HEATING AND AIR-CONDITIONING
OF BUILDINGS

Plate I. Belfast City Hospital. 95 m multi-flue chimney (see Fig. 4.8, p. 86)

HEATING AND AIR-CONDITIONING OF BUILDINGS

(BY OSCAR FABER AND J. R. KELL)

SIXTH REVISED EDITION (IN SI UNITS)

BY

J. R. KELL

C.B.E., F.I.MECH.E., F.C.I.B.S., F.INST.F.

PAST PRESIDENT, INSTITUTION OF HEATING AND VENTILATING ENGINEERS

HOLDER OF I.H.V.E. GOLD MEDAL

AND

P. L. MARTIN

F.I.MECH.E., F.C.I.B.S., F.INST.F., A.M.R.AE.S., M.CONS.E.

PAST PRESIDENT, INSTITUTION OF HEATING AND VENTILATING ENGINEERS

VISITING PROFESSOR, UNIVERSITY OF STRATHCLYDE

HOLDER OF I.H.V.E. GOLD MEDAL

THE ARCHITECTURAL PRESS: LONDON

First published 1936
Second and revised edition 1943
Reprinted 1945
Reprinted 1948
Reprinted 1951
Third and revised edition 1957
Reprinted 1958
Reprinted 1961
Fourth and revised edition 1966
Fifth and revised edition 1971
Reprinted 1974
Sixth and revised edition 1979

PRINTED IN GREAT BRITAIN BY
ROBERT MACLEHOSE AND CO. LTD, PRINTERS TO THE UNIVERSITY OF GLASGOW

PREFACE TO SIXTH EDITION

WHEN THE FIFTH EDITION WAS in preparation some eight years ago, this was at the time of transition from Imperial to SI units. Metric units have now become familiar and therefore the older equivalents, previously given in brackets, have been dropped for the present revision. At the same time the pascal and the kelvin have been substituted, as appropriate, for the units they succeeded. The 'Introduction to SI', previously taking pride of place, has been relegated to an Appendix where it may serve as a refresher for those who were brought up with the Btu, the pound of steam and so on.

The necessity for revision has been brought about by various factors, among them the current emphasis on conservation of energy. This new concept is affecting building design and thus involves the architect as much as the building services engineer. Further text has been included to deal with this important subject, albeit briefly.

The changed and changing relationship of energy costs from all the different sources constitutes a quicksand where nothing seems stable. In the last reprint, a caveat was included to cover the re-evaluation of any figures previously given. Since then, the matter seems even more fluid and therefore any costs hinted at here can only be treated as a reflection of trends now seen. Nevertheless, electrical heating, whether off-peak or on-peak, must always suffer from the inefficiency of the Rankine cycle and hence, in the interests of energy conservation, continues to be at a disadvantage compared with other fuels converted to heat energy on-site, quite apart from any question of economics.

Other revisions have proved necessary because staple articles of ironmongery, such as the cast-iron radiator, have virtually disappeared from the market. Changing patterns in building design and demands by occupants for improved working conditions have led to advancement in the science and practice of air conditioning. The technique of the variable volume system has developed, and this is treated more fully than before as are methods for heat recovery applied to ventilation systems.

As in previous editions, the text of this book refers again and again to factual data taken from the *IHVE Guide.** In most cases extracts and condensations only are given and the reader should refer to the source for

* Published by the Chartered Institution of Building Services, 49 Cadogan Square, London SWIX oJB. Available in booklet form for many of the Sections within Volumes A, B and C. Prices on application to the Assistant Secretary (Technical).

more complete information. The Authors wish, once again, to offer their thanks for permission to quote from the *Guide*, but in this case the acknowledgement is made to the Chartered Institution of Building Services and is coupled with good wishes for the future of this successor to the now laid-down Institution of Heating and Ventilating Engineers.

Further thanks are due to many colleagues and friends who have provided willing assistance and also to patient secretaries who have transcribed obscure manuscript into legible copy.

St Albans, 1978   J. R. K.
                                                                 P. L. M.

CONTENTS

LIST OF PLATES

ACKNOWLEDGMENTS

The Authors wish to express their thanks to the following for their kind permission to reproduce illustrations:

Plate VIII: Basari Ltd.
Plate IX: Allen Ygnis Ltd.
Plate XXXIX: Runtalrad Ltd.
Plate XXV: Colt Heating & Ventilation Ltd.
Plate XXVIII: Thorn Benham Ltd.
Plate XXXIII: Moducel Ltd.
Plate XLI: Henry Hargreaves & Sons Ltd.
Plate XLIII: Haden Carrier Ltd.
Plate XLV: Nicholas Horne Ltd.
Plate XXVIII: Thorn Benham Ltd.
Figs. 10.5, 10.13 and 10.14: Spirax Sarco Ltd.

CHAPTER 1

Heat

KEEPING WARM IS A primitive instinct in man, and modern life still depends on warmth for existence.

The more civilized we become, the more sophisticated are the demands—the fashion is to wear lighter clothes and hence to require warmer environments.

Buildings can be warmed by the sun in hot countries and, even in temperate climates, given sufficient insulation, they can be warmed by the internal heat from occupants and lights. But it is necessary, as a rule, to warm buildings by the burning of some sort of fuel, or by consuming electrical energy which may in turn be derived from fuel or from nuclear energy or water power. It is to the means for performing these functions and distributing the heat to places where it is wanted that the heating engineer directs his attention.

But warmth alone is not the sole criterion of comfort. There must also be a pleasant atmosphere, which involves considerations of ventilation for the supply of fresh air and the removal of foul air, and air-conditioning for cooling in summer as well as warming in winter. Modern buildings with their large expanses of glass and high concentration of occupancy, coupled with the noise and dirt of large cities, create an increasing demand for such systems.

Apart from comfort conditioning, there is the further field in industrial applications of providing carefully controlled temperatures and humidities for process requirements and for the storage and manufacture of a wide variety of materials.

Fundamentally, all these and other related problems may be resolved simply into questions of how to add or remove heat, how to add or remove moisture and how to move air. Air will generally be involved somehow or another in all these mechanisms. It will be as well to keep this clear perception in mind in all that follows, but in the first place it is necessary to define some of the terms commonly used, with special reference to the particular meaning which the engineer normally applies to them.

DEFINITIONS

Temperature—Temperature is that state of a substance that determines the direction of heat flow to or from the substance. Heat will flow from a warm body to a cold body, and the rate will be directly proportional to the temperature difference. Temperature is akin to potential or pressure and is a relative term. The temperature of boiling water is higher than that of cold: the temperature of cold water is higher than that of ice. Ice may be said to be hot, or at a high temperature, compared with liquid air at − 190° C.

The scale of temperature now being used in the United Kingdom is (as explained in Appendix I) the *Celsius Scale* (° C). If a mercury-in-glass thermometer is so calibrated that the freezing point of water is marked 0° and the boiling point is marked 100°, then any intermediate temperature is given a figure between these two, proportionate to the length of the mercury column.

The *absolute zero* has been established by studying the characteristics of expansion of gases, and other phenomena, and is assumed to be the temperature of space. The absolute scale is referred to in *kelvins* or units of thermodynamic temperature (Symbol K); an interval of 1° C is the equivalent of an interval of 1 kelvin. Absolute zero = − 273·15° C or 0 K.

To avoid confusion in terminology, it was the accepted convention in Imperial units that temperature *level* (i.e. thermal head) should be expressed in terms of ° F whereas temperature interval or difference was in terms of deg. F. Under strict SI rules, confusion is avoided by expressing temperature level in degrees Celsius (° C) and temperature interval or difference in kelvins (symbol K). Whilst this convention is not obligatory

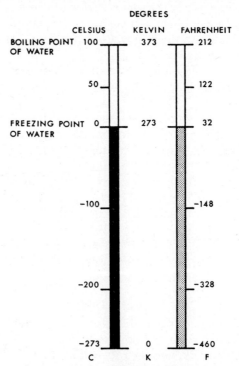

FIG. 1.1.—Temperature scales.

and degrees Celsius may be used in either context, for this edition it is intended that the strict SI rules will be followed.

The Fahrenheit scale, having been ingrained in the mind for so long, will no doubt die hard; hence conversion tables will continue to be useful and one is produced in Appendix II, being in whole numbers of degrees Celsius. Fig. 1.1 also serves to illustrate the comparisons.

To convert Celsius degrees into Fahrenheit, multiply by 9, divide by 5 and add 32, and conversely, to convert Fahrenheit into Celsius, deduct 32, multiply by 5 and divide by 9. A convenient short cut for mental arithmetic, reasonably accurate in the range of British weather charts, is to multiply degrees Celsius by 2 and add 30 to obtain degrees Fahrenheit.

It is interesting to note that the intermediate temperatures as measured by a mercury-in-glass thermometer do depend a little on the expansion characteristics of mercury and glass, and that thermometers which depend on the expansion of other materials do not necessarily give an exact subdivision between freezing and boiling point when checked against a mercury thermometer. For very accurate scientific work this has to be taken into account, but is not necessary for the purposes of this present treatise.

Fig. 1.2 illustrates various forms of thermometer in common use.

The reading of an ordinary mercury thermometer is said to give the temperature of a gas, liquid or solid in which it is immersed, but in reality what it gives is something considerably more complicated when applied to air in buildings. It gives a temperature at which the heat received from the surrounding objects exactly equals the heat given off to other surrounding objects when a perfect balance is obtained, and this heat transference may be partly by conduction, partly by convection and partly by radiation (defined later).

It is a matter of considerable interest that an ordinary mercury thermometer gives approximately the same reading in a room which has been allowed to reach stable conditions as regards temperature, whether it is screened from the source of radiant heat or not, provided that the source of radiant heat is one of relatively low temperature. This is because glass is practically impervious to the radiation from low-temperature sources.

This, however, is not at all true where high-temperature radiation, such as that received from the sun and from incandescent surfaces such as open fires, gas and electric fires is concerned. This is really the explanation of the curious phenomenon of the ordinary glass greenhouse, which, as everyone knows, gets extremely hot inside when the sun is pouring down on it, and retains its heat long after the sun has ceased to shine upon it.

The air outside the greenhouse in the sun receives exactly the same amount of radiation from the sun as the air inside the greenhouse. Why, therefore, does that outside remain, say, at 20° C when the air inside may be at 40° C? In both cases the temperature of any body subject to the radiation goes up until the total quantity of heat received from the sun and all other surrounding objects is balanced by the heat lost to the other

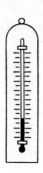

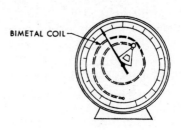

BIMETAL COIL

DOMESTIC THERMOMETER

**BIMETAL COIL
THERMOMETER**

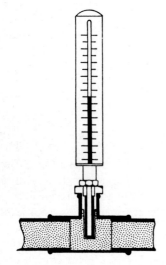

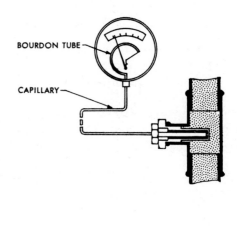

BOURDON TUBE

CAPILLARY

PIPE THERMOMETERS

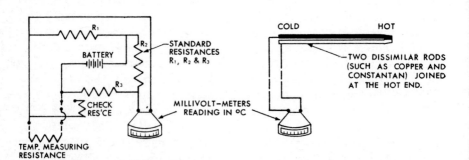

R₁

BATTERY

R₂

STANDARD
RESISTANCES
R₁, R₂ & R₃

R₃

CHECK
RES'CE

MILLIVOLT–METERS
READING IN °C

TEMP. MEASURING
RESISTANCE

COLD   HOT

TWO DISSIMILAR RODS
(SUCH AS COPPER AND
CONSTANTAN) JOINED
AT THE HOT END.

**ELECTRICAL RESISTANCE
THERMOMETER**

**THERMOCOUPLE
THERMOMETER**

FIG. 1.2.—Types of thermometer.

surrounding objects, and at first there appears to be no reason why the temperature reached in the two cases should be in any way different. The explanation of this is to be found in the peculiar property of glass in being pervious to high-temperature radiation and impervious to low-temperature radiation (a phenomenon sometimes referred to as *dia-thermancy*). The heat from the sun (which is a source of high-temperature radiation) passes through the glass and warms the objects therein contained, which are unable to radiate this heat back through the glass.

A thermometer known as a *solar thermometer* is employed when it is desired to obtain a reading which includes the effect of solar radiation (see Fig. 1.3). This thermometer con-sists of a glass bulb containing a vacuum, in the middle of which is an ordinary thermometer bulb with a blackened surface. The idea is that the thermometer bulb will not be affected by the air tempera-ture, from which it is insulated by the intervening vacuum, but will receive only radiation, and so will measure the radiant heat only. This gives quite satisfactory readings when applied to the measurement of sun temperatures, but it is of little use when we wish to measure radiation from low-temperature surfaces for the reason stated above.

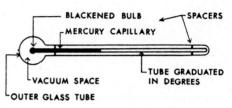

Fig. 1.3.—Solar thermometer.

TABLE 1.1
MOISTURE IN SATURATED AIR AT
VARIOUS TEMPERATURES.

Temperature ° C	kg moisture per kg of dry air
0	0·0038
5	0·0054
10	0·0076
15	0·0107
20	0·0148
25	0·0202
30	0·0273
35	0·0367
40	0·0491
45	0·0653
50	0·0868

Humidity—The total pressure of the air is the sum of the partial pressures of the separate constituents. Water vapour, or the constituent of air which we describe as *humidity* exists in the space independently of the oxygen, nitrogen and other gases forming the atmosphere. When the highest partial pressure of water vapour appropriate to the air temperature exists, the air is said to be saturated. At lower partial pressures, the air is unsaturated, and the ratio of the two pressures is known as *relative humidity*. The greatest mass of moisture which can obtain in a given space is dependent only on temperature, and Table 1.1 expresses this in terms of kilogram per kg of dry air. Such a condition is stated to be saturated. If air which is unsaturated is lowered in temperature, the relative humidity will rise until it reaches 100 per cent. and is saturated. Any further lowering causes condensation. Another term used is *percentage saturation* which for practical purposes is almost the same as relative humidity (see p. 409 for a precise definition).

The relative humidity is the condition of air upon which depends, more than anything else, the rate at which evaporation from a moist surface will occur. Thus at 100 per cent. relative humidity no evaporation can occur, while at 50 per cent. relative humidity the evaporation will be rapid. The human body provides such a moist surface, and the evaporation which takes place from it with low relative humidities produces various physiological effects, such as a parchiness in the throat and a cooling effect on the face and hands, especially in the presence of air movements, thus producing discomfort.

Too high a relative humidity produces other signs of discomfort, such as lassitude, caused primarily by the inability of the skin to rid itself of moisture and heat.

The normal evaporation from the skin is accompanied by a cooling effect on the body owing to the great heat required (equivalent to approximately 2300 kJ/kg) to convert water into water vapour (see also *latent heat*, p. 13). It is on this that the cooling effect of a current of air principally depends. With excessive humidity, when this evaporation does not occur, the cooling effect clearly disappears also, and discomfort arises at high temperatures.

This explains what travellers in tropical climates have experienced for many years, that a high air temperature with a low relative humidity may be borne more easily than a lower air temperature in very humid conditions.

In cold climates, however, the opposite is the case. Air at low temperatures has a greater feeling of coldness at high relative humidities than at low. The reason for this effect on the sensation of cold in cold weather and of heat in hot weather, being both greatly accentuated in a humid atmosphere, is due to the changes which high humidity induce on the human skin. In a dry atmosphere, the skin dries up and hardens and becomes more insulating, and so feels the cold or heat less.

Conversely, in a moist atmosphere the skin swells, the pores open, the skin becomes more conducting and sensitive. This is an important effect, which no instrument so far devised takes into account.

It therefore appears that high relative humidity causes discomfort both at high and low temperatures, in the former case by producing a sensation of extreme heat, and in the latter case of extreme cold. The temperature at which no change in sensation occurs with change of humidity is stated to be 8° C for still air and 10° to 13° C for air moving at speeds of 0·5 m/s to 2·5 m/s respectively.

Humidification and De-humidifying—From what has been said it will be clear that in hot summer weather the hot air from outside coming into a cool building will have its relative humidity raised, and will therefore produce oppressive conditions.

Comfortable conditions may be produced by some device whereby the relative humidity may be reduced, and this is known as de-humidifying. Most de-humidifying processes depend on allowing the air to come in contact with cold surfaces where the excess moisture is condensed by being lowered below saturation temperature, and then warming it again to a comfortable temperature before distribution.

In cold winter weather the opposite occurs. Cold air from the outside comes into a warm room, and is warmed by the heating system to a higher temperature accompanied by a drop in the relative humidity. In such cases there is need for humidification, i.e. for adding to the moisture content of the air. One method of effecting this is to pass the air through a washer in which it is intimately mixed with water in the form of a very fine spray or mist.

In general it may be said that to most people a comfortable humidity is between 40 and 70 per cent. combined with a temperature of somewhere between 18° and 24° C.

Conduction, Convection and Radiation—It will be important, in the discussions that follow, to understand clearly the difference between the three main methods of transferring heat from one body to another.

Conduction may be described as a heat transfer from one particle to another by contact. If, for example, a hot lump of copper is placed in contact with a cold lump of copper, as in Fig. 1.4, heat is said to be conducted from the hot to the cold, until the two will finish at an intermediate temperature which may be calculated by equating the

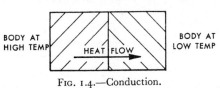

FIG. 1.4.—Conduction.

total mass times the final temperature with the mass of the hot body times its temperature added to the mass of the cooler body times its temperature. If the two bodies are of different material, then the mass of each has to be multiplied by its specific heat capacity before this calculation can be made.

Conductivity is the property of bodies of being able to conduct heat, and the measure of conductivity is the number of units of energy transmitted per degree difference per unit thickness and per unit face area, in unit time.

Table 1.2 gives conductivities of various metals, building materials and insulators, and it will be seen that metals have a high conductivity, while materials like brick have a low conductivity. Certain bodies have such a low conductivity that they are known as *insulators*, and are used for covering warm surfaces to reduce heat losses from them. For convenience Table 1.2 also gives the density, specific heat capacity and coefficient of expansion of these substances. These terms will be referred to later.

Good conductors of heat are generally good conductors of electricity, though there is no rigid relationship between the two.

The conductivity of many materials varies considerably with temperature, and therefore figures should only be used within the range to which they apply. The insulation value is inversely proportional to the conductivity.

Porous materials are bad conductors when dry and good conductors when wet, a fact which is sometimes lost sight of when attempting to warm a newly-constructed building, where the heat losses may be far higher in the early months than they will be when it has properly dried out.

Convection is a transfer of heat which involves the movement of hot particles of a fluid medium from a hot body to a cold. A common illustration may be found in the ordinary radiator (so-called), see Fig. 1.5. This warms the air immediately in contact with it, which expands and so becomes lighter than the rest of the air in the room. It consequently rises, forming an upward current from the radiator, and travels to the colder portions of the room, to which it gives up its heat and eventually returns to the radiator for the process to be repeated.

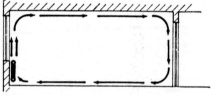

FIG. 1.5.—Convection in air.

Another example is the heating of water in a boiler, as shown in Fig. 1.6. The water in contact with the hot surfaces over the fire becomes heated, expands, and produces an upward circulation, exactly as in the case of the air, eventually returning to the boiler for reheating. Convection, therefore, implies a medium capable of movement from the hot body to the cool body to be heated, and cannot occur in vacuum when no such medium exists.

Radiation is a phenomenon with which we are most familiar in its application to the problem of light.

In Newton's time the phenomenon of radiation was explained as a bombardment of infinitesimal particles from the source of heat, which were supposed to impinge on the cool body to be heated. At a later date radiation, whether of light or of heat, was supposed to be a wave action in an imaginary medium known as the ether, which was invented by mathematicians to account for this phenomenon. It is now known that all radiation is an electromagnetic process be it light, heat, X-rays or radio

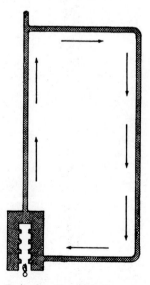

FIG. 1.6.—Convection in water.

TABLE 1.2

PROPERTIES OF MATERIALS

Material	Density (Specific Mass)	Specific Heat Capacity	Coeff. of Linear Expansion per K $10^{-6} \times \ldots$	Thermal Conductivity (k)
	kg/m³	kJ/kg K		W/m K
Metals				
Aluminium (sheet) - - - - -	2 700	0·98	25·5	238
Brass (Cast) - - - - -	8 100	0·36	18·8	109
Copper (sheet) - - - -	8 800	0·39	17·5	385
Iron (Cast) - - - - -	7 400	0·51	10·2	47
Lead - - - - - -	11 400	0·14	29·0	35
Magnesium - - - - -	1 700	1·05	25·5	157
Mercury (0° C) - - - -	13 600	0·14	60·0	7
Mild Steel - - - - -	7 800	0·48	11·3	48
Tin - - - - - -	7 300	0·23	21·4	64
Zinc (sheet) - - - - -	7 200	0·39	26·1	112
Building Materials				
Asbestos cement (sheet) - - -	1 550	0·84	9·9	0·45
Asphalte - - - - - -	2 250	1·68	—	1·2
Brick (exposed) - - - -	1 800	0·79	2·2	1·07
Concrete (exposed) - - - -	2 400	0·84	9·9	2·55
Firebrick (at 400° C) - - -	2 000	0.84	4·9	1·0
Glass (sheet) - - - - -	2 500	0·84	8·4	1·05
Granite - - - - - -	2 650	0·90	7·9	2·9
Limestone - - - - -	2 200	0·86	6·3	1·5
Marble - - - - - -	2 700	0·90	11·0	2·0
Plaster - - - - - -	1 300	0·84	—	0·46
Plaster board - - - - -	950	0·84	—	0·16
Slate - - - - - -	2 700	0·75	19·6	1·9
Tiles (burnt clay) - - - -	1 900	0·84	—	0·85
Timber				
Deal - - - - - -	600	1·21	4 to 8 along grain	0·13
Oak - - - - - -	750	1·88	20 to 80 across grain	0·16
Pitch pine - - - - -	650	2·30	(when dry)	0·14
*Insulating Materials				
Asbestos Millboard - - - -	700	0·82	—	0·11
Lightweight Concrete - - -	600	0·84	1·4	0·18
Cork board - - - - -	150	1·80	—	0·04
Diatomaceous Brick - - - -	500	0·80	1·4	0·09
Fibreboard - - - - -	380	—	—	0·05
Glass Fibre (quilt) - - - -	80	0·82	—	0·04
Calcium Silicate - - - -	200	—	—	0·07
Polystyrene (expanded) - - -	15	—	—	0·04
Vermiculite (loose) - - - -	100	—	—	0·07
Wood wool (slab) - - - -	600	—	—	0·11
Miscellaneous				
Water 4° C - - - - -	1 000	4·205	0	0·6
15° C - - - - -	998·5	4·186	65	0·6
100° C - - - - -	958·4	4·214	250	0·67
Ice - - - - - -	920	2·1	52	2·2
Air (at normal 20° C - - -	1·205	1·012	—	0·027
pressure) 100° C - - -	0·88	1·012	—	0·027

*For fuller list see Table 2.2 (p. 25).

waves. Heat radiation occupies a band between the red end of the visible spectrum and the shortest of radio waves. Fig. 1.7 gives a pictorial representation of the order of the various forms of radiation.

Confining our attention to heat radiation, this may be regarded as a transference of energy which takes place in rays in such a way that the intensity varies inversely with the square of the distance, and is independent of any substantial medium such as air, i.e. it occurs just as readily across a vacuum as across a room filled with air, and does not depend on warming the medium through which it travels.

The so-called 'radiator' transfers its heat partly by convection and partly by radiation, the proportion depending mainly on the shape of the radiator. Those surfaces which face one another radiate very little to the rest of the room and depend principally on convection. Also, if anything is done to obstruct the free flow of air over the radiator, its proportion of convection is reduced. Flat surfaces such as a warmed wall or ceiling, however, may have a relatively high proportion of radiation as compared with convection.

In very rough figures, the proportion of radiation to the total heat emission in ordinary radiators may be about 20 per cent, in a wall panel about 50 per cent., and in a ceiling panel about 90 per cent. This shows that the term 'radiator' is a misnomer, since approximately 80 per cent. of its heat is transmitted by convection.

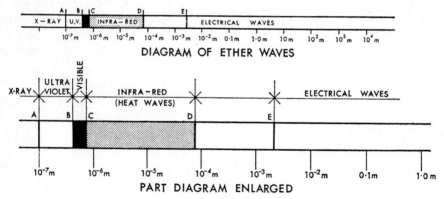

FIG. 1.7.—Types of radiation. (Wave lengths are in metres.)

Fig. 1.8 shows the relatively concentrated radiant beam emitted from an electric bowl fire and the diffused radiation from a warmed ceiling panel.

The amount of radiation emitted from a surface depends on its texture and colour. Dead black is the best radiator and polished metal the worst. This effect is dealt with more fully later. (See p. 143.)

Surfaces which are good radiators of heat are found to be the best absorbers also: thus a black-felted roof is often found to be covered with hoar frost on a cold night, due to its good radiation into space, whereas

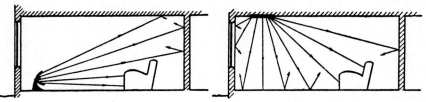

Fig. 1.8.

(a) Visible heat radiation from electric   (b) Invisible radiation from heated
bowl-fire.   ceiling panel.

other surrounding objects may be apparently unaffected. Similarly, a black suit of clothes on a warm day is found to absorb much more heat from the sun than one of white material such as is worn in tropical countries. Again, asphalt on a flat roof does not become nearly so hot in summer if painted white.

Unit of Heat—As explained in Appendix I, it is fundamental in SI to treat heat as a form of energy. The unit is the *joule* (J), which is the energy of a mass of 1 kg accelerated to a speed of 1 metre per second in 1 second, i.e. a force of 1 newton (N) acting through 1 metre.

Heat flow is measured in joules per second, but 1 J/s = 1 watt and this latter unit or its multiples will be used wherever heat energy flow in unit time is the criterion.

Specific Heat Capacity—The old term *specific heat* related to the quantity of heat required to raise unit mass through unit temperature, water being taken as unity. In SI, the unit of specific heat capacity is kilojoules per kilogram per kelvin (kJ/kg K). Thus heat flow to or from a substance is taken as

$$\text{kJ} = (\text{kg}) \times \left(\frac{\text{kJ}}{\text{kg K}}\right) \times (\text{K})$$

If the problem involves heat flow in unit time (i.e. one second), then this is calculated as

$$\text{kW} = \frac{\text{kJ}}{\text{s}} = (\text{kg}) \times \left(\frac{\text{kJ}}{\text{kg K}}\right) \times (\text{K}) \times \left(\frac{1}{\text{s}}\right)$$

Table 1.2 gives specific heat capacities for materials in common use in buildings. For water the figure is approximately 4·2 kJ/kg K.

Expansion—It will be found that Table 1.2 gives a column for the *Coefficient of Linear Expansion* of various materials.

Materials, with very few exceptions, expand on being warmed, and contract on being cooled. The Forth Bridge expands approximately 0·75 m as between warm and cold weather.

It is found that for most materials the expansion varies directly with the length and with the difference of temperature.

The coefficient of expansion for a material is defined as the proportion of its original length which it lengthens with a temperature rise of 1 kelvin. Thus the coefficient of expansion of mild steel is approximately 0·000011, which means that if a bar is say 1 m long at 15° C, it will be 1·000011 m long at 16° C, or, if raised through 100 K it will be 1·0011 at the higher temperature. A bar 100 m long will expand for the same temperature rise 0·11 m (110 mm).

It will be seen from this how necessary it is to make due allowance for the expansion and contraction of such things as long straight pipes, where the movement may be quite considerable. In practice, pipes beyond a certain length of straight have to be provided with expansion joints—a matter which will be dealt with in the appropriate place later. 'Invar' steel, a material much used in instruments, does not expand over a wide range of temperature.

The coefficients of expansion given in Table 1.2 are average figures within the range of normal temperatures—that is, between 15° C and 100° C—but a word of warning is necessary in applying these figures, as in some cases they vary considerably beyond this range. Water is an example of this, since near the freezing point is has a negative coefficient of expansion, i.e. it actually expands on *lowering* the temperature from about 4° C down to 0° C, the coefficient at 4° C being zero.

Superficial Expansion is the increase in area due to increase of temperature. The Coefficient of Superficial Expansion is taken as twice the linear coefficient. For an expansion of 'α' on the sides of a square of 1 unit initial length, increase in area $= (1+\alpha)^2 - 1 = 2\alpha + \alpha^2$. α^2 is negligible, hence coef. sup. exp. $= 2\alpha$. *Cubic expansion* is the increase in volume due to increase of temperature. The Coefficient of Cubic Expansion is taken as three times the linear coefficient (by a similar argument to the above).

Specific Heat Capacity and Expansion of Gases—The perfect gas conforms to Boyle's and Charles' Laws, which state that the volume varies inversely with the pressure when the temperature is constant, and the volume varies directly with the absolute temperature when the pressure is constant. In other words, PV/T is a constant where P is the pressure, V is the volume and T is the absolute temperature.

Now the number of kelvins between freezing point and absolute zero is 273 (see Fig. 1.1). It therefore follows that, if a certain volume of perfect gas is contained in an envelope and its temperature is raised through 1 kelvin from 0° to 1° C without altering the pressure, the increase in volume will be $\frac{1}{273}$ of its original volume. This gives a coefficient of cubic expansion of $\frac{1}{273}$.

If, however, a volume of gas at the boiling point of water is raised through 1 kelvin, then its absolute temperature being 373 K, its coefficient of expansion is $\frac{1}{373}$.

Most gases conform very closely to the properties of the perfect gas when at a temperature remote from their temperature of liquefaction.

Ordinary air conforms very closely indeed, but CO_2, which can be liquefied at only moderate temperature reductions, does not conform quite so well.

Gases have two specific heat capacities according to whether the pressure or the volume is kept constant while the temperature is raised. If the pressure is kept constant, the increase in temperature is accompanied by an increase in volume, and, in the case of air, the specific heat capacity at 20° C is 1·012 kJ/kg K. If the volume is kept constant, the increase in temperature is accompanied by an increase of pressure and, again for air, the specific heat capacity is 0·72 kJ/kg K at this temperature.

Power—Power and energy being synonymous terms within the present context, the old term *horsepower* is no longer used and power is expressed in terms of the watt or the kilowatt. (The hp was equivalent to 746 watts.)

It is found that an ordinary human being, in maintaining an internal temperature of approximately 37° C, emits approximately 120 watts when at rest at normal temperatures. This heat goes out partly as radiation and convection, partly as warm air containing water vapour from the lungs, and partly as evaporation from the skin. A man doing heavy physical work, such as carrying a 50 kg mass up a 20 m ladder in five minutes, is producing energy at a rate of 30 watts. In so doing his heat emission will rise to about 400 watts, which shows an efficiency of only $7\frac{1}{2}$ per cent! This is lower than that of the worst engine in common use, and should conduce to humility. A man can exert still more energy for short periods, and will then emit correspondingly greater quantities of heat.

Electrical Energy—It follows that electrical energy, whether in the form of power taken by a motor or heat due to passage through a resistance, is measured, as it always has been, in watts or multiples thereof. In the past, the *output* of a motor was given in horsepower but is now quoted in watts as for the input, and some confusion may arise unless it is made quite clear as to which is referred to.

The Unit of Electricity is 1 kilowatt maintained for 1 hour, and may represent 10 amperes flowing in a circuit at 100 volts pressure for one hour or 100 amperes flowing in a 10 volt circuit for an hour, and so on.

Latent Heat—If water is heated from freezing point to boiling point (i.e. through 100 K), $4 \cdot 2 \times 100 = 420$ kJ are added per kg; and, if the supply of heat be continued, it is found that the temperature does not increase (assuming the vessel is open to atmosphere) but the water is gradually converted into steam at the same temperature. The experiment will show quite clearly that the amount of heat absorbed to effect this conversion is extremely large in comparison with the heat required to warm the water, and in fact it requires 2300 kJ to convert 1 kg of water at boiling point into 1 kg of steam at the same temperature.

It will be seen that this is approximately 5·4 times as great as the heat required to warm the water from freezing point to boiling, and this gives some idea of the importance of this phenomenon. This large quantity of heat required to convert a unit weight of any substance from the liquid to the vapour state is known as the *latent heat of evaporation.*

The reverse is equally true; when 1 kg of steam at 100° C condenses so as to produce 1 kg of water at the same temperature, 2300 kJ are liberated. Of all known substances, water has the greatest latent heat.

The boiling point of water varies with the pressure: at high pressures the boiling point is greatly raised and, conversely, it is reduced at pressures below atmospheric. It is often convenient to add the heat required to raise the water from freezing point to the temperature at which it is vaporized to the latent heat of vaporization, and this quantity is known as *specific enthalpy* at that temperature.

There is also a *latent heat of fusion* when ice is converted into water, or the reverse, and this has the value of 330 kJ/kg.

In all problems of humidification and de-humidifying, where water is either evaporated or condensed, the latent heat is the principal factor concerned.

All materials have a latent heat of fusion and one of vaporization, but these have lower values than in the case of water, and, of course, the temperatures at which the changes of state occur may be widely different.

Availability of Heat—The 'value' of heat depends on its temperature. Thus a certain quantity of heat at a high temperature may be useful for warming rooms or heating water. The same quantity of heat at a low temperature may be useless for these purposes. Heat is degraded in flowing from a high to a low level of temperature—increasing thereby in 'entropy.' This is perhaps a subject which need not concern us here, but it is as well to remember that such a process exists in all heat-flow phenomena.

CHAPTER 2

The Building and the Heating System

THE BUILDING

As A GENERAL PRINCIPLE when approaching the problem of the heating of a building, it is desirable that we should consider the building and its heating system as one entity. The form and construction of the building will have an important effect on any question of how to heat it. The amount of heat required to maintain a given internal temperature can be greatly reduced by insulation. The recent fashion to use vast areas of glass runs counter to this and imposes considerable loads on any heating system. The lightness or heaviness of the structure has a direct bearing on the type of heating system and on its control and running costs. A lightly constructed building requires a heating system which is quickly responsive to changes of temperature, whereas a heavy, massive building is probably better served by a system with a slow steady output.

Tall multi-storey blocks of offices and flats introduce fresh problems due to exposure to wind and sun as well as chimney effects within the building itself due to height. These again have an important bearing on the form and design of the heating installation.

In considering the problem of how to keep a building warm, we cannot overlook the fact that the people, electric lighting and other heat-producing items of equipment within the space contribute in some measure to the maintenance of the desired temperature. This is particularly so in the case of buildings with large expanses of glass, where the heat of the sun, even in winter, may often be sufficient to maintain the desired temperature without any additional aid. Whilst this may be so on one side of a building, however, it will not apply to all aspects at the same time, and therefore the system of heating has to be so devised that it can be shut off automatically in those areas where fortuitous heat gains of this kind are sufficient in themselves, whilst other areas continue to be supplied with warmth for the maintenance of the desired temperature.

An important question arises as to how far the designer of a heating installation for a given building should take into account fortuitous heat gains. If they are not allowed for, the system may be unduly large and unwieldy, and rarely, if ever, run at its full output. On the other hand, if the designer has made certain assumptions that heat gains will occur which in practice, due possibly to change of usage, do not actually arise, he might be faced with the building being inadequately heated.

The tendency in the past, when designing heating installations, has been to ignore entirely the effect of internal heat gains, in so far as the calculation of the heat necessary to maintain a given internal temperature is concerned; but, when it comes to a question of running cost and heat consumption per annum, the importance of such effects has been brought out by the study of actual consumptions of buildings heated electrically, and, by the same token, similar calculations for other sources of heat should make the same allowances.

CONSERVATION OF ENERGY

The past age of a seemingly limitless supply of fuels of one sort or another has come to an end. The new age just beginning comes about due to a world-wide realisation that the supply is not inexhaustible, a situation well understood by engineers for some decades but unpalatable in political circles.

Furthermore, international economic forces are causing a startling rise in the cost of energy with the result that old standards of comparison no longer apply.

It has therefore come about that energy conservation in the sense of fuel saving is now a doctrine of prime importance. In the context of heating and air-conditioning of buildings the scope for economies is considerable such as by structural insulation, double glazing, restriction of unwanted ventilation, accurate control of internal temperature, reductions in plant operating hours and recovery of heat from lighting, industrial processes or exhaust air. These are dealt with in the text under the various headings.

Taking a broader view, the use of solar energy has yet to be developed on any considerable scale. The use of geothermal heat may or may not be possible in this country. Energy contained in tides and waves has yet to be exploited.

It follows that the first step in embarking on the assessment of heat requirements for a building should be to ensure that they have been cut down to an economic minimum. This may involve persuading the architect to reconsider building orientation, to change materials, to add thermal barriers, or even to reduce window areas.

But the architect is involved, even before this stage is reached, in his basic planning for the shape of building, bearing in mind that the major component of the total thermal load is through the peripheral surfaces.* The most economical shape for maximum volume with minimum surface area is a sphere; this being impracticable for a building, however, the nearer the plans approach to it the lower will be the heat demand. A tall shallow slab building is obviously one of the worst in this respect.

---

* Page J. K., *Energy Requirements for Buildings* Public Works Congress, 1972.
   Jones W. P., *Designing Air-Conditioned Buildings to Minimise Energy Use* RIBA/IHVE Conference 1974.

HEAT LOSSES

The conventional basis for design of any heating system is the estimation of heat losses. It is assumed that a steady state exists as between inside and outside temperatures and that air temperature is the sole criterion.

In fact, steady-state conditions rarely exist and, as will be seen later, the radiant temperature of the enclosure if taken into account may call for higher or lower air temperatures for equal comfort. However, it will be best to deal with the conventional method first and to discuss these other effects subsequently. The method may be explained as follows.

Each room of a building is taken in turn and an estimate made of the amount of heat necessary to maintain a given steady temperature within the space, assuming a steady lower temperature outside. The calculation is in two parts: (a) that due to conduction through the materials of the walls, floor and ceiling, known as *transmittance*, and (b) the heat necessary to warm infiltration-air up to room temperature, this air escaping by some means or other to the outside—an effect which is known as *air-change*. This air-change is, in fact, ventilation, and without it a space would quickly become uninhabitable; but knowledge of what exact air-change rate to allow in any given case is debatable and hence this part of the calculation tends to remain empirical.

It might be thought that, with so many assumptions and 'rule of thumb' estimates, heat losses were little better than guesswork; but in practice they appear to give a satisfactory basis, partly no doubt due to the fact that all areas within the building are treated in a like manner and should therefore be consistent, and secondly due to the flywheel effect of the structure itself which, even with a lightly constructed building, still has floor slabs, partitions, furniture, etc. to absorb and emit heat and so avoid violent fluctuations.

Adjacent rooms maintained at the same temperature will have no heat transfer through partitions or other surfaces, and these are therefore ignored. All that concerns us is the leakage of heat from a room at one temperature to outside, or to other rooms at a lower temperature. If certain areas of the room are used for heating, such as the ceiling or the floor, then these, being warmer than the room, will not be taken into account in so far as the heat losses from the room are concerned; but they will have inherent losses upward or downward, dependent on the amount of insulation, and these will have to be taken account of separately from the heat loss calculations.

Transmission Losses—The transmission of heat through any material depends on the temperature difference between the two surfaces and on the conductivity of the material itself. We are concerned with building materials, and with the transmission of heat from air in a room to air outside. The air temperatures will not be the same as the surface temperatures of the material, due to a dead layer of air, which may be supposed to exist in contact with the surface, retarding the flow of heat. Fig. 2.1 illustrates this as it might be for a thin material such as a sheet of glass. The air inside the room being relatively still offers higher resistance to heat flow than that outside, where the effect of wind is to be considered.

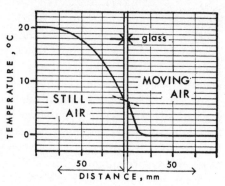

FIG. 2.1—Heat transmission gradient through glass.

The effects of these boundary layers are defined as *surface resistances*, and values are put upon them as given later. In the case of the outside surface resistance, the value depends on the degree of exposure, such that a sheltered surface has a higher surface resistance than one with a severe exposure. In the extreme case, indeed, of say a very tall building, it might well be supposed that the force of the wind was such that the boundary layer completely disappears and the surface resistance is equal to zero. This would be the limiting case.

In order to calculate the transmission losses from a room, it is necessary to measure from drawings, or from the building, the area of each type of exposed surface, i.e. window, wall, roof, floor, etc. The measurements will be in square metres, and we then require to know the amount of heat which will be transmitted per square metre of surface, and this will depend on the material itself.

Conductivities of materials (k) are stated in watts for 1 metre thickness per square metre per kelvin, i.e.

$$k = \frac{Wm}{m^2 \, K} = \frac{W}{m \, K}$$

The resistivity of a material (r) is the reciprocal of the conductivity, thus

$$r = \frac{1}{k} = \frac{m \, K}{W}$$

Conductivities of a few of the more common building materials are given in Table 1.2, but the *Guide, Section A3* should be consulted for fuller information.

Note that the suffix '. . ivity' in these expressions signifies per unit thickness; the suffix '. . ance' signifies the value overall for the thickness stated.

U **Values**—The heat transmitted in watts per square metre per kelvin (W/m² K) for the purposes of heat loss calculations is termed *the trans-*

mittance coefficient U, which is the reciprocal of the sum of all the resistances. Thus:

$$U = \frac{1}{R_{s1} + R_{s2} + r_1 L_1 + r_2 L_2 \text{ etc. } + R_a + R_h}$$

R_{s1} = surface resistance internal,
R_{s2} = surface resistance external,
r_1, r_2, etc. = resistivity of individual materials of which element of structure is composed, e.g. brick, plaster, etc.
L_1, L_2 = thickness of individual materials,
R_a = resistance of air gap (if any),
R_h = resistance of hollow block (if any) for which resistivity per unit thickness does not apply.

The values of the surface resistances R_{s1}, R_{s2}, vary with the type of surface, degree of roughness, and with the air movement over the surface. For the purpose of calculating U values they may be taken as follows:

Internal Resistance R_{s1}
Walls - - - - - - 0·13 m² K/W
Ceilings and Floors
    upward heat flow - - - 0·11 m² K/W
    downward heat flow - - - 0·15 m² K/W

External Resistance R_{s2}
'Normal Exposure'
Walls - - - - - 0·06 m² K/W
Roof - - - - - 0·05 m² K/W

The resistance of an air gap R_a in normal material such as brickwork, wall-board, etc., 20 mm wide or over, may be taken at the value of 0·18. If the air gap is lined with bright metallic surfaces, the value becomes 0·35. If, as in a recent development, the cavity is filled with insulating material to increase its resistance, it ceases to behave as an air gap and the wall becomes a sandwich construction. But see p. 26 for further reference to cavity fill.

If the transmittance coefficients multiplied by the surface areas of all the exposed surfaces in the room are added together, we then have the total quantity of heat which will be lost through the fabric per kelvin, inside to outside. Table 2.1 gives a range of transmittance coefficients from the *Guide* by way of example of various forms of construction, but there are nowadays so many composite elements involved, that it is usually necessary to calculate the U factor from first principles.

For example, consider a curtain wall construction comprising:

Outside - - - 40 mm thick concrete panel
Air gap - - - 20 mm
Inside - - - 75 mm lightweight concrete block
Plaster - - - 20 mm

the resistances are summated thus

	k	$r=\dfrac{1}{k}$	thickness m		R
Concrete Panel	1·40	0·72	× 0·04	=	0·028
Air gap					0·18
Concrete Block	0·18	5·55	× 0·075	=	0·415
Plaster	0·46	2·18	× 0·02	=	0·044
Surface R_{s1}					0·13
R_{s2}					0·06
			Total R	=	0·857

$$U = \frac{1}{R} = \frac{1}{0·857} = 1·17 \text{ W/m}^2 \text{ K}$$

It should be noted that the transmittance coefficients U quoted in Table 2.1 are for normal exposure. The *Guide* gives further values for sheltered and severe exposure, as follows:

'Sheltered'—including—the first two storeys above ground in the interior of towns;

'Normal' —the 3rd, 4th and 5th storeys of buildings in the interior of towns and most suburban and country premises;

'Severe' —the 6th and higher storeys of buildings in the interior of towns and all storeys of buildings on hill, coast or riverside sites.

U **Values and Moisture**—A review of basic data on which U values have been based has been made by Loudon,* comparing calculated values with experimental data derived for BRE wall and roof laboratories. It has been shown that, in general, the conductivities of all masonry materials bear the same relationship with dry density and follow similar increases proportionately with increasing moisture content. The *Guide* is based on corrected practical values of moisture content thus:

Brickwork, protected from rain   1% Moisture Content by volume

Concrete,	,,	,,	,,	3%	,,	,,
Brickwork, exposed to rain				5%	,,	,,
Concrete,	,,	,,		5%	,,	,,

It is pointed out that driving rain and condensation may give higher moisture contents.

Thus, calculation of U values from conductivities should take account of moisture content where the material is exposed to weather. Inner skins of building elements should remain dry, as in the case of the inner material of a cavity wall, and the protected value is then used.

Table 2.1 from the *Guide* is on this basis.

* *U values for the 1970 Guide,* J.I.H.V.E., 1968, 36, 167.

U **Values for Floors**—Transmittance coefficients for solid floors on earth as given in Table 2.1 need some explanation. Where the building covers an extensive area the loss will be chiefly around the perimeter, as the earth temperature near the centre will have built up to nearly that of the room. In a small or narrow building the perimeter loss will be proportionately greater. The *Guide* gives the appropriate values for floors of varying extent of which the coefficients given in Table 2.1 are samples. They may be multiplied by the full temperature difference inside to outside.

TABLE 2.1

TRANSMITTANCE COEFFICIENTS U
(W/m^2 K) FOR 'NORMAL' EXPOSURE

(*including allowance for moisture as appropriate*)

WALLS

	Thickness, mm*				
	105	220	260	335	375
Brickwork					
Solid					
Unplastered - - - - - - - - - -	3·3	2·3	—	1·7	—
Plastered - - - - - - - -	3·0	2·1	—	1·7	—
Cavity (unventilated) brick outer skin					
Brick inner skin, plastered - - - - - -	—	—	1·5	—	1·2
Lightweight concrete inner skin, 13 mm expanded polystyrene in cavity, plastered - - - - -	—	—	0·7	—	—
Brick inner skin, plastered, cavity filled with urea formaldehyde or mineral wool - - - - -	—	—	0·5	—	—

	Thickness, mm		
	150	165	200
Concrete			
In situ			
Cast, unplastered - - - - - - - - - -	3·5	—	3·1
Cast with 50 mm woodwool on inside, plastered - - - -	1·1	—	1·1
Pre-cast			
Panel with 50 mm cavity and lined with 25 mm expanded polystyrene plus plasterboard finish - - - - -	—	0·80	—

Framed Constructions	
5 mm asbestos cement sheet	
Bare on frame - - - - - - - - - - -	5·3
On frame with cavity and aluminium foil backed plasterboard - - - -	1·8
Double skin with 25 mm glass fibre insulation in between - - - - -	1·1
Tile hanging on timber battens and building paper, 50 mm glass fibre in cavity, plasterboard finish - - - - - - -	0·65
Weather boarding on building paper and timber frame, 50 mm glass fibre in cavity, plasterboard finish - - - - - - -	0·62
Curtain Walling (typical examples)	
With 5% bridging by metal mullions	
Mullions projecting outside - - - - - - - -	1·2
Mullions projecting inside and outside - - - - - - -	1·8
With 10% bridging by metal mullions	
Mullions projecting outside - - - - - - - -	1·5
Mullions projecting inside and outside - - - - - - -	2·8
GLAZING (*measured over wall opening*)	
Single	
Metal frames - - - - - - - - - - - -	5·6
Timber frames - - - - - - - - - - -	4·3
Double, air space 20 mm or over	
Metal frames - - - - - - - - - - -	3·2
Timber frames - - - - - - - - - - -	2·5

TABLE 2.1 *continued*

Double, air space 3 mm
  Metal frames - - - - - - - - - - - -   4·8
  Timber frames - - - - - - - - - - -   3·9
Roof Skylight
  Open - - - - - - - - - - - -   6·6
  With laylight under, ventilated - - - - - - - - -   3·8

DOORS
  Solid timber, 25 mm thick - - - - - - - - - -   2·6

ROOFS
Flat
  Concrete 150 mm thick, covered asphalt or felt
    Uninsulated - - - - - - - - - - - -   3·4
    With 100 mm (average) lightweight concrete screed to falls, plaster soffite - -   1·8
  Hollow tiles 150 mm thick, covered asphalt or felt
    Uninsulated - - - - - - - - - - - -   2·2
    With 100 mm (average) lightweight concrete screed to falls, plaster soffite - -   1·4
  Woodwool slabs, 50 mm thick with asphalt or felt on 13 mm screed, on timber joists
    Aluminium backed plasterboard ceiling - - - - - - -   0·9
    25 mm glass fibre above ceiling - - - - - - - -   0·6
  Hollow or cavity asbestos cement decking with asphalt
  or felt on 13 mm insulating board
    Cavity as void - - - - - - - - - -   1·5
    Cavity with 25 mm glass fibre - - - - - - - - -   0·73

Pitched
  Tiles on battens roofing felt and rafters, ceiling below joists
    Aluminium backed plaster board - - - - - - - -   1·5
    50 mm glass fibre over plaster board - - - - - - - -   0·5
  Corrugated asbestos cement or plastic covered steel sheeting
    Uninsulated - - - - - - - - - - -   6·7
    With cavity and aluminium backed plaster board lining - - - - -   2.0
  Corrugated double-skin asbestos cement sheeting with 25 mm glass fibre insert -   1·1

	4 edges exposed	2 edges exposed
FLOORS		
Solid Floors in contact with earth		
Very long × 30 m broad - - - - - -	0·16	0·09
Very long × 7·5 m - - - - - - -	0·48	0·28
60 m × 60 m broad - - - - - - -	0·15	0·06
60 m × 15 m broad - - - - - - -	0·32	0·18
30 m × 30 m broad - - - - - - -	0·26	0·15
30 m × 7·5 m broad - - - - - - -	0·55	0·32
15 m × 15 m broad - - - - - - -	0·45	0·26
15 m × 7·5 m broad - - - - - - -	0·62	0·36
7·5 m × 7·5 m broad - - - - - -	0·76	0·45
3 m × 3 m broad - - - - - - -	1·47	1·07

Timber Floors
  With ventilated airspace below - - - - - - - -   1·3
  With 25 mm fibre board below floor boarding over ventilated airspace - -   1·08

	Heat Flow	
	Upwards	Downwards
Intermediate Floors		
Timber on joists, plaster ceiling - - - - - -	1·7	
Concrete		
150 mm thick, 50 mm screed - - - - - -	2·7	2·2
150 mm thick, 20 mm woodblocks - - - - - -	2·0	1·7
Hollow tile		
150 mm thick, 50 mm screed - - - - - -	2·0	1·7
150 mm thick, 20 mm woodblock - - - - - -	1·6	1·4

BUILDING INSULATION

Building insulation has now come to be regarded as an essential element in construction and an inherent part of the building. It is indeed frequently possible, by the selection of the right materials, to achieve a high degree of insulation with little or no extra cost. Insulating materials are generally porous by nature, and hence weak structurally, so are more commonly used as inner skins. When used on flat roofs, they are prone to absorb condensation, and hence some vapour seal surrounding them, or means of venting, is needed.

The Thermal Insulation (Industrial Buildings) Act (1957) requires industrial buildings to be insulated in so far as the roofs are concerned, and certain maximum figures of U values according to temperatures are stated in the Schedule under the Act, in effect setting the maximum at 1·7 W/m² K. The Health and Safety at Work Act (1974), Part III, provides power whereby the Secretary of State may make Regulations with respect to the design and construction of (all) buildings for the purpose of 'Furthering the conservation of fuel and power'. This represents a considerable advance in that previous powers existed only under the Public Health Acts, were specific to the construction of dwellings and were not related to energy conservation.

For dwellings, the Building Regulations 1976 (Second Amendment) apply. These stipulate that:

1. The U value of any part of any wall, floor or roof shall not exceed the values stated:

External wall	1·0 W/m² K
Internal wall	1·7 ,,
Floor to ventilated void	1·0 ,,
Roof	0·6 ,,

2. The calculated average U value of perimeter walling shall not exceed 1·8 W/m² K.

3. For the purpose of calculating the average U value, that for windows shall be taken at 5·7 W/m² K for single glazing and 2·8 W/m² K for double glazing. Other openings to be taken at a U value equivalent to the wall.

   The value of walls between one dwelling and another, or an enclosed unventilated space in the same dwelling, shall be taken at a U value of 0·5 W/m² K.

Considering a room with one exposed wall and a window area of 25 per cent. with single glazing, we have

$$(5\cdot7 \times 0\cdot25) + (U_w \times 0\cdot75) = 1\cdot8.$$

Hence, $\qquad U_w = 0\cdot5.$

Some form of wall insulation would be required to achieve a value of 0·5 W/m² K. (See note on cavity fill, p. 26.)

Alternatively, if double glazing were used in this example, it would be found that a wall U value of 1·46 W/m² K would satisfy the average and that a normal uninsulated cavity construction would be adequate were it not for the overriding requirement for a maximum U value of 1·0 W/m² K. Some insulation to the wall would therefore still be required.

Burberry* has shown that the Regulations may be presented graphically for a variety of dwelling types and plan ratios. Fig. 2.2 illustrates, for a detached building, the maximum glazed areas permissible for two levels of wall insulation. Higher percentages are allowable in the case of semi-detached and terraced houses.

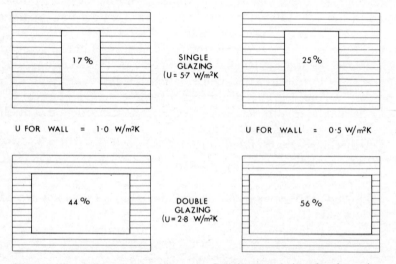

Fig. 2.2—Permissible window areas under Building Regulations for alternative wall and window constructions.

A draft document, currently in circulation for comment,† proposes that the Building Regulations (1976) be amended under the Health and Safety at Work Act (1974) such that:

1. Buildings in purpose groups II to VIII, i.e. other than dwellings, fall within the scope of the Regulations.
2. Maximum U values for walls and floors are 0·7 W/m²K for factories and heated storage buildings and 0·6 W/m²K for all other buildings.
3. Maximum wall areas of single glazing are 15 per cent for places of assembly, factories and heated storage buildings; 25 per cent

---

* 'The New Insulation Standards for Houses', *Architects' Journal*, 1975, pp. 161, 235.

† Now published as Building Regulations (First Amendment) 1978 and due to be applied from 1st July 1979.

for institutional and residential buildings and 35 per cent for offices and shops.

4. Maximum roof areas of single glazing are 20 per cent.

5. As an alternative to meeting the prescribed U values and limiting the glazed area, the building designer may use alternative construction methods provided that the heat loss then arising will not exceed that which would have obtained had the rules in (2) (3) and (4) been followed.

A number of the more common types of insulating material are listed in Table 2.2, together with their conductivities.

TABLE 2.2

THERMAL INSULATING MATERIALS FOR BUILDING

Material	Bulk Density kg/m³	Conductivity W/m K
Asbestos insulating board	700	0·110
Asbestos, sprayed	240	0·075
Concrete, Lightweight block*	600	0·180
Corkboard	145	0·042
Fibreboard (insulating)	380	0·060
Glassfibre, quilt	80	0·040
Kapok, quilt	30	0·030
Mineral wool, loose mat	180	0·042
Polystyrene, expanded, board	15	0·037
Polyurethane board	30	0·023
Pumice, loose granules	350	0·070
Sawdust, loose	145	0·080
Thatch, straw	240	0·070
Urea formaldehyde foam	10	0·031
Vermiculite, granules	100	0·065
Wood-wool, slabs	600	0·110

* Protected, moisture content 3% by volume

With regard to glass, double-glazing reduces the heat loss from this element by half, and is desirable on its own account in cases of severe exposure. The cost of double-glazing is generally high, and it is often difficult to make a case on economic grounds. If windows are to be open for ventilation, the value of double-glazing as a noise barrier is much reduced. Hence double windows go with some means of mechanical ventilation.

Where translucence only is necessary, a sandwich form of insulating glazing may be used containing glass-fibre in the centre.

Figs. 2.3 and 2.4 show methods of insulating walls and roofs.

Another reason for the use of insulation in buildings is for the prevention of condensation. If the internal surface temperatures of walls and/or roofs drop below the dew point of the air within the space, condensation is bound to occur. It is necessary then to calculate the thickness of insulation necessary to avoid this condition arising. Condensation on windows, where it is likely to occur through crowded occupancy or pro-

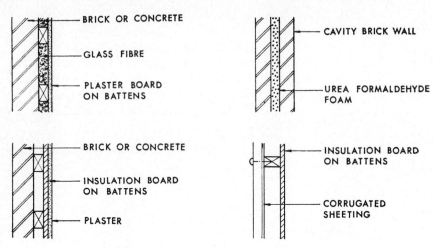

FIG. 2.3—Methods of applying insulation to walls.

duction of steam or vapour within the space, can be dealt with by double-glazing, or by heating the surface at the bottom of the glass.

Apart from the benefit of insulation as a means of reducing heat losses in winter, it will equally be effective in reducing heat gains in summer. Where a building is air-conditioned, as will be discussed later, double-glazing can generally be shown to be economically justified, owing to the much higher cost of cooling than of heating.

Where a concrete floor is suspended with a cavity below or indeed possibly an open space below, condensation may occur on the floor surface above unless insulation is provided of adequate thickness. It may well be found that this is best placed on the top of the slab rather than underneath, due to edge conduction. Edge conduction from exposed balconies can also be troublesome where the floor slab and the balcony are one homogeneous unit. This has been a fruitful cause of mould growth in multi-storey blocks of flats.

Cavity Fill—The resistance R_a of an airgap, as stated earlier, may be taken as 0·18 m²K/W. If the cavity is filled with an insulating foam such as urea formaldehyde resin for a 50 mm gap, the resistance becomes

$$\frac{1}{0\cdot031} \times \frac{50}{1000} = 1\cdot6 \text{ m}^2 \text{ K/W}$$

an increase of 1·42 m² K/W.

Taking a normal cavity wall of 260 mm thickness overall, from Table 2.1 (U value = 1·5 W/m² K) but with foam insulation in the cavity, the U value becomes 0·48 W/m² K, i.e. roughly ⅓ of the normal. U values with other materials and widths of filled cavities may be similarly calculated using the appropriate factors.

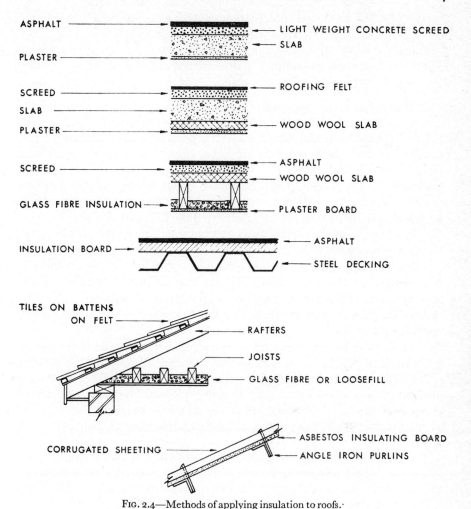

FIG. 2.4—Methods of applying insulation to roofs.

Features to be noted concerning the filling of cavities include the following:

(*a*) Savings in overall fuel or energy consumption over a period will depend on area of wall relatively to total heat loss.

(*b*) Theoretical savings may not be reached if heating is intermittent due to heat absorbed by the inner skin being dissipated during the non-heating period.

(*c*) When a steady state is reached, the inner wall surface will be at a higher temperature than that of an unfilled cavity thus adding to comfort in the room and also reducing possibility of condensation.

(d) Any insulated construction enables a saving to be made in the initial capital cost of the heating installation.

(e) Filling of cavities can seriously reduce sound insulation between dwellings such as by transmission through a U turn at the end of a party wall.

Safeguards in this method of treatment are necessary owing to the fact that rain penetration through the outer skin must not be allowed to enter the foam or other added material, otherwise the insulating value will be lost and furthermore damp patches may occur on the inner wall face. It has always been a cardinal rule in the building of cavity walls that mortar droppings and the like which could bridge the gap, particularly by lodging on wall ties, must be avoided. For this reason Building Regulation C9 has in the past not permitted the filling of cavities.

However, with the significant potential energy savings accruing, the Department of Environment in Circular 105/75 allows local authorities to relax regulations so as to permit such insulation of buildings, subject to BRE Agrement Board approval of the material and method of application. For instance, the material must be water resistant and not subject to rotting or mould growth: only approved firms may be employed to apply the treatment.

The chief danger is that of driving rain penetrating the outer skin of the wall, if of porous brick, and finding a channel through the insulating material if this contains voids or cracks due to shrinkage. Once a cavity is filled there is no known method of clearing it. Some literature has referred to the necessity of waterproofing the external skin of the wall.

Materials used for the purpose of cavity fill include:

(a) Urea formaldehyde which is injected as a plastic foam under pressure through holes drilled in an existing wall at intervals checks being made to ensure that the cavity is completely filled; new construction may be filled from the top as work proceeds. The material has a short 'shelf-life' and frequent quick deliveries from works are arranged.

(b) Mineral wool, treated with a water repellant, blown in to the cavity by air-jet in a manner similar to that used with foam.

(c) In new construction, other materials used include resin bonded rockwool slabs, resin bonded glass fibre slabs and polystyrene slabs.

Alternatively, for new construction, the use of a lightweight inner skin (Table 2.1) achieves a U value of $1 \cdot 0$ W/m² K compared with $1 \cdot 5$ uninsulated and roughly $0 \cdot 5$ for cavity fill, and the question whether the added complications and risks of insulation in the cavity are worth the extra savings must be considered for each separate application.

Further detailed information on this subject may be found in BRE publication TIL. 41 of January, 1972, and the addendum of February, 1974. Many articles have appeared in *The Architects' Journal*, particularly in the issues of 11 August 1971, 25 August 1971, 29 September 1971 and 12 November 1971.

Heat Required for Air-Infiltration—What has been said so far relates to heat lost from a room through the fabric. There is an additional loss due to movement of air through the room, such air generally entering from outside through the lack of seal of the structure itself. If there is a flue in the room, this will frequently create a suction by which air is drawn into the room, sometimes at an excessive rate. Leakage through windows may be greatly reduced by weatherstripping or by the use of double windows. In single-storey sheeted factory construction, air leakage often occurs through joints in the sheeting, particularly at eaves and ridges. Large door openings may involve considerable air change. In tall buildings the chimney effect will produce strong inward air currents at doors at ground level.

The importance of air-infiltration is that it frequently accounts for as much as one-third or a half of the heat losses in total, and yet it is the most indeterminate. It is this question of the indeterminacy of air-infiltration losses that renders temperature guarantees for buildings difficult to sustain, since any test is more a test of the building itself than of the heating installation.

There are two methods of assessment for air infiltration. One is empirical and is based on the allowance of an assumed number of changes per hour of the content of the space in question. The first column of Table 2.2 lists the commonly accepted rates for various types of buildings.

Where leakage is confined to windows, as in a multi-storey office block or block of flats, another more precise method has been devised. To use this method it is necessary to know the building exposure, height, type of window and opening areas. The *Guide*, *Section A4* gives in detail the graphs and factors applying to this technique, but it is too involved to summarise here. By way of example, however, taking a city centre site and a building height of 30 m with weather stripped sliding windows, the infiltration rate is given as 0·45 litre/s per metre of opening joint; without weather-stripping, the figure is 0·9. Values range between 0·1 and 4·0 litre/s m. Having arrived at the air-flow rate in this manner, it may be converted into an hourly quantity for comparison with the air-change method previously referred to.

The quantity of heat required per m^3 to warm the infiltration air is obtained from the specific heat capacity of air (at constant pressure) at 20° C = 1·01 kg/kg K (from Table 1.2) and the mass per m^3 = 1·2 kg. Thus the heat required to raise 1 m^3 by 1 kelvin = 1·01 × 1·2 = 1·212 kJ.

To be consistent with the units for heat loss through the building fabric (watts = joules per second) the hour is an unacceptable time interval and, of course, the use of air changes per second would lead to unwieldy values over the accepted range. Using the heat-requirement conversion, however, for one air change we have

$$m^3 \text{ (the volume of the space)} \times \frac{1 \cdot 212 \times 1\,000}{3\,600}$$

$$= 0 \cdot 33 \text{J/sm}^3 \text{ K } (\text{W/m}^3 \text{ K})$$

Having regard to the arbitrary nature of the definition of air-change rate this may be rounded to $0 \cdot 3$ W/m³ K. The term *ventilation loss* has been adopted to define the SI 'air-change' units which have values as listed in the right-hand column of Table 2.3.

TABLE 2.3

NATURAL AIR INFILTRATION FOR HEAT LOSSES

(Air Changes and Specific Ventilation Loss)

Room or Building	Air Change per hour	Ventilation Loss (W/m³ K)
Large spaces (e.g. Factories, Assembly Halls, Churches Canteens) Solid Construction		
up to 5000 m³	1	0·3
5000 to 25 000 m³	$\frac{1}{2}$	0·2
over 25 000 m³	$\frac{1}{4}$	0·1
Light Construction		
up to 5000	3	1·0
5000 to 25 000 m³	$1\frac{1}{2}$	0·5
over 25 000 m³	$\frac{3}{4}$	0·3
Living spaces, offices, libraries		
Windows exposed one side	1	0·3
Windows exposed two sides	$1\frac{1}{2}$	0·5
Windows exposed more sides (if windows are weather-stripped, halve the above rates.)	2	0·7
Circulating Spaces Generally	2-3	0·7-1·0
Lavatories	2	0·7
Laboratories	3	1·0
Hospitals, Schools etc See Department of Health and Social Security, Ministry of Education and Science publications setting standards		

NOTE: This Table refers to air-change rates for heat loss calculations. Where mechanical inlet and extract ventilation is provided, heat for additional air-change must be added.

Whilst the heat quantity for air loss arrived at by either method will be that to be offset by heat emission from the appliances room by room, the sum total of all the rooms will not necessarily represent the total heat input required for the building. Windows on the windward side will be subject to inward leakage, whereas those to rooms on the leeward side will be the means by which warm air escapes. If, therefore, rooms are disposed equally on two sides of a corridor, the total air loss will be 50 per cent of the sum of the individual room air losses; this would not of course apply to a building one room wide with windows on both sides.

The precise method given in the *Guide, Section A4*, shows how the overall building-air loss can be derived from the room totals for various dispositions and proportions of glazing. It is noted that the building-air loss may be as little as one third of the sum of the individual room-air losses. If the air change or ventilation-loss method has been employed, some similar allowance can be made by assessment from the drawings. It will be clear that this procedure applies not only to the sizing of boiler plant, but also to the estimation of fuel consumption.

It is clear that in the interests of conservation of energy and economy of installation and running costs for all time, money spent on weather-tightness of windows is a real economy, since uncontrolled ventilation goes on for 24 hours a day, whereas ventilation is necessary only during the occupied period of a matter of a few hours as a rule, and can usually be dealt with by some controlled means, such as the opening of a window. During occupation, the heat of occupants alone is sufficient to allow of some additional ventilation; for instance, in the case of school classrooms it is recognized that the heat of the children is equivalent to three air-changes per hour and so the heating system is designed accordingly.

If ventilation is necessary at a known rate, such as in workshops and factories coming under the Factories Act, or in places of public entertainment coming under local authority regulations, mechanical ventilation is required (dealt with in later chapters); in which case, so far as the heating system is concerned, the designer has to make a choice as to whether he will install heating equipment to cope with the fabric losses only, or whether he will include this heat in the ventilation air, bringing it in somewhat warmer to allow for this loss. What has been said earlier relates to buildings and rooms where no such mechanical ventilation is involved.

TEMPERATURE DIFFERENCE

The sum of the transmission losses and the air-infiltration loss, both in W/K, requires to be multiplied by the difference of temperature between that of the room inside and some assumed temperature outside.

The inside temperature depends on the purpose of the various rooms and the kind of building, and a list of conventional temperatures is given in Table 2.4.

With regard to the outside temperature, in the British Isles this is commonly taken at $-1°$ C or $0°$ C, but such a basis, whilst it has sufficed for the types of buildings and central heating systems of the past, may not get by during periods of colder weather with lightly constructed buildings and heating systems with no inherent margin.

A committee set up in 1950 by the various Institutions, in its report

'Basic Design Temperatures for Space Heating'*, examined the frequency of cold spells and the likely number of days per annum on average when the specified internal temperatures would not be met by designs based on − 1° C. The argument brought out in this report is that a heavily constructed building has sufficient thermal time lag to tide over a period of a few days of exceptionally cold weather, whereas a lightly constructed building has not this thermal capacity, and so will suffer more quickly.

TABLE 2.4

ROOM TEMPERATURES
(Generally accepted levels—rounded)

Space or Area	Temp. °C
Swimming Pool Halls	26
Art Classrooms, Bathrooms, Changing Rooms, Hotel Bedrooms	22
Banqueting Halls, Hotel Lounges, Operating Theatres, Residential Living Rooms, X-Ray Rooms	21
Banking Halls, Canteens, Laboratories, Libraries, Offices, Reading Rooms	20
Assembly Halls, Auditoria, Churches, Domestic Bedrooms, Exhibition Halls, Factories (sedentary work), Hospital Day Rooms and Wards, Restaurants, School Classrooms, Shops	18
Cloakrooms, Factories (light work), Gymnasia, Lobbies and Circulation Spaces, Sports Halls, Stores (working areas)	16
Factories (heavy work), Warehouses (storage areas)	14

TABLE 2.5

EXTERNAL DESIGN TEMPERATURES

Multi-storey buildings with solid
   intermediate floors and partitions - Type A
Single storey buildings   -   -   - Type B

Type of Heating System	Thermal Time Lag of Building	
	Type A	Type B
Hot water with overload capacity 15–20% (25% if no tolerance on indoor temperature allowable) - - - - -	− 1° C	− 3° C
System without overload capacity (e.g. electric, steam unit heaters) - - -	− 4° C	− 5° C

* *Post-War Building Study* No. 33, H.M.S.O., 1955.

Furthermore, central heating systems with boilers of some sort usually have some overload capacity which can be brought into use during periods of exceptional cold, whereas certain other systems such as electrical floor panels and free standing storage units or those relying on heat emitters of a fixed output capacity, such as unit heaters, have no such reserve. The *Guide* recommends that systems should be designed to the criteria listed in Table 2.5.

Arising from an analysis of background data to international norms, Billington* has proposed that the United Kingdom be divided into three climate zones for the purpose of determining standards for structural insulation and selection of external design temperatures. In place of the datum level of $-1°$ C quoted in Table 2.5, the alternatives shown in Fig. 2.5 are proposed.

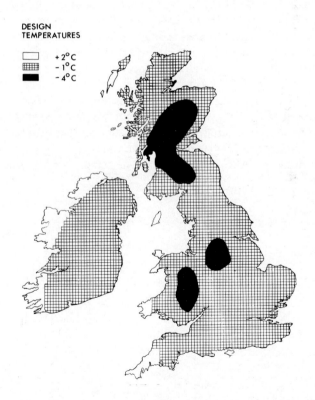

DESIGN TEMPERATURES

+ 2° C
– 1° C
– 4° C

FIG. 2.5—Proposed variation of outside design temperature by geographical area in the U.K.

* *Thermal Insulation of Buildings*, J.I.H.V.E., 1974, 42, 63.

It should perhaps be mentioned that, in calculating total heat losses, the temperature difference across inner walls or partitions or from one floor to the next (where the adjacent rooms are warmed to a different temperature or are unwarmed) are best taken off separately, and the loss per degree difference multiplied by some lesser assumed temperature difference. Transmittance coefficients should also be adjusted in such cases, as both surfaces are internal.

ALLOWANCE FOR HEIGHT

It appears reasonable to make allowance for the height of a room bearing in mind that warm air rises towards the ceiling. Thus, in a room designed to keep a comfortable temperature in the lower one and a half or two metres, a higher temperature must exist nearer the ceiling, which will inevitably cause greater losses through the upper parts of windows walls, and roof. This effect is greatest with a convection system, i.e. one which relies on the warming of the air in the room for the conveyance of heat. This would occur in the case of conventional radiators and convectors.

In the case of radiant-heated rooms the same tendency does not occur, and a much more uniform temperature exists from floor to ceiling. In the case of floor heating there is virtually no temperature gradient whatever. This effect is illustrated in Fig. 2.6 which gives the vertical temperature gradients which may be expected with radiators and warmed-floor and ceiling systems. Fig. 2.7 gives the same for warm-air systems and shows how a low-level inlet produces a much smaller temperature gradient than a high-level inlet due to better mixing with room air.

In the design of a heating installation, much can be done to reduce

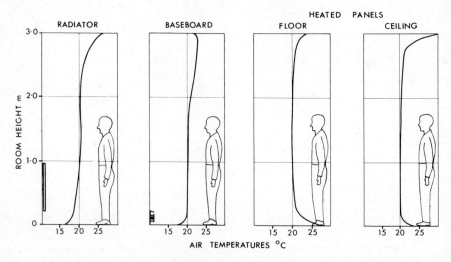

FIG. 2.6—Vertical temperature gradients in a room with various types of heating, assuming rooms above and below heated by similar means. See also Fig. 2.7.

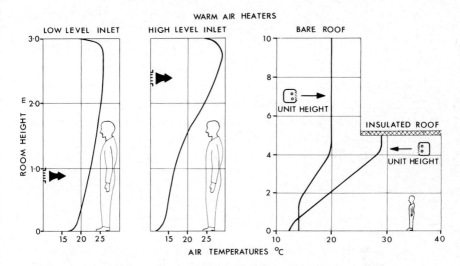

FIG. 2.7—Vertical temperature gradients with warm-air heating and unit heaters.

temperature gradient by the correct disposition of the heating surfaces under large cooling surfaces such as windows. In fact it might be said that a system producing a big temperature gradient is a bad system.

Table 2.6 gives height factors proposed in the *Guide, Section A9*, for various types of system.

TABLE 2.6

HEIGHT FACTORS

Type of Heating	Type and distribution of Heaters	% addition for height of heated space		
		5 m	5-10 m	over 10 m
Mainly Radiant	Warm floor	nil	nil	nil
	Warm ceiling	nil	0–5	*
	Medium and high temperature cross radiation from intermediate level	nil	0–5	5–10
	Ditto, downward radiation from high level	nil	nil	0–5
Mainly Convective	Natural warm air convection	nil	0–5	*
	Forced warm air crossflow from low level	0–5	5–15	15–30
	Ditto, downward from high level	0–5	5–10	10–20

* not applicable

HEAT LOSSES GENERALLY

Fig. 2.8 and the following heat loss calculation provide a simple example of the method normally adopted. The roof is partly exposed to the outside and partly has a warmed room over; so the latter has no loss.

The floor being solid, on earth, is taken at the appropriate coefficient for an area of this kind from Table 2.1.

CONDITIONS OF COMFORT

It has been mentioned already that the ordinary thermometer measures air temperature. If placed in a room in which the air, walls, floor and ceiling are all at the same temperature, and if this temperature is one in which we say we are comfortable, then the thermometer may be regarded as giving a satisfactory index of comfort. That is to say if under the same conditions a lower temperature is registered, we shall no doubt say it is too cold, and likewise if a much higher temperature is registered that the room is too warm.

Such conditions of uniformity of temperature of air and surfaces, however, rarely pertain. In a warm room there is 'cold radiation' from the window, meaning in effect that heat is escaping from the body through the glass at a greater rate than to other surfaces, and, in order to counter-balance this, either the air temperature has to be raised or some form of radiant heating surface be provided. Under these conditions, where air temperature and surface temperatures of the enclosure are different, the ordinary thermometer fails as a true index of comfort. Furthermore, comfort is concerned with air movement. Even if air is warm, at a speed of much over 0·15 m/s most people would express discomfort.

Numerous attempts have been made to devise a scale of comfort of which the following deserve attention.

The Equivalent Temperature Scale. This combines the effect of air temperature, radiation and air movement, as measured by an instrument named the *Euphatheoscope*, a laboratory instrument. The late Mr A. F. Dufton, who developed this scale and instrument, defined equivalent temperature as that temperature of a uniform enclosure in which, in still air, a sizeable black body would lose heat at the same rate as in the environment, the surface of the body being one-third of the way between the temperature of the enclosure and 38° C (i.e. approximately blood heat). The instrument contained an electric heater, to maintain the surface temperature at a fixed point—the consumption of the heater being a measure of the Equivalent Temperature. Being a warm body it was sensitive to air movement.

Resultant Temperature is more easily measured than Equivalent Temperature. It is used as the most convenient index of warmth in the *Guide*. At 0·1 m/s the Resultant Temperature is the mean of air and mean-radiant temperature.

Environmental Temperature. This is not a comfort scale but is held to correlate more closely with measured heat losses at BRE wall and roof

EXAMPLE: HEAT LOSS
(See Fig. 2.8, all dimensions in m)
Cube (allow 1½ air changes)
14·5 × 8·5 × 4·0 = 493
6·0 × 8·5 × 4·0 = 204

697 × 0·5 = 398·5

Windows
16·0 × 2·0 = 32 × 5·7 = 182·4

Wall
48·5 × 4·0 = 194
less windows = 32

162 × 1·5 = 243·0

Floor
14·5 × 8·5 = 123
6·0 × 8·5 = 51

174 × 0·62 = 108·0

Roof
6·0 × 8·5 = 51 × 1·8 = 91·8

1023·7 W/K

Allow 0° C outside, 20° C inside = 20 K rise
∴ Loss = 1023·7 × 20 = 20 474 W (say 20·5 kW)
(This approximates to 30 W/m³)

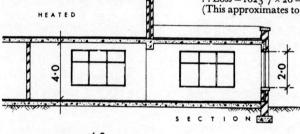

SECTION

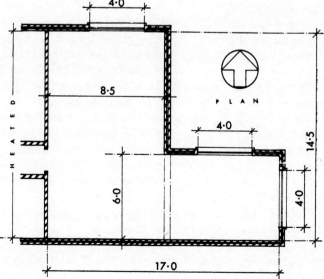

PLAN

FIG. 2.8—Heat Loss Example.

laboratories than other scales. For practical purposes, except where extremes of air temperature or radiant intensity occur, it can be regarded as equal to Equivalent or Resultant Temperatures. Environmental Temperature is a weighted mean between mean-radiant and air temperatures such that it is equal to $\frac{2}{3}$ mean-radiant temperature plus $\frac{1}{3}$ air temperature.

Globe Thermometer. This instrument comprises a thermometer inserted into a matt black sphere of copper or other thin material about 150 mm in diameter (Fig. 2.9). When suspended in a room, the black sphere will absorb heat radiation or emit radiation to cold surfaces, and so achieve some kind of balance, comparable with what is termed the *mean radiant temperature* of the enclosure: that is, the weighted sum of all the surfaces at their varying temperatures. In still air, or with air movements of a low order, up to say 0.15 m/s, the Globe thermometer temperature in conjunction with the air temperature may be taken as an indication of the equivalent temperature and will be closely akin to Resultant Temperature.

If strong draughts prevail, this thermometer tends towards air temperature. It is sluggish in response and, in normally heated rooms

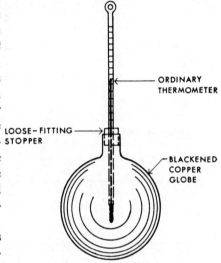

ORDINARY
THERMOMETER

LOOSE-FITTING
STOPPER

BLACKENED
COPPER
GLOBE

FIG. 2.9—Globe thermometer.

(such as offices) even with radiant ceiling heating, rarely reads more than 1 kelvin above air temperature. In large spaces served with high temperature radiant surfaces, such as in factories, a Globe thermometer reading as much as 5 kelvins in excess of air temperature may be found. In spaces heated by means of warm floors, with the thermometer 1 m above the floor, an excess temperature of about 2 kelvins may be expected.

Effective Temperature. This is a scale which has been devised and developed in the United States, mainly applicable to air-conditioning, and combining the effects of air movement, air temperature and humidity, but having no reference to radiation. The scale is the outcome of a long series of subjective tests carried out on a number of people in a great variety of conditions, and has been revised from time to time. The results are expressed in a series of curves known as *Effective Temperature Lines.* This scale has little application where heating of a building alone is being considered, since, within the range of 15–20° C, variations in relative humidity between 40 per cent. and 70 per cent. have small effect on comfort. (See Fig. 2.10.)

A Corrected Effective Temperature Scale has been devised to include the

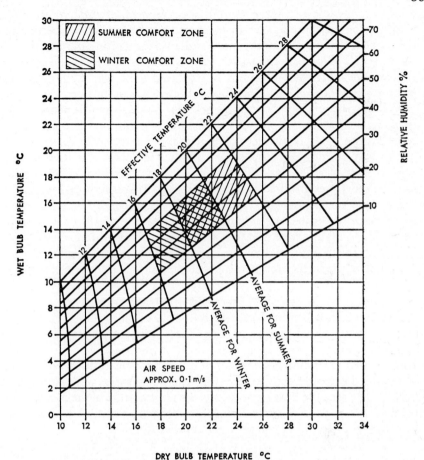

FIG. 2.10—'Effective temperature' chart. This chart is postulated from a variety of sources as applicable to Great Britain only. It gives a winter and summer range of conditions within which most people would feel comfortable, provided they are not transient.

effect of radiation, and, for this, four measurements are required: the Globe thermometer temperature, the wet bulb temperature, the air speed and the dry bulb temperature. By means of a nomogram, given these four measurements, the Corrected Effective Temperature can be obtained. Its reliability is restricted to the central part of the scale.

Application—It will be clear that, whilst the various methods of measuring comfort may be of interest and concern as a matter of research or appraisal of one system against another, in so far as the design of heating systems is concerned, their application is difficult. In a space heated by means of radiant surfaces, the radiation falling on the various surfaces of the room and on the objects within the room, such as furniture, causes them to be warmed slightly so they in turn warm the air in contact by convection. The air temperature and radiant temperature thus tend to approach one another and eventually become nearly equal. With radiant heating, the air temperature as measured by an ordinary thermometer will be slightly cooler than the mean radiant temperature derived from the reading of a Globe thermometer and, conversely, if the heating is convective, the air temperature may be expected to be higher than the mean radiant temperature. The differences will be slight in the normal case.

Where the problem concerns heating by high-temperature radiant surfaces, such as in a factory, it is possible to estimate the mean radiant temperature, and a convenient method is given in the *Guide, Section B1*.

By way of example, if the design intent is to achieve an environmental temperature of 18° C and the air temperature is taken at 13° C, it will be found that a mean radiant temperature of 20° C is required which calls for radiant heating in some form. If the heating were to be by warm air, to produce the same environmental temperature of 18° C it would be necessary to maintain an internal air temperature well in excess of 18° C, since the mean radiant temperature would be lower, depending as it would upon the surface temperatures of the building structure alone.

INCIDENTAL SOURCES OF HEAT

Reference has been made to fortuitous heat gains and their effect on the estimation of heat requirements. These heat gains may be evaluated from the following:

Heat From Occupants—Heat is emitted from the human body, partly as sensible heat and partly as latent heat in perspiration and vapour in the breath. The sensible heat alone affects the temperature of the room.

Table 2.7 gives average values at different states of activity in an atmosphere at 20° C and 60 per cent. relative humidity.

The sensible heat dissipated is reduced at higher temperatures, becoming zero at 37° C (blood temperature), and increases at low temperatures (e.g. at 0° C it is 70 per cent. greater than at 18°). The latent heat follows the opposite course, decreasing at lower temperatures and increasing at higher.

The sensible heat of one person at rest is sufficient to heat 15 m³ of air per hour through 18 kelvins. If the space per person allocated in a room

TABLE 2.7

HEAT FROM OCCUPANTS

Activity	Heat emission per occupant Watts		
	Total	Latent	Sensible
At rest - - - - -	115	25	90
Sedentary worker - - -	140	40	100
Walking - - - - -	160	50	110
Light manual work - - -	235	105	130
Heavy work - - - -	440	250	190

approximates to this figure, and if the infiltration rate is the same, the air-change loss is covered by the heat from the occupants. If a building is designed so thermally perfect that fabric losses are insignificant, we can use the electric light to bring it up to temperature and thereafter rely on the occupants to keep it warm. This is the basis of the Wallasey School example. Limitations, however, arise when attempting to extend this principle to other types of building.

Heat from Machines—Energy expended by machines within a building results for the most part in the production of heat. In a lathe, for instance, in which metal is turned, the heat liberated is divided between the tools, the metal and the lubricant, which subsequently dissipate their heat in the room. The friction losses of shafting, belt drives, etc., are converted into heat. An electric motor driving machines transmits dynamic energy perhaps to a number of points within a building, where each machine in turn converts it into heat. The efficiency losses within the motor itself similarly come out in the form of heat. The exception is where work is done to produce potential energy, as in the pumping of water to a height, or in the lifting of goods in a hoist; in such cases the losses only are converted into heat.

Energy in the form of mechanical work being the same as in the form of heat, it follows that, for every kilowatt of power expended in a space, there is liberated 1 kilojoule of heat per second. If the power derives from an electric motor of 90 per cent. efficiency, 10 per cent. will come out as heat in the motor casing and air loss, and 90 per cent. as work which in turn will be degraded into heat—except where energy is conveyed outside of the space concerned.

Heat from Lighting—The energy consumed in the lighting of a space comes out as heat. Thus, if a space of 100 m² is lit to an intensity of 30 watts/m², the load will be 3000 watts, or 3 kW, and will be the heat energy involved.*

Heat from Process Equipment—The liberation of heat within the enclosure from equipment such as steam presses, hot plates, drying ovens, gas or electric furnaces, etc., may be determined by summating the consumptions of all the items in watts or kilowatts, less any flue loss.

Heat from Sun—While the heat from the sun absorbed by a building is most important when considering cooling plants under summer conditions (and is treated later in Chapter 18), it is so unreliable in winter that no allowance can be made for it when designing a heating plant, as in the coldest weather it may be entirely absent.

Nevertheless, there are other days, even in winter, when its effect on the southerly aspect may be quite important, while rooms facing north receive no such benefit. Hence, there is much to be said for having rooms with south aspect separately zoned, so that they can have heat reduced without affecting rooms facing north.

* See also p. 407.

Annual Total—A variety of interesting facts regarding the probable *annual* total of such adventitious heat input are set out in *Digest* 94 published by the Building Research Establishment. These relate to a semi-detached house of 100 m² area and are summarized in Table 2.8. The approximate total of these items represents something over a third of the theoretical *annual* heat requirement of such a dwelling.

TABLE 2.8

APPROXIMATE ANNUAL HEAT
FROM INCIDENTAL SOURCES

Source	Heat input per annum in GJ
Two adults, one child (body heat) - - -	2·5
Radiation from Sun - -	7·0
Cooking, gas - - -	6·0
Cooking, electric - - -	3·5
Lighting, radio, TV and miscellaneous - -	1·0
Losses from water-heating appliances, circulating pipes etc. - - - -	10 to 30

TEMPERATURE CONTROL

From what has already been said, it is clear that the ideal system is one in which each room has separate automatic control. It is then supplied with a heating element sufficient to warm it up to temperature, and then, as the heat from occupants, machines, light and sun begin to have effect, the heat input is cut down to prevent overheating and to save consumption. These factors probably operate at quite different times and in different degrees in all the rooms of a building.

A system which automatically allows for all such combinations is the ideal, other things being equal. Any departure from this ideal involves loss of comfort and waste of heat through the opening of windows in rooms which are otherwise too hot. In practice, a compromise is to be effected between the ideal, which may be expensive in first cost, and the primitive, which may be expensive in fuel. The more costly the fuel, the more important this item becomes.

The thermal characteristics of the building, its heating system and the system of temperature control are closely interlinked.

TIME LAG OF BUILDING—CONTINUOUS VERSUS INTERMITTENT HEATING

Continuous heating is necessary in buildings occupied for twenty-four hours during the day, like hospitals, police stations and three-shift factories.

Most buildings are, however, occupied for only a limited period during the day, such as schools (which are probably the lowest) at about 20 per cent. of the weekly hours. It is a matter of considerable importance when considering running costs to know whether, by operating the heating system intermittently, savings in fuel can be made whilst still maintaining

TABLE 2.9

PERCENTAGE OF FUEL CONSUMED OVER YEAR TO MAINTAIN INTERNAL TEMPERATURES
STATED, BY DAY AND BY NIGHT

Period	Internal temperatures maintained					
	20° C for 24 hrs		20° C for 10 hrs. 15° C for 14 hrs.		20° C for 10 hrs. 10° C for 14 hrs.	
	Day	Night	Day	Night	Day	Night
January	51	49	60	40	72	28
February	50	50	60	40	73	27
March	49	51	60	40	76	24
April	48	52	61	39	86	14
May	44	56	65	35	99	1
September	43	57	78	22	100	0
October	47	53	66	34	100	0
November	50	50	62	38	80	20
December	51	49	60	40	73	27
Average over season	48	52	64	36	84	16

satisfactory comfort conditions in the building (see Table 2.9).

The time lag of the building is perhaps best visualized by considering two cases. One, a shed of extremely light construction and having, in effect, no time lag whatever. A heating installation in such a building would, as soon as the heat is turned on, raise the internal temperature up to that desired, with a very small lag. The second kind of building— one with thick, massive walls, negligible windows and extremely heavy construction—involves a very great time lag. It might take a whole week to warm it up to the desired temperature. In the lightly constructed building, it is obvious that intermittent heating would achieve great economy, as heat would be supplied only during the hours of occupancy and no more. In the other case of the massive building, intermittent heating would be impracticable and thus no saving by such a means would be possible.

Between the two lie all the practical buildings of light, medium and heavy construction; and, according to the thermal time lag involved, so will intermittent heating show greater or lesser economy. Similarly, according to the hours of occupancy per week, the possibilities of savings are greater or lesser in proportion to the hours when not occupied.

Colthorpe* summarizes these effects in a paper on the subject, by means of a graph, from which Fig. 2.11 has been prepared. This was based on a series of extensive tests on a number of buildings of various types in which savings compared with continuous heating were achieved, ranging from 5 per cent. to 50 per cent.

* B.S.R.I.A. (late H.V.R.A.) Lab. Report No. 7, August 1962.

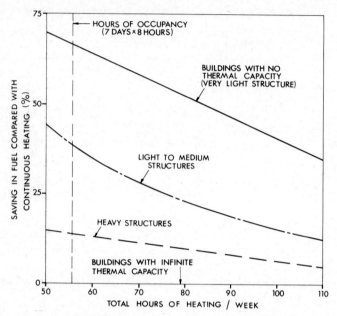

FIG. 2.11—Comparative fuel savings for different weekly heating periods indicating effect of thermal capacity of structure.
(With acknowledgment to B.S.R.I.A.)

With an intermittently heated system, the building temperature will
drop during the unoccupied period and, in order to bring this back to
normal by the time the occupants enter in the morning, a preheating
period is necessary. Some excess capacity in the heating system would be
desirable to bring this temperature up quickly; but what appears to suffice
in practice in most cases is to start up the heat input at full rating,
irrespective of the weather, so drawing upon the spare capacity in the
heating plant in the manner indicated in Fig. 2.12, which represents a
normal cooling and heating curve for a building heated intermittently.
The preheating time will be greater or less according to the outside
temperature, and control systems have been devised to take account of this
and adjust the preheating time automatically.

As the weather becomes colder, so the preheating time will become
longer, until in extremely cold weather, or with outside temperatures
below the basic design temperature, continuous operation of the heating
system may be the only way of maintaining the desired internal temperature.

It has been shown that there is little advantage to be gained by installing a heating system of much greater capacity than that necessary to meet
the hourly heat losses in an attempt to accelerate the warming up period,
where such a system has its own time lag. This is due to the fact that as the
system, boiler plant, etc., is increased in size, so its own time lag becomes

longer and the advantages in a quicker heat-up are almost cancelled out.

Where systems with a fixed input rate are involved, such as electric systems having fixed resistance elements, and gas systems using a set gas-rate, some additional capacity is necessary for a quick heat-up. Where an electric system is of the floor-warming type, operation so far as heat output is concerned is continuous even though the heat input off-peak is intermittent.

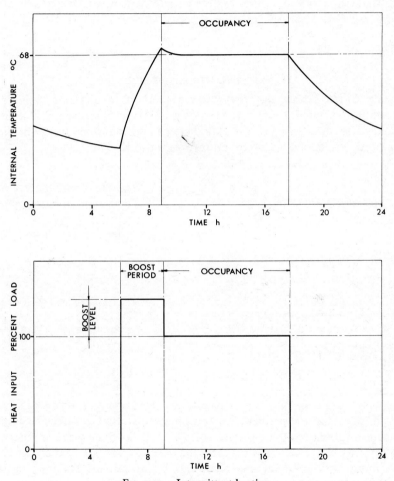

FIG. 2.12.—Intermittent heating.

With regard to the type of system and its effect on fuel consumption under intermittent operation, it has been shown by Dick* and others that systems with small heat capacity and quick heating-up rate achieve greater economy than ones with a long time lag. This is an obvious con-

* *Experimental and Field Studies in School Heating*, J.I.H.V.E., 1955, 23, 88.

clusion, and points to the generalization that the greatest economy in buildings of light construction will be to run the system intermittently, the system itself to be of the shortest possible time lag. This accounts for the greater use of systems using warm air in some form or another in the present vogue of light construction. Older and more solid buildings achieve a satisfactory condition and reasonable economy with a steady supply of heat and with no fine degree of control. The light building with a quick heat-up system can obviously easily overrun its temperature, and hence a highly responsive control system is imperative.

PREHEAT AND PLANT SIZE

The *Guide, Section A9,* postulates a method of assessing plant size according to preheat time, plant response and type of structure. The basis is 'plant size ratio' which is the ratio between normal maximum output of plant and design load for a temperature rise of 20 K.

Rounded up and in brief this shows:

	Preheat time	Plant size ratio required
for Heavy Structure		
Quick response plant	1 hr	3
	3 hr	2
Slow response plant	3 hr	3
	6 hr	$1\frac{1}{2}$
for Light Structure		
Quick response plant	1 hr	2
	3 hr	$1\frac{1}{2}$
Slow response plant	2 hr	$2\frac{1}{2}$
	4 hr	$1\frac{1}{2}$

The same table in the *Guide* gives the corresponding daily fuel consumptions taking 100 per cent. as continuous heating. For heavy structures the consumptions for the various preheat times range between 80 per cent. and 90 per cent., for light structures between 54 per cent. and 60 per cent. Exceptions are stated in respect of highly intermittent systems, radiant systems and storage systems.

Any approach such as this to the subject of intermittent heating must remain theoretical, and there is no point in attempting meticulous accuracy. It makes clear, however, that it is useless in a heavy structure to attempt a rapid heat-up as it is shown that for a one-hour preheat a plant capacity some three times normal is required, which would be uneconomic and often impracticable. This is why in church-heating a preheat time of twelve hours or more is required.

CHAPTER 3

Choice of Heating System

HAVING ARRIVED AT THE quantity of heat energy required for each room and thus for the building as a whole, as described in Chapter 2, it is now necessary to consider how this heat shall be supplied. It is proposed, therefore, to take a quick look at the whole range of available systems and to proceed thence to see how a choice can be made.

The number of heating systems is almost unlimited, if every combination of fuel, method of firing, transmission medium, and type of emitting element is considered. It is therefore quite useless to attempt to describe them at all clearly or systematically unless they are classified under the headings of the two main groups into which they all fall. This classification is indicated in Table 3.1, below and on the following page.

TABLE 3.1

CLASSIFICATION OF HEATING SYSTEMS

DIRECT SYSTEMS

Fuel or Source of Energy		Example
Solid	Open fire	Coal / Coke / Peat
	Stove	Coal / Coke / Anthracite
Liquid		Oil Stove / Oil-burning Radiator / Oil Warm Air
Gas	Primarily Radiant	Gas fire / High-Temperature Gas Panel / Low-Temperature Gas Panel / Gas-fired radiant tube
	Primarily Convected	Gas Radiator / Gas Convector / Gas Unit-heater / Gas Warm Air
Electricity	Primarily Radiant	Luminous Radiator / High-Temperature Panel / Low-Temperature Panel / Electric embedded Floor Heating
	Primarily Convected	Low-Temperature Tubes / Electric Convector, natural and forced / Electric Unit-heater / Electric Night Storage-heater / Electric Warm Air / Electric Oil-filled Radiator

TABLE 3.1 (*Continued*)

CLASSIFICATION OF HEATING SYSTEMS

INDIRECT SYSTEMS

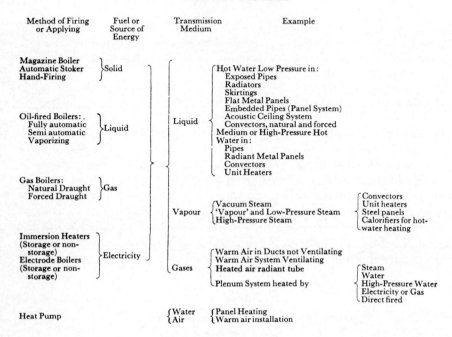

Method of Firing or Applying	Fuel or Source of Energy	Transmission Medium		Example
Magazine Boiler Automatic Stoker Hand-Firing	} Solid	Liquid	{	Hot Water Low Pressure in: Exposed Pipes, Radiators, Skirtings, Flat Metal Panels, Embedded Pipes (Panel System), Acoustic Ceiling System, Convectors, natural and forced Medium or High-Pressure Hot Water in: Pipes, Radiant Metal Panels, Convectors, Unit Heaters
Oil-fired Boilers: . Fully automatic Semi automatic Vaporizing	} Liquid			
Gas Boilers: Natural Draught Forced Draught	} Gas	Vapour	{ Vacuum Steam, 'Vapour' and Low-Pressure Steam, High-Pressure Steam	Convectors, Unit heaters, Steel panels, Calorifiers for hot-water heating
Immersion Heaters (Storage or non-storage) Electrode Boilers (Storage or non-storage)	} Electricity	Gases	{ Warm Air in Ducts not Ventilating, Warm Air System Ventilating, Heated air radiant tube, Plenum System heated by	Steam, Water, High-Pressure Water, Electricity or Gas, Direct fired
Heat Pump		{ Water Air	{ Panel Heating, Warm air installation	

The two groups into which all systems may be divided are:

(*a*) '*Direct*' systems, in which the fuel or energy purchased is consumed in the space to be heated.

(*b*) '*Indirect*' systems, in which energy is transferred from some more or less centralized point outside the space to be heated to equipment within the space for liberation as heat.

The *direct systems* are separated into four main subdivisions, according to whether *the fuel* (*or energy source*) is solid, liquid, gaseous, or electrical. Each of these is again subdivided according to the nature of the emitting or radiating element. This group is simplified by the fact that there is no transmitting medium.

The *indirect systems* are subdivided according to the kind of *transmitting medium,* which may be liquid, vapour, or gaseous, and these in turn are subdivided according to their emitting element and other characteristics.

Within this large group, the fuels may be interchanged in any combination of transmitting medium and emitter, and a place is found for them in the Table.

We may here make the following comments on the different systems.

DIRECT SYSTEMS—SOLID FUEL

Open Fires—Open fires were, of course, the primitive source of heat, greater economy being obtained in mediaeval times by having the hearth in the centre of the room and allowing the gases to escape through a hole in the roof. Little heat was lost, but the system was not without its disadvantages.

Later, flues were invented, whereby economy was sacrificed for the greater advantage of smoke elimination.

The ordinary open fire is notoriously extravagent in fuel for the heat it gives out. Furthermore, it has a big open flue which draws in sometimes as much as eight or ten changes of cold air per hour into the room: five times as much as is necessary for good ventilation.

The modern version of the open fire is usually arranged for continuous burning, with means for controlling the air inlet, and with a restricted throat to the flue. It is designed to burn coal or smokeless fuel and efficiencies are of the order of 30 to 40 per cent. as compared with 15 to 20 per cent. for the old-fashioned type of grate.

Some forms of modern grates provide for warming convection air, and others for heating hot water.

Open fires of any sort are, of course, impracticable for buildings of any considerable size, partly owing to the labour and inefficiency already referred to, and partly owing to the great waste of space and cost of the multiplicity of flues required.

Stoves—Anthracite stoves in this country, and the large coal stoves so common on the Continent, are much more efficient, labour saving, and draught reducing than the open fire, and will keep alight all night. The fire is not visible in them, however, and for this reason their use in this country is generally confined to halls and corridors. Their efficiency may be as high as 50 per cent.

Again there are many modern forms of 'closable' stoves for burning coal, coke, anthracite, or one of the processed fuels. Sometimes these combine convection heating or water heating.

Coke stoves of the so-called slow-combustion type were a common method of heating churches but owing to the intense local heating which they produce, are often the cause of bad draughts. Their use is tending to die out although some have been converted to burn oil.

DIRECT SYSTEMS—LIQUID FUEL

Oil Stoves—The liquid fuel commonly used for room heaters is paraffin, burned either on a wick or on a vaporizing plate. The sulphur content of this fuel is as low as 0·04 per cent. so that the products of combustion are unobjectionable except in so far as they contain moisture.

Such heaters are commonly used in houses for background heating in cold weather, their merit being cheapness and portability. The efficiency can be as much as 100 per cent provided no flue is connected to the heater and provided the water vapour formed due to the combustion of the hydrogen in the fuel is condensed—as is frequently the case—causing moisture, however, to collect on the windows and cold walls.

It is this last factor—the condensation of the moisture in the products of combustion on windows and walls—together with the potential hazards of fire and asphyxiation which are now leading to a decrease in the popularity of such units.

Direct-Fired Warm Air System—A variation of the warm air system, used for industrial space heating, is one in which the air is heated by direct firing. The heater takes the form of a combustion chamber and passages through which the gases pass from an oil-fired combustion chamber over which the air is impelled by a fan. Oil is pumped from a central storage to a number of such units.

It should be mentioned, incidentally, that this type of heater is economical in first cost and it is also simple to run. An example is illustrated in Fig. 3.1.

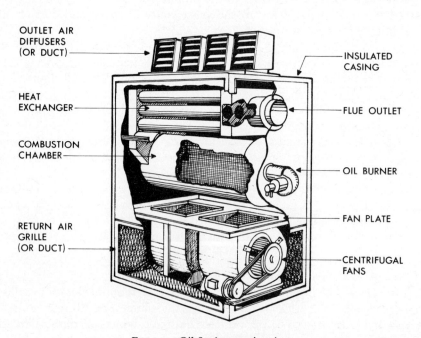

FIG. 3.1.—Oil-fired warm air unit.

Direct-Fired Warm Air in Overhead Ducts—In this system, a unit similar to that just described is used to circulate hot air through ducts overhead which, in turn, radiate heat to the space below. This system is also applicable generally to industrial buildings, the radiant heating ducts being large and unsightly and, of course, necessarily visible.

DIRECT SYSTEMS—GAS FUEL (PRIMARILY RADIANT)

Gas Fires—Gas fires have been greatly improved over the last thirty years by increasing the radiation effect with special forms of radiant elements, and by combining convection surface for warming the air, see Fig. 3.2. Their efficiency is about 60 per cent. They have rapid response, but some people find the 'drying' effect of the radiation unpleasant for comfort.

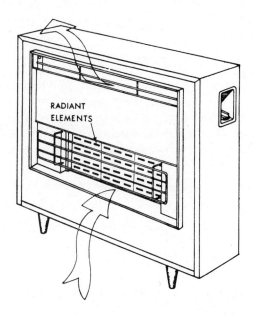

FIG. 3.2.—Radiant-Convective Gas Fire. The heating elements provide radiant heat whilst convected heat is supplied by the cold air which is drawn in at low level and then warmed and delivered from the top outlet.

A non-luminous form of radiant heater burning gas is shown in Fig. 3.3. This is suitable for industrial applications. Efficiencies as high as 92 per cent. are claimed for this system.

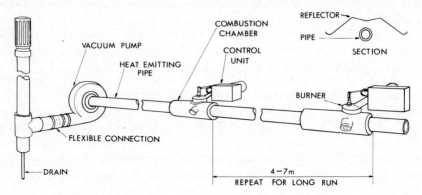

FIG. 3.3.—Radiant tube gas-fired heater (Nor-Ray-Vac).

DIRECT SYSTEMS—GAS FUEL (PRIMARILY CONVECTED)

Gas Convectors—Consist of a gas-heated element contained in a metal box with an opening at the bottom through which the air enters and another at the top through which it escapes after being warmed. The convection currents are then confined and a higher air temperature is the result. They are used chiefly in halls, shops, etc. and they have an efficiency of about 90 per cent. assuming that the vapour in the gases is not condensed.

Two types are shown: Fig. 3.4 shows an improved type in which the air for combustion is taken from outside and burnt gases returned to outside on a balanced flue principle; Fig. 3.5 is free-standing, drawing in

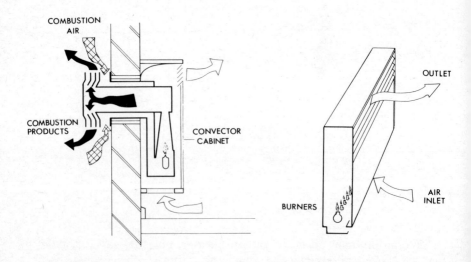

FIG. 3.4.—Balanced flue convector, gas fired.   FIG. 3.5.—Free standing gas convector.

air from the room at the base and discharging it with the burnt gas back into the room.

Gas Unit Heaters—Gas-fired heaters are similar to their steam counterparts described later (p. 216), and may be convenient if no steam or hot water for heating is available. The removal of fumes through a flue is normally essential. Gas-fired unit heaters are useful in factories, workshops, garages and other large spaces.

Gas Warm Air—The gas warm-air system is half way between a direct and an indirect system; it is much used in housing and is discussed in Chapter 6.

Flueless Heaters Generally—The discharge of products of combustion into the air of a room is only allowable if there is a good rate of ventilation, otherwise the concentration of CO_2 and moisture in the atmosphere will in time cause lassitude and headaches even if not being injurious to health.

DIRECT SYSTEMS—RADIANT ELECTRIC SYSTEMS

Luminous Fires—These comprise a wire element forming a resistance which becomes incandescent on being switched on and rapidly achieves working temperature. A great variety of types and designs are available ranging in capacity from about $\frac{1}{2}$ kW to 3 kW. A particularly useful type has a polished reflector at the rear projecting a beam forward, and giving a concentrated band of radiation. The effect of any heater of this type is localized, and they are not generally applicable to the heating of large spaces.

High-Temperature Electric Panels—Consisting of a tile of ceramic material (often about 650 mm by 350 mm), enclosing a resistance element. They have been used for meeting halls, etc. (see Fig. 3.6). They are generally placed high up on the walls facing diagonally downwards, and reach a temperature of about 250° C with an emission of about 1·25 kW for the size stated. The radiation is about 80 per cent. These panels are much below red heat.

This system is naturally of limited application but it is an interesting example of what can be done by radiation alone.

Low-Temperature Electric Panels—These may consist of a flat plate of metal (Fig. 3.7), hardboard or other conducting material, to which is attached, or embodied, a heating element of nickel chromium, or other resistance wire or, in some special instances, a graphite coated mesh of synthetic fibre. The panels generally operate at about 70° C. One application is in the heating of churches, where they may be fixed in the pews, thus giving a quick effect of warmth to the congregation without fully heating the church.

Another system consists of electric resistance elements embedded in a kind of wallpaper strengthened by fabric, which can be stuck to walls or ceilings with heat insulating material behind. The system generally

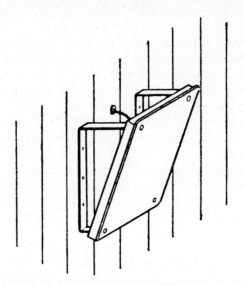

FIG. 3.6.—High-temperature electric radiant panel.

runs at about 32° C and 180 watts per sq. metre with about 90 per cent radiation in still air when used on ceilings.

Tubular Electric Heaters—Are commonly of a round or oval shape (Fig. 3.8). The elements, enclosed in a thin steel tube for protection, are run at a temperature of 70° to 90° C. They are frequently fixed to skirtings

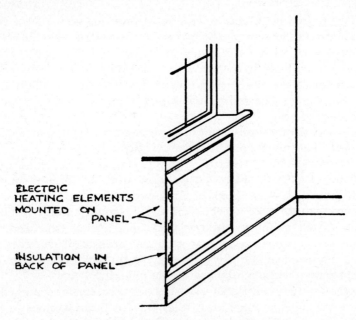

ELECTRIC
HEATING ELEMENTS
MOUNTED ON
PANEL

INSULATION IN
BACK OF PANEL

FIG. 3.7.—Low-temperature electric wall panel.

DIRECT SYSTEMS—ELECTRIC HEATING (PRIMARILY CONVECTED)

under windows. They have also been applied to churches, placed under the pews, in which position they tend to keep the lower air warm without heating the whole building, and this naturally leads to economy for intermittent use. Downdraughts from the upper windows are prevented by more tubes on the sills at high level.

The tubes emit about 70 per cent convection and 30 per cent radiation. Tubes 40 mm to 50 mm diameter are rated at 180 to 240 watts per metre run.

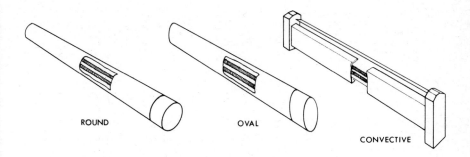

ROUND   OVAL   CONVECTIVE

FIG. 3.8.—Electric tubular heaters.

Electric Convectors—Generally consist of high-temperature elements in a casing with an opening below and a grating above through which an air current is induced. The warming effect of convectors is slower than with the radiant type of heater. A type of convector containing a fan gives a more rapid circulation of warm air in the room, one with a tangential fan being extremely compact.

Electric Oil-Filled Radiators—Are a convenient form of providing electric heating in buildings of the type otherwise furnished with central heating such as offices, residences etc. They usually comprise a steel radiator of neat design filled with oil so as not to be subject to freezing, and with an electric immersion element thermostatically controlled. Such units commonly have a maximum surface temperature of less than 90° C and are thus less potentially dangerous than are luminous fires.

Electric Warm Air (commonly 'Electricaire')—This system is very similar to gas warm air in that the air is warmed and distributed by a fan to various rooms in a house. It is usually a storage unit taking off-peak current, and is dealt with in Chapter 12.

Electric Unit Heaters—These consist of a series of wire coil heating elements, electric fan and adjustable louvres to direct the air (Fig. 3.9). They are usually fixed overhead and find a use in work-rooms, workshops etc. where no centralized system is possible.

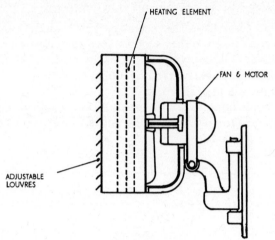

FIG. 3.9.—Electric unit heater.

DIRECT SYSTEMS—ELECTRIC STORAGE

Where off-peak supplies remain available, use may be made of suitable equipment which must, necessarily, involve storage of heat by some means at night for use during the day. Systems of this type include *Electric Floor-Warming* and the *Block Storage Heater* either uncontrolled or with controlled output. These systems are dealt with in Chapter 12 (*Electric Heating*).

INDIRECT SYSTEMS—LIQUID MEDIUM

Distribution of Heat—Hot water is the only liquid medium which need be considered under this heading. Water is cheap and has high heat capacity per unit volume. Its greatest merit is flexibility in control of temperature to meet variations in weather.

Its application is most common at low pressure: that is, with the system filled from a tank above the highest point of the system open to atmosphere. The water temperature is then controlled at some level below atmospheric boiling point, such as 80° C. Such a system is what is popularly called 'Central Heating'.

For large installations where considerable distances have to be covered, the system is closed, so allowing pressure to be raised above boiling point. A medium-pressure system may operate up to about 130° C, and a high-pressure system up to 180° C or above. By this means the quantity of heat carried in a given size of pipe can be greatly increased. This subject is dealt with in detail in Chapter 11 on *High-Pressure Hot Water*.

Group heating or district heating exploits the possibilities of distribution of heat still further by serving a number of buildings or a whole town from one heat station. Chapter 25 explains the advantages of this. Plate II, facing p. 86, shows a pipe tunnel for distribution of services.

Whatever the method of distribution, the emission of heat requires apparatus to be fitted into the rooms to be heated. The following are representative types for use with low-pressure hot water.

Exposed Pipes—This is an old form of heating surface, and was much used in factories for economy in the past. The present tendency is to use small piping for conveying heat rather than as a heat-emitting surface. The downward radiation from overhead piping may be increased by the fitting of flat plates attached to the pipes. These are known as 'strip heaters'. (Plate III, facing p. 86.)

Hot-Water Radiators—Radiators are too well known to need detailed description. Previously made of cast iron, they are now usually of steel which, though probably neater, is more subject to corrosion unless the water is suitably treated. Radiators of cast aluminium are a new development. The word 'radiator' is a misnomer, since the heat is only about 20 per cent radiant, the rest being convected. A simple form of radiator system is shown in Fig. 3.10. For domestic use, radiators and

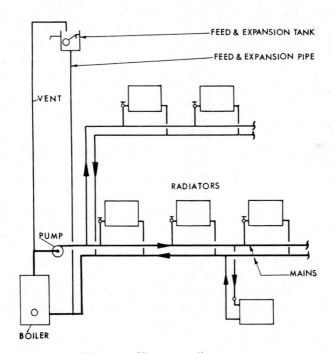

FIG. 3.10.—Hot-water radiator system.

other forms of heat emitters are often served from small-bore or 'mini-bore' piping with pump circulation as described on p. 179.

Flat Panels—Heating of a mainly radiant character is achieved by the use of metal panels comprising a large surface of flat plate heated in some way by the attachment (or otherwise) of waterways at the back. One of the advantages of this type of surface is its neat appearance in a room, since the surface may be treated to match the walls etc. and be inconspicuous.

Flat plate panels are generally of steel. Various types have been produced with ribs or other convolutions at the back so as to increase the convection component and increase output for a given size.

Skirting Heating—The object of using skirtings for the heating of a room is to provide a well-distributed source of heat, unobtrusive in appearance. Developments by various makers make this type of system available in a convenient form. The units may be either flat-fronted to provide largely radiant heat, have apertures at the base and above to allow a certain amount of convection heat in addition or be of a type providing convection heating alone, i.e. a convector heater of small height.

Ceiling Heating—The use of the ceiling for heating originated in the 'panel' system in which pipes of 15 mm bore are embedded in the structural slab and plastered over. These were fed by hot water at low temperature (40 to 50° C). This system, however, is sluggish in response and unsuitable for buildings of the current, lighter, forms of construction.

The most common forms of ceiling heating in use today make use of light metal trays perforated for acoustical effect and insulated above, being either clipped to a grid of heating pipes (Fig. 7.15 p. 149) or independently suspended below a series of pipe coils. In another type, use is made of plaster panels perforated acoustically (Fig. 7.16 p. 149).

Characteristics of ceiling heating are cleanliness of decorations, uniformity of warmth, freedom in planning and economy in fuel consumption.

Floor Heating—Hot-water coils may be embedded in the floor screed. The floor surface being warmed to 24 to 27° C emits heat largely in radiant form. This system is commony used in entrance halls, churches, banking halls and the like where occupancy is transient. Another valuable application is to the floors of buildings suspended over open-air car parks etc. where, whatever the air temperature maintained internally, discomfort can arise due to cold feet.

Hot-Water Convectors—A hot-water convector generally consists of a finned tube heater so placed as to induce an air current, as shown in Fig. 3.12. It is a neat and compact form of heating, but being entirely convective suffers from the disadvantages associated with the ordinary radiator as regards the carrying of dust and dirt by the air from the floor to the walls and breathing zone.

The absence of the radiant component makes it unsuitable for places requiring good through-ventilation.

An interesting feature of the convector is that the emission increases rapidly with the height of the flue, being nearly doubled when this is increased from 0·3 m to 1 m.

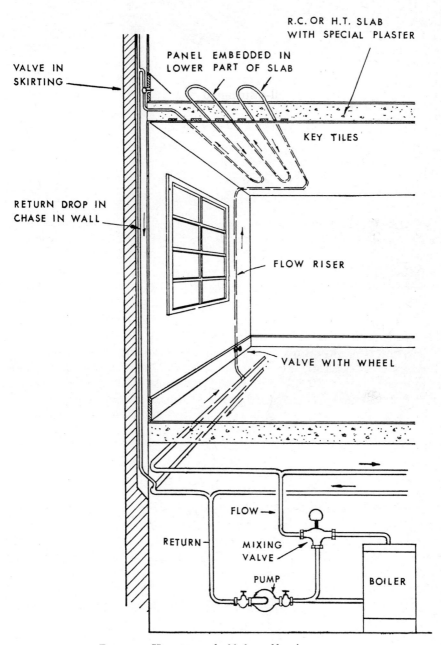

VALVE IN
SKIRTING

PANEL EMBEDDED IN
LOWER PART OF SLAB

R.C. OR H.T. SLAB
WITH SPECIAL PLASTER

KEY TILES

RETURN DROP IN
CHASE IN WALL

FLOW RISER

VALVE WITH WHEEL

FLOW →

RETURN

MIXING
VALVE

PUMP

BOILER

FIG. 3.11.—Hot-water embedded panel heating system.

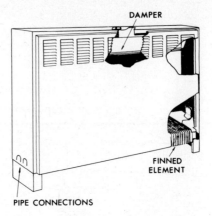

FIG. 3.12.—Hot-water natural convector.

Forced Convectors—Forced (fan-) convectors are similar to natural convectors except that the air flow is produced by electrically operated fan or fans (see Fig. 3.13). The output is thus much increased in a given space, and they find particular application in schools, large halls, entrance spaces and the like. The fans are particularly quiet and are usually controlled thermostatically. A similar type of equipment has been developed for domestic use either in the form of units suitable for a single room, or for central installation to serve a whole dwelling when supplied with hot water from a centralised group or district-heating system.

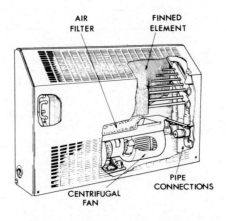

FIG. 3.13.—Hot-water fan convector.

INDIRECT SYSTEMS—VAPOUR MEDIUM

In earlier editions of this book, steam was considered in some detail, but its use as a heating medium has died out in favour of hot water, except in industrial applications. Thus, in the context of the present chapter, it need have no more than a passing reference.

Steam was used at various pressures—under vacuum to preserve a low temperature in the emitting surfaces—at atmospheric pressure, and at low, medium and high pressures. For industrial purposes steam is commonly used in unit heaters, one type of which is shown in Fig. 3.14. Steam is also used in calorifiers (or heat exchangers) to heat hot water which is then used in low-pressure hot-water equipment.

Further consideration to this subject is given in Chapter 10 on *Heating by Steam*.

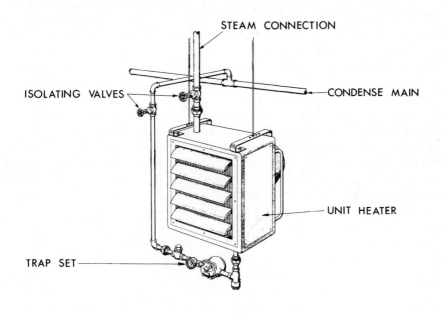

FIG. 3.14.—Steam unit heater, suspended type.

INDIRECT SYSTEMS—GASEOUS MEDIA

Air is the only gaseous medium now used in heating distribution, but due to its low specific heat capacity for a given volume it is only practicable for limited distances; otherwise the ducts sizes and heat losses therefrom become disproportionately large.

Flue Gases in Ducts (Roman System)—The culture of the Romans developed a type of heating which produced much the same effect as our radiant systems of to-day by warming the floors and walls from a furnace in the basement. Examples are brought to light from time to time in this country, and those at Bath are quite well known. Excavations at Verulamium and elsewhere show clearly that this form of central heating was an essental item of all the better-class houses of the time; yet the art was evidently completely lost when the Romans departed, and it has taken some sixteen hundred years or more to get back to their state of civilization in this respect.

Their system comprised a heating chamber below the ground from which the hot gases were conveyed in ducts under the floor to flues in the walls, emerging to the atmosphere at various points around the building. Sometimes proper ducts were formed, and in other cases the whole space under the floor, known as a 'Hypocaust', appears to have been used. The wall flues were invariably of hollow tiles, very similar to their modern counterparts. The floor consisted of a slab corresponding to our concrete slabs supported on 'pilæ', and over this a mosaic was laid. The heat was furnished by wood or charcoal. Probably the distribution of heat was not uniform, but there can be no doubt that the general effect was highly successful.

Warm Air in Floor Ducts—Part of the Anglican Cathedral at Liverpool is heated by the same means as used by the Romans; except that air, not flue gases, is circulated through ducts beneath the floor by means of a fan.

Warm Air Natural Circulation—In this system air is heated in a furnace in the basement, and rises through a grille in the floor over, by natural means, pervading the whole interior; return air is brought back to the furnace through other grilles. This system is still to be found in churches and may be fired by coke, oil or gas.

Gas Warm Air, Electricaire, Oil Warm Air—As mentioned under *Direct Systems*, these belong to a class of system which might equally be considered an indirect system as the heat is usually produced outside the rooms to be heated. Their use is preferably confined to domestic applications where the plan is such that ducts are short or are non-existent. In instances where long distribution ducts have been required, conditions of instability can occur due to differential wind pressures. In any event, the economics of the systems demand compactness.

Warm Air Forced Circulation (Plenum System)—The Plenum system of heating comprises a centrifugal fan, a heater warmed by steam, hot water, electricity or gas, and a system of ducts distributing the air to the points required, as in Fig. 3.15.

Air is drawn in at the fresh-air intake and is warmed at the heater to a temperature generally between 40° and 55° C. When this air is discharged into the room to be heated it cools to the room temperature, so giving up its heat to the walls, floor, ceiling, etc., just as the heated air from a convection radiator system cools down in its circulation through a room. There is, however, an important difference between these two systems in

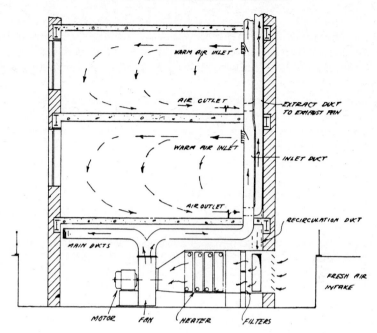

FIG. 3.15.—Plenum heating system.

that the warmed air of the Plenum system is under a slight pressure, so that inward leakages of air through windows, etc., usual with radiator systems, are non-existent, and all air leakages are outward.

This air leakage means that a great loss of heat is continually taking place, since all air supplied by the fan has been warmed in winter from say 0° to 50° C, and has cooled only from say 50° to 18° C, so that one-third of the heat input is entirely wasted. It is therefore usual for such systems to incorporate a re-circulating arrangement by which air from the building is returned to the system for re-heating during the winter months. This is usually effected by including a damper at the inlet, which can be set to give any proportion of fresh to re-circulated air. When 100 per cent. re-circulation takes place the pressure in the building is lost and infiltration will then occur, and so it is common practice to allow a maximum of 75 per cent. of air to be returned to the system (see reference on p. 559 to the potential for heat recovery).

Strong winds have the effect of counteracting the minute internal pressure generated by a Plenum system, so that on the windward side of a building air leakages inward through windows, etc., may still occur, with a corresponding drop in temperature.

Owing to the dirt brought in by the fresh air stream an air filter is generally interposed with the object of cleaning the air supply.

The system was in the past much used for the heating of factories but other systems have now largely replaced it.

Ventilating Systems—A ventilation system is not truly a heating system, and is mentioned here only to point out differences from the Plenum system.

When a building is ventilated mechanically it has a fan, heater, filter, and duct system similar to that described above, but the air is warmed only to room temperature or a few degrees below. The warming of the rooms is accomplished by direct radiation, or some other system entirely independent of ventilation.

It is necessary to remove the air from the room, and a system of exhaust ducts and fans is provided for this purpose.

The great advantage of such a combined system is that a feeling of freshness can be maintained in the rooms, and the objectionable features of a heated air supply are avoided. The subject of ventilation is dealt with in Chapter 16.

FACTORS AFFECTING CHOICE

The choice can only be related to the type of building, since what may be suitable for a factory would be out of the question say for a block of flats.

Home Heating—Choice here may be influenced by personal preference, by sales pressure of various fuel interests, by close regard for economy or, where a public authority is concerned, by what is permitted by various local or national regulations.

Again, where the property exists with no more than a few open fireplaces, the desire for some form of central heating may limit the choice to what is practicable.

Direct-heating systems using solid-fuel fires and stoves involve labour, dust and dirt and, with modern habits of families being out all day, are often inconvenient. Gas and electric fires are then preferred, but tend to be expensive in running cost. Gas Warm Air and Electricaire using off-peak electricity, are then to be favoured. They can be controlled to give a set temperature at times when the house is occupied, with some minimal background warmth during the rest of the day to counteract, among other things, condensation.

Indirect systems for home heating can give whole-house comfort as well as hot water probably more consistently than any direct system. The capital cost may be higher but the running cost less, subject to the vagaries of the various fuel tariffs.

The question of fuel for central heating in the home, however, may not always be a matter of cost, even supposing that relative prices remain stable. Thus, the saving of space for an oil tank or coal store may give the advantage to gas. The saving of labour with gas or oil may rule out solid fuel. Maintenance of any fuel-burning appliance is small in comparison with fuel cost and, on the domestic scale, insignificant.

If central heating is decided upon, probably a small-bore, mini-bore, or even micro-bore system is to be recommended. The old gravity systems with large pipes are out-dated.

Some may prefer floor heating in a new house, avoiding convection-current marking of walls. In an old house, skirting heating may be less obtrusive than radiators, although the latter will be cheaper. If floor heating is preferred, electric off-peak may be a simple answer, but great care with insulation and operational routine will be required to minimise consumption.

Control systems for home heating have been developed to a high degree of sophistication, being adjustable as to times and temperature in a variety of ways. Intending purchasers should not be misled by claims of one particular fuel or other interest, since what can be done by one can probably be done by all.

Thus, as will be seen, the choice for house heating is a bewildering one, there being in effect many answers to what, in all the circumstances, is the best. It is impossible to generalise even in the matter of cost of heat, for this again is dependent to some extent on the system and how it is run. Further reference to this is made later.

Flats—Multi-storey—Systems of heating in common use and from which a choice would no doubt be made are:

Electric floor-warming off-peak;

Electric warm air;

Gas warm air;

Oil warm air from common storage;

Central heating by low-pressure hot-water radiators;

Hot-water warm air from a boiler per block;

Group or district heating by hot water to a number of blocks.

The housing authority or estate developer in making a choice will be influenced by the capital cost, fuel cost, maintenance costs and labour to run. There is also the question of how the heat is to be charged—whether by the public utility reading its own meters (the consumer paying direct), or whether the landlord is responsible for reading meters and collecting the money, or, again, whether the cost of heat is included in the rent.

All these matters have to be considered as well as the type of tenant and what kind of expenditure can be afforded before a recommendation can be made.

Commercial—Offices being usually in blocks are most economically heated by a central system in some form. Choice is then confined to the kind of emitting surface, radiators, convectors, ceiling heating etc., and to the kind of fuel.

For reasons already explained, an office block with large expanses of glass and light construction will be subject to rapid swings of temperature; hence a system which is quickly responsive to change of output is needed. Thus, embedded-coil systems, or electric-floor warming, are not to be recommended although they may find a place in buildings of heavy construction.

Small blocks, where a central plant may be unsuitable for a variety of reasons, will probably best be served by one of the direct systems such as electric block-storage heaters, gas fires or convectors.

A study of probable temperature rise in modern office blocks reveals a tendency to overheating in summer, due to solar heat gains in addition to heat from office machinery and personnel. It is then necessary to consider air-conditioning instead of heating alone, which matter is developed in Chapter 17.

Public Buildings and Schools, Universities, Halls of Residence, Swimming Pools, Hospitals, Hotels etc.—This general class of substantial buildings will in most cases be in the hands of a consulting engineer, or public authority engineer, who can be expected to advise which system and fuel should be used. Matters to be reported on will cover, among other things, architectural appearance, the need for ventilation, capital cost, effect on redecoration costs, fuel, plant space, controls, labour costs, maintenance costs, noise and amenities generally.

Industrial—Here the choice will be largely governed by economics. Which system will produce adequate conditions with minimum upkeep, fuel consumption and labour?

If steam is produced for process work, the choice may fall in the direction of using steam for heating also—unit heaters being then one answer. If no steam is required, hot water at medium pressure or high pressure is to be preferred. For heat emission, unit heaters may again be selected on account of low cost, but they involve maintenance which does not apply with radiant panel or strip heating.

Direct oil or gas fired air heaters meet certain cases where a boiler plant is not desired or where extensions cannot be dealt with from existing boilers. Furthermore, such units can be moved to suit any change in building plan or workshop layout. Radiant tube systems, heated either by combustion products from gas burners or by hot air, should also be considered having regard to the high efficiency of heating effect and operating economy inherent to these arrangements.

Choice of Fuel—There has in the past been complete freedom of choice but it is not known how long this may continue. Any kind of control would necessarily limit this freedom. The choice in the past has largely been determined by cost, but the present instability of prices tends to make any such comparison unrealistic, however generalized this may be.

Thus, it is proposed here to do no more than indicate how comparisons may be made so that, in any particular case, given the basic cost or tariff for alternative fuels, an answer may be derived.

The unit quantity of heat for calculation purposes may be taken as 100 MJ, this being of a convenient magnitude and of the same order as the previously familiar 'therm' (1 therm = 105·5 MJ). A table may be set down to show that, at 100 per cent. efficiency, the heat content per quantity as sold would be:

	Unit sold	*Unit heat content in 100 MJ*	*To produce 100 MJ*
Gas	therm	1·055	0·948 therms
Oil	litre	0·40	2·5 litres
Solid fuel	tonne	300	3·3 kg
Electricity	Unit (kWh)	0·036	27·8 kWh

If unit prices, roughly representative of the situation at the time of going to press, are put against these figures then the equivalent unit costs per 100 MJ, at 100 per cent. efficiency, are:

	Unit price as sold		*Cost per 100 MJ*
Gas (North Sea)	17·9p/therm × 0·948	=	17p
Oil (Class D)	8·4p/litre × 2·5	=	21p
Solid fuel	£30/tonne × 0·33	=	10p
Electricity			
Off-peak	1·2p/unit × 27·8	=	33p
On-peak	2·3p/unit × 27·8	=	64p

These figures do not represent the cost of the heat arriving in the building, except with direct electric heating. In an indirect central system there is a loss from piping—mains loss—and a loss in combustion in the boiler due to hot gases leaving by the flue. The first loss may be 10 per cent., the second loss 25 per cent. with gas and oil, and somewhat more with solid fuel. If, for simplicity, we take the overall loss as 30 per cent.—a system efficiency of 70 per cent.—the costs above become approximately as shown in Table 3.2.

Other factors affect fuel consumption, such as controllability, running hours and no-load losses. For instance an off-peak electric system may have losses during the night when otherwise heating would be off; heating using heavy oil may involve electrical consumption for pre-

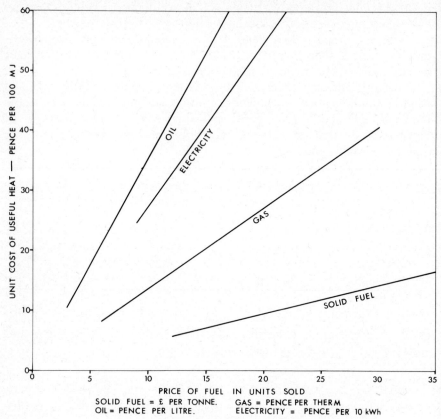

FIG. 3.16.—Comparative fuel costs at normal efficiency of utilization.

heating or fuel pumping and gas firing may carry the penalty of pressure boosting as well as that of a permanent pilot flame for ignition.

Each case merits separate evaluation. As a guide, a plot such as Fig. 3.16 may be used to determine the cost of useful heat per 100 MJ arising from a range of unit selling prices for fuel. This is, of course, a simplification, being based upon the same approach as that used for Table 3.2.

TABLE 3.2
NOTIONAL COST OF HEAT FROM VARIOUS FUELS

Type of fuel	Cost of heat per 100 MJ
Electricity (direct heating at 100% efficiency) off-peak -	33p
Gas (70% efficiency) - -	24p
Oil (70% efficiency) - -	30p
Solid fuel (70% efficiency) -	14p

CHAPTER 4

Combustion of Solid and Liquid Fuels

IT IS DESIRABLE THAT the principles of combustion for solid and liquid fuels be understood before considering boiler types and methods of firing; the subject is involved and the treatment herein should be regarded as an outline with reference to heating boilers only.

Fuels encountered in practice are mixtures or compounds chiefly of carbon, hydrogen and oxygen. In addition, there are generally small quantities of sulphur, nitrogen and—in the case of solid fuels—ash. The carbon burns to form carbon dioxide (and possibly small amounts of carbon monoxide), the hydrogen to form water, and the sulphur to form sulphur dioxide which later may combine with more oxygen to form trioxide.

Combustion Process—The combustion of fuel is a physico-chemical process which is accompanied by the liberation of heat. Combustion reactions can only take place at a high temperature known as the *ignition temperature*, which varies between $400°$ C and $600°$ C according to the fuel.

The elements combine with oxygen in proportion to their molecular weights, which are as follows: $O_2 = 32$, $C = 12$, $H_2 = 2$, $S_2 = 64$, $N_2 = 28$. Air contains $23·15$ per cent. of oxygen by mass and $76·85$ per cent of nitrogen, etc. (by volume: $O_2 = 20·8$ per cent, N_2 etc. $= 79·2$ per cent). The basic combustion reactions are given in Table 4.1.

The Calorific Value of a fuel is the quantity of heat released on the complete combustion of unit mass. There are two such values always given—

TABLE 4.1

COMBUSTION REACTIONS

Reaction	Products	Requirement for combustion kg/kg of combustible		Heat Liberated MJ/kg of Combustible
		Oxygen	Air	
$C + O_2 = CO_2$ $12 + 32 = 44$		2·7	11·6	34
$2C + O_2 = 2CO$ $24 + 32 = 56$		1·3	5·8	10
$2H_2 + O_2 = 2H_2O$ $4 + 32 = 36$		8·0	34·6	145
$S_2 + 2O_2 = 2SO_2$ $64 + 64 = 128$		1·0	4·3	9

gross or higher, and *net* or lower. The *gross calorific value* includes the heat given up in the reaction towards the supply of the latent heat of vaporization to any water forming part of the products of combustion. The *net calorific value* is the gross value minus the latent heat referred to. The greater the amount of hydrogen in the fuel the greater the difference between the two values.

The calorific value of a fuel may be calculated approximately from a knowledge of its analysis and the heat due to the reaction of oxygen with each element, but it is necessary to check any such computation by experimental determination.

Solid Fuel—Solid fuels contain carbon as *fixed* or uncombined carbon, or as *volatiles* or hydrocarbons. The former burn direct to form CO and CO_2 giving a short blue flame, the volatiles which occur in coal give a yellowish flame which is liable to cause smoke if cooled too quickly on water-backed boiler surfaces or by cold air entering above the fire. As the volatiles are consumed, the residue is largely coke.

Solid fuels also contain hydrogen not combined with carbon, referred to as *inherent moisture*. *Free moisture* is that on the surface or in the interstices often augmented by rain in open trucks. Whilst the moisture content in both forms constitutes a loss in available heat, it appears to have some beneficial effect on combustion with some coals such that arrangements

TABLE 4.2
CHARACTERISTICS OF TYPICAL SOLID FUELS

Description	Anthracite (102)*	Coal			Coke
		Medium Volatile (301*b*)	High Volatile (502)	General Purpose (802)	
Typical ultimate analysis, %					
Carbon	84·3	81	75·5	67·8	82
Hydrogen	3·4	4·5	4·8	4·4	0·4
Nitrogen	1·2	1·4	1·6	1·4	
Oxygen	2	3·8	6·2	8·5	0·9
Sulphur	1·1	1·3	1·9	1·9	1·7
Ash	5	5	5	5	8
Moisture	3	3	5	11	7
Calorific Value, MJ/kg					
Gross	32·8	32·9	31·4	27·9	28·6
Net	31·9	31·8	30·1	26·6	28·3
Bulk Density, kg/m³					
Graded	750		640		450
Small	800		750		560
Angle of repose					
Graded	40		40		45
Small	55		55		

* Numbers quoted are N.C.B. Rank Codes.

are sometimes made to spray the coal before firing, or to admit steam below the fire bars.

The *proximate analysis* gives the percentage mass of fixed carbon, volatiles, moisture and ash, as above. From this the probable characteristics of combustion can be forecast. The *ultimate analysis* gives the percentage mass of the various elements or compounds in the sample. From this the theoretical air for combustion may be estimated and the combustion efficiency. Table 4.2 lists the principal characteristics of a few typical solid fuels.

Assume that a sample of coal has the following analysis by mass:

Carbon 80%   Hydrogen 4%   Oxygen 1·1%
Sulphur 1%   Ash and moisture etc. 13·9%

The theoretical air for combustion would then be (using Table 4.1):

$$\begin{array}{lccc}
 & \textit{kg air/kg} & & \textit{air, kg} \\
 & \textit{combustible} & & \\
C\text{---}CO_2 & 11\cdot6 \times 0\cdot8 & = & 9\cdot28 \\
\left.\begin{array}{l} H^* \\ O \end{array}\right\}H_2O & 34\cdot6\left(0\cdot04 - \dfrac{0\cdot011}{8}\right) & = & 1\cdot33 \\
S\text{---}SO_2 & 4\cdot3 \times 0\cdot01 & & \\
\end{array}$$

Total air kg/kg fuel   10·61

Proportion of O_2, 23·15% mass   =   2·46
Density of air at 20° C   =   1·2 kg/m³

Thus volume per kg fuel   = $\dfrac{10\cdot61}{1\cdot2}$ = 8·8 m³

The theoretical calorific value would be:

$$\begin{array}{llcl}
C & 0\cdot8 \times 34 & = & 27\cdot2 \\
H & 0\cdot04 \times 145 & = & 5\cdot8 \\
S & 0\cdot01 \times 9 & = & \text{---} \\
\end{array}$$

33·0 MJ/kg

The hydrogen combining with oxygen to form steam is condensed out before the CO_2 is sampled and therefore does not enter into the assessment of CO_2 percentage. The maximum CO_2 content possible from combustion of this sample with no excess air is:

Air for combustion   10·61 kg/kg
   of this, O_2 2·46 ,,
   thus, N_2 8·15 ,,
C   =   0·80
O_2 2·7 × 0·80   =   2·16
          2·96 ,,

* The oxygen burns one-eighth of its equivalent in hydrogen:
i.e. $4 - \dfrac{1\cdot1}{8} = 3\cdot875$ per cent hydrogen is left requiring air for combustion.

by volume $\quad$ 8·15 N_2 ÷ mol wt 28 $\quad$ = $\quad$ 0·290 $\quad$ 81 %

$\qquad\qquad$ 2·96 CO_2 ÷ ,, ,, 44 $\quad$ = $\quad$ 0·068 $\quad$ 19

$\qquad\qquad\qquad\qquad\qquad\qquad\qquad\qquad$ 0·358 $\quad$ 100

The maximum CO_2 from combustion of this sample of coal will thus be 19 per cent, assuming no CO or other unburnt products. If the air supply is in excess of the theoretical requirement:

	50%	100%
CO_2 as above	2·96	2·96
N_2 8·15 × 1·5 etc.	12·23	16·30
O_2 2·46 × 0·5 etc.	1·23	2·46
by volume		
CO_2 ÷ 44	0·07	0·07
N_2 ÷ 28	0·44	0·58
O_2 ÷ 16	0·08	0·16
	0·59	0·81
in per cent.		
CO_2	11·9	8·6
N_2	74·5	71·6
O_2	13·6	19·8
	100·0	100·0

Liquid Fuel—Oil firing of boilers for central heating has for long taken a predominant place, being automatic and labour free. The arrival of North Sea gas as a serious competitor and the dramatic rise in world oil prices have, however, changed the picture.

Furthermore, concern at the pollution caused by the liberation into the atmosphere of noxious effluents militates against the use of oils of high sulphur content except for power stations or industry where tall chimneys can be used. Thus for heating purposes, where stack heights are comparatively low, Class D oil of low sulphur content will generally be used or, indeed, alone permitted.*

Table 4.3 lists the various grades of oil and their characteristics; Class D requires no pre-heating whereas the heavier grades must be stored at elevated temperature and further heated for handling and atomization (see Chapter 5).

* For a detailed study of Atmospheric Pollution in Great Britain, see *Conservation and Clean Air*, I.H.V.E., 1973.

A fifth grade, kerosene, should be mentioned and is referred to under *vaporising burners* later.

TABLE 4.3

CHARACTERISTICS OF TYPICAL FUEL OILS

Description	Class D Gas Oil	Class E Light	Class F Medium	Class G Heavy
Specific Gravity at 15·6° C - - - - -	0·835	0·93	0·95	0·97
Flash Point (closed) min. - - - - -	66° C	66° C	66° C	66° C
Viscosity				
Redwood No. 1 at 38° C secs. - - - -	34	250	1000	3500
Kinematic, centistokes, at 38° C secs. - -	4	62	247	864
Pour Point, °C - - - - - - - -	− 18	− 7	21	21
Calorific Value MJ/kg, gross - - - -	45·5	43·4	42·9	42·5
net - - - -	42·7	41·0	40·5	40·0
Storage Temp., min. °C - - - - -	−	7	20	32
Handling Temp., min. °C - - - - -	−	7	27	38
Sulphur Content, max. % by weight - - -	0·75	3·2	3·5	3·5
Ash Content, % by weight - - - - -	0·01	0·05	0·12	0·2
Mean Specific Heat Capacity, kJ/kg 0–100° C -	1·93	1·82	1·78	1·78

With reference to the characteristics of the oils shown in Table 4.3, the following should be noted:

The Flash Point, Pensky Marten (closed), is the minimum temperature at which a flash can be obtained on the apparatus when the oil in it is heated. While the flash point limits the amount of low boiling-point materials which can be incorporated in the oil, from a combustion point of view it has little or no significance.

The Viscosity, which is given above as Redwood number 1 at 38° C, is generally determined nowadays in a U tube viscometer; it is also usual to quote values for kinematic viscosity in centistokes ($1 cSt = 10^{-6}m^2/s$).

The Pour Point generally follows the viscosity, and is the temperature at which the oil ceases to run freely. For practical reasons this point must be below the normal temperature of the storage vessel.

The Calorific Values (gross and net) have already been defined.

The Sulphur Content determines the quantity of sulphur trioxide which is produced on combustion. It is the chief cause of pollution from chimneys and corrosion in boilers.

In normal boiler plant, where high superheat temperatures are not employed, the *ash* in the fuel is of little importance.

A sample of oil may have the following analysis by mass:

Carbon 86% Hydrogen 11% Sulphur 3%

The theoretical air for combustion would then be (using Table 4.1):

	$kg\ air/kg$ combustible		$air,\ kg$
C—CO_2	$11 \cdot 6 \times 0 \cdot 86$	$=$	10
H—H_2O	$34 \cdot 6 \times 0 \cdot 11$	$=$	$3 \cdot 79$
S—SO_2	$4 \cdot 3 \times 0 \cdot 03$	$=$	$0 \cdot 13$

$$\text{Total air kg/kg fuel} \qquad 13 \cdot 92$$

$$\text{Proportion of } O_2\ 23 \cdot 15\% \text{ mass} \quad = \quad 3 \cdot 21$$
$$\text{Density of air at } 20° \text{ C} \quad = \quad 1 \cdot 2 \text{ kg/m}^3$$
$$\text{Thus volume per kg fuel} \quad = \quad \frac{13 \cdot 92}{1 \cdot 2} = 11 \cdot 6 \text{ m}^3$$

The theoretical calorific value would be:

C	$0 \cdot 86 \times\ \ 34$	$=$	29 MJ/kg
H	$0 \cdot 11 \times 145$	$=$	16 ,,
S	$0 \cdot 03 \times\ \ \ 9$	$=$	—
			45 ,,

The actual calorific value would be obtained experimentally, as mentioned earlier, typical values being given for various grades of oil in Table 4.3.

Assume sample as before contains 86% carbon:

Air for combustion			$=$	$13 \cdot 92$ kg/kg	
of this O_2			$=$	$3 \cdot 21$,,	
thus N_2			$=$	$10 \cdot 71$,,	
C		$=$	$0 \cdot 86$		
$O_2\ 2 \cdot 7 \times 0 \cdot 86$		$=$	$2 \cdot 3$		
				$3 \cdot 16$	

(The balance of O_2 $(3 \cdot 2 - 2 \cdot 7)$ is in the hydrogen and sulphur reactions)

				%
by volume $10 \cdot 71\ N_2$	$\div$ mol wt 28	$=$	$0 \cdot 38$	$84 \cdot 5$
$3 \cdot 16\ CO_2 \div$	,, ,, 44	$=$	$0 \cdot 07$	$15 \cdot 5$
			$0 \cdot 45$	100

The maximum CO_2 from combustion of this sample of oil will thus be $15 \cdot 5$ per cent, assuming there is no CO or free O_2 in the products. If the air supply is in excess:

	25%	75%
CO_2 as above	$3 \cdot 16$	$3 \cdot 16$
$N_2\ 10 \cdot 7 \times 1 \cdot 25$ etc.	$13 \cdot 4$	$18 \cdot 7$
$O_2\ 3 \cdot 2 \times 0 \cdot 25$ etc.	$0 \cdot 8$	$2 \cdot 4$

by volume

$CO_2 \div 44$	0·07	0·07
$N_2 \div 28$	0·48	0·64
$O_2 \div 16$	0·05	0·15
	0·60	0·86

in per cent.

CO_2	11·7	8·2
N_2	80·0	74·5
O_2	8·3	17·3
	100·0	100·0

Flue Gas Analysis—Analysis of flue gases resulting from combustion of fuel is the way used to assess excess air and thus 'efficiency' of burning.

The standard apparatus is the *Orsat*, by which a sample of the gas is first dried to remove water vapour, and is then subjected to adsorption by three liquids in turn. The first is caustic soda which adsorbs CO_2. The second is an alkaline solution of pyro-gallol which adsorbs O_2. The third is acid cuprous chloride which reacts with CO. Measuring burettes enable the proportion of gas adsorbed in each case to be measured and hence the volumetric content of the gases present.

Maximum CO_2 contents may be set down approximately thus:

Pure carbon	21·0%
Coke	20·0%
Coal 80% carbon	19·0%
Oil 86% carbon	15·5%

The presence of CO, as mentioned earlier, indicates incomplete combustion and, if present in any quantity, steps should be taken to reduce it to zero.

More convenient instruments than the Orsat for day to day use are now available for determining CO_2 content when setting up and adjusting automatic boilers. For permanent indication and recording in large boilerhouses, various types of automatic instruments are used, some depending on relative electrical or thermal conductivity of gases, some on relative mass, and some on chemical reaction as in the Orsat.

Flue Gas Temperature—Gases at boiler exit are measured by pyrometer or high temperature thermometer, which may be nitrogen filled mercury in glass, mercury in steel, or electrical type.

Chimney Loss—The chimney loss is directly proportional to the mass gas flow, the temperature of gases leaving the boiler and the specific heat capacity of the gases. The loss also includes the latent and sensible heat of the steam content due to combustion of hydrogen in the fuel, the steam being in a superheated condition. The mass gas flow varies according to

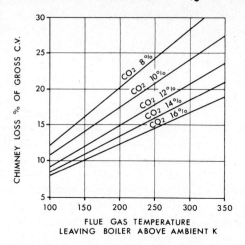

FIG. 4.1.—Chimney loss for coal firing.

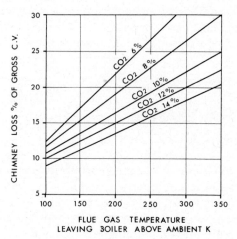

FIG. 4.2.—Chimney loss for oil firing.

the excess air quantity, which in turn can be assessed from the CO_2 analysis. The formulae for chimney loss are involved, and hence it is usual to refer to graphs prepared for particular fuels such as are to be found in the *Guide*; one for coal is as Fig. 4.1, and one for oil is as Fig. 4.2.

Unless some excess air is admitted, combustion will be incomplete, producing CO and soot. Fuels containing hydrocarbons (coal and oil) will produce smoke in such circumstances.

If the flue gas temperature is too low, there will be condensation, with serious corrosion hazard in the boiler. There is a water dew-point at which any steam in the gases from the combustion of hydrogen is deposited. This is generally in the region of 50° C. In the case of sulphur-bearing fuels (coal, oil) there is an 'acid dew-point' in the region 100° to 130° C, where the sulphur being converted from SO_2 to SO_3 or SO_4 combines with water, forming sulphurous or sulphuric acid. There is a great volume of research on this topic which may be studied by those interested.

In the practical low-pressure hot-water heating boiler, the endeavour is to keep the temperature of the water-backed surfaces up, so that they are above the water dew-point but below the acid dew-point—generally about 80° to 85° C. Boilers serving high temperature water systems are best operated at 140° C or over.

Flue gases, on leaving the boiler, may condense in the flue system (which is referred to later), but, so far as the boiler is concerned, the practical lower limit of exit temperature is generally about 200° C.

SMOKE INDICATION

The *Ringelmann* charts are a series of four grids of increasing darkness marked on cards 100 mm square, against which the smoke from a chimney may be compared. The Clean Air Act prohibits smoke darker than No. 2 on the Ringelmann scale but where smokeless zones are in force, smoke of any shade is prohibited.

Another scale is the *Bacharach*. A controlled volume of flue gas at a constant rate is passed through a filter paper and the discoloration is compared with standard shades 0 to 9 smoke numbers. This is clearly a laboratory instrument.

Smoke-density indicators consist of a light source on one side of a flue and a light sensitive cell on the other. Any shade causing a diminution of the light received by the latter is caused to register electrically on a dial in the boilerhouse from which the smoke number (Ringelmann) is read. In addition, a recorder may be incorporated for use as evidence in case of dispute. Also, an alarm bell is incorporated to give audible warning if smoke is being produced above the limit prescribed.

BOILER EFFICIENCY

The efficiency of a boiler expressed as a percentage is

$$\frac{\text{Heat output} \times 100}{\text{Heat input}}$$

Heat Input—With steam boilers, the water pumped into the boiler and evaporated is conveniently measured by a water meter. From the inlet-water temperature and outlet-steam pressure, the total heat per unit quantity of water evaporated may be determined, and hence the total heat output. With hot-water boilers, measurement of water flow through the boiler is more troublesome and costly due to the large volumes involved and is not commonly included except in large installations. Where such measurement is made, the quantity, multiplied by the temperature difference inlet to outlet, gives the total heat output per unit of time.

Heat input is measured according to the fuel: solid fuel by weighing and deducting the weight of ash and liquid fuel by meter or tank gauge. Calorific values of fuel and ash must be determined. Detailed methods of carrying out tests are covered by B.S. 845. Duration of tests is generally six hours with one control hour before and after. From such tests, including combustion analysis, a heat balance may be struck. The following is an example with a certain class of coal:

Overall thermal efficiency	78%
Loss due to sensible heat in chimney gases	15 ,,
Loss due to unburnt CO	1 ,,
Combustible matter in ash and clinker	2 ,,
	96 ,,
Radiation and other losses	4 ,,
	100%

Boiler test efficiencies with modern automatically-fired and controlled plant range between 75 and 85 per cent. Hand-fired boilers are not capable of sustained high efficiencies: no more than 65 per cent may be expected. Seasonal efficiencies may be 10 or more per cent. less than values obtained under test conditions.*

AIR SUPPLY TO BOILER HOUSES

It has been shown above how the theoretical air supply for combustion of fuel is determined. Table 4·4 gives a summary.

TABLE 4.4

THEORETICAL AIR FOR COMBUSTION PER kg OF VARIOUS FUELS

Fuel	Theoretical air required	
	in kg	in m³ at 15° C
Oil	13·9	11·5
Bituminous Coal	10·6	8·8
Anthracite	9·6	8·0
Coke	9·1	7·6

In the design of a boilerhouse, provision must be made for the theoretical air to enter plus excess air, which, due to lack of adjustment, can be as much as 100 per cent. Thus the theoretical quantities may well be doubled. The entering velocity through grilles or openings of various sorts may be taken at between 1 and 2 m/s. For example, for combustion of 100 kg of oil

* *Part load efficiencies of gas and oil fired boilers.* B.S.R.I.A. Technical Note 1/76.

per hour, the free area of inlet-air openings should not be less than:

$$\frac{100 \times 11 \cdot 5 \times 2}{1 \cdot 5 \times 3600} = 0 \cdot 43 \ m^2$$

In addition to combustion requirements, an allowance should be made for an air supply to ventilate the boiler house and the area calculated as above is commonly doubled in order to cater for this further need. The *Guide, Section B13*, recommends that an equivalent free area of $1 \cdot 2 \ m^2/MW$ boiler capacity should be allowed to serve the dual purpose. It should be borne in mind that boiler plant may be noisy and that a nuisance may arise from air inlet grilles to adjacent properties: acoustic treatment to the louvres may be necessary in consequence.

CHIMNEY SIZING

The calculation of the size and height of a chimney can be a very involved matter, but for the purposes of heating boilers using solid or liquid fuel it may be simplified by reducing the number of variables. The following notes show such a method. Flues for gas appliances are discussed in Chapter 6.

1. *Products of combustion. Quantity*

Assume excess air - - - - 75%
Average temperature of products of
combustion (flue gases) - - - 200–300°C
Boiler efficiency - - - - 75%

It can be shown that the products of combustion per MJ of boiler output are, by volume for coal, coke or oil, between $1 \cdot 1$ and $1 \cdot 3 \ m^3/MJ$ when these assumptions are made, say $1 \cdot 2 \ m^3/MJ$ mean.

The temperatures assumed cover the range of most heating boilers but volumes at other temperatures may be derived in proportion to absolute temperature in kelvins. Since the remaining variables, including the calorific value of the fuel used, will almost certainly vary during the life of the chimney, there is no purpose to be served by undue precision.

2. *Time*

Bringing a time scale into consideration, the above (for one second) become the products deriving from fuel burnt at a rate to give an output of $1 \ MJ/sec$:

$$1 \ MJ/sec = 1 \ MW \ (1000 \ kW)$$

3. *Velocity of flue gases in chimney*

Assume the range of velocity with natural draught from 4 to 8 m/s; with mechanical draught, from 10 to 12 m/s.

4. *Area*

The cross-sectional area of the chimney may then be derived according to boiler duty, fuel and type of draught. Thus, if the boiler duty is 200 kW, oil-fired, with natural draught (5 m/s) chimney temperature average at 250° C,

$$\frac{1 \cdot 2 \times 200}{1000 \times 5} = 0 \cdot 048 \text{ m}^2$$

$$= 250 \text{ mm diam.}$$

Fig. 4.3 gives areas and diameters for round chimneys direct, having selected a velocity. The 'equivalent diameter' of a square or rectangular chimney, in mm, is 1000 × the square root of the area as read from Fig. 4.3.

5. *Efflux Velocity*

In order that the plume of gas should rise clear of the chimney top and not flow down the outside of the chimney (down-wash), the diameter at the

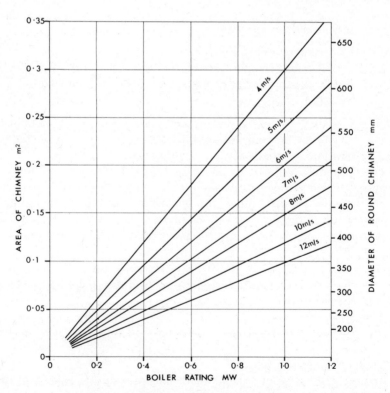

FIG. 4.3.—Chimney areas for stated velocities.

top should be reduced so as to maintain as high a velocity as practicable. For small boilers with natural draught a velocity of 6 m/s is advised. Larger boilers with mechanical draught should achieve 7·5 to 15 m/s.

The velocity head corresponding to these rates may be read from the bottom line of Table 4.5 as appropriate to a flue gas temperature of 250° C.

6. Draught required
The following may be calculated from the data on duct sizing in Chapter 19, adjusted for temperature:

At boiler exit, consult makers' data.

Oil-fired boilers vary from	7 to 50 Pa
Solid fuel fired, if burning rate is 5 kg/m² grate area	70 Pa
Flue connection, boiler to chimney, depending on number of bends and other losses, average	15 to 30 Pa
Efflux velocity head for 6 m/s	12·1 Pa
The total for an oil-fired boiler may then be from about	40 to 100 Pa
„ „ „ a solid fuel „ „ „ „ „ „	100 to 120 Pa

7. Draught produced by Chimney
The theoretical draught of a chimney at the two temperatures named varies with the external ambient temperature. Assuming that this is 20° C in summer and 0° C in winter, unit values are:

Winter
per metre height at 300° C	6·7 Pa
„ „ „ „ 200° C	5·5 Pa

Summer
per metre height at 300° C	5·8 Pa
„ „ „ „ 200° C	4·5 Pa

Fig. 4.4 is drawn on this basis.

Thus a chimney of 30 m height will in summer produce theoretically a draught of 155 Pa with flue gases at 250° C.

8. Draught loss in Chimney
The pressure loss per metre height may be taken from Table 4.5. Values for other velocities may be interpolated. Fig. 4.5 gives the pressure loss for chimneys of given heights having determined the loss per unit length from Table 4.5.

9. Rectangular Equivalents
For square or rectangular chimneys, the effective areas are those of the circle or ellipse which may be inscribed within them. The 'equivalent

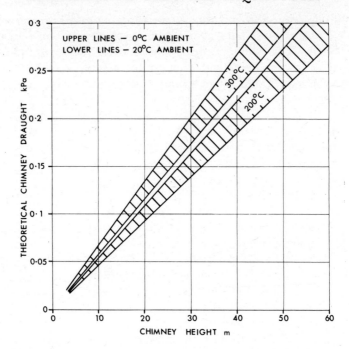

FIG. 4.4.—Theoretical draught for a chimney of given height.

TABLE 4.5

PRESSURE LOSS PER METRE HEIGHT OF SMOOTH CHIMNEY (MULTIPLY VALUES
BY 4 FOR BRICK OR ROUGH CEMENT RENDERING)

Effective Chimney Diameter (mm)	Pressure loss, Pa per m height or run for gas flow at the following velocities, m/s								
	4	5	6	7	8	9	10	11	12
200	0·45	0·71	1·02	1·39	1·81	2·29	2·82	3·41	4·07
250	0·34	0·52	0·76	1·03	1·34	1·70	2·10	2·53	3·04
300	0·27	0·41	0·60	0·81	1·06	1·35	1·66	2·00	2·39
350	0·22	0·35	0·49	0·68	0·88	1·12	1·38	1·67	1·99
400	0·18	0·29	0·41	0·56	0·73	0·93	1·14	1·38	1·65
450	0·16	0·25	0·36	0·49	0·64	0·81	1·00	1·21	1·44
500	0·14	0·22	0·32	0·44	0·57	0·72	0·89	1·07	1·28
550	0·13	0·20	0·28	0·38	0·50	0·64	0·78	0·95	1·12
600	0·11	0·18	0·25	0·35	0·45	0·57	0·71	0·85	1·02
650	0·10	0·16	0·23	0·32	0·41	0·52	0·64	0·78	0·92
700	—	0·14	0·21	0·28	0·37	0·47	0·57	0·69	0·83
750	—	0·13	0·19	0·26	0·34	0·43	0·53	0·64	0·76
Velocity Pressure	5·4	8·4	12·1	16·5	21·5	27·3	33·6	40·6	48·4

diameter' of such flues is therefore the square root of the square or rectangular area. If the flue must be rectangular, it is a general rule that a ratio of sides of 3:1 should not be exceeded. The pressure loss arising from gases flowing in brick or rough rendered concrete chimneys will be greater than for relatively smooth sheet steel.

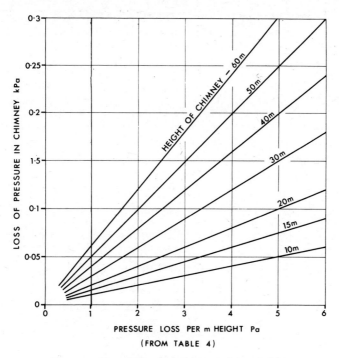

FIG. 4.5.—Draught loss for a chimney of given height.

10. *Procedure*

(1) Using Fig. 4.3 select a velocity according to whether draught is natural or mechanical and find chimney area and diameter.

(2) From Table 4.5 determine pressure loss per metre of height for this diameter.

(3) Make an assumption as to chimney height and hence from Fig. 4.5 note loss in chimney. Add for loss through boiler and flue connection and efflux velocity head.

(4) Using Fig. 4.4, the available draught may be found for the same assumed chimney height. If this is equal to or exceeds the sum of the losses, the assumption as to height may stand. If the available draught is insufficient, either the height may be increased or the velocity reduced, or both. If the calculated draught is much in excess of requirements, a smaller chimney or less height may be the solution, or if neither is possible, or desirable, a damper may be used.

11. *Clean Air Act*

The chimney height obtained, as above, is that necessary for combustion. It is necessary to consider it next in relation to mandatory requirements of the Clean Air Act, 1968. For this purpose reference should be made to the second edition of *Memorandum on Chimney Heights*, 1967.

The *Memorandum* is concerned to limit the SO_2 contamination near the chimney by making the chimney higher as the rate of SO_2 increases. The type of locality is taken into account as well as the relationship of building height to chimney height.

The *Memorandum* excludes installations releasing less than 1·5 kg of SO_2 per hour (about 200 kW with fuel oil and 400 kW with solid fuel). Further reference to this matter occurs in Chapter 5.

The Local Authority needs to be consulted in every case of a new boiler installation. The *Memorandum* is intended as a guide and calls for intelligent interpretation: the *Guide, Section B 13*, provides a digest of the text and rather clearer graphical data.

BOILERS WITH MECHANICAL DRAUGHT

Any boiler may be fitted with an induced-draught fan for exhausting the products of combustion and discharging them up the chimney, as in Fig. 4.6. Many boilers are obtainable with such a fan fitted as part of the standard unit. In others, the fan supplying combustion air has sufficient power to expel the products under pressure.

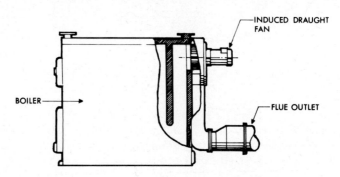

FIG. 4.6.—Sectional boiler with induced draught fan.

The advantages of mechanical draught are: first, that the boiler can be designed for higher velocities over the heating surface, so giving a higher rating for a given size; second, that draught produced by stack height is unimportant, and the chimney may thus be short (subject to the Clean Air Act), or, if the boiler is on the roof, notional only.

Induced-draught fans usually take in some diluting air and makers' data should be consulted for the volume to be handled by the chimney, the area of which will then be determined by the velocity decided upon,

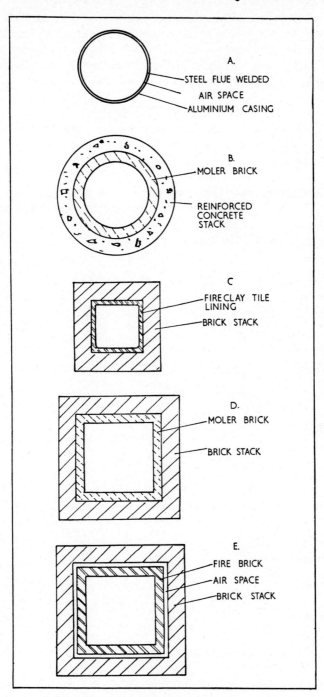

FIG. 4.7.—Various methods of chimney construction.

such as 10 to 12 m/s. The resistance will be calculated as before—it is usual, where the fan is part of the boiler, for about 60 Pa surplus pressure to be available for chimney loss and efflux velocity. A purge period before and after firing is incorporated in the control system to clear the boiler of stray products.

CHIMNEY CONSTRUCTION

Domestic Chimneys—*Building Regulations*, 1965, requires that all chimneys be lined with some impervious material such as tile. For advice on details of construction, reference may be made to the Building Research Station's *Digest No. 60.*

Larger Installations—The importance of keeping the products of combustion in the chimney warm is explained in the next chapter. With this in mind some form of insulation is a necessity and various forms of construction are illustrated in Fig. 4.7. They are:

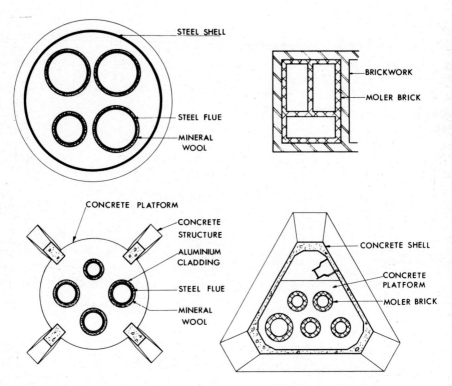

Fig. 4.8.—Multiple flues—alternative methods of construction.

(A) steel, enclosed in an aluminium outer shell having about 6 mm air gap. The air gap constitutes the insulation. These chimneys may be self-supporting;

Plate II. Pipe tunnel for service distribution (see p. 56)

Plate III. Continuous strip heating at high level in a small workshop (see p. 57)

Plate IV. View at rear of boilers showing coal feed from structural bunkers to automatic stokers (see p. 97)

Plate V. Hopper-type automatic stokers (see p. 98)

(B) pre-cast concrete in sections lined with insulating moler concrete;
(C) brick-lined with fire-clay tile;
(D) brick-lined with moler brick;
(E) brick or concrete outer stack, and independent lining of fire brick or moler brick with air gap between.

The relative heat losses from these constructions may be calculated as for U values—the inner surface resistance being assumed as zero. Cost, permanence and appearance if free-standing, will always be determining factors.

Multiple Flues—It will be apparent that, where more than one boiler occurs and there is a common chimney, it would be impossible to maintain the design efflux velocity with anything less than the full number of boilers in use. Furthermore, where mechanical draught is used and the flue is under pressure, back draught might occur to any boilers idle. Hence, it is now advocated that each boiler should have its own individual flue connection and chimney, and that they should not be combined into one large stack as has been the practice in the past. The vertical flues may be grouped into one stack, as in Fig. 4.8. (See Plate I, frontispiece.)

CHIMNEYS FOR OIL-FIRED BOILERS

The corrosive effect of sulphurous gases on the heating surfaces of a boiler has been alluded to previously (page 77). If the products of combustion on entering the chimney are allowed to cool to the region of the acid dew point, a new hazard is set up with oil-firing, namely acid smuts.

It appears that the minute unburnt particles of carbon always present in flue gases are liable to form nuclei on which condensation takes place, and they then agglomerate into visible black oily specks. In a chimney they are particularly liable to collect on any roughnesses, or at points of change of velocity. In so doing, on starting up, the sudden shock may cause them to be discharged from the top—often giving rise to complaints from surrounding property.

It has come to be recognized that, in order to avoid this trouble, the following principles should be adhered to:

(*a*) to keep the velocity up the stack as high as possible with the draught available;

(*b*) to keep the gases hot by insulation of the chimney as referred to in the previous chapter;

(*c*) to avoid cooling the gases by a draught stabilizer or air leaks;

(*d*) to design flue connections with easy bends and gradual changes of velocity;

(*e*) to keep the chimney smooth inside;

(*f*) on multi-boiler plants, to use one flue per boiler; or, if a common flue is unavoidable, the dampers on idle boilers must be gas-tight.

Acid smut formation is less of a problem with a low sulphur oil such as Class D, even though draught stabilizers are often part of standard boilers on a domestic scale.

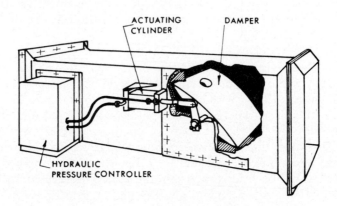

Fig. 4.9.—Automatic draught control damper.

For the control of draught on plants burning the heavier oils, in lieu of a draught stabilizer admitting cold air, some form of damper control is necessary which may be automatic, of a type such as Fig. 4.9.

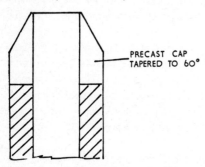

Fig. 4.10.—Form of chimney top.

Treatment of flue gases by means of a proprietary additive to the fuel or in the hot-gas outlet may be considered where smut nuisance is particularly troublesome.

The top of a chimney for an oil-burning installation for good dispersal is preferably of the form shown in Fig. 4.10. It will be noted that the cap is tapered at 60° to give a thin edge which has the effect of reducing the tendency to down-wash.

CHAPTER 5

Boilers and Firing Equipment for Solid and Liquid Fuels

THIS CHAPTER IS DEVOTED, first, to the sizing of the boiler or boilers for low pressure hot water heating and to the location of the boiler plant, and, second, to methods of firing of solid and liquid fuels and their handling. Firing by gas is considered in the next chapter.

Boiler Power— The sizing of the boiler plant is to be such as to meet the heat losses from the building under basic design-temperature conditions. As mentioned already, the total of the maximum heat losses may exceed the actual peak demand, due to the fact that infiltration air entering rooms on one side of the building leaves by rooms on the other side, so that a correction should be made to avoid taking the same air-change twice. In addition, where heating is continuous, it has been proposed that diversity factors apply as follows:

Single space	1·0
Single building or zone, controlled centrally	0·9
Single building, individual room control	0·8
Group of buildings with similar use	0·8

The boiler output must, however, be in excess of the corrected total in order that the design temperature can be achieved in a reasonable time. If heating is continuous, the excess will be a minimum; the more intermittent the usage, the greater excess capacity needed.

There is a further case for some margin of capacity, where a boiler is thermostatically controlled, in order to provide what may be termed 'acceleration'—the ability of the boiler to surmount the load under peak conditions and still be under control. A still further purpose of a margin is to deal with the declining efficiency of a boiler due to fouling of the heating surfaces by soot or ashy deposit, especially where solid fuel is used.

The reduction on air change may be set against the increase required for a margin for the reasons stated, and to some extent these cancel out. In the past, these refinements of calculation have generally been ignored and a margin of around 25 per cent added to the total of the heat losses and mains losses to cover a multitude of sins. Where time is short for design and estimating, this rough and ready method may suffice, but it would be worth while in due course checking with the data now available.

By way of example, using the approximate method, let us
assume that the heat loss total for the building is   500 kW
Also let us assume the heat losses from main piping in
trenches etc. is 10 per cent of this figure   50 kW
                                                                        ————
                                                                        550 kW
                    Allow margin of 25 per cent   137 kW
                                                                        ————
                                                                        687 kW

Reference to makers' catalogues may show that the nearest boiler size
up is 750 kW. If this is chosen, the margin would be 36 per cent and might
be considered excessive. The next size down might be 650 kW, still giving
a margin of 18 per cent which would be more reasonable.

In the old days of hand-stoked boilers a big boiler was considered an
advantage as it meant longer periods of run on one charge of fuel. Nowadays, with automatically-fired boilers burning oil or gas, it is often a disadvantage to overdo the boiler power as overall efficiency may be lowered.
If the requirement has been accurately calculated, a margin of 10 to
15 per cent may be adequate. Any margin is likely to be a boon during
exceptionally cold spells.

Boiler Ratings and Margins—Boilers are normally rated in kW and fall
broadly into the following ranges:

> Small   10–50 kW, mainly domestic
> Medium   50–500 kW
> Large   500–2500 kW

In any sizeable installation over say 200 kW there is a case for providing
more than one boiler. If there are two boilers, one will suffice in mild
weather working near its full output, which has advantages in efficiency
and avoidance of corrosion. The second is brought in during cold weather
and can also act as a standby during breakdown or cleaning. In the past,
it was customary with two boilers to arrange each to take about two-thirds
of the net heat load subject to selection from standard units as mentioned
above; thus, together, they had $33\frac{1}{3}$ per cent margin. More modern
practice suggests that this might well be reduced due to the diversity which
undoubtedly exists, the bigger the installation.

It will be obvious that for very large installations three, four or more
boilers may be used, giving greater flexibility than with fewer larger units.
Several boilers each having a small margin may then give almost a
complete one-boiler standby.

Selection of the size of individual units for a multi-boiler installation is
a matter for compromise. Limitation of the number of spare parts to be
stocked suggests the use of equally sized units (e.g. three boilers at 166 kW
each for a total requirement of 500 kW). Conversely, to obtain maximum
efficiency whilst meeting a load which varies from day to day with the

external temperature, there is a case for sizing the units unequally in order to provide 'steps': e.g., for the same total requirement, one boiler at 100 kW and two at 200 kW each would allow an output at 100, 200, 300, 400 and 500 kW as required (five 'steps' as against the three provided by equal-sized boilers).

Modular Boilers—The modular approach to boiler power is completely different. It comprises the use of an array of small boilers. Makers' ratings offer three or four sizes within the range 40 to 70 kW. These units are connected in parallel, thus ten units would produce 400 to 700 kW, and if only one is fired the turn-down ratio is 10:1, which is well below what can be achieved by conventional large units. Essential to the system are the controllers which bring on the appropriate number of boilers to match the load, and the sequencing step controller which varies the order in which the units are fired.

Fig. 5.1 shows a diagrammatic arrangement of a modular system.

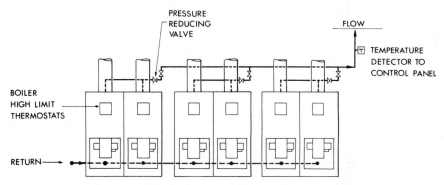

FIG. 5.1.—Modular boilers, principles of pipe connection.

Advantages claimed are:

(a) Higher efficiency than conventional large units especially under conditions of low load.
(b) Reduction in standby capacity.
(c) Lower cost and ease of installation.
(d) Reduction in weight. This renders them particularly suitable for roof-top boiler houses for use with oil or gas.
(e) Low thermal capacity.

There are a number of points of design which need care such as the arrangement of mixing or diverting valves, positioning of control thermostats and interaction of outside compensator control with load matching controller. A report on research carried out on the modular system is given in a paper by Spivey and Jarvis.* This contains reference to a field study

* Institution of Gas Engineers, No. 917, Nov. 1973.

on two identical multi-storey blocks of flats, one with conventional boilers and one with modular, and concludes that 'for normal operation over a heating season the differences (in gas consumption) will be of the order of 10%'.

Hot-Water Supply—Although this is dealt with in a later chapter, it is customary where a central system is installed to couple with it the supply of hot water. The same boiler or boilers can serve both heating and hot water by means of a heat exchanger or calorifier, so that, during the heating season, the hot-water load is added to that for heating. Bearing in mind that hot-water demand is spasmodic, it is possible to supply it by slight 'robbing' of the heating for short periods.

In summer, when heating is 'off' in a single-boiler installation, it becomes a question of proportion as to whether the hot-water supply load is sufficient to warrant the running of the boiler; if not, then a small boiler for summer use is required. In a multi-boiler installation it is likely that one boiler may be of a smaller size than the others for the same purpose.

BOILERHOUSES: SIZE AND LOCATION

Size—Various attempts have been made to give guidance to architects for planning a boilerhouse in advance of design being worked out. There are so many designs of boilers and kinds of arrangement that any advice can only be in very general terms. Some boilers are long and narrow, others short and tall. Access for installation may decide whether a sectional type must be used. A cylinder for hot-water supply must often be housed in the same space.

Table 5.1 may be used as a rough guide.

TABLE 5.1
APPROXIMATE BOILERHOUSE SIZES

Boiler Rating kW	Length m	Width m	Minimum Height m
30	2·5	3·5	2·3
75	3	4	2·3
100	3	4·5	2·5
150	4	5	2·5
200	5	5	2·5
300	6	5	2·8
500	7	6	3
750	7	7	3
1000	7	8	3

Location—It was usual to house boiler plant in the basement, a relic of the days when coke was delivered by road trucks. Construction below ground is expensive and the flue must be accommodated up to the roof, which is costly and takes up space. Hence, where conditions permit, there is often a case for putting the boilerhouse on the roof. With mechanical draught the flue may be quite short. Ventilation and access are easy.

Roof boilerhouses are really only applicable with oil or gas although prototype installations exist where small coal has been delivered to roof level plant by delivery vehicles equipped with pneumatic blowers. Oil is pumped-up from storage at ground level as illustrated in Fig. 5.17. Considerations of weight, noise and vibration will affect the structure and acoustical treatment, but are usually capable of being simply resolved (see Chapter 6 also).

Alternatively, boiler plant may be at ground level—possibly detached from the building with an independent stack. A group-heating scheme would most likely require this sort of planning.

BOILER TYPES

Cast Iron—By far the most common type of heating boiler is the cast-iron sectional, which has been in use for nearly a century. Originally designed for burning solid fuel, particularly coke, on a set of fire bars, it has been

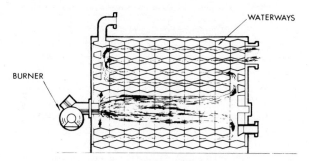

FIG. 5.2.—Cast-iron sectional boiler (Beeston).

developed and refined for firing by automatic underfeed stokers and for application of oil firing or gas firing at high efficiency. Sizes range from the smallest up to 3000 kW.

The boiler is made up of sections usually connected together by means of three nipples, one at the top and two at the bottom, all pulled together and held water-tight by steel tie-bars externally. Some of the larger types are in two halves, having four nipples. Back and front sections differ from the intermediate ones as they make provision for firing, cleaning and out-let for products of combustion. Sections may be added for extensions or replaced on failure.

With the decline in availability of coke and of labour for stoking, this method of firing has become a thing of the past. Fig. 5.2 shows a type specially designed for oil or gas firing in which the flue passages are arranged for higher gas velocities and hence greater output for a given size. This type has a waterway bottom instead of being open, and hence the base of brick or concrete on which it stands is subject only to the relatively low temperature of the water, otherwise an insulating base is necessary to prevent undue temperature build-up in the floor structure.

The normal cast-iron sectional boiler is designed for working water pressures up to about 40 m head of water (400 kPa), which covers most normal heights of building; but certain makes for larger sizes, using a special grade of cast iron, are suitable for heads up to 100 m of water (1 MPa). Test pressures are in excess of these working levels.

Cast iron is less prone to attack from sulphurous corrosion products arising from combustion of fuel, especially oil, than is steel.

Steel—Wrought iron was at one time a favourite material for the manufacture of boilers, being very ductile and resistant to corrosion. The art

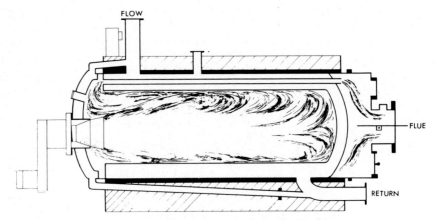

FIG. 5.3.—Welded mild-steel boiler (Allen Ygnis).

of 'puddling' of iron being largely a manual process, however, wrought iron has disappeared from the market and steel has taken its place. Steel is homogeneous in structure and, unlike wrought iron, is very liable to attack by sulphurous corrosion products.

For long life, a steel boiler requires the system which it serves to be designed to operate at a temperature outside the region where severe attack is liable to occur. Given these conditions, steel is probably more versatile than cast iron—which feature has been taken advantage of in the wide range of designs. One example is illustrated in Fig. 5.3. In this the combustion chamber is a welded cylinder, whilst the secondary heating surface consists of fire tubes.

Sizes in a variety of forms range from about 30 to 3000 kW, above which types referred to in Chapter 11 are applicable to higher pressures.

Concrete—Experiments have been made in France using concrete as a material for sectional-type boilers. A special aggregate is used and each section is cast in halves to a pattern, including flue-ways for the passage of the combustion gases. Within the concrete, an extended coil of steel pipe is embedded, and this has terminal ends brought outside the half section at top and bottom. A great advantage would seem to be that no refractory lining is required to the combustion chamber for firing with oil or gas and, since the actual concrete reaches red heat, combustion is complete and virtually no cleaning of the flue-ways is needed.

PACKAGED-BOILER UNITS

As applying to hot-water boilers, the term 'packaged unit' implies a boiler specially designed for oil- (or gas-) firing, having all the necessary equipment fitted to it of the type best suited to the characteristics of the boiler.

These units generally operate with high combustion rates, the flue passages being designed for high velocity and, hence, for maximum heat transfer for a given heating surface.

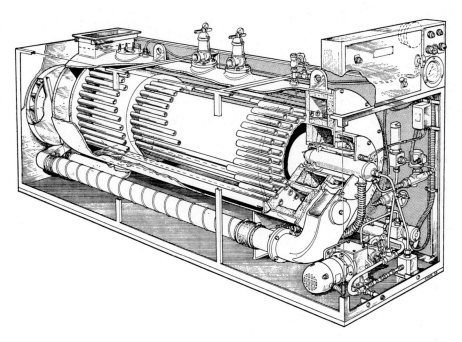

FIG. 5.4.—Packaged boiler unit, forced draught.

The burner will have been selected to give the best flame shape for the combustion chamber and may be arranged to pressurize the chamber so that flue pull is not relied upon. Fig. 5.4 illustrates a unit of this type. Alternatively, an induced-draught fan may be included as part of the equipment.

The unit will be complete with control gear for fully-automatic operation, including safety controls. Certain types include a boiler-circulating pump.

The chief merit of this kind of approach is that all parts are suitably matched and pre-set, so that the task of setting-up and commissioning is reduced to a minimum.

METHODS OF FEEDING BOILERS WITH SOLID FUEL

Solid fuel is fed into boilers in one of three ways:

(a) by hand;
(b) by gravity from a magazine;
(c) by automatic stoker.

Hand Feeding—The vagaries of hand feeding have already been mentioned briefly. The high cost and the difficulty of obtaining labour for the hand-firing of boilers, apart from other considerations, is causing automatic firing methods to be generally adopted.

Magazine Boilers—The fuel is fed into a magazine and descends by gravity into the burning zone where it is consumed, the ash ending up as clinker to be removed from the base. Fig. 5.5 illustrates a magazine boiler for burning anthracite grains.

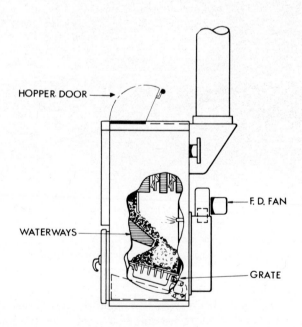

FIG. 5.5.—Magazine boiler (Trianco).

Automatic Stokers—One of the most usual forms of automatic stoker for heating boilers operates on the worm-feed principle. Fuel is fed into a hopper, at the bottom of which is situated a worm or screw, rotated at a slow speed through reduction gearing from a motor drive. The worm is enclosed in a tube beyond the hopper, and serves to convey the fuel into the firepot, which is built into the firebrick inside the boiler, as in Fig. 5.6. The fuel is bituminous coal of small size.

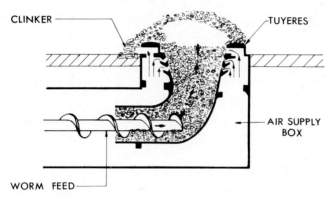

Fig. 5.6.—Firepot and fuel bed for underfeed stoker.

A second tube delivers air into the firepot from a fan driven from the same motor that operates the worm. This air is discharged through a series of slots or openings in the firepot (tuyeres), so disposed as not to be closed up by ash or coal.

Thus, forced draught is provided, and a very high combustion rate is possible, so high, in fact, that a grate is unnecessary. The fuel is burnt as it passes over the edge of the firepot, and all the ash is reduced to clinker in the process. This, however, does not impede combustion, as the fresh coal brought in by the worm pushes the waste material to one side where it remains for periodical removal.

Safeguards are generally provided to prevent jamming of the worm from damaging the mechanism, also to prevent the fire going out if the machine is shut off for lengthy periods by its thermostat. The former is accomplished by a shearing pin, or slipping clutch. The latter, by arranging the motor to start up every hour or so for a few minutes, whether required by the thermostat or not.

Thermostatic control is applied either as stated above, by the stop-and-start method, or by a variation in the rate of the fuel feed. With the latter method it is necessary to vary the air supply at the same time, or considerable excess air will result when operating at low outputs.

Another method of feeding is direct from bunker. In this, the worm extends into the main fuel storage and the hopper is eliminated. Fig. 5.7 illustrates the two alternative arrangements. (Plate III facing p. 86.)

For smokeless combustion, makers provide a method of supplying preheated secondary air.

The advantages of automatic stokers are that labour is reduced and higher efficiency is obtainable than with hand-stoking. Provision for grit arresting may be required under the Clean Air Act. Plate III (facing page 86) illustrates an example of under-feed stokers fed by gravity chutes from coal hoppers overhead. The boilers serve a small group heating scheme.

For larger plants, coal elevators may be used such as screw type, pneumatic, rubber belt, or endless chain bucket. Similarly ash may be handled by mechanical means to a hopper for removal from site, but these subjects are beyond the scope of the present treatise.

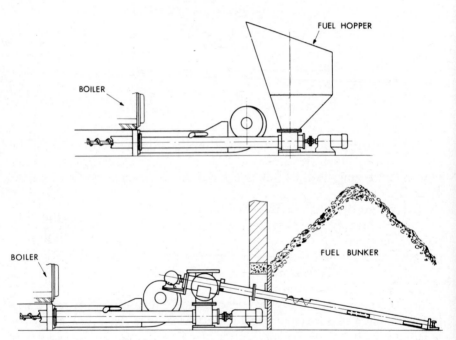

FIG. 5.7.—Alternative fuel supply and storage methods for underfeed stoker, hopper and bunker type.

Vekos Boiler—Another type of automatically-fired boiler for burning coal is the Vekos, shown in section in Fig. 5.8. The coal is delivered pneumatically or by worm to the top of the boiler and is then 'sprayed' into the combustion chamber. Ash and unburnt fine fuel is removed by a grit collector, which is integral with the flue system, and is conveyed back to the combustion chamber for re-burning. Ultimately, ash is burnt into clinker and this may be removed daily through access at the boiler front. It is common practice to incorporate a clinker-crusher system with the boiler plant so that the waste matter may be raked directly into a hopper at the boiler front and then crushed and conveyed pneumatically to a storage cylinder, from which it may be removed with minimal effect upon amenities. Plate VI (facing page 118) illustrates this type of boiler applied to a District Heating System.

The Vekos is also available as a multi-fuel boiler capable of being changed over from coal to oil or gas by the simple manipulation of levers. Possible sudden changes in cost and availability of fuels may render such a facility a worthwhile 'insurance policy' for future security of energy supply.

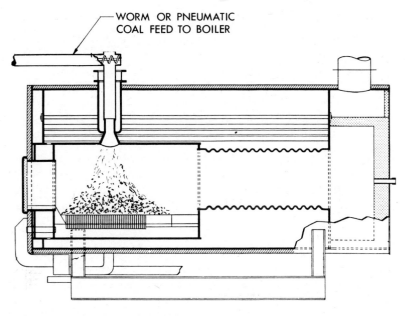

FIG. 5.8.—Overfeed fuel supply to GWB Vekos boiler, worm type coal feed.

METHODS OF FIRING BOILERS WITH OIL FUEL

Atomization—The whole problem of burning oil efficiently and without smoke is one of atomization, that is to say the intimate mixture on a molecular scale of carbon in the fuel and oxygen in the air. All kinds of attempts to solve this problem have been made over the years, but for heating boilers they have now been resolved into relatively few. Systems have included:

Vaporization—The oil is heated as in a blow lamp or Primus stove. Vapour is formed which is ignited. The heat of the flame continues the process. It is suitable only for very light oils such as kerosene.

Pressure jet—The earliest system simply pumped oil at high pressure through a fine jet giving it a swirling motion. Natural draught air entered through a front register giving a swirl in the opposite direction. Used mainly in marine practice, it was also used for large industrial plants in the early days. Skill was required in adjustment of the burner and the air to achieve reasonable efficiency and to avoid smoke. Heavy crude oils were used, requiring considerable preheating.

Rotating cup—The atomization in this system is caused by a spinning cup. Little or no pressure is required to deliver the oil to the cup, which has a serrated edge. The cup in some of the methods is driven by the primary-air blast, in others by mechanical means from the motor-driven fan and the oil pump.

Compressed air—The principle of atomization in this case depends on admission of a jet of compressed air around the nozzle, delivering oil under pressure.

Steam—This is the same as the last except that steam is used in lieu of the air blast. Its use is confined to steam boilers.

Emulsifying—The emulsifying burner relies on the principle of pre-mixing air and oil before delivery to the burner. A reduction in preheat requirements results, but the system has not been developed extensively.

Combustion Air—Natural draught as a means of introducing air for combustion is used in some of the small vaporizing burners. Apart from this, mechanical draught is employed, either supplying the primary air to the burner—the remainder entering by natural draught as secondary air—or supplying 100 per cent of the combustion air. The tendency now is to adopt the latter method, in which case the combustion chamber of the boiler is sealed and pressurized. Alternatively, the burner delivers the oil and air mixture into the combustion chamber and relies on natural draught or mechanical-induced draught to remove the products of combustion.

It will be seen that there are various possible combinations of atomizing methods with systems of air introduction, and hence as many variations in design of the burner. Of these the following are described.

Vaporizing-Type Burner—This is a burner for small boilers up to about 20 kW and uses kerosene marketed by the oil companies for this purpose and known as Class C. Its properties are: Calorific Value 43·6 MJ/kg net, Flash Point 38° C, Sulphur 0·2 per cent. One form of vaporizing burner is illustrated in Fig. 5.9, being the pot type. Oil enters the chamber under control of a ball float and is thence fed to the burner pot by gravity where it is ignited manually. Air is supplied by means of a small fan. The flame is controlled by a thermostat actuating the magnetic valve, but a small flame is always maintained sufficient to preserve ignition.

Other forms of vaporizing burner are fully-automatic and therefore include means for ignition and safety devices for flame failure. One such is the *Wall-flame* which is available up to about 50 kW. The vaporizing burner is relatively cheap, quiet in operation and suitable for small domestic applications.

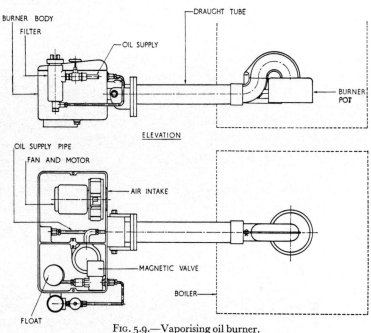

FIG. 5.9.—Vaporising oil burner.

Gun-Type Burner—This is by far the most common type of oil burner, ranging in size from a boiler output of 15 kW to 2·5 MW. It contains an electrically-driven fan delivering 100 per cent combustion air, see Fig. 5.10. Oil is delivered under pressure from an oil pump coupled to the spindle driving the fan. The oil is sprayed through a fine calibrated jet, the size of which controls the output of the burner and is fixed. Different jets can be used according to the duty of the boiler. Air from the fan, delivered *via* swirling vanes in the nozzle, is controlled as to quantity by adjustable slots or dampers so as to give the desired oil/air mixture. This adjustment is pre-set when the burner is tested by CO_2 measurement to give say $10\frac{1}{2}$ per cent by flue gas analysis.

Ignition is by electric spark from two electrodes located near the nozzle tip, the high-tension supply at about 8·5 kV being from a transformer which is energized when the burner starts up. The spark is maintained for a sufficient period to cause ignition, and is then cut off by a time delay in the control system.

The burner is under the control of a thermostat in the case of a water boiler, or pressure device in the case of a steam boiler. In addition, a high-limit thermostat is incorporated to cut off the burner if an unduly high temperature results—this being of a manual reset type.

In the smaller sizes, control is usually on–off and above this, either high–low or modulating control is incorporated to reduce the shock on the boiler of a sudden big flame. In the high–low method the burner starts up on one-third load, and after a delay changes to full load, being cut down to one-third as the temperature is reached and finally to off. In the modulating system, the burner starts up on one-third load and changes to a modulated condition between that and full load, according to demand. On the downward run, if temperature is satisfied at one-third load, the burner becomes on–off below that point.

The flame shape of this burner may be varied to suit the geometry of the fire-box by selecting the angle of divergence of the jet and by adjusting the angle of the directional vanes in the nozzle.

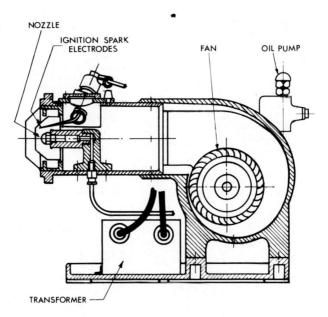

FIG. 5.10.—Gun type pressure jet burner.

When Class D oil is used, no preheating is necessary. For heavier grades, an electric preheater is incorporated as part of the burner unit. The heater contains a thermostat set to maintain the required temperature for atomization of 60 to 70° C. The control system is interlocked so that the burner cannot start until this temperature is reached.

Rotary-Cup Burner—The main elements of a burner of this type are illustrated in Fig. 5.11. The burner is mounted on the front of the boiler. Oil is delivered from the pump at low pressure through a hollow spindle to which is attached the fan and spinning atomizing cup. The periphery of the cup is surrounded by an annular air nozzle with swirling vanes. The air from this nozzle may represent 20 per cent of the air for combustion. The remaining (secondary) air for combustion is introduced at low pressure by a separate fan, being preheated to some extent by its passage over the primary-air and secondary-air quarls.

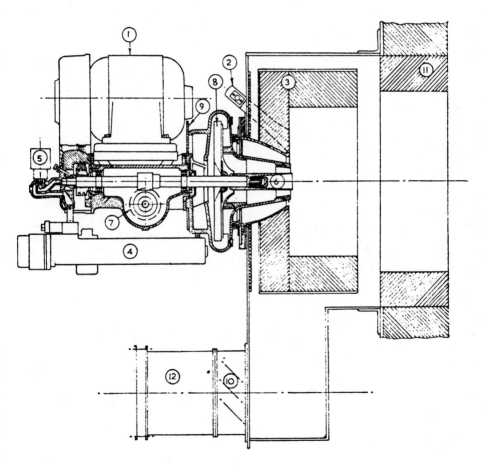

(1) Motor	(5) Solenoid valve	(9) Primary air control valve
(2) Photocell	(6) Atomizer	(10) Secondary air control damper
(3) Primary air quarl	(7) Pump Drive	(11) Secondary air quarl
(4) Heater	(8) Fan	(12) Forced draught fan

FIG. 5.11.—Rotary cup burner (Hamworthy).

Ignition is usually by gas, ignited by electric spark. Control of output is on a modulating basis from one-third to full load, being on–off below one-third, or on high–low.

Control of output is by adjustment of oil delivery linked with damper control of primary and secondary air.

This type of burner is applicable to larger outputs from 150 kW to 5 MW, with oils of Class E, F or G.

FURNACE LININGS

The firebrick of oil-fired combustion chambers calls for a material which will withstand temperatures of 1400 to 1600° C without fusing or premature disintegration. The brick, therefore, requires to be burnt to a temperature in excess of this, and several standard brands are available for the purpose.

One important point in the building-in of this brickwork, particularly where a sectional boiler is concerned, is to watch that the material is not carried solid up to the metal without an air-gap of half an inch or so being left around it. The great temperature in the combustion chamber causes considerable expansion of the bricks, and the air-gap allows freedom for this to take place without any strain being put on the boiler plates or joints. Many of the packaged type boilers, however, dispense with firebrick linings except for the quarl.

Under-Hearth Insulation—Where a boiler does not have a waterway bottom, it is necessary to insulate beneath the firebrick forming the hearth of the furnace, so as to avoid cracking the concrete floor which, in turn, might cause damage to any waterproof membrane that may exist. For boilers of any size up to about 600 kW, this insulation may comprise, say, 150 mm of Moler insulating brick. For larger boilers it is desirable to construct a sub-base of honeycomb form to allow air-cooling and, with some types of boiler and burner, this must be so arranged that the air drawn in by the forced-draught fan for combustion is drawn through these passages, so serving as a means of preheating the air. Alternatively, the sub-base may be cooled by circulating return water through a pipe coil embedded therein. It is desirable that the concrete below a boiler should not exceed a temperature of 70° C. (Plate VII facing p. 119 and Fig. 11.4.)

REGULATIONS AND SAFETY PRECAUTIONS

Owing to the fact that the burning of oil contains the elements of a fire hazard, various bodies have drawn up regulations governing the installation of oil-firing equipment, oil tanks, etc. A British Standard *Code of Practice* has also been produced, No. CP. 3002, Part 1 being in respect of oil-firing installations burning Class D fuel oil. Part 2 covers vaporizing burner installations and Part 3 installations using preheated fuels, Classes E, F and G.

The various local authorities, fire authorities and insurance companies have their own regulations and these should be ascertained when any new installation is under consideration. Local authorities also have responsibilities under the Clean Air Act which may affect the grade of oil selected and the dimensions of the chimney, referred to later.

Fire Valve—The outlet from a tank should include a fire valve. This generally takes the form of a lever-type valve heavily weighted, kept open by a taut wire stretched across the boiler house and having a fusible link of low melting-point alloy over each boiler front.

Should a fire occur, the rising flames or hot gases will melt the link and the valve will shut immediately, thus cutting off the supply of oil and minimizing further damage. A hand release is also provided for testing and this may incorporate a solenoid release in circuit with the sump float referred to later. Since such mechanical equipment is rarely, if ever, called upon to act, friction may prevent its functioning at the crucial moment. An electrical system may alternatively be used in which a solenoid valve in the main oil feed is kept open only so long as current passes. The circuit is broken by a heat sensitive device over each burner. Such devices require to be hand re-set.

Controls—The ideal always aimed at is to make an installation completely automatic and thus dispense with the labour of attendance altogether. Fully-automatic types of burner have been referred to and it will be obvious that they inherently include means for self-igniting and self-extinguishing, the rate of firing or the period of operation being controlled by temperature, i.e. thermostatically in the case of a hot-water boiler or air-heater, or by pressure in the case of a steam boiler. One thermostat is provided for control purposes and a second high-limit stat to cut off the burner in the event of over-run of temperature. The latter is for hand re-set.

In self-contained unit type burners, in which one motor drives fan and oil pump, there is little likelihood of oil being delivered into the boiler without the fan running and supplying the necessary air; hence safeguards necessary with the old tailor-made systems to guard against this possibility are not required. The worst that can happen, however, is that the flame is not established due to failure of ignition or, having been established, is subsequently extinguished from some cause such as water in the oil or maladjustment of the burner. To guard against this hazard the flame failure cut-off device has been developed and will be found to be incorporated in one form or another in most of the self-contained burners referred to earlier. The most reliable and rapid device of this nature depends on a light-sensitive element sometimes called an 'electric eye' so positioned as to receive light from the burner flame. When starting up, this device is shorted out for a period and, if by the end of this period the flame is not established, the burner will cut off automatically. In some types of control, after a delay period to allow any gases from the first attempt to have cleared from the boiler, a second attempt at ignition is made. If the second attempt fails, the burner is locked out permanently and it is then necessary for an attendant to discover the fault and reset the control.

In the normal course, the flame having been established, the light-sensitive cell is actuated and maintains the burner in operation. Should the flame fail, however, under this normal running condition, the burner is caused to shut down immediately.

Complete automation of a heating system, whether the source of heat is oil, gas or electricity, brings with it control of operation according to time, involving clock control. In the case of oil firing, the burner may be switched off completely at night or may change over to another thermostat with lower setting. This is sometimes further extended on the clock switch to remain at the lower level over a week-end when applied to offices, shops etc. Where a boiler or boilers are supplying heating only, it is tempting to allow the control to vary the temperature of the water delivered from the boiler, but this may result in unduly low water temperatures at times of mild weather with danger of condensation forming within the boiler causing corrosion. Excepting in small installations it is therefore preferable to run the boiler at constant temperature and to rely on a mixing valve to give the variable temperature for the heating system, as referred to on page 181. Where boilers are supplying, in addition to heating, hot-water supply or hot water for air-heater batteries in a ventilation system, this method will be adopted in any event for obvious reasons.

Further Safety Precautions—Further devices for protection against oil fires, particularly on large plants, are often called for. These include the following:

(a) Boiler dampers to be removed or locked open with an indicating plate showing position of vane.

(b) Alternatively, the burner valve and damper must be provided with an interlock operated by a common key which cannot be withdrawn from the damper lever until this is in the open position.

(c) A foam chemical extinguisher of portable or permanent type to be installed, serving both the boiler room and oil-tank chamber.

(d) Alternatively to (c), foam pipes to boiler house and tank chamber to which the fire brigade can connect their apparatus in the street.

(e) A sump in the boiler house provided with a ball float and electrical contact so as to cut off the oil supply to the burner(s) should there be an oil leak.

LIQUID FUEL STORAGE AND HANDLING

It is usual to provide a storage based on two to three weeks' running on full load. Except in special cases, the full load would not apply over twenty-four hours, and it is generally reasonable to assume the equivalent of about twelv.e hours per day. In the case of small installations, such as in domestic premises, certain oil companies now deliver oil metered through a pump, according to requirements, in which case a tank of about 1300 litres may serve. Otherwise, it is usually considered desirable to have a tank large enough to take at least a 2500 litre load, which, allowing for some reserve, gives a minimum tank capacity of say 3500 litres.

In larger installations, it is frequently necessary to provide two or more tanks. This has the advantage that one tank can be filled while the others are in use, thus allowing sludge and water, if any, to settle.

TABLE 5.2

OIL STORAGE CAPACITY

21 days × 12 hours = 252 hrs. Boiler efficiency 75%. For 1 kW output heat

$$\text{input to boiler} = \frac{252 \times 3600}{0.75} \text{ kJ/s}$$

$$\text{Oil: take 41 MJ/kg} \frac{252 \times 3600}{41\,000 \times 0.75} = 29.3 \text{ kg}$$

Take sp.gr. 0·9 = say 35 litres per kW

Thus:

Boiler Rating kW	Storage for 3 Weeks Litres
20	700
40	1400
60	2100
80	2800
100	3500
150	5250
200	7000
300	10 500
400	14 000
500	17 500
750	26 250
1000	35 000

Note: The thousands column gives storage in m³. (m³ × sp. gr. = tonnes).

Table 5.2 gives the consumption of various sized plants based on three weeks' run at full load (twelve hours per day), from which the appropriate storage may be estimated.

Storage Vessels in Buildings *—Oil-storage vessels are usually of steel construction and may be either rectangular or cylindrical. They may be brought into the building in one piece, or welded *in situ*. The latter is

* See B.S. 799 and C.O.P. 3002 for detailed recommendations.

necessary in confined situations. Table 5.3 gives capacities of tanks, cylindrical and rectangular.

Tanks deeper than 1 m often have iron access ladders inside. To allow proper inspection outside of the tanks, they should have a walking-way of 0·5 m clear all round. Cylindrical tanks are supported on steel cradles, rectangular tanks on steel joists, in either case bearing on sleeper walls about 0·3 m high with bituminous felt or lead packing on top.

TABLE 5.3

OIL-TANK CAPACITIES

Cylindrical

Diameter m	Length m	Gross Capacity Litre	Net Capacity* Litre
1	2	1570	1300
	2·5	1960	1600
1·5	2	3530	3200
	2·5	4420	4000
	3	5300	4800
2	3	9430	8800
	3·5	11 000	10 200
	4	12 570	11 800
2·5	3·5	17 180	12 900
	4	19 640	14 800
	4·5	22 090	16 600
3	4	28 280	27 500
	5	35 350	34 200
	6	42 420	40 200

Rectangular

Length m	Width m	Depth m	Gross Capacity Litres	Net Capacity* Litres
1	1	1	1000	700
1·5	1	1	1500	1100
2	1	1	2000	1500
3	1·5	1	4500	3300
3	1·5	1·5	6750	5600
3	2	1·5	9000	7500
4	2	2	16 000	14 000
4	3	2	24 000	21 000
4	4	2	32 000	28 000
6	4	2	48 000	42 000

* Allowance 150 mm up to outlet and 100 mm ullage (at top), rounded to nearest 100 litres, lower.

Pipe-work and tanks used in connection with oil should not be galvanized.

Tank Rooms, separated entirely from the boiler house by a brick or concrete construction, are called for in C.O.P. 3002. An oil-storage room should have the lower portion oil-tight to hold the full capacity of the tanks, should they leak; this calls for an access door at a higher level with steps outside and inside. The door must be fire-resisting; also, separate inlet and outlet ducts direct from outside are required for the ventilation of the tank room.

Fig. 5.12 shows this arrangement, together with other details which will be referred to later.

Tank Fittings—Various fittings are required in connection with storage tanks, the principal ones being given below.

A *manhole* must be provided to each tank, a common size being 0·5 m diameter. The joint with the tank top should be made gas-tight.

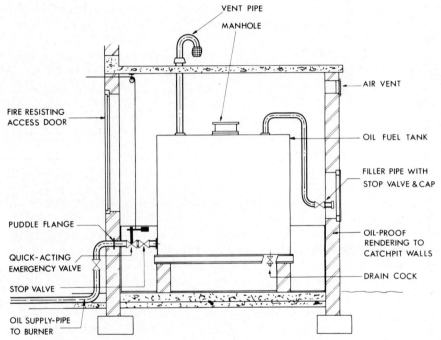

FIG. 5.12.—Oil tank and connections, above ground.

The filling terminal should be standard 50, 65, or 80 mm, according to length and grade of oil; male gas thread for hose coupling with gunmetal cap. From this, the pipe connects to the tank with a steady fall. Where more than one tank exists, a 3-way cock or set of valves will be required, or one filling pipe per tank.

Vent pipe. A vent pipe from each tank is required. It should be 80 mm in diameter, be carried up to the roof of the building or to some point where smell will not be troublesome, and be fitted with a wire guard. It should be separate from the filling pipe.

In the event of the overfilling of a tank from a road wagon which is delivering with compressed air or by pump, a combined vent pipe arrangement allows oil to flow into the next tank, through the pipe, without obstruction.

In the event of overfilling, oil may rise in the vent pipe to a considerable height if the building is tall, so placing an undue pressure on the tank. An alarm device is desirable, therefore, to give warning of this condition, and, in addition, an oil seal or one of the proprietary unloading devices should be branched from the vent pipe.

It is unnecessary to point out that vent pipes should have a steady rise to the top, as any dip which might become filled with oil would obstruct the free passage of vapour.

The outlet connection should have a valve next to the tank and should be 100 to 150 mm above the tank bottom. The size is determined by the number and size of burners connected, each of which should again be separately valved.

If a number of boilers is being served from the oil-storage tank, some designers favour the use of a daily service tank. This enables the boiler plant to be started up before steam or electric current is available for heating the fuel in the main storage tank. The capacity of the service tank is limited by C.O.P. 3002 to 900 litres per boiler.

The daily service tank also serves to limit the amount of oil which could escape in the case of pipe-fracture. If an electric pump is provided for filling the daily service tank, it is necessary to arrange for hand-starting and automatic stopping.

Sludge Outlet. Water, being denser than oil, collects at the bottom of the tank, generally in the form of mud, since it mixes with the sludge and solid matter which gradually settles out from the oil. After a few years of partial emptying and refilling of the tanks the accumulation increases, and, before there is any chance of its reaching the outlet, it must be removed. A sludge outlet is therefore provided at the lowest point and is fitted with a valve. The end may terminate in the tank room, provided it is easily accessible. Oil companies generally make arrangements for the collection and removal of sludge, as it cannot, of course, be put down the drains. A small oil sump with hand pump in the tank chamber is an advantage.

Level indicator. Gauge glass indicators, being fragile, are unsatisfactory. Direct reading dial gauges for mounting in the side of the tank are available, or there are various forms of remote reading gauges operating by sealed pressure system (see Fig. 5.13). There are also pneumatically and electrically operated types.

Filters. To protect the burners and pumping equipment, it is desirable to fit a coarse filter, either immediately on the outlet or at the burner installation, the size being such that it causes no excessive resistance to the flow of the oil. The object of this filter is to stop any extraneous matter, which may have gained access to the storage tank, from reaching the transfer or burner pump, and therefore protect them from mechanical damage.

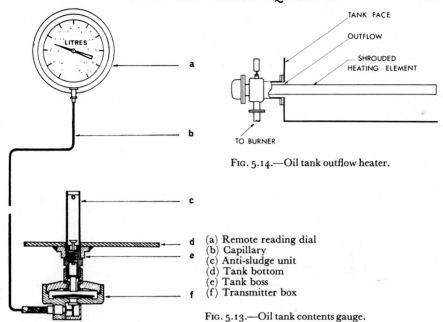

FIG. 5.14.—Oil tank outflow heater.

(a) Remote reading dial
(b) Capillary
(c) Anti-sludge unit
(d) Tank bottom
(e) Tank boss
(f) Transmitter box

FIG. 5.13.—Oil tank contents gauge.

Heaters. In the case of heavy oils it is necessary to provide means of heating in the main storage tanks, either by steam or hot-water coils, or by electric immersion heaters. Electric heaters are usually confined to the proximity of the outlets the object being to heat the oil at the point of exit rather than the whole tank. This device is termed an *out-flow heater*, one form of which is illustrated in Fig. 5.14, being so designed that oil cannot be withdrawn below the level of the heating element.

In addition, steam or hot-water tracer lines are run in contact with the outlet pipe to the pumping and heating unit, or, alternatively, electric heating cable is wound round the pipe. The heating lines and pipe are then lagged in one envelope.

Underground Storage Tanks—Where it is possible, tanks may be placed out of doors underground. This economizes building space and is a very safe and practical arrangement. If buried without an enclosing pit, the tank should be properly protected with bituminous paint on the outside and buried in concrete. It is generally necessary to anchor the tank to a block of concrete to overcome buoyancy when empty. A better arrangement, as in Fig. 5.15, is to construct a pit, so that inspection may be made all round.

Underground oil tanks may also be of concrete, lined with special tiles, this being a proprietary design. (Plate VIII facing p. 119.)

Oil Preheating—For the larger installations using Class F and G oils, it is the usual practice to circulate hot oil from a heating and pumping unit mounted in the boilerhouse separately from the burner, as in Fig. 5.16.

Heating is by means of a hot-water or steam heat exchanger with electric immersion heaters for start-up. The circulation is taken right up to the burner so as to avoid any cold oil which might cause smoke on start-up. Oil lines are kept warm by heated piping or electric tracing. Un-used oil from the ring is returned to the pump suction. Pumps are usually in duplicate, one being a stand-by. Where the type of burner does not have a self-contained pump, a slightly different arrangement applies. The burner

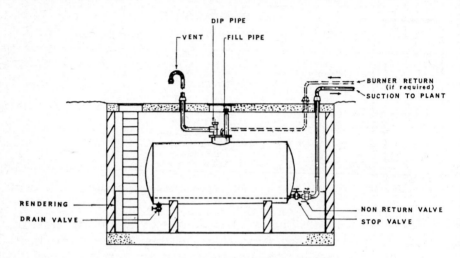

Fig. 5.15.—Underground oil storage tank.

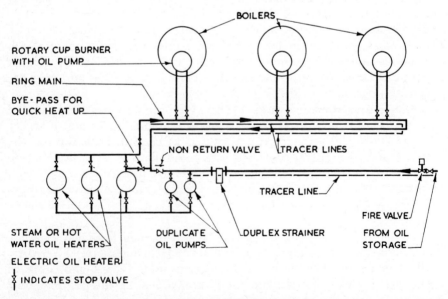

Fig. 5.16.—Hot oil circulating ring main.

connections are then in parallel across the flow-and-return mains and, at the end of the ring, a pressure relief valve is inserted thus setting up a pressure at the burner for atomization.

Roof-top Boilerhouse—The advantages of a roof-top boilerhouse have already been referred to, particularly for multi-storey construction. The arrangement for oil supply from main tankage at ground, or below ground, is shown in Fig. 5.17. (Plate IX facing p. 150.)

PIPED OIL SUPPLIES

An interesting development in the use of oil for heating is the provision of a central oil-storage to serve a group of flats or residences each of which has its own oil-burning appliance. The oil is continuously pumped from a main storage on the estate and fed to each consumer by meter. Piping is laid underground.

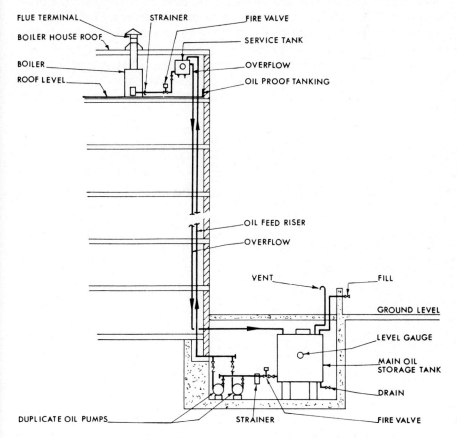

FLUE TERMINAL   STRAINER   FIRE VALVE
BOILER HOUSE ROOF   SERVICE TANK
BOILER   OVERFLOW
ROOF LEVEL   OIL PROOF TANKING
OIL FEED RISER
OVERFLOW
VENT   FILL
GROUND LEVEL
LEVEL GAUGE
MAIN OIL STORAGE TANK
DRAIN
DUPLICATE OIL PUMPS   STRAINER   FIRE VALVE

Fig. 5.17.—Roof top boiler house.

CHAPTER 6

Heating by Gas

NATURAL GAS FROM THE North Sea is currently stated to be supplying 97 per cent. of the national consumption and thus manufactured gas need no longer be considered in great detail.

A high pressure trunk main system has been constructed for distribution to the various Area Boards, this being also capable of receiving liquefied natural gas from the Sahara imported under long term contract by tanker from Algeria. The distribution system includes underground storage and, in addition, the familiar local gasometers remain for local low pressure storage. The storage capacity of the high pressure distribution pipework system is not inconsiderable.

To cater for exceptional peak demands, special 'interruptible' tariffs have been offered to large users under which their supply may be cut off for specified periods. In such instances some alternative fuel for heat

TABLE 6.1

PROPERTIES OF NORTH SEA AND MANUFACTURED GAS

Description	North Sea Gas	Manufactured Gas
Specific mass, kg/m³	0·78	0·6
Relative to air = 1·0	0·60	0·48
Calorific value, MJ/m³	38·6	18·6
Wobbe number	54	29
Burning velocity, m/s	0·34	1·0
Max. flame temperature, °C	1930	1960
Sulphur compounds, mg/m³	0–20	120–390
Toxicity	Nil	Toxic
Constituents, typical, % volume		
Methane	92·0	33·5
Ethane	3·5	—
Propane	0·7	—
Butane	0·3	—
Benzene etc.	0·3	—
Carbon Dioxide	0·3	13·7
Carbon Monoxide	—	4·9
Hydrogen	—	47·9
Nitrogen	2·9	—

supply is needed in the form of individual bulk stores of LPG or fuel oil. For the latter, dual fuel burners would be required.

In general terms, however, gas has the advantage of requiring no storage facilities at the point of usage and, by virtue of its composition with negligible sulphur content, the products of combustion are principally steam and carbon dioxide, both relatively harmless and pollution free. The principal properties of North Sea gas are listed in Table 6.1, those for manufactured gas being given for comparison.

The principal differences are in burning velocity, and calorific value. The slow-burning rate of natural gas is coupled with a tendency for the flame to lift, for which reason the neat-gas or flat-flame burner no longer functions. The flame must be aerated and the burner so designed as to remain stable. One method of achieving this is by means of a small keeping-flame near the base of the burner. By way of compensation for this difficulty is the fact that the natural-gas burner has little inclination to light-back.

The calorific value of natural gas is double that of towns gas; hence for a given duty only half the volume is required. This means less ability to entrain air for aeration, yet twice the quantity is needed. In consequence, it is necessary to increase pressure to produce a higher velocity whilst reducing jet size.

COMBUSTION OF NORTH SEA GAS

The principles set out in Chapter 4 apply equally to the combustion of North Sea gas, the basic equations appropriate to the constituents being as given in Table 6.2, p. 131. For the proportions listed in Table 6.1, the theoretical air for combustion may be calculated as:

	m^3 air/m^3 combustible	air, m^3
Methane	0.92×9.5	$= 8.74$
Ethane	0.035×16.7	$= 0.58$
Propane	0.007×23.8	$= 0.17$
Butane	0.003×30.9	$= 0.09$
Benzene etc.	0.003×29.3	$= 0.09$

Total air, m^3/m^3 of gas $= 9.67$

Similarly, the theoretical calorific value would be calculated as 40.79 MJ/m^3, using individual values from the last column of Table 6.2. The actual measured calorific value of natural gas varies little from this calculated figure. The maximum CO_2 arising from combustion is calculated, as has been shown in Chapter 4, using mass values and for the analysis quoted would be 11.9 per cent.

BURNERS

Burners for the firing of boilers with natural gas fall under two broad classifications: one is *natural-draught or 'atmospheric'*, the second is *forced-draught*.

The natural-draught or atmospheric burner is usually quiet in operation but is limited in application to boilers rated at up to about 1 MW. It is dependent on flue conditions which may affect burner performance: it is usual to provide the boiler outlet connection with a

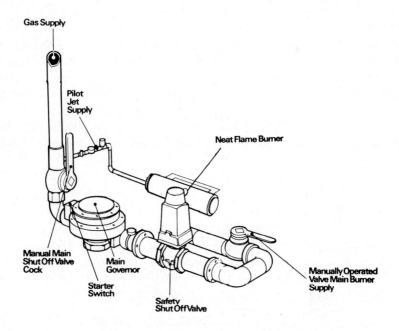

Fig. 6.1.—Natural draught (atmospheric) burner.

draught-diverter which provides inlet for dilution air at the base of the vertical flue and thus stabilizes draught conditions. One type of natural-draught burner, applied as a conversion set to a sectional boiler, is shown in Fig. 6.1. The burner control system is relatively simple, usually consisting of an electrical ignition device which operates in conjunction with a gas pilot ladder, a flame failure sensor connected to a safety shut-off valve plus control and overheat thermostats, the latter being of a manual re-set type. A gas pressure of 1·5 to 2 kPa is required at the burner.

The forced-draught burner is supplied as a factory-tested unit complete with all safety and other controls, and can be expected to achieve the maximum efficiency of combustion with or without a flue. It is noisier than the natural-draught burner, and often worse in this respect than any oil burner. Its capacity is virtually unlimited. A gas pressure of about

2 kPa is required and where this cannot be made available, boosting equipment will be required as described later.

The forced-draught or 'automatic' burner-control system is electrical and comprises a pre-purge period; pilot-flame ignition by spark; a period for pilot-flame proving; followed by main flame establishment and control. A safety lock-out is provided in the event of malfunction and this is arranged such that it may only be re-set manually; on shut-down, a post firing purge is often provided after which the control system re-cycles ready for the next start.

In addition there is a flame-failure control which will shut off the gas supply in one to two seconds: types at present in use include either a flame-failure probe or an ultra-violet scanner. The luminosity of a gas flame is not adequate to operate a light-sensitive cell as in an oil burner. The arrangement of main and pilot governors, safety shut-off valves, flame probe and other features for one particular make will be apparent from Fig. 6.2.

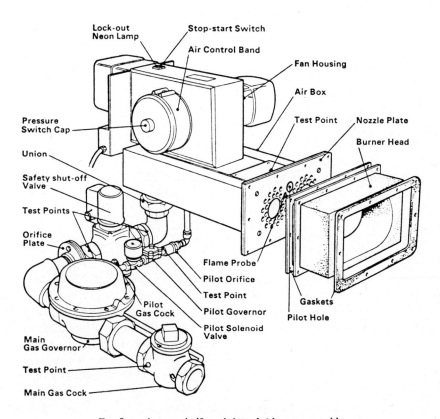

Fig. 6.2.—Automatic (forced-draught) burner assembly.

Boiler/Burner Units—Burners such as those described may be applied to existing boilers but where a new installation is concerned it is obviously preferable to adopt a packaged boiler/burner unit. The flame shape, combustion-chamber volume, application of controls and so on will have been designed to be compatible one with the other and the whole unit will have been factory tested. Site work therefore will be reduced to the minimum and, after connection, the combined unit should require little more than tuning to suit the actual gas pressure and flue conditions prevailing. Sizes range from the smallest for domestic use to about 1 MW in sectional form, and to 6 MW or over in steel-shell or tubular form.

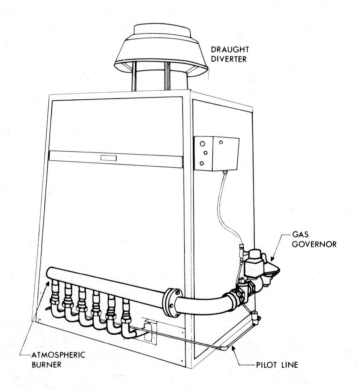

DRAUGHT
DIVERTER

GAS
GOVERNOR

ATMOSPHERIC
BURNER

PILOT LINE

FIG. 6.3.—Packaged boiler unit with atmospheric burner (Potterton).

Fig. 6.3 shows the exterior appearance of one make of natural-draught gas fired boiler; Fig. 6.4 shows a similar view of a forced-draught boiler. It will be found that makers' data usually give input ratings alongside output ratings. From these may be deduced the design efficiency. For instance for the largest size, as Fig. 6.3, output = 593 kW, input = 791 kW and thus efficiency = 75 per cent.

Gas Burner Controls—British Gas have published comprehensive guidance material, with particular regard to safety provisions, and

Plate VI. Vekos boiler with pneumatic coal feed. Front hopper serves clinker crusher and conveyor (see p. 99)

Plate VII. La Mont water tube boilers mounted on raised base which forms passage for combustion air to burner fans (see p. 104)

Plate VIII. Underground oil storage tank with special tiled lining (see p. 111). Photograph by courtesy of Borsari & Co. Zurich

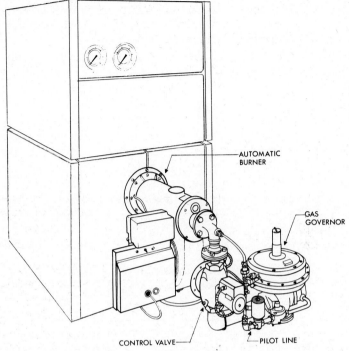

FIG. 6.4.—Packaged boiler unit with automatic burner (Ideal-Standard).

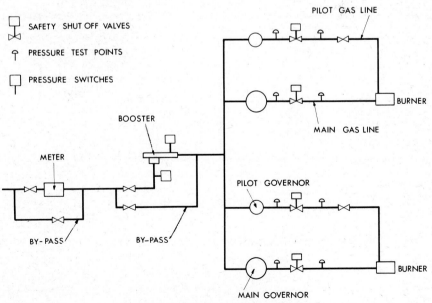

FIG. 6.5.—Connections to gas booster unit.

K.H.A.C.

reference should be made to this for further details. Additional information of a general nature is included in B S 1250 and 3561. Alarm equipment calibrated to sense a gas concentration of 20% of the Lower Explosive limit is available for mounting in larger boiler houses.

Gas Boosters—In certain instances, as has been mentioned previously, it is necessary that pressure boosting equipment be provided between the gas meter and the point of firing. The booster sets are operated automatically in conjunction with the burner control system and may take the form of one centrifugal unit per boiler/burner or a common unit serving a range of burners. When boosters are used, a pressure switch is introduced at the boiler to ensure that a satisfactory supply is available before the burner operates and, in addition, a minimum pressure switch is fitted at the booster inlet to ensure that a mains supply is available. This latter switch is set to cut out the booster if the inlet pressure falls to less than 250 Pa. Fig. 6.5 shows how a booster is connected into the gas supply piping.

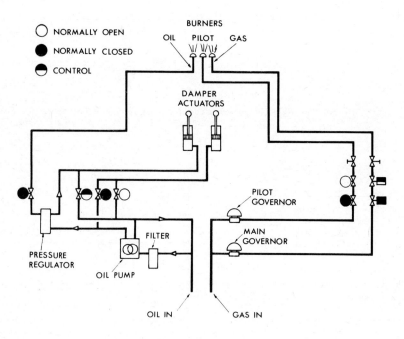

FIG. 6.6.—Dual Fuel burner (Dunphy).

Dual Fuel Burners—Mention has previously been made of such equipment which, in essence, consists of a combination unit capable of firing either fuel oil or gas, as shown in Fig. 6.6. Both fuel supplies are commonly connected and change-over is a simple matter of shut-down, isolation of one supply, activation of the other and start-up.

AIR FOR COMBUSTION
With Natural draught—
  Assume 30% excess air (over the theoretical 9·67 m³/m³).
  Volume of air required, including that admitted at diverter (100%)

Air m³/m³ gas (theoretical × 2·6)	= 25
CV—MJ/m³	= 39
Air per MW input—m³/s	= 0·64
Air per MW of boiler rating at 75% efficiency	= 0·85 m³/s

With Forced draught—
  Assume 30% excess air.

Air m³/m³ gas (theoretical × 1·3)	= 13
Air per MW input—m³/s	= 0·33
Air per MW boiler rating at 75% efficiency	= 0·44 m³/s

Inlet area—
  Taking air-inlet velocity at 2 m/s.
  Free area required:

Natural draught	0·43 m²/MW boiler rating
Forced draught	0·22 m²/MW boiler rating

According to BS *Code of Practice* 332, *Part* 3, the required free area for admission of air for combustion for a gas appliance should be the equivalent of

  at low level   1100 mm² per 1 kW (1·1 m²/MW) of boiler output;

and for ventilation

  at high level   550 mm² per 1 kW (0·55 m²/MW) of boiler output.

This applies up to boiler ratings of 600 kW, and it will be found to agree as to the low-level inlet with the foregoing for natural draught, after allowing for inlet of ventilation air. For larger duties, the area should be calculated on the lines indicated. The area given for ventilation outlet as distinct from combustion inlet is no doubt safe enough for removal of heat, but an accurate rather than empirical approach is to be preferred.*

CHIMNEYS AND FLUES FOR GAS BOILERS
Natural Draught—The requirements for gas boilers differ from those discussed in Chapter 5 for oil and solid fuel, due to the air admitted by the

* *Combustion and ventilation air for boilers having a rated output exceeding* 45 kW. I.H.V.E. Practice Note No. 2. 1975.

draught diverter acting as a dilutant to the combustion gases. The CO_2 content of the latter may be 9 per cent with a gas temperature of 240° C, but the normal for the 'secondary' flue is about 4 per cent and a gas temperature of about 120° C. Curves for sizing and data as to resistance factors are given in the *Guide*, and a British Gas publication* provides comprehensive tables for a wide range of circumstances. In brief, the required chimney height is that which will provide for adequate dispersion of the products of combustion such that their concentration at ground level does not exceed a critical value. The negligible sulphur content of natural gas means that the provisions of the Clean Air Memorandum (p. 84) apply only to the very largest installations. In terms of height, the flue terminal need be only about 1·5 to 2 m above roof level of a building or 3 to 4 m above ground level where free-standing, for outputs of up to 5 MW. For natural-draught burners with gas dilution at a diverter, the area of the flue for each unit may be read from Fig. 6.7 where the value of the factor F has been calculated from:

$$F = \frac{\text{Height of flue above boiler outlet (m)}}{K + 2 + b/2}$$

where

K = $\frac{2}{3}$ any suction required at the flue base (Pa)
b = the number of bends in the route of the flue.

For a forced draught burner, the capacity of the burner fan must be considered and the *Guide* should be consulted.

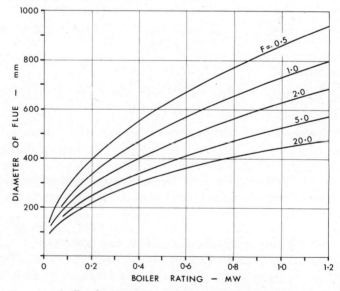

FIG. 6.7.—Chimney areas for a given height.

* *Technical Notes on the Design of Flues for Larger Gas Boilers, Dec., 1971.*

Fan-Diluted Draught—In this system, one form of which is shown in Fig. 6.8, the combustion products, together with a quantity of diluting air, are exhausted to the atmosphere by a fan. By this means it is possible to dispense with a chimney entirely by discharging through the wall of the boilerhouse. The diluting-air quantity is such as to bring the CO_2 content of the mixture down to 1 per cent, which involves a fan to handle approximately 100 m³ per m³ of natural gas burnt. Included in the system is a fan-failure device to shut off the burners in the event of draught failure.

The duct sizing with this system should be on the basis of a gas velocity of 3 to 5 m/s.

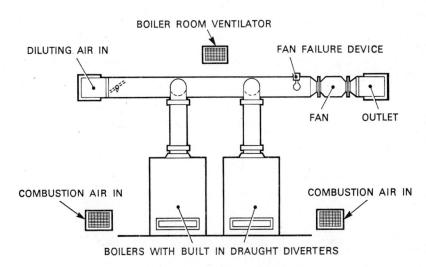

BOILER ROOM VENTILATOR

DILUTING AIR IN

FAN FAILURE DEVICE

FAN   OUTLET

COMBUSTION AIR IN

COMBUSTION AIR IN

BOILERS WITH BUILT IN DRAUGHT DIVERTERS

FIG. 6.8.—Fan diluted draught.

Materials for chimneys and flues—Condensation is likely to occur in gas flues and chimneys, particularly on start-up; hence they should be lined with an impervious and acid-resisting material such as glazed stoneware or asbestos cement. It is desirable to keep the chimney warm to assist draught, either by enclosing it in brickwork or concrete, or by insulation. Connecting flues between boiler and chimney are usually of asbestos cement. In the fan-diluted system they may be of galvanized steel with sealed joints and painted edges.

Explosion Doors—In the event of the malfunctioning of safety controls, the remote contingency of an explosion must be faced and means provided for its relief. If the boiler has an open skirt, this will serve to relieve pressure. Where, however, the combustion chamber is sealed, as when a boiler has been converted from another fuel or where a forced-draught burner is fitted, other provision must be made, such as by relief doors on the boiler or panels in lieu of doors which will shatter under pressure.

Terminals—Care must be taken to see that flues discharge the products of combustion freely to the atmosphere. In buildings with return walls, areas, etc., peculiar atmospheric pressure conditions occur and it is advisable to extend the flue to a height of at least 500 mm above the eaves of the roof. The possibilities of downdraught are then much reduced.

Every flue should be fitted with a terminal of approved design to prevent birds nesting, etc. Bafflers, or more correctly 'draught diverters', should be fitted to the flues of all natural-draught gas boilers and heaters if these are not incorporated in the design of the heater.

Balanced Flues—In difficult situations where the flue terminal cannot be

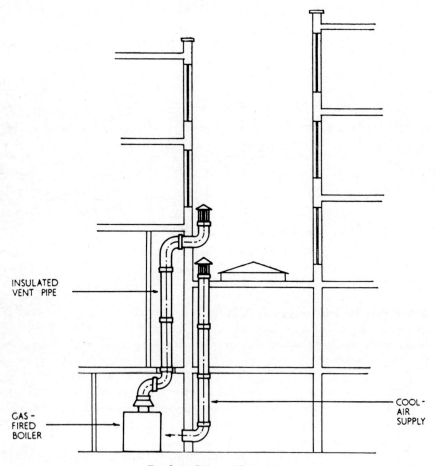

INSULATED
VENT PIPE

COOL-
AIR
SUPPLY

GAS-
FIRED
BOILER

Fig. 6.9.—Balanced flue system.

carried above roof level, but must terminate in some position such as in an internal light well, differences of pressure are found to cause back draughts. These may be overcome by the provision of a balanced flue, as in Fig. 6.9,

and comprises a return flue similar to the rising flue except that it is carried
down to near floor level in the boiler chamber. The sizing of this flue must
be generous, as it has to convey not only the air for combustion, but also
that drawn in by the baffler; and the resistance should, in any event, be
kept as low as possible.

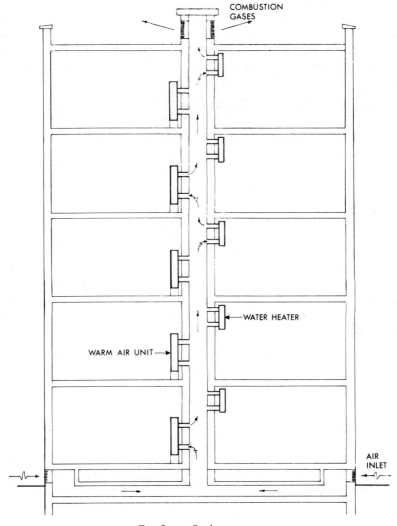

FIG. 6.10.—SE duct system.

Another flue system is the SE, as shown in Fig. 6.10. This is suitable
for multi-storey buildings and, as will be seen, air enters at the base and the
products of combustion from the various appliances discharge into the
same duct. Thus the gas combustion space is virtually sealed off from
the occupied space.

An alternative to the Se duct, for use where a bottom inlet is not possible, is the U-duct shown in Fig. 6.11. Fresh air taken in at the roof is conveyed down a shaft running parallel with the rising shaft, which acts as a shared flue as in the Se duct, the products of combustion being exhausted at the roof from a terminal adjacent to the intake, but at a

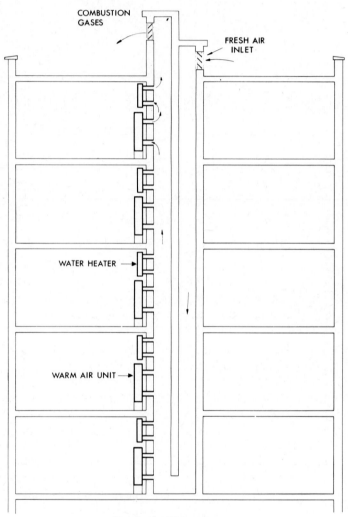

COMBUSTION GASES

FRESH AIR INLET

WATER HEATER →

WARM AIR UNIT →

Fig. 6.11.—U duct system.

slightly higher level. All gas burning appliances take their combustion air from, and return the products of combustion to, the rising duct.

Roof-top Boilerhouse—The flue problem is avoided altogether in a boiler installation if the boilers are on the roof. Fig. 6.12 shows such an arrangement with short outlets from the boilers carried through the roof of the penthouse to the atmosphere. Gas boilers on the roof are obviously simpler to deal with than oil—there is no delivery pumping and negligible fire risk. Thus, in many examples of multi-storey buildings, the use of gas in this location for boilers has much to commend it. As referred to earlier, the modular boiler system is particularly appropriate to roof-top application.

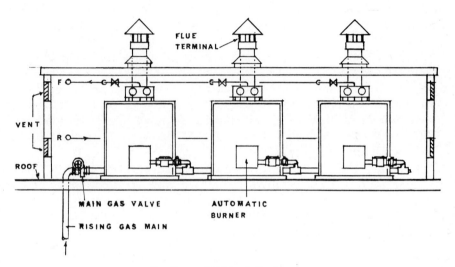

Fig. 6.12.—Roof-top boiler house.

GAS RADIANT TUBE

The principle on which this system operates is that of infra-red space heating and is suitable for industrial and commercial buildings.

Nor-Ray-Vac—In one system, Nor-Ray-Vac using the Phoenix burner, the chief elements comprise the burner, the radiant tube 65 mm dia, a reflector over the tube, and a vacuum pump. Products of combustion are drawn through the tube which achieves a temperature of about 540° C being below luminosity; the heat radiation is reflected down to warm the occupants and objects in the space. As with any form of radiant heating, the temperature of the air in the heated space is raised by the surfaces becoming warmed and themselves acting as convectors. (See Fig. 3.3.)

The system may be designed to provide an environmental temperature of 15 to 20° C, the actual air temperature to satisfy such conditions being in the region of 10 to 14° C. This, being much lower than would be comfortable with convection heating alone, leads to a considerable saving in heat input due to reduced heat losses, particularly in high buildings. High efficiency of combustion is achieved by a regulated gas/air ratio and exhaust temperatures are low, thus overall efficiencies of the order of 90 per cent are stated to be achieved.

The products of combustion are normally discharged outside the building. Where there are a number of units in line, they may be connected in cascade, the gases from the first joining those from the second and so on, finally being exhausted by one vacuum pump. Air for combustion is normally taken from the space but where this is dust-laden or contains explosive gases the air may be drawn from outside via ducting.

Units are in lengths of 4 to 7 m, having a heat output of 12 to 14 kW. Of this, radiation accounts for over 80 per cent. The units are suspended overhead lengthwise down a shop bay normally at a height of 3 to 5 m. Additional units may be desirable near large glass areas, doors, or cold walls, to off-set cold radiation. A minimum clearance above combustible materials of 1·2 m is stipulated.

The system may be thermostatically controlled, preferably by a thermostat susceptible to radiation. It may also be controlled by time switch and frost thermostat. The burner is ignited by electric spark, full safety controls are included and, should the vacuum pump fail, the system instantly shuts down.

Gas-Rad—The principle on which this system works is exactly the same as that just described. It differs in that two tubes are used instead of one, in U formation, the second tube acting as a return so that the suction fan is adjacent to the burner box, each unit being separated and not interconnected, as shown in Fig 6.13.

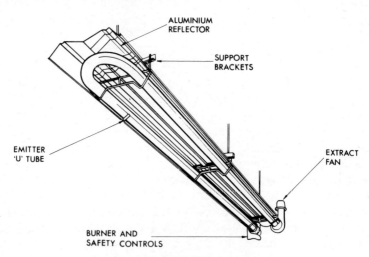

FIG. 6.13.—Radiant tube heater (Gas-Rad).

The unit is 5 m overall in length, width over the reflector is 533 mm, tubes are 62 mm dia and the output about 15 kW.

For design purposes the numbers of units may be determined for a given area against average U-values and air changes. Efficiencies of the order of 77 to 82 per cent are claimed for this system by the manufacturers.

GAS WARM AIR

This system has been developed for residential use, i.e. for houses and flats.

The air-heating unit is shown diagrammatically in Fig. 6.14 and comprises a combustion chamber formed of some non-corrodible metal, a fan for blowing air over the combustion chamber, and a casing into which air is drawn from the rooms and from which it is returned after warming.

Units range from 8 to about 24 kW, and are suitable for heating a small 2-bedroom flat up to a 4-bedroom house.

As generally planned in a flat, the unit has short duct connections to discharge the warmed air near floor into two or three rooms, the return being at high level from the hall, or from one of the rooms. When applied to a house, delivery ductwork carries the warm air *via* the under-floor

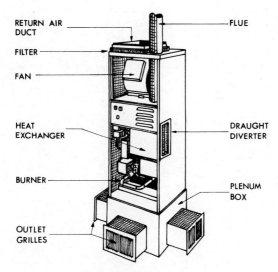

RETURN AIR DUCT

FLUE

FILTER

FAN

HEAT EXCHANGER

DRAUGHT DIVERTER

BURNER

PLENUM BOX

OUTLET GRILLES

Fig. 6.14.—Warm air unit.

space or roof space to floor grilles in the various rooms to be heated, the return being via the hall. So as to allow free movement of return air, some gap below the door or a transfer grille is necessary. It is important that the ducts are airtight, that all ducts are well insulated, and that gas pressure is according to maker's requirements.

The control of gas warm air units is generally by stopping and starting the fan and a magnetic gas valve so arranged that the gas can only be 'on' when the fan is running. A thermostat is provided in one of the rooms, together with a clock switch and often a low limit in order to maintain some background warmth such as 10° C even when 'off'.

The gas warm air system is economical in first cost, readily responsive to demands for warmth and has become a popular feature in multiple housing schemes.

Industrial Gas Warm Air—For industrial applications the type of air-heater shown in Fig. 3.1, page 50 (but with gas burner in lieu of oil) is applicable. The absence of storage is an advantage and, like the oil version, the unit is mobile if change of lay-out is required.

With the advent of North Sea gas and taking advantage of the associated lack of pollutants in the products of combustion, a number of 'direct fired' air heaters have been developed. These operate to circulate outside air to replace that exhausted through process extract fans, the ventilation air stream passing directly over the burner with no separate

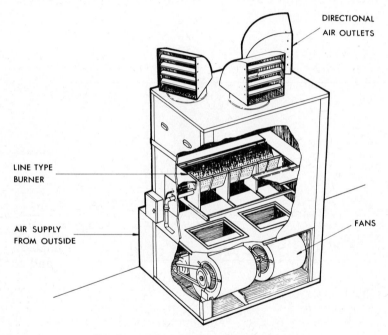

DIRECTIONAL
AIR OUTLETS

LINE TYPE
BURNER

AIR SUPPLY
FROM OUTSIDE

FANS

FIG. 6.15.—Direct fired air heater (Wanson).

combustion circuit or flue. The purity of the output, *with an outside air supply*, is within accepted threshold values as laid down by HM Factory Inspectorate. Unit ratings range up to 650 kW and one such is illustrated in Fig. 6.15. Since all the energy of the fuel passes to the air circulated, efficiency approaches 100 per cent.

GAS-SUPPLY PIPES

At the point of entry there will be a main gas cock, a filter, a governor, and a meter.

TABLE 6.2

COMBUSTION REACTIONS

Reaction Products	Requirements for Combustion by unit proportion of combustible				Heat Liberated MJ/m³ of Combustible
	kg/kg		m³/m³		
	O_2	Air	O_2	A	
Methane $CH_4 + 2O_2 = CO_2 + 2H_2O$ $16 + 64 = 44 + 36$	4	17·3	2	9·5	40
Ethane $2C_2H_6 + 7O_2 = 4CO_2 + 6H_2O$ $60 + 224 = 176 + 108$	3·7	16·0	3·5	16·7	69
Propane $2C_3H_8 + 10O_2 = 6CO_2 + 8H_2O$ $88 + 320 = 264 + 144$	3·6	15·6	5	23·8	99
Butane $2C_4H_{10} + 13O_2 = 8CO_2 + 10H_2O$ $116 + 416 = 352 + 180$	3·6	15·6	6·5	30·9	129
Benzene etc. $C_5 +$	3·1	13·4	8·3	39·3	164

TABLE 6.3

FLOW OF NATURAL GAS IN STEEL PIPES
(MEDIUM GRADE TO BS 1387)

Pa perm run	Litres per second in pipes having stated nom. bore—mm										
	15	20	25	32	40	50	65	80	100	125	150
0·5	0·07	0·24	0·47	1·02	1·6	3·0	6·1	9·5	20	35	57
1·0	0·14	0·37	0·71	1·54	2·4	4·5	9·1	14·1	29	52	84
1·5	0·21	0·48	0·90	1·96	3·0	5·7	11·5	17·9	37	65	106
2·0	0·25	0·56	1·08	2·31	3·5	6·7	13·6	21·0	43	76	124
2·5	0·29	0·64	1·23	2·64	4·0	7·6	15·4	23·9	49	86	140
3·5	0·34	0·78	1·50	3·20	4·8	9·2	18·6	28·9	59	104	169
5·0	0·43	0·96	1·83	3·92	5·9	11·2	22·7	35·1	72	127	205
7·0	0·51	1·17	2·23	4·75	7·2	13·6	27·4	42·3	86	152	246
10·0	0·64	1·44	2·72	5·81	8·8	16·6	33·4	51·5	105	185	298
15·0	0·81	1·82	3·43	7·30	11·0	20·7	41·8	64·3	130	230	370
20·0	0·95	2·14	4·04	8·57	12·9	24·3	48·9	75·3	152	269	432
25·0	1·09	2·43	4·58	9·71	14·7	27·5	55·2	84·9	172	303	487
	0·5	0·5	1	1	1·5	2	2·5	3·5	5	7	8

1. *Single Resistances*
   For approximate purposes equivalent length may be taken as length in metres listed in bottom line of table for each bend, tee etc. For accurate values see *Guide*.

2. *Heat equivalent*
   Natural Gas 39 MJ/m³ = 39 kJ/litre
   thus   1 litre/s = 39 kW

From this point onwards the supply piping is to be sized to suit the demands of the apparatus. The *Guide* gives Tables for flow of gas against pressure drop for pipes, from which Table 6.3 is extracted for steel pipes.

It will be noted that, retaining the kilowatt as the basis of calculation, a flow of gas of 1 litre/sec is equivalent to 39 kW (strictly 38·6) with natural gas. Thus, the input rating of any boiler or other appliance in kW may be converted into gas flow by dividing these equivalents. Alternatively, makers' lists usually give gas rates volumetrically, from which litre/sec may be derived.

In terms of pressure loss, the pipe loss from point of entry to terminal may be 100 Pa.

Then, taking as an example,

Input rating of boiler, 750 kW

Natural gas ÷ 39   = 20 litre/s

Pressure drop assume available   100 Pa

Run from intake to boiler,
    including single resistances   30 m

per m run $\dfrac{100}{30}$ = 3·3

From Table 6.3, size   = 65 mm

The pressure available for distribution within the building is bound up with governor settings and burner pressures and with requirements for boosting equipment. All these are matters on which the technical departments of the Area Gas Boards will advise.

CHAPTER 7

Low-Pressure Hot-Water Heat Emitters

In this country, with its unpredictable climate, hot water is usually the best medium for heating on account of its simple temperature control to meet variations in weather, and the absence of the parching effect commonly associated with steam and hot air. Warming by hot water therefore deserves detailed consideration.

Low pressures are generally employed—which means temperatures below boiling point, usually at 82° C maximum. The use of higher pressures is considered in Chapter 11.

The calculation of the amount of heat necessary to maintain the desired internal temperature has previously been explained and it now remains to consider how that heat input shall be supplied.

EMISSION FROM HEATED SURFACES OF VARIOUS FORMS

The total emission from a heated surface may be divided into radiant and convective components, of which the radiation is proportional to the difference of the fourth powers of the absolute temperatures of the radiating and absorbing surfaces. The convection varies considerably with the form and height of the surface, but if the heat emitted by convection at any one temperature be known, that at some other temperature will be proportional to the temperature difference to the power of 1·25.

Due to the convection currents set up by any heated element in air, no surface gives 100 per cent radiation, though a heated ceiling closely approaches it (about 90 per cent). Panels on walls may have 60 per cent radiation and 40 per cent convection, and in floors 50 per cent radiation and 50 per cent convection. Radiators of the ordinary type vary according to their convolutions, but transmit commonly 20 per cent by radiation and 80 per cent by convection. Convectors give almost 100 per cent of their output by convection and only an accidental fraction by radiation.

It has in the past been usual to evaluate heating surface in terms of area and to give a coefficient of heat emission per unit area based on makers' tests. The coefficient multiplied by the temperature difference mean water to air gives the transmission per m², and hence the heating surface for a given heat loss in a room may be derived. Whilst the heat emission from a plane surface or a simple shape such as a cylinder can be calculated on theoretical grounds, conventional radiators and the like are not susceptible to such treatment.

It is not possible, in any event, to apply this method to a convector which contains some form of finned surface in a metal box, and the same applies to many other forms of heating appliance. Thus a British Standard has been produced, No. 3528, which stipulates a standard method of test according to which actual emissions are to be stated per section of radiator or per unit length.

Normal Design Temperatures—For low-pressure hot-water systems, recommended operating temperatures are given in Table 7.1 (see page 150)* when the external temperature is assumed to be − 1° C. As explained in Chapter 3 this does not preclude higher flow temperatures being used under colder weather conditions.

EMISSION FROM PIPES

Exposed piping as a form of heating surface is rarely used in current practice but in any system of distribution, main and branch piping running through the spaces to be heated may contribute something thereto. For instance, if there is a system of overhead radiant panels or unit heaters in a factory, the piping serving this equipment may be run below north-light glazing, so serving to prevent downdraught.

Any piping so disposed in the heated space gives useful heat and the amount emitted may be deducted from the total required.

The same applies where piping serving radiators under a range of windows is left exposed. Greater economy is achieved than if the pipework were buried in a trench, though with some sacrifice in appearance.

The theoretical heat emission from piping may be calculated from formulae given in the *Guide, Section C3*. Based on these formulae Table 7.2 (page 151), is extracted from the *Guide*.

The reduction in emission, where pipes are used in coil form vertically over one another, may be taken approximately as follows:

2 pipes	5% reduction	
4 „	15%	„
6 „	25%	„

In order to reduce the heat loss from mains and other piping, insulation is applied in one of the forms discussed later (see page 191). The loss is dependent upon the conductivity of the insulation, its thickness and the surface finish: a bright metal cladding to insulation will reduce the heat loss by up to 10 per cent. below that resulting from a dull painted finish. Table 7.3 (page 151) gives the loss expressed as a percentage of the bare-pipe loss for painted sectional glass fibre insulation, having a conductivity at 80° C of about 0·04 W/m² K.

By way of example in the use of these Tables, assume a two-pipe pumped system and main piping 30 m in length, average size 50 mm.

---

* Table 7.1 and all ensuing Tables dealing with heat emissions are grouped together at the end of this Chapter, on pages 150 to 158

From Table 7.1, flow at 80° C   return at 65° C

air 15° C   15° C

difference 65 K   50 K

bare-pipe loss from Table 7.2,

$$(180 \text{ W/m} \times 30 \text{ m}) + (130 \text{ W/m} \times 30 \text{ m}) = 9300 \text{ W}$$

If insulated with 25 mm thick glass fibre sections, the loss will be

Mean temp. diff. hot surface to air	=	57·5 K
Interpolating from Table 7.3, per cent. of bare pipe loss	=	13
The loss for the two pipes insulated		

$$= 9300 \times \frac{13}{100} \qquad = \qquad 1209 \text{ W}$$

The thickness of insulation to be applied to pipes is determined by a calculation which takes into account the capital cost of the material over its expected life and balances this against the saving in heat loss. Recent increases in energy costs have led to a re-examination of the insulation thicknesses which were 'traditionally' applied to hot water piping.

RADIATORS

Radiators are generally best placed under windows, both for architectural and technical reasons. Architectural, because here they do not mar an otherwise unblemished wall surface, and technical, since this is the point of maximum heat loss and dirty marks on the walls due to convection currents are minimized. Furthermore, cold 'negative radiation' from the window is counter-balanced directly by radiant heat from the heating surface.

The placing of radiators in relation to cooling surfaces also has a bearing on temperature gradient. Thus a radiator on an inner wall will deliver a strong current of warm air to the ceiling, whereas under a window the cool downward currents from the glass will lower the rising air temperature and hence the force of the upward current. For the same reason, long windows may be better dealt with by long radiators rather than short concentrated ones.

In a high building, such as a church, radiators or pipe coils under high level glass will prevent strong downward draughts, and, under a factory north-light roof or any roof-light, the same applies.

Radiator Types—Cast iron radiators have been in existence for 100 years or more but they are no longer being manufactured in this country. In some respects this situation is to be regretted in that there seems to be no modern substitute for the 75 mm deep 'Hospital' type radiator which, in terms of neatness, cleanliness and ability to withstand maltreatment, by physical impact or by untreated water, represented the ultimate of its type.

The place of cast iron has been taken by radiators fabricated from light gauge mild steel pressings welded together, a technique which originated in Scandinavia and elsewhere. Steel is more susceptible to

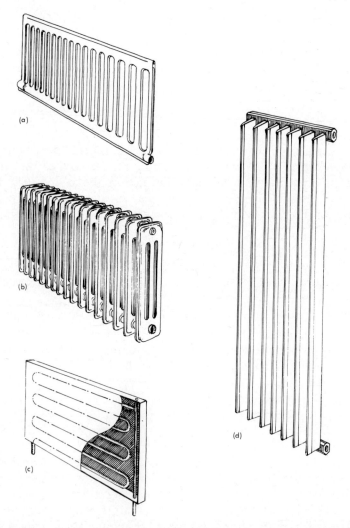

Fig. 7.1.—Steel radiators (a) panel type, (b) column type, (c) sinuous coil, (d) flat oval tubes.

corrosion than cast iron and thus some form of water treatment for heating system contents is advisable; this subject is discussed later.

The great merit of steel radiators is that they are generally much neater in appearance, and smaller for a given heat output than their cast iron equivalent; furthermore, they are lighter for handling on the site. Their water content is less than in the older types and a quicker response to control thus follows as an added advantage.

Due to the versatility of the manufacturing process, there is an almost infinite variety of shapes and patterns of steel radiator available. These fall, however, into a number of clearly recognizable categories: the convoluted panel which may be single, double or more; the column type which in effect reproduces the appearance of the early 1930 cast iron equivalent; the simple tubular which utilizes headers top and bottom with vertical pipe arrays between; the panel with added convective surface which may vary from the simple to the complex and, lastly, the complex tubular which surrounds the heating element with a louvred enclosure welded to it. It will be noted that, from the panel to the enclosure, the differences in design place progressively less emphasis upon the radiant component of output, the last named type being for practical purposes a pure convector.

Typical of these five basic types are the examples shown in Figs. 7.1 and 7.2. The first of these illustrates the simple panel radiator, available in various forms, heights and lengths. Plate X shows an example of this type with extensive length in a hospital ward. Also available are extra slim patterns especially suitable for domestic application. The same figure illustrates one pattern of column steel radiator, at (b). Fig. 7.1 also shows two examples of the tubular type, the first, at (c), incorporating a steel front plate spot-welded to a sinuous pipe coil and the second consisting of profiled tubes between box headers at (d). Plate XI shows how the latter may be used as a screen/room divider to suit a special application.

Where added convective surface is applied to panels, this may be in one of two forms as illustrated in Fig. 7.2 (a) and (b). In the first case, the convective surface is supplementary to that of the panels but in the second case it predominates. The radiators illustrated in Fig. 7.2 (c) and (d) are of a type which nears the dividing line between what is known commercially as a radiator and what is clearly a purpose designed (and designated) convector. As may be seen, the radiant component of the total output will be very small in each case.

Aluminium Radiators—A clear departure from pre-conceived ideas is the use of die-cast aluminium for radiators. Fig. 7.3 shows this type which, like its cast iron forebears, is sectional. The material and production technique clearly leads to a fine finish. It also follows early patterns of so-called 'ventilating radiators' in that it has fins cast integrally with the section, so increasing heat output. A disadvantage of the old ventilating radiator was its propensity to collect dirt, but the smooth surface of

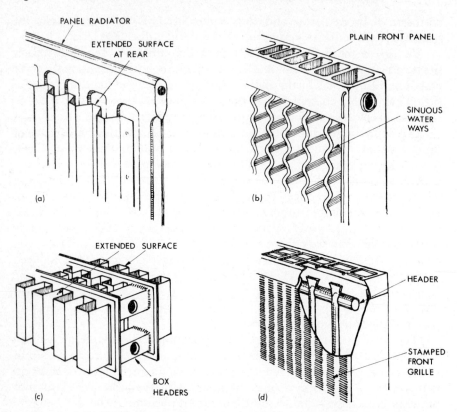

FIG. 7.2.—Steel radiators (a), (b) panels with simple extended surface, (c) panel with complex extended surface, (d) grid waterway in convector casing.

FIG. 7.3.—Cast aluminium radiators.

aluminium no doubt largely overcomes this problem.

Ratings of Radiators—With so many forms of radiators, of which those mentioned are only a few samples, it is necessary to refer the reader to makers' lists for output data. There are in each type various heights, widths, and lengths, also temperature differences, water to air and temperature drops inlet to outlet, within which the designer has to make a selection.

By way of example, however, Tables 7.4, 7.5 and 7.6 have been prepared for the various types illustrated, giving a limited range in each case with rated emissions per metre length. Also given is the equivalent rating per square metre of front face from which a comparison may be made of the space saving of the alternative types. For instance a plain single panel radiator has an output at the stated temperature difference of 1200–1400 W/m² front face, whereas that shown in Fig. 7.2 (*b*) will provide an output of up to 6500 W/m² front face.

Temperature Difference—The basis of emissions in Tables 7.4, 7.5 and 7.6 previously referred to is 56 K between air and mean water. Where some other temperature difference is adopted, the factors given in Table 7.7 (page 154) apply. These are based on $\left(\dfrac{\text{Temp. diff. K}}{56}\right)^{1\cdot3}$ which is the proven radiant/convective ratio of interaction between this type of heating equipment and room conditions.

CONVECTORS

A 'natural' convector comprises a finned tubular element mounted near the bottom of a casing, such that the 'flue effect' of the rising column of heated air within the casing causes a flow of warm air to issue from the top, whilst cooler air is drawn in near the floor. The greater the height of the casing, the greater is the emission of heat for a given size.

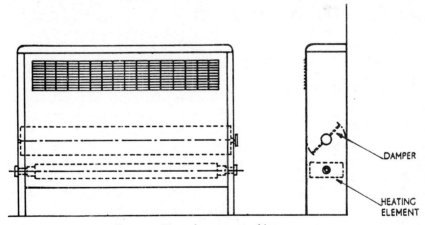

Fig. 7 4.—Natural convector, cabinet type.

Convectors may be free-standing as in Fig. 7.4, or recessed. A damper for hand control is incorporated in some types. Convectors should only be used with pumped water circulation.

Where control is by adjustment of water temperature, it is to be noted that convector emission falls off more rapidly as water temperature is lowered than does that from a radiator, due to the absence of the radiant component as well as to the drop in flue effect with the decline in temperature. Hence, control of water temperature with convectors should be over a more limited range than with radiators. Furthermore, it is inadvisable to mix the two forms of equipment on the same circuit. Separate circuits under different control characteristics are usually necessary.

The velocity of water through the tubes also has a bearing on transmission rates; thus, where the element has two or more tubes, a series arrangement offers a greater output than a parallel arrangement.

Typical heat emissions for a few sizes and heights of convector are given in Table 7.8 (page 155) based on 56 K temperature difference air to mean water. A correction for other temperature differences may be based on $\left(\dfrac{\text{Temp. diff. K}}{56}\right)^{1\cdot5}$; but see makers' lists also.

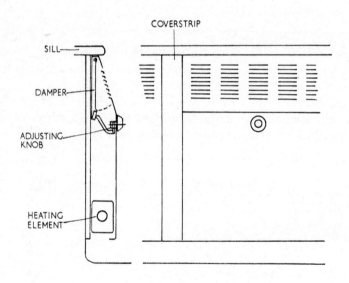

FIG. 7.5.—Continuous convector.

Continuous Convectors—This application of the convector is specially suited to the concept of continuous glazing in a building designed on a modular basis. The finned heating element is continuous from end to end of the building (subject to certain provisions for expansion) and the steel casing is likewise continuous, see Fig. 7.5. The louvre outlets for warm air at the top are in sections, one per module, and each section is provided with a damper for control of heat. The heating element naturally runs continuously uncontrolled. Where partitions occur, a sound-proof barrier is necessary inside the casing.

An example of the use of continuous convectors treated architecturally, as a barrier rail, may be seen in Plate XII (facing page 182).

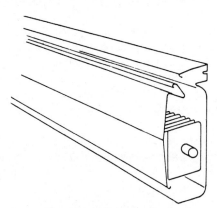

FIG. 7.6.—Skirting heating.

Skirting Heating—This provides a neat and unobtrusive system especially for domestic use. The skirting is heated by pumped hot-water circulation and may run round one or all sides of the room.

One form is in effect a miniature convector and comprises a finned element enclosed in a steel casing, as in Fig. 7.6. Masking and dummy sections are available to fit corners, etc.

Emission for this type is given in Table 7.8 (see page 155).

Fan Convectors—One make is illustrated in Fig. 3.13 (page 60) and another is shown in Fig. 7.7. This is the basic unit having an outlet grille on the front. A base section may also be added, if required, allowing fresh air to enter from the back, as well as recirculated air from the front. The unit shown also has a filter. In some applications a high-level outlet is required, for which purpose an extension is added to the casing.

Two-speed or three-speed control of the fan is usually included, the

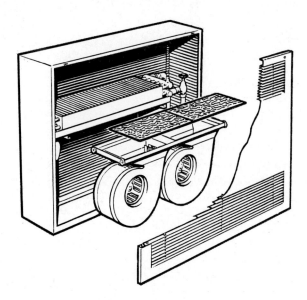

FIG. 7.7.—Fan convector: note the fans below the filter, the heating coil above and the inlet and outlet grilles on the face.

top or boost speed being useful for quick heat-up when the noise level is unimportant. At the lower speeds these units are generally acceptably quiet. Thermostatic control is by on–off switching or by change speed often with a low-limit water temperature cut-out.

Typical emissions are given in Table 7.9 (page 156), from which it can be seen that relatively large spaces may be heated with few units, for instance, one unit per classroom suffices in a school and two or three units suffice in a hall or gymnasium. The system has been applied successfully in churches. The emissions in Table 7.9 are based on 56 K temperature difference air to mean water. Outputs for other temperatures follow a linear relationship.

Fan convectors or 'apartment heaters', shown in Fig. 7.8, can be used for the heating of multi-storey flats. Hot water is circulated from a boiler

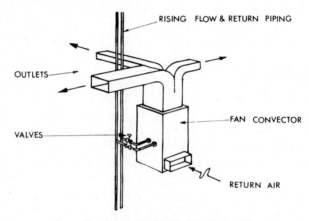

FIG. 7.8.—Apartment heaters in a block of flats.

serving a unit in each flat. This then functions as any other gas, oil or even electric warm-air system, air being drawn into the unit and delivered through short ducts to the various rooms. Charge for heat may be on the basis of fan-running hours, for which a simple meter suffices.

Miniature-size fan convectors have been developed, such as in Fig. 7.9, suitable for domestic use and taking the place of radiators.

The small water content of these convectors means that they are very rapid in response and give the greatest economy with intermittent heating.

Treatment for Radiators—The old treatment for radiators to make them architecturally more acceptable was often to mount them in recesses with a more or less elaborate grille in front, which, however, reduced their emission considerably with the result that they had to be made larger to provide the required heat output. The small radiant component, of course, was lost altogether.

The present trend is to accept the radiator for what it is and not attempt to disguise it. This is no doubt in part due to the advent of simpler, cleaner designs, and in the case of convectors and convector panels, due to the fact that they already have a casing, no further treatment being necessary.

Most forms of radiator require painting except those already supplied with a stove enamelled or similar finish. The colour of the painting at these wavelengths is immaterial so long as the base is not metallic, which would reduce the radiation effect. The convective component is not of course affected.

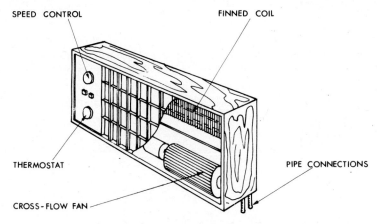

SPEED CONTROL   FINNED COIL

THERMOSTAT   PIPE CONNECTIONS

CROSS-FLOW FAN

FIG. 7.9.—Domestic fan convector.

Where radiators are so positioned that convection currents rising therefrom are liable to make dirty marks on the wall surfaces above, the fitting of a shelf over is common practice. This must be tightly fitted to the wall and have end shields. Even so, after a time, some discoloration may occur, but it will be much reduced. Convectors and some of the panel convectors with front grille outlets largely overcome this problem without the fitting of shelves.

Another adjunct to the older systems was the fresh air inlet. This comprised a grating to outside to admit fresh air which became heated in passing through the radiator. To stop wind blowing directly in, a baffle plate was fitted to the front of the radiator. The air admitted without filtration naturally brought with it dust and dirt, quickly fouling the heating surface and choking the grille. This method is now a thing of the past. Fan convectors with fresh air inlets would now be used to meet this kind of application, the filter in the unit serving to arrest entering particulate matter.

RADIANT PANELS (METAL)

Steel plate panels (of which there are various makes comprising, in principle, a flat steel plate with tubes attached or welded to the back) are adaptable to almost any shape, size and position. They are best concentrated under or over the windows, where the major heat loss occurs. This type has also been developed for warming skirtings, floor-borders and cornices, or they may be suspended overhead in work-rooms or factories—but these are matters dealt with in Chapter 11.

It is important that where panels of any type back onto exposed walls, the rear surface should be insulated to reduce direct transmission losses.

EMBEDDED PANELS

Ceiling Panels—The practice of heating rooms by means of embedded ceiling panels was common in the older type of building with heavy construction, where a long time lag was unimportant. Present-day light forms of construction and extensive glass areas require systems with rapid response. Furthermore, the greater use of false ceilings, often with some form of acoustical treatment, renders the embedded system impracticable. It is proposed, therefore, to make only brief reference to it here.

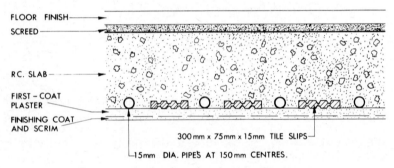

FIG. 7.10.—Section through embedded ceiling panel.

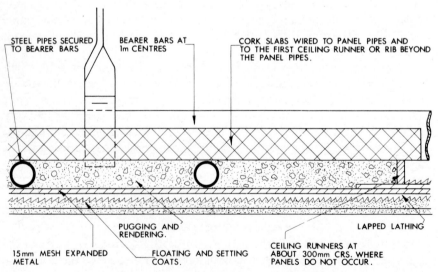

FIG. 7.11.—Section through suspended ceiling panel.

The sinuous coils, usually of 15 mm bore steel or copper at 150 mm centres, were laid on the shuttering of a concrete slab prior to concreting. For a good key, slip tiles were placed between the pipes, see Fig. 7.10.

The soffit was plastered to a special specification now available from B.S.R.I.A. Ground floors having no heat below may require some floor heating and roof slabs must be well insulated.

A technique for incorporating heating panels in suspended ceilings is shown in Fig. 7.11. Emissions from ceiling panels are given in Table 7.10 (page 157).

Floor Panels—Floor panels consist of 15 or 20 mm bore pipes, usually at 225 to 300 mm centres in coil form, laid in the floor screed which is usually of 50 to 75 mm thickness. A variety of floor finishes is now commonly used over floor panels, and, subject to appropriate measures being taken to render material suitable for the temperatures involved, almost any normal floor finish may be used. Thus, wood blocks adequately kilned and bedded in suitable mastic, not softening at the temperatures involved, are found to be satisfactory. Cork tiles and a variety of thermoplastic and rubber tiles have also been used.

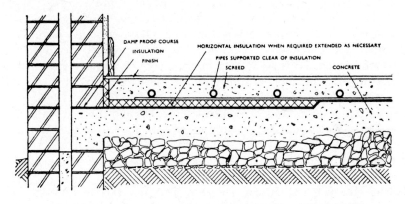

FIG. 7.12.—Floor panel heating: section showing pipe coils and edge insulation.

Most of the harder types of material for floor finishes are equally satisfactory for floor warming. Thus, marble, stone, terrazzo, Granwood and various similar proprietary materials are suitable, subject to due precautions being taken, bearing in mind the possibility of expansion which might lead to cracking. Advice on precautions and details associated with floor heating are available from the Building Services Research and Information Association. For instance, in the laying of stone or marble paving over floor-warming panels, it is necessary to bed the material all over and not by dabs as is sometimes done, otherwise an air gap is introduced and the transmission seriously retarded.

Another important matter for consideration, where floor panels occur on the ground floor, is the insulation of the floor from the earth below. It is found that where insulation is provided a quicker heating-up rate takes place, and the downward loss, which may be as much as 25 per cent is largely eliminated. Fig. 7.12 shows a method of insulation in which it will be noted that the insulating material is carried up at the perimeter of the ground floor slab, where it abuts on the wall, so as to form an edge insulation.

The form of the material must be suitable for the conditions, and involves the choice of such materials as are capable of withstanding loads likely to occur as well as being immune from rot or other deterioration. A damp-proof course is usually carried under the insulating material. It will also be noted that the pipes forming the panel coils do not lay directly on the insulating material, which might be liable to cause external corrosion through moisture contained therein.

It is found to be unpleasant and tiring to sit for long with the feet in contact with a warm floor. The maximum floor-temperature to give comfort with continuous use is about 24° C, and, where occupants are not stationed for long periods, 27° C. In entrance halls and the like, and within 1·6 to 2 m of the walls of a room, a temperature of 30° C is permissible.

Where floor heating is adopted, it is generally found necessary to cover the whole area of the floor with embedded pipes, varying the spacing according to the amount of heat required. Such systems have been successfully used in schools and churches (see Plate XIV, facing page 183).

Floor panel coils are sometimes made of copper, to avoid rusting due to water from washing of floors, etc., entering through cracks.

Emissions from floor panels are given in Table 7.11 (page 157).

Wall Panels—Pipes may be embedded in walls and covered with plaster, as in the case of ceilings, or with marble or some other similar finish. They are, however, liable to be obscured by furniture and to cause convection-current markings. Where pipes are in external walls, insulation is desirable.

Ceiling Heating and Room Height—It is necessary to consider, in the case of ceiling heating, whether the radiant heat on the head will be unpleasant. In the case of a low ceiling height there is obviously a greater danger than in higher rooms. At the same time, by limiting the width of a panel the effect of the direct radiation can be reduced. The *Guide* gives the relationship between maximum desirable temperatures of ceiling panels at different heights and of different sizes in relation to the comfort of the occupants. From this it is clear that, for any room height of 3 m or over, the panel dimensions common in practice will not produce conditions of discomfort. For rooms of less height it may be desirable to reduce the width to 0·6 m or less, which would involve the spreading of the panel surface in a narrow band round the perimeter of the room.

The effectiveness of ceiling panels is reduced with increasing height of rooms. An arbitrary allowance for this is to add 1 per cent to the heating surface for every 0·3 m of height above 5 m. In addition it is desirable to bring the panels for unusually high rooms nearer to the centre of the ceiling so as to reduce the amount of heat falling on the exposed walls and windows. High rooms commonly have tall windows, and it is desirable to check the downdraught from these by convection heaters under the sills.

'PANELITE'

A proprietary version of panel heating under the above trade name consists of pipe coils enclosed in a split tubing of asbestos cement as shown in Fig. 7.13. This arrangement allows freedom for expansion and contraction; consequently, higher water temperatures may be used than in the solid embedded type. The coils can be installed in concrete or hollow tile slabs, or laid in floor screeds, or recessed into walls. Emissions are similar to the embedded panel type.

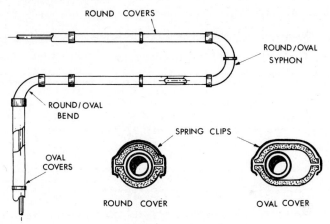

FIG. 7.13.—Panelite coils in floor.

HEATED ACOUSTIC CEILINGS

This type of heating combines acoustical treatment with radiant warming. In one of the proprietary makes, the ceiling is composed of thin aluminium plates (0·6 metres square) having perforations, and above this is placed a layer of thermal and acoustical insulating material, such as a 25 mm blanket of glass silk or slag wool. Fig. 7.14 shows a typical layout of a ceiling. The plates are clipped to the piping. It is usual to cover the whole ceiling of a room with this ceiling treatment, and only such parts of the coils as are necessary for heating are made alive, the remainder being blanked off or plugged during the construction.

It is usual to equip systems of this kind with thermostatic control, which may be of the individual room-control type comprising a thermostat in each room which controls, by means of a motorized valve, the circulation of water through the coils serving that particular room. It should be noted that as the metal ceiling replaces the normal sub-ceiling, no plastering is involved. (See Fig. 7.15 and also Plate XIII, facing p. 182.)

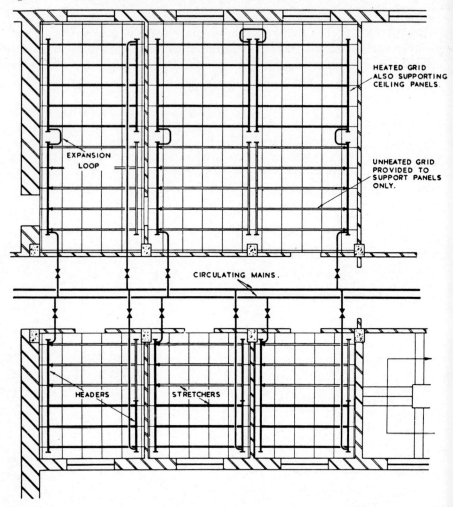

HEATED GRID
ALSO SUPPORTING
CEILING PANELS.

EXPANSION
LOOP

UNHEATED GRID
PROVIDED TO
SUPPORT PANELS
ONLY.

CIRCULATING MAINS.

HEADERS   STRETCHERS

FIG. 7.14.—Example of Frenger ceiling heating applied to a number of rooms: heated sections shown in solid line.

In another proprietary form a continuous hot-water pipe coil is supported above and independently of the fibrous plaster panels forming the acoustical ceiling, as shown in Fig. 7.16. Above the pipes is placed a thermal and acoustic insulating pad supported on wire mesh. The construction may be applied directly below a floor slab, or with a space as in the case of a normal sub-ceiling and as shown in the Figure.

The system is of low thermal capacity and therefore lends itself to individual room thermostatic control.

Typical emissions from the above types of acoustic ceilings are given in Table 7.12 (page 158).

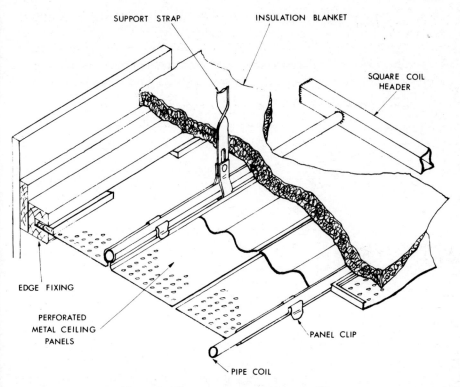

FIG. 7.15.—Component parts of a Frenger ceiling.

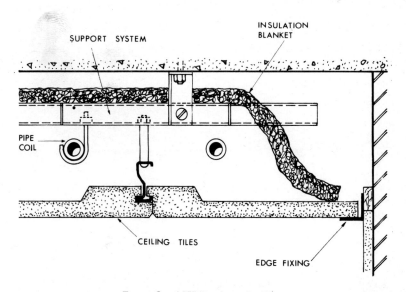

FIG. 7.16.—ASH-Ray heated ceiling.

A recently developed system of overhead radiant heating (Andrews Weatherfoil) uses the space above the false ceiling to house fan-convector units heated by hot water. The units circulate warm air in the ceiling void to heat the ceiling tiles which then emit low temperature radiation. Other fans discharge warm air through grilles to the space beneath, thus combining radiant and convective heating. The fan-convector units are thermostatically controlled, the object being to achieve a balance between radiation and convection so that a satisfactory environmental temperature is achieved. Fig. 7.17 illustrates the principles of this system.

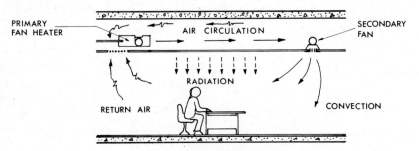

FIG. 7.17.—Andrews-Weatherfoil heated ceiling.

TABLES: CHAPTER 7

TABLE 7.1*

DESIGN WATER TEMPERATURES FOR LOW-PRESSURE
HOT-WATER SYSTEMS

System	Temps. °C at Boiler or Calorifier	
	Flow	Return
Radiators (including pipe coils)		
Gravity circulation - - - - - -	80	60
Pump circulation 2-pipe - - - - -	80	65
Pump circulation 1-pipe - - - - -	80	70
Metal Plate Panels		
Pump circulation 2-pipe - - - - -	80	65
Convectors and Unit Heaters		
Pump circulation - - - - - - -	80	70
Embedded Panel Coils		
Pump circulation 2-pipe		
Ceiling and wall - - - - - -	53	43
Floor - - - - - - - -	43	35
Sleeved panel coils, all positions - - - - -	70	60

Where radiators are served off a single pipe, successive radiators receive water at a lower temperature than those preceding. The following temperatures are recommended for design purposes:

Flow temperature at radiator	Minimum temperature at outlet from radiator
80	70
75	65
70	60
65	55

* See page 134.

Plate IX. Delivery by crane of oil fired boiler to roof-top plant room (see p. 113). Photographs by courtesy of Allen Ygnis Boilers Ltd.

Plate X. Long steel panel radiator in rehabilitated hospital ward (see p. 137)

Plate XI. Tall tubular steel radiator used as screen to large glazed area (see p. 137). Photograph by courtesy of Runtalrad Ltd.

TABLES: CHAPTER 7

TABLE 7·2*
THEORETICAL EMISSIONS FROM HORIZONTAL STEEL PIPES†

Nominal Bore (mm)	Heat emission Watts/metre run												
	Temperature difference, K, surface to air. Ambient air temperature between 10° C and 20° C												
	40	45	50	55	60	65	70	75	80	85	90	95	100
8	29	33	36	41	45	48	53	57	62	71	74	79	83
10	35	39	43	48	52	58	63	68	72	83	88	94	99
15	42	48	55	62	69	77	84	92	100	109	118	127	135
20	51	59	67	75	84	93	103	112	122	132	143	154	164
25	62	71	81	92	102	114	125	137	149	161	175	188	200
32	75	87	99	112	125	138	152	167	181	196	213	229	244
40	84	98	111	125	140	155	170	186	203	220	238	259	273
50	102	118	135	152	169	188	206	226	246	267	289	310	331
65	125	145	165	186	207	230	253	277	301	327	354	380	406
80	143	166	189	213	238	263	290	317	345	378	411	444	466
100	179	207	236	266	297	329	362	396	431	469	507	544	582
125	214	247	281	317	354	392	432	473	515	560	605	650	695
150	248	287	327	368	411	456	502	549	598	650	703	756	808

* See page 134.
† Steel pipes to BS 1387 and ISO/R 65.

TABLE 7.3*
EMISSIONS FROM INSULATED PIPES EXPRESSED AS A PERCENTAGE OF BARE PIPE LOSS
For rigid glass fibre sections at temp. differences stated

Temperature Diff. Hot surface to ambient K	Thickness of Insulation	Nominal bore of steel pipe to BS 1387—mm								
	mm	15	20	25	32	40	65	100	150
80	19	17	17	16	15	15	14	14	14
	25	15	15	14	13	13	11	11	11
	38	13	12	11	10	10	9	8	8
	50	10	10	9	9	8	7	7	6
60	19	18	18	16	16	16	15	14	14
	25	15	15	14	13	13	12	11	11
	38	13	12	11	10	10	9	9	8
	50	11	10	9	9	8	8	7	7
40	19	18	18	17	17	17	16	16	16
	25	16	16	15	14	14	13	13	13
	38	14	13	12	11	11	10	9	9
	50	13	11	10	10	9	8	8	7

* See page 134.

K.H.A.C.

TABLE 7.4*

EMISSIONS FROM STEEL RADIATORS OF VARIOUS TYPES FOR TEMPERATURE
DIFFERENCE AIR TO MEAN WATER OF 56K (20° C TO 76° C)

Make and Type	Pattern	Dimensions mm		Emission W		Remarks
		Depth	Height	per m length	per m² front face	
Myson (Hullrad) Panel type as Fig. 7.1(a)	Single panel	33	300	430	1430	Also made in 4 panel form, and in 440 mm height
		33	590	760	1280	
		33	740	930	1250	
	Double panel	60	300	670	2230	
		60	590	1190	2010	
		60	740	1450	1960	
	Triple panel	95	300	910	3050	
		95	590	1630	2760	
		95	740	1980	2670	
ABK Holdings† (Fixschafer) Column type as Fig. 7.1(b)	2 column	110	450	1160	2570	Also made in 600 mm height (except 5 column) and in 75 mm depth
		110	1000	2580	2580	
	3 column	160	450	1560	3460	
		160	1000	3320	3320	
	4 column	220	450	2100	4660	
		220	1000	4320	4320	
	5 column	250	300	1640	5470	
Runtal† Flat oval tubular elements: headers top and bottom (or either end) as Fig. 7.1(d)	Elements at 40 mm crs.	98	400	1150	2870	Also made with elements at 50 mm crs. And in heights from 400 mm up to 6 m 50 mm steps
		98	1000	2700	2700	
		166	400	1950	4870	
		166	1000	4750	4750	
	Elements at 60 mm crs.	98	400	770	1920	
		98	1000	1800	1800	
		166	400	1300	3250	
		166	1000	3160	3160	
Special (Weldex) Sinuous coil as Fig. 7.1(c)	Coil at 100 mm crs.	35	500	560	1110	Made in 2 lengths, 900 and 1200 mm
		35	700	750	1070	
		35	900	950	1060	
Sensotherm† Sinuous coil with extended Surface in casing	Single depth	78	300	420	1400	Made in 5 lengths only. Low surface tempr. (42° C) Tested to 2 MPa
		78	450	800	1780	
		78	600	1130	1880	
	Double depth	156	300	820	2730'	
		156	450	1430	3180	
		156	600	2070	3450	

* See page 139.   † Type tested by B.S.R.I.A.

TABLE 7.5*
EMISSIONS FROM STEEL RADIATORS OF VARIOUS TYPES FOR TEMPERATURE
DIFFERENCE AIR TO MEAN WATER OF 56K (20° C TO 76° C)

Make and Type	Pattern	Dimensions mm		Emission W		Remarks
		Depth	Height	per m length	per m² front face	
Barlo†	Single panel	30	320	520	1650	Also made in
		30	520	810	1560	220, 420
Panel type with		30	720	1070	1490	and 620 mm
extended surface						height
at rear as Fig.	Double panel	64	320	1020	3190	
7.2(a)		64	520	1530	2950	
		64	720	1990	2770	
	Triple panel	108	320	1480	4650	
		108	520	2130	4100	
		108	720	2820	3920	
Trianco†	Single panel	40	300	630	2110	Also made in
(Hudevad)		40	1000	1770	1770	200, 400, 600,
						700 and 800
Flat front plate		82	300	1020	3410	mm height,
with rear water		82	1000	2780	2780	and in other
and air surfaces						depths
as Fig. 7.2(b)	Double panel	80	300	1120	3750	
		80	1000	2720	2720	
		164	300	1960	6540	
		164	1000	4570	4570	
Thor-rad†	Single panel	75	100	410	4100	See makers'
		75	150	550	3660	data for
Panel type		75	200	630	3150	use in floor
core(s) with						channels
extensive added	Double panel	150	100	720	7200	
surface each		150	150	1020	6800	
side (and between)		150	200	1230	6150	
as Fig. 7.2(c)	Triple panel	225	100	1020	10200	
		225	150	1430	9530	
		225	200	1820	9100	
Anglo-Nordic†	Top outlet	70	370	840	2280	Also made
(Thermalrad)		70	620	1440	2320	in 450 and
		70	1020	2380	2330	750 mm height
Grid pattern						
waterway with	Top outlet	100	370	1000	2700	
perforated casing		100	620	1750	2820	For front
as Fig. 7.2(d)		100	1020	2800	2750	outlet, rating
						is reduced
	Top outlet	172	370	1110	3010	by about 10%
		172	620	2010	3250	
		172	1020	3400	3330	

* See page 139.   † Type tested by B.S.R.I.A.

TABLE 7.6
EMISSIONS FROM CAST ALUMINIUM RADIATORS FOR TEMPERATURE DIFFERENCE
AIR TO MEAN WATER OF 56 K (20° C TO 76° C)

Make	Type	Dimensions mm		Emission W		Remarks
		Depth	Height	per m length	per m² front face	
Faral (Tropical) Fig. 7.5(a)	Sectional with flat front and rear and internal fins	95 95 95 160 160	430 580 690 285 435	1840 2560 2880 1750 2700	4280 4410 4160 6150 6200	Enamelled finish at works in range of colours
Aelrad Fig. 7.5(b)	Similar to above	90 90 90 90 135 135	422 572 622 722 322 422	1490 2270 2520 3010 1490 2410	3530 3960 4050 4170 4630 5710	Tested to 1 MPa for working pressure of 600 kPa

Note: 1. Both types manufactured in 60 mm wide sections.
2. Weight of highest type in each case is approximately 1·8 kg per section (empty).
3. Water content for 690 mm Faral is stated as 0·55 litres per section.

TABLE 7.7

Values of $\left(\dfrac{\text{Temp. Diff. K}}{56}\right)^{1\cdot3}$

Temp. Diff. K	0	1	2	3	4	5	6	7	8	9
20	0·27	0·28	0·30	0·32	0·34	0·35	0·37	0·39	0·41	0·43
30	0·45	0·47	0·49	0·51	0·53	0·55	0·57	0·59	0·61	0·63
40	0·65	0·67	0·70	0·72	0·74	0·76	0·78	0·81	0·83	0·85
50	0·87	0·90	0·92	0·94	0·96	0·99	1·00	1·03	1·06	1·08
60	1·11	1·13	1·15	1·18	1·20	1·23	1·25	1·28	1·30	1·33
70	1·35	1·38	1·40	1·43	1·45	1·48	1·50	1·53	1·55	1·58
80	1·61	1·63	1·66	1·69	1·71	1·74	1·77	1·79	1·82	1·85
90	1·87	1·90	1·93	1·95	1·98	2·01	2·04	2·06	2·09	2·12
100	2·15	2·18	2·20	2·23	2·26	2·29	2·32	2·35	2·37	2·04

TABLE 7.8*

EMISSIONS FROM NATURAL DRAUGHT CONVECTORS, CABINET, CONTINUOUS AND SKIRTING TYPE, FOR TEMPERATURE DIFFERENCE AIR TO MEAN WATER OF 56K (20° C TO 76° C)

Make and Type	Pattern	Dimensions mm		Emission W		Remarks
		Depth	Height	per m length	per m² front face	
Dunham-Bush	2 row element	110	400	680	1700	Made in units
		110	500	790	1580	at lengths of
Cabinet type		110	600	860	1430	600 to 1800
Top-front outlet		110	700	920	1310	mm in 100
grille, open bottom						mm steps.
inlet	3 row element	160	400	1000	2500	Also with
		160	500	1150	2300	front bottom
		160	600	1270	2120	grille inlet at
		160	700	1360	1940	10% reduced
						output
	4 row element	210	400	1260	3150	
		210	500	1460	2920	
		210	600	1600	2670	
		210	700	1720	2460	
Myson	Single tube,	73	406	560	1380	
(Copperad)	22mm	73	508	610	1200	Single tube
		73	610	640	1050	type also
Continuous type		73	711	660	930	made at
Sloping front at						305 mm high
outlet grille	Single tube,	133	406	910	2240	
(Sill-line)	35mm	133	508	1010	1990	22 mm tube
		133	610	1070	1750	type may
		133	711	1110	1560	have 28 mm
	Double tube,	73	406	680	1670	tube
	22mm	73	508	790	1560	
		73	610	860	1410	
		73	711	920	1290	
	Double tube,	133	406	1110	2710	
	35mm	133	508	1280	2520	
		133	610	1390	2280	
		133	711	1550	2180	
Myson	Single tube,	65	210	440	2090	With damper
(Copperad)	15mm					closed, output
						reduced to
Skirting type						30%
(Wallstrip)						

* See page 140.

TABLE 7.9

Make and Type	Dimensions mm		Ratings for alternative fan speeds						
			Emission kW			Air Supplied litre/s			
	Depth	Width	Low	Med.	High	Low	Med.	High	
Biddle	200	750	1·84	2·18	3·46	34	49	91	Heights are
(Force flo)	200	1000	4·06	5·43	6·97	80	122	196	standard for
	200	1250	5·97	7·39	8·70	120	161	209	all sizes
Commercial									Front outlet
Fan									= 690 mm,
Convectors	250	1000	7·51	9·13	12·11	164	212	326	Top outlet
	250	1250	8·15	10·92	13·75	181	266	365	= 560 mm
	250	1500	10·92	14·27	16·78	242	344	431	Extended
	250	1750	13·53	18·39	25·17	273	413	580	case = 1850 mm
Flexaire	356	610	—	3·30	—	—	95	—	Height =
(Tempaflex)	356	610	—·	4·83	—	—	140	—	760 mm
	356	610	—	5·97	—	—	165	—	
Apartment	356	760	—	7·41	—	—	195	—	
Heaters									
	432	815	—	9·31	—	—	235	—	
	432	915	—	10·95	—	—	285	—	
Fenton-Byrn†	127	597	1·50	1·91	2·32	37	46	55	Height =
Domestic use									346 mm
Fan									
Convectors	127	771	1·72	2·61	3·50	38	56	82	

Note: 1. This type of equipment is normally selected using the ratings at medium speed. Noise ratings are commonly NC 25, NC 35 and NC 40/50 for low, medium and high speeds. Domestic units are somewhat quieter.
2. Emissions with fan 'off' vary with construction in the range of 10–15 per cent. of medium speed rating.
3. Emissions stated are for 10 K temperature drop in water.

† Type tested by B.S.R.I.A.

TABLE 7.10*

EMISSIONS FROM CEILING PANEL HEATING WITH WATER AT 50° C MEAN

Construction	Downward emission W/m²		Upward emission W/m²	
	Room Temperature °C		Room Temperature °C	
	15	18	15	18
1. Hollow Tile 150 mm				
(a) cement screed and 25 mm hard finish above	175	160	80	70
(b) Screed and wood block finish above	190	175	70	55
2. Solid Concrete 150 mm				
(a) Screed and 25 mm hard finish above	175	160	100	80
(b) Screed and wood block finish above	175	160	80	70
3. Suspended ceiling 50 mm pugging, 25 mm insulation over	190	175	55	40
4. Roof. Hollow Tile or concrete with 50 mm insulation over	190	175	30	

Note: 1. Where upward emission is 80 W/m² and over with room temperature of 18° C, the floor area over the panel may be too warm for comfort.
   2. Roof loss is to be taken in estimating boiler power and pipe sizing but not as contributing to room heating.
   3. See *Guide, Section B1* for more comprehensive data.

* See page 145.

TABLE 7.11*

EMISSIONS FROM FLOOR-PANEL HEATING

Floor Finish	Centres of 15 mm nom. bore pipes mm	Upward emission W/m² floor surface for temperature differences air to mean water, K				
		20	25	30	35	40
Screed 50 to 75 mm thick and 25 mm tile or other hard material finish	150	90	112	136	155	—
	225	83	104	124	142	—
	300	77	97	112	130	—
Screed as above but with 25 mm wood block finish	225	—	—	75	85	95
	300	—	—	70	80	90

Note: See *Guide, Section B1* for more comprehensive data.

* See page 146.

TABLE 7.12

EMISSIONS FROM HEATED ACOUSTICAL CEILINGS

Metal plates clipped to pipes (as Frenger):

Air temperature	20° C
Mean water temperature	80° C
Emission	175 W/m²

Metal plates with pipes above, but not attached
(Burgess Sulzer or Ash-Ray)

Temperatures	(as above)
Emission	170 W/m²

Upward emission (heated room over):

Both types	15 W/m²

* See page 148.

CHAPTER 8

Pipe Sizing for Hot-Water Heating

HAVING SELECTED THE TYPE, position, and extent of the heating surface, and marked this on the plans of the building in question, the next stage is to determine how this shall be connected to the boiler with pipes in the most economical and effective manner, and how the water shall be circulated.

Water Circulation—Water circulation in a system of pipes may either be by thermo-syphon (gravity), or by mechanical means (pump). A thermo-syphon or gravity system is one in which the circulation is produced by difference of temperature. Water at a higher temperature, such as issues from a boiler (flow), is less dense than the cooler water which has passed round a circulation (return), and advantage is taken of the difference of mass of the rising warm-water column and the dropping cool-water column to cause the circulation through the system.

In the case of circulation by mechanical means this usually implies the employment of a centrifugal pump, which is a convenient method of impelling water since it may be easily connected to an electric motor. A centrifugal pump is preferable to a plunger type of pump owing to the absence of pulsations whereby noise is produced. The centrifugal pump can be rendered to all intents and purposes noiseless. Furthermore it is more foolproof; for instance, even the closure of valves in the circulation will cause no harm.

Pump circulation has supplanted gravity circulation for all medium and large installations, and is now common even for the smallest installations, such as in houses. The advantages of pump circulation are:

(*a*) The circulation is independent of temperature.

(*b*) It gives a quicker response.

(*c*) Pump circulation allows the use of smaller bore pipes, thus saving in first cost and enabling pipes to be fitted inconspicuously more easily. The thermal capacity of a system using smaller pipes is naturally less than one with large pipes, and this again affects the rate of heating up and response to controls.

(*d*) Many forms of heat emitter are now suitable only for pump circulation, such as forced air convectors and metal plate radiant panels. Other types such as skirting heating, and normal convectors, are also much more advantageously served by mechanically circulated water. Temperature drops may be less, and hence the size of equipment may be reduced due to the higher mean-water temperature in the apparatus.

New kinds of pumps have become available (as described later) designed to reduce the maintenance and cost of fixing, and some are particularly suitable for small installations.

Although the cost of the pump or pumps and the cost of power for running have to be allowed for, the advantages of forced circulation greatly outweigh the consideration of cost, which is usually a minor item compared with the overall capital and running cost of the installation.

In view of the importance of forced or pump circulations, it is proposed to consider the design of these in greater detail first, and to refer to gravity circulation second. Before doing so, however, it is necessary to describe some of the systems of piping commonly used.

SYSTEMS OF PIPING

Fig. 8.1: Single-Pipe Circuit—In this case the circulating piping leaves the boiler and returns to it in one continuous loop. Heat emitters are teed off the single-pipe flow and return and are therefore in shunt with a short section of the piping. The circulation from the pipe into the heat emitter and out again is thus largely independent of the circulation produced by the pump, and it is necessary in this case to size the connections of the branches as if they were gravity operated. The first radiator receives water at full boiler temperature, and the cooler return water from the outlet mixes with the water in the main which has bye-passed the radiator. So radiator No 2 receives a mixture at some lower temperature than that received by radiator No 1. The temperature drop from radiator to radiator is progressive, the last one receiving water at little more than boiler-return temperature. To preserve heat output, radiator sizes are therefore graded according to their calculated mean temperature. A small

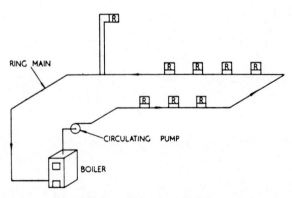

FIG. 8.1.—Single-pipe circuit.

overall-system temperature drop—boiler flow to boiler return—reduces the effect. The pump is shown in this case in the flow, but may equally well be in the return to the boiler.

Fig. 8.2: Parallel Single-Pipe Circuits—In this type of system there are a number of single-pipe circuits arranged in parallel, either in the form of a ladder or in the form of a series of single-pipe drops. Such an arrangement is frequently the most economical for serving a multi-storey building, or a school, as there are any number of ways in which similar parallel circuits may be arranged. Like the previous case, radiators are only slightly assisted as to their circulation by the action of the pump, being in shunt from a short section of the pipe in each case, and they are similarly subject to progressive mean temperature decline.

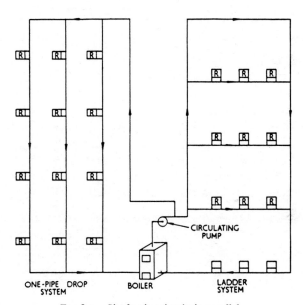

FIG. 8.2.—Single-pipe circuits in parallel.

Fig. 8.3: Two-Pipe System—This system is sometimes called a 'tree' system, as it starts from a main near the boiler and branches out in all directions. Each branch conveys water at boiler temperature to the heat emitter, less any temperature drop due to heat given out by the mains, and returns the water direct to the boiler. Thus, the progressive lowering of temperature to successive heat emitters of the single-pipe system is largely avoided and, due to this, the two-pipe system is of greater application, being widely used on extensive systems. It may be applied to multi-storey buildings with risers, or to single-storey factories etc. with long horizontal flow and return mains, and drop-pipe connections to radiators or heat emitters in any one of a vast variety of ways. All radiators or other heat emission surfaces connected to the system, receive the full benefit of the pump in promoting circulation through the equipment in question. Many types of heater offering a high resistance to flow are only suitable for two-pipe application.

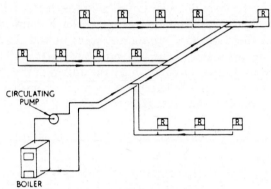

FIG. 8.3.—Two-pipe system.

Fig. 8.4: Reversed Return System—This is a modified form of the two-pipe system and, as will be noted from the Figure, the pipework is arranged such that the total travel from the boiler to any one heat emitter and back to the boiler is the same in each case. This is clearly a method of simplifying the balancing-up of resistances, and is an admirable system where conditions permit. It is, however, usually more expensive than the previous tree system.

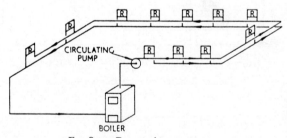

FIG. 8.4.—Reversed return system.

Fig. 8.5: Panel Systems—These are always designed on the two-pipe principle. Panel coils offer a considerable resistance to flow, and assist in establishing a differential throughout the system, which aids in balancing up. The resistance of the water flowing through the panel coil must be allowed for in establishing the pump head of the system. It will be noted that the circulation is upwards through the panels, so as to remove accumulation of air to the highest points where it may be easily vented. An alternative method of piping panel systems on the drop-system has been used, but has fallen into disfavour due to difficulties of air venting.

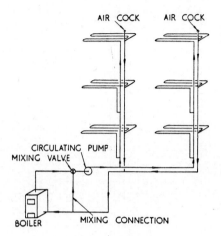

Fig. 8.5.—Panel system.

FLOW OF WATER IN PIPES

The mass flow of water to convey the required quantity of heat from the boiler or water heater to the heat emitters, in unit time, depends on the temperature drop from flow (t_f) to return (t_r) and the specific heat capacity of water $(J/kg\ K)$ at that temperature. Thus

$$\text{mass flow kg/s} = \frac{J/s}{(t_f - t_r) \times J/kg\ K}$$

Over the range of low-pressure hot-water systems, the specific heat capacity may be taken as $4 \cdot 2\ J/kg\ K$.

This is a tedious sum to work out a great many times for various sections of a pipework complex and shows up the disadvantage of the SI system compared with a calorimetric unit, such as the British Thermal Unit, in which the specific heat of water is taken as unity.

Tables of mass flow in piping are to be found in the *Guide, Section C4*, based on kg/s. Heating loads, on the other hand, are worked out in kW; but 1 kW = 1 kJ/s, so it would be more convenient if the data were presented direct in kW per kelvin, thus:

$$kW/K = \text{mass flow kg/s} \times kJ/kg\ K = kJ/sK$$

It then only remains to multiply by the temperature difference flow to return to obtain the energy flow in question. The properties of water do not enter into the matter.

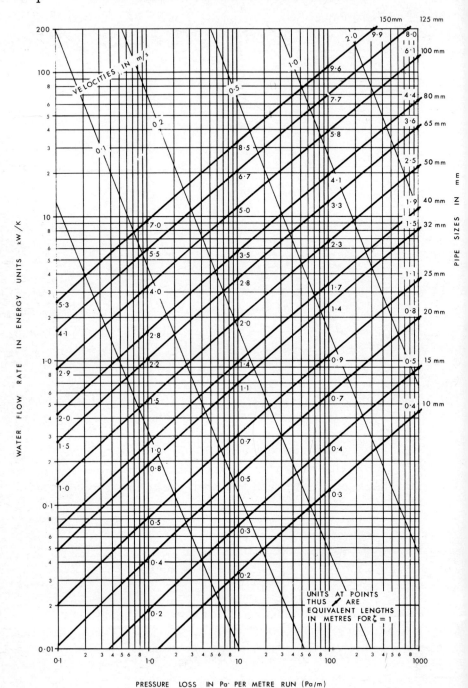

FIG. 8.6.—Pipe-sizing chart for water at 75° C (steel).

This book presents Fig. 8.6, which has been prepared from *Guide* data by computer* in the form of a chart. The pipe size lines give kilowatts of flow per kelvin temperature drop. Pressure loss is in Pa per metre: i.e., Pa/m. The chart covers the range 10 mm to 150 mm. Larger sizes are less commonly used and the *Guide* Tables are available.

By the use of this chart it is possible to retain the kilowatt throughout all heating calculations: heat loss, proportioning of heat emitters, summation of boiler power and pipe sizing.

It is as well to remember that the kilowatt is no more than a measure of energy flow. Thus we see in the above steps, first, how the rate of energy dissipation through the fabric is calculated in heat losses, and second, the output of energy to counteract this loss room by room or space by space. Third, all come together at the boiler, or other heat source, where energy is to be produced in sufficient quantity to supply all demands. Finally, the system of piping has to be sized so that water circulating through the boiler and carrying the energy which it generates is split up into perhaps a great many sub-circuits and branches, and hundreds of terminal connections, to heat-emitting appliances. The object of pipe sizing is to achieve a balance throughout such that each appliance receives its proper share of the total energy distributed.

It is as well to keep this picture in mind and to remember that pipe sizing is only a means to an end and not an academic exercise. Thus, extreme meticulosity is entirely unnecessary and a waste of time. The limitations of commercial pipe sizes and the tolerances to which they are made have to be accepted; furthermore, for one reason or another the practical fitting of the system into the building may not go exactly to the drawings: more bends may be introduced, tees may branch in a different way and rough joints may occur, particularly if welded. A separate study might be made of the relative importance of things—some surprises might be in store!

Single Resistances—The method of making allowance for the resistance of bends, tees and other fittings is to add to the measured length of piping an *equivalent length* (*EL*) for each fitting. The values of *EL* are marked on the graphs in Fig. 8.6 at intervals and, as will be seen, they vary according to velocity and pipe size. They are, in fact, velocity-pressure equivalents and require to be multiplied by a factor, ζ. As given in the *Guide*, the factors vary from 0·3 for a long radius bend over 100 mm bore to 0·9 for a return bend of 25 mm or under. Tees and normal bends and elbows are in the range $\zeta = 0·5$ to 0·7.

Bearing in mind the above dictum on the relative importance of things, and that resistances are of minor significance, it is excusable in the interests of simplicity to allow all such fittings as having $\zeta = 1$.

---

* The Authors are indebted to Oscar Faber and Partners for the use of their computer for this purpose.

This is certainly on the right side and contributes something to unknown factors such as have been mentioned.

The allowance for single resistances thus becomes simply a matter of counting up bends, tees, elbows, reducers and the like, and multiplying by the *EL* for the particular size taken off the chart. As a first shot, when neither size nor velocity are known, no more than a guess can be made, but this can be checked later.

PIPE SIZING

Single-Pipe Circuit—Referring again to Fig. 8.1 (page 160), the single pipe must carry sufficient energy, via the water flowing in it, to supply the total emission from all the radiators plus the emission from the main. This total will equal the output of the boiler.

Assume Radiator emission	= 60 kW
Pipe emission (taken as 50 mm)	= 10 kW
	70 kW

Temp. drop flow to return at boiler   12 K

$$kW/K = \frac{70}{12} \qquad = \quad 6$$

Length of main	90 m
Allow for single resistance	10 m
	100 m

There is now a choice, using Fig. 8.6:

50 mm	85 Pa/m × 100 m =	8500 Pa (8·5 kPa)
40 mm	300 Pa/m × 100 m =	30 000 Pa (30 kPa)

If 40 mm is used, radiators would be increased, as emission from 50 mm is assumed as contributing to room heating. On the other hand, 50 mm gives a very low pump head, but there would be little significant difference in the cost of a pump if the higher pressure loss arising from 40 mm is selected.

Probably, overall, 40 mm is preferable: the energy-flow rate is quite out of the range of 32 mm.

Having established the size, the allowance for single resistances can now be revised. The pressure to be set up by the pump will be the sum of the circuit-pressure loss and the boiler resistance. The latter may be found from makers' data (where high velocities may be used to promote efficiency).

The radiator connections, being in shunt from a short length of main, derive a pressure difference corresponding to that of the length of pipe carrying the bye-pass water. This can be evaluated and added to the gravity head referred to later.

Parallel Single-Pipe Circuits—It will be apparent from Fig. 8.2 (page 161) that each single-pipe element can be treated as set out above, as if a boiler served it individually. The mains to which they connect will, however, be sized as the two-pipe system considered next. Note that the pump pressure will be that of the longest or index circuit plus boiler resistance. To adjust for out-of-balance of parallel circuits, offering less resistance than the index circuit, regulating valves are desirable.

Two-Pipe system—By way of a simple example, the system shown in Fig. 8.3 (page 162) is reduced to one element of the circulation, as in Fig. 8.7.

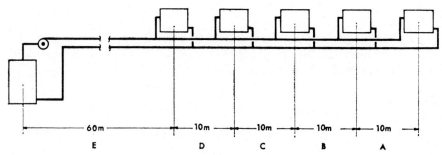

FIG. 8.7.—Apportionment of mains emissions.

First approximation—Assuming the emitting units are 20 kW fan convectors:

Five fan convectors of 20 kW each	=	100 kW
The mains loss may be say 10 per cent	=	10 kW
		110 kW

The overall temperature drop may be taken at 12 K. Thus:

$$kW/K = \frac{110}{12} = 9 \cdot 1$$

From Fig. 8.6, assume a unit pressure drop of say 100 Pa/m. The flow for various pipe sizes can then be set down thus:

	kW/K × 12 = kW	
15 mm	0·25	3
20 mm	0·6	7
25 mm	1·2	14
32 mm	2·4	29
40 mm	3·4	41
50 mm	6·7	80
65 mm	13·0	156

Using the same overall mains heat loss we may then give progressive totals thus:

	Section A	22 kW
	B	44 kW
	C	66 kW
	D	88 kW
	E	110 kW

Sizes will then be:

	A	32 mm
	B	40 mm
	C	50 mm
	D	50 mm
	E	65 mm

These sizes being generally larger than strictly needed, the unit pressure loss will be less than the 100 Pa/m³ assumed. If taken at 90 average, we can arrive at the mains pressure loss thus:

Total length flow and return	200 m
Allow single resistances	20 m
	220 m

$$90 \times 220 = 19\,800 \text{ Pa } (= 19 \cdot 8 \text{ kPa})$$

A fresh run-through with a higher pressure loss could be tried:

e.g. if the unit pressure loss is taken at 200 instead of 100 Pa/m³, the largest pipe size will be 50 mm not 65 mm, but the overall mains pressure loss will be about 40 kPa; which, with loss through boiler and heater added, may be considered a rather high pump head for an installation of this modest size.

'Accurate' Sizing—The previous example may serve to show a quick method of pipe sizing, sufficient perhaps for an estimate of cost; but, when the project is to proceed further, a more accurate approach is necessary. The following steps, related to Fig. 8.8 (which is a diagram of the system shown in Fig. 8.3 with dimensions added), will illustrate one method by which 'accurate' pipe sizing may be accomplished.

1. The process of allocating guessed preliminary pipe sizes to a system in order to establish the heat emitted by each section, is tedious and use may be made of Fig. 8.9 as a short-cut. From this figure, a near approximation to heat emission for a pipe may be read from the right-hand scale, for a number of system-temperature drops, against the energy-flow

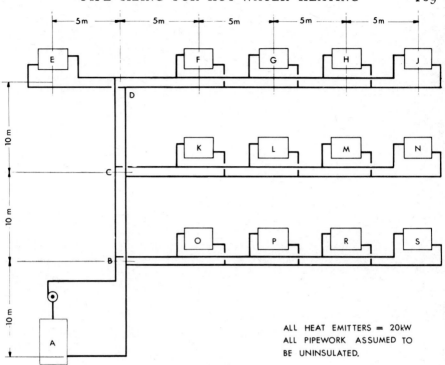

FIG. 8.8.—Example of two-pipe sizing.

rate shown on the bottom scale. Thus, for the system in Fig. 8.8, a list may be made as follows for a 12 K temperature drop:

Pipe section key	Nett Energy flow (kW)	Pipe heat emission W per (m) Fig. 8.9	length (m)	Total (kW)
AB	260	188	30	5·64
BC	180	178	20	3·56
CD	100	148	20	2·96
DE	20	92	10	0·92
DF, CK, BO	80	135	10	1·35
FG, KL, OP	60	125	10	1·25
GH, LM, PR	40	112	10	1·12
HJ, MN, RS	20	92	10	0·92

2. The apportionment of these heat emissions, section by section, must now be carried out systematically on a 'compound interest' basis. A first step would be to consider the three identical main circuits on the right-hand side of the Figure, taking the upper identifications as follows:

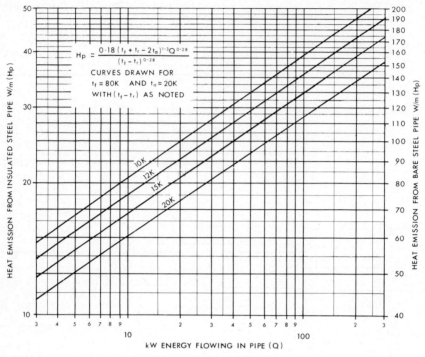

FIG. 8.9.—Heat emission related to energy flow.

	Heaters			
	F	G	H	J
Rated emission, kW	20·00	20·00	20·00	20·00
Pipe HJ is allocated to heater J alone				0·92
	20·00	20·00	20·00	20·92
Pipe GH is allocated to heaters H and J			0·55	0·57
	20·00	20·00	20·55	21·49
Pipe FG is allocated to heaters G, H and J		0·40	0·41	0·44
	20·00	20·40	20·96	21·93
Pipe DF is allocated to all four heaters	0·32	0·33	0·34	0·35
	20·32	20·73	21·30	22·28

3. Proceeding similarly, pipe CD is allocated to heaters E, F, G, H and J, pipe BC to heaters E, F, G, H, J, K, L, M and N, and lastly, pipe AB is allocated to all the heaters shown. This process produces the total loadings shown below and it will be seen that, although the *average* mains loss is just over 10 per cent, that for heater J is 18·9 per cent whilst that for heater O is only 3·5 per cent.

Heater	Gross load (kW)	Mains loss (%)
O	20·71	3·5
P	21·15	5·7
R	21·74	8·7
S	22·73	13·6
K	21·09	5·4
L	21·53	7·1
M	22·12	10·6
N	23·14	15·7
E	22·35	11·7
F	21·69	8·4
G	22·14	10·7
H	22·76	13·8
J	23·79	18·9

The effect of the mains losses upon the heat emitters is that the full *system* temperature drop of 12 kelvins will not be available at the individual flow and return connections. Taking the two extreme examples:

Heater O, temperature drops*

$$\text{In flow main} \quad = 12 \times \frac{0·35}{20·71} = 0·25\text{K}$$

$$\text{Across heater} \quad = 12 \times \frac{20·00}{20·71} = 11·50\text{K}$$

$$\text{In return main} = 12 \times \frac{0·35}{20·71} = 0·25\text{K}$$

Heater J, temperature drops*

$$\text{In flow main} \quad = 12 \times \frac{1·89}{23·79} = 0·95\text{K}$$

$$\text{Across heater} \quad = 12 \times \frac{20·00}{23·79} = 10·10\text{K}$$

$$\text{In return main} = 12 \times \frac{1·89}{23·79} = 0·95\text{K}$$

* These are not strictly correct since the flow pipe is hotter than the return, and will thus have a slightly higher heat loss.

The object of the exercise is to maintain the *mean* temperature at the heat emitter constant throughout the system. Naturally, this small simple example does not show up the relative importance of mains loss as would be the case in an extensive plant, but no doubt it will serve to show the principle.

It will be seen that, for a large installation involving a great many branches of different lengths and sub-circuits of varying size, the apportionment of mains losses, if pursued to its ultimate refinement, can be a very laborious process. Various methods have been devised to simplify this task, such as have appeared in previous editions of this book. By way of rough compromise, if the total mains loss of one circuit from the boiler is calculated and divided by the number of branches on that circuit, even if not of uniform load, some attempt at apportionment can generally be made by sight to achieve a percentage basis which is probably not far from reality.

However arrived at, the mains loss for each section is added to the emitter load of the branch, and these are added progressively back to the boiler, or to the headers if there are several main circuits.

4. We now have loads which each section of main must carry, and the size can be judged from a starting basis of say 100 Pa/m unit pressure drop. The length of each section, flow plus return, plus single resistances can be set down and a table prepared, thus, for heater J in the example:

Section of Pipe (1)	Load Carried kW/K (2)	Total Length (L + EL) m (3)	Pipe Size mm (4)	Unit Pressure loss Pa/m (5)	Section Pressure Loss (3) × (5) (6)
AB	23·85	41·1	80 (3″)	140	5760
BC	16·70	23·6	80 (3″)	70	1652
CD	9·37	21·3	65 (2½″)	58	1235
DF	7·52	11·1	50 (2″)	120	1332
FG	5·72	11·2	50 (2″)	70	784
GH	3·88	11·1	40 (1½″)	120	1332
HJ	1·98	14·0	32 (1¼″)	60	840
Emitter	1·98	—	—	Catalogue	5000

$$\text{Total for circuit} = 17\ 935 \text{ Pa}$$
$$= 18 \text{ kPa}$$

5. Next come the other branches and sub-circuits. At each off-take from the index circuit there will be some surplus pressure available. This must be dissipated in the branch connections, otherwise short circuiting will occur.

Each branch taken in turn then becomes a fresh exercise to be re-tabulated as above, sizes being adjusted to absorb surplus pressure.

There often comes a stage where this is impossible within the limits of commercial pipe sizes, and hence all such branches are provided with regulating valves which can be adjusted to take up over-pressure. Alas, in the hurry of completion of site work this final regulation is often scamped and the circuits are never properly balanced. The saving grace, however, is a generous margin on pump capacity (but *not* on pump pressure) above that strictly necessary so that the ill-effects of the short-circuiting hazard are minimized.

6. Having established the total pressure loss of the system based on a number of initial assumptions, there is always the question as to whethe**r**, by selecting a different temperature drop or a greater or lesser pump pressure, a more economical solution would have emerged. Furthermore, as main sizes are changed, so are their heat losses, and the question of optimum insulation thickness also arises where appropriate.

Pipe Sizing for the Reversed-Return System—It will be obvious that with this system (see Fig. 8.4) the loads to be carried by the piping are added progressively forwards for the return and back from the index run to the boiler on the flow. Branches will be dealt with, however, in the normal way as for a two-pipe system. A word of caution is necessary, calling for a careful checking of the available pressure (flow and return) at each branch, as a condition may be found where a greater than average pressure drop has occurred in, say, the flow and a less than average one has occurred in the return. In such a case there might be no differential pressure available for the branch, which would then fail to work.

Pipe Sizing for the Embedded-Panel System—The normal two-pipe method of sizing (see Fig. 8.5) is used except that, due to low water temperatures, mains losses are low and may be taken on an overall basis of say 10 per cent. Furthermore, as the pressure loss in each coil is high relative to the pressure loss in mains and the temperature drop in the system is low, involving large water quantities, the system tends to be self-balancing.

The task of pipe sizing may be simplified by noting that, as panel coils are all usually of much the same length, each may be assumed to take the same quantity of water. The progressive totalling of loads for the mains is then only a matter of adding up the numbers of coils branch-by-branch and length-by-length.

It is necessary to recall that the heat emitted by an embedded panel has an upward and a downward component, whether in ceiling, roof or floor. Thus the total in each case must be allowed for, and not simply the emission to the room which the panel is designed to heat.

Design by Computer—To obtain the most refined solution to the problem of pipe sizing by long-hand arithmetical methods, having regard to the large number of variables each of which is mutually interdependent, is virtually impossible. Methods in the past have followed well-defined principles and limitations which, in the main, have worked in practice.

With the terrific advantage of a new tool, namely the computer, techniques have quickly been developed to assimilate all these variables and to evolve therefrom solutions to problems which are impossible to solve by any other means. For all but the smallest systems, therefore, the future of pipe sizing will no longer be in the sweating-through innumerable pages of tedious calculations, and the pushing of a slide rule until it is red hot, but in the careful completion of machine input-sheets.

A number of suitable *programs* (to use the transatlantic spelling unfortunately adopted by the computer industry) are already in existence. Experience is showing that a project of suitable size can be processed in one-quarter of the time and at one-third of the cost which would apply to the equivalent manual exercise. Pipe sizing *per se* is, however, only a part application of a fully developed computer approach: a full suite of programs will accept building dimensions and system criteria as input and will provide, as output, not only pipe sizes but also radiator dimensions and bills of quantities. Mention has already been made, however, of the limitations imposed upon any pipe sizing calculation by the limited range of commercial pipe sizes available. The computer may tick and whirr to ten decimal places but, in the end, circuits must be balanced one with the other by Fred's skill in adjustment of regulating valves.

PIPE SIZING WITH GRAVITY CIRCULATION

Circulating Pressure—Fig. 8.10 shows a simple single-pipe gravity cir-
culation. The force creating circulation is that due to the difference in weight of positive column P_1 at temperature t_2 and negative column N_1 at temperature t_1.

The force available will be that of unit mass, subject to acceleration due to gravity, which may be taken as 9·81 m/s. If ρ_2 and ρ_1 are densities of water at temperature t_2 and t_1 respectively in kg/m³, H is the height of P_1 and N_1 in metres, and CP is the circulating pressure in Pa/m, then

$$CP = H \times 9·81 \ (\rho_2 - \rho_1)$$

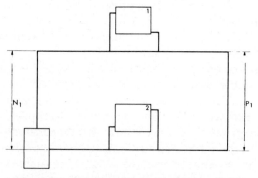

FIG. 8.10.—Simple single-pipe gravity circulation.

For convenience, the *Guide, Section C4,* gives a Table of circulating pressures for various flow temperatures and temperature drops, from which Table 8.1 is extracted.

TABLE 8.1

CIRCULATING PRESSURES FOR GRAVITY HOT-WATER SYSTEMS

Flow Temperature °C	CP in Pa per metre height for following temperature differences, K, flow minus return						
	8	10	12	14	16	18	20
40	27·63	33·73	39·50	44·90	49·94	54·59	58·83
50	33·35	41·01	48·37	55·44	62·20	68·64	74·74
60	38·38	47·36	56·10	64·57	72·78	80·72	88·37
70	42·96	53·14	63·09	72·80	82·28	91·52	100·50
80	47·57	58·54	69·61	80·46	91·09	101·50	111·68
90	51·25	63·57	75·64	87·61	99·32	101·82	122·11

In order to determine the temperature drop from N_1 to P_1, it is necessary to take the heat emission of the top radiator 1 and the top length of the circulating pipe and that of the bottom radiator 2 and the bottom length of pipe. The overall temperature drop from boiler flow to return, multiplied by the ratio of emissions top: total, gives the temperature drop to P_1. If Q_1 is emission at the upper level and Q_2 that at the lower level, then

$$\text{Temp } P_1 = \text{Temp } N_1 - \left\{ \frac{Q_1}{(Q_1 + Q_2)} \right\} \times \text{overall temp. drop.}$$

Heat loss from riser and drop are ignored.

Resistance to Flow—The circulating pressure assessed in the manner described is the means of creating and maintaining a circulation through the system, and if this is a closed circuit, as in Fig. 8.10, it will cause just such a velocity that its force is balanced by the resistance or friction encountered in the pipes and boiler.

For a given quantity of heat to be transmitted from the boiler to radiators 1 and 2, water must flow through the system, the mass depending on the temperature drop. For a high temperature drop t_1 to t_2 each kg of water will carry more heat and, therefore, less water will be required than at a lower temperature. At the same time, the greater the difference between t_1 and t_2 the greater will be the circulating pressure, and obviously the greater the flow of water.

With a constant heat output from the boiler, these effects strike a balance in such a way that the temperatures t_1 and t_2 adjust themselves to produce just that circulating pressure which will be absorbed in impelling through the circuit a quantity of water to an amount equal to the energy delivered, divided by the temperature difference $t_1 - t_2$.

Thus the circulating pressure balances the sum of the resistances, or in other words:

$$CP = \Sigma R$$

where ΣR is the sum of all the resistances to flow of water throughout the circuit.

Of these, the most important is the resistance of the piping, bends, tees, and valves. The lesser resistance is that in the boiler.

Available CP per Metre Run—It is necessary to determine the total metres travel 'T' of the circulation, including an allowance for the single resistances, which must therefore be assumed beforehand, both for size and velocity and corrected later. In the case of a branched system we take the travel for which CP/T is the least, and this is called the *index circulation*. The flow and return temperatures will be as selected from Table 7.1 (page 150), such as 80° C flow and 60° C return in a typical case.

Centre of Boiler—In measuring the height of the column producing circulation it is convenient to assume the boiler to be concentrated at one central point. This is the mid-point between flow and return connections.

Radiators on Single Pipe—The matter of radiator connections from a single pipe has been mentioned. Fig. 8.11 illustrates the method often

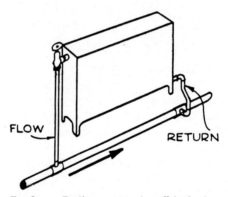

Fig. 8.11.—Radiator connection off single pipe.

employed. The gravity circulation through the radiator is calculated as if there was a boiler at the centre of the main pipe. The *CP* available is then that from the centre of the radiator to the centre of pipe. The *T* will be say one metre in the connecting pipes plus the equivalent length for the valve and radiator resistance and other fittings. The ζ factor for an angle valve is 5·0 and for a radiator also 5·0. The bends in the return and the tees may total three: a grand total of thirteen. The EL for $\zeta = 1$ may be taken from Fig. 8.6 and the total length assessed. CP/T is then obtained and hence the kW carried per kelvin. The actual temperature drop across the radiator must be decided, trying this perhaps on 8 K at first. The lower the drop the higher the mean-radiator temperature, but, on the other hand, the larger the connections.

Where the system is pumped, as mentioned earlier, some additional pressure is available corresponding to the differential loss through the length of main between the flow-and-return connections carrying the bye-pass water. This extra pressure is added to the gravity CP.

Two-Pipe (Gravity) System—In a two-pipe system, as Fig. 8.12, the CP (centre of radiator to centre of boiler) will obviously be greatest for radiator 1 and least for radiator 3, which is the index and the worst placed. This latter circuit must then be sized first and the resistance to the branches to radiators 2 and 3 deducted from the CP in each case. The risers to radiators 2 and 3 are then sized to absorb the available CP.

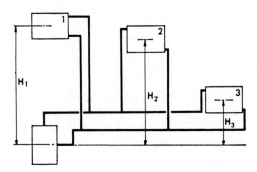

Fig. 8.12.—Example of two-pipe gravity circulation.

Having established $\dfrac{CP}{T}$ for the index and other circuits, sizing then proceeds as with a pumped system by reading, from Fig. 8.6 (page 164), the kW of energy flow for each pipe size for the available pressure drop per metre run. The loads will have been totalled progressively, including mains losses as before in the case of the two-pipe system duly apportioned.

Sizes of pipes necessary for the flow in question will then be selected and, if different from the first assumption, some revision of emissions may be necessary followed by a final run-through again.

Gravity Systems in Perspective—The above brief reference to sizing of gravity systems is intended to give some idea of the principles involved. Being no longer in common use for anything but the smallest installations, there is no point now in pursuing the subject in great detail. Those who wish to do so for historical interest may consult earlier editions of this book and other works. There were, for instance, many more types of system— the one-pipe drop, the two-pipe drop, the irregular and so on. With the tiny forces available, large pipes were the rule and, indeed, considerable

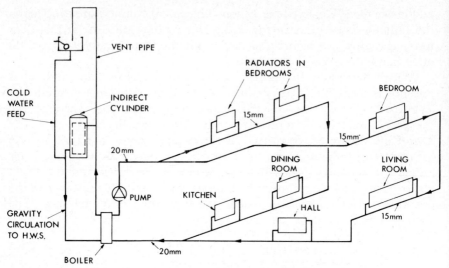

FIG. 8.13.—Diagram of small-bore installation.

experience was called for to ensure that a system would circulate in all its parts. For one reason or another some circuits might be sluggish, only responding when the boiler was forced unusually hard. Then there was always the menace of an air lock. All this is a thing of the past, due to the great flexibility now offered by the pump.

THE SMALL-BORE SYSTEM

This is a special case of a single-pipe system developed for residential property from tentative experiments made twenty-five years ago in commercial premises but taking advantage of the later development of a silent submerged rotor pump. Fig. 8.13 is a diagram of a typical layout; in this case two single-pipe loops are in parallel, which is necessary where the number of radiators becomes greater than can be carried on one loop, or there may be three or four loops. Also, it will be noted that an indirect hot-water supply cylinder is served by gravity circulation from the same boiler, Fig. 8.14.

Use is made of small-bore copper piping, usually 15 mm nominal bore for the loops and 20 mm or over where serving several loops, thus enabling it to be unobtrusively fitted into new or existing houses. Radiators are served by tees from the single pipe as shown.

Pipe sizing is a matter of assessing the maximum heat output which can be carried by a 15 mm loop, devising further loops as necessary. Piping common to the multiple circuits is then sized to carry the total load.

As copper piping is not covered by Fig. 8.6 (page 164), reference may be made to Fig. 8.15 which, as before, is based on *Guide* data but in terms of kW per kelvin temperature drop. A temperature drop of 10 K is commonly adopted. Water velocity is limited by what is permissible without creating noise, usually considered to be about 1 m/s. Single resistances are dealt with as already stated, the values for equivalent length being given in Fig. 8.15.

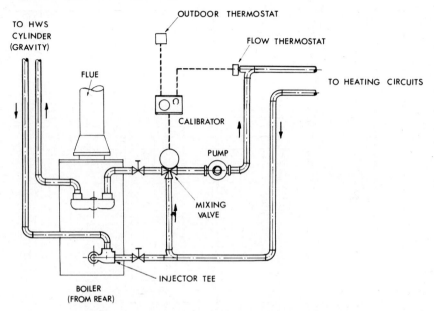

FIG. 8.14.—Small-bore system: pump and mixing arrangement.

THE MICRO-BORE OR MINI-BORE SYSTEM

In this system, the possibilities of small-bore piping are further exploited with water velocities up to 2 m/s and sometimes more. Each radiator, convector or other form of heat emitter receives a separate flow and return from headers or manifolds strategically placed, as in Fig. 8.16 (a). (See also Plate XV, facing page 183.)

It is found under these conditions that copper piping no more than 6 mm outside diameter may often suffice or, for larger duties, 8, 10 or 12 mm outside diameter. The common header or manifold may be 20 mm, or as required, and a number of types of prefabricated unit are currently available, as shown in Fig. 8.17.

Radiators or other emitting elements may be connected with two small-bore valves as normally, or use may be made of a special single valve carrying an extension pipe which runs inside the waterway to the far end, the valve having two connections—one for flow and one for return. In one derivative of this type of system, a single-pipe circuit is used as shown in Fig. 8.16 (b). Here a proprietary type of valve is used and fixed at the bottom centre of the radiator to act as a divertor or a by-pass according to setting.

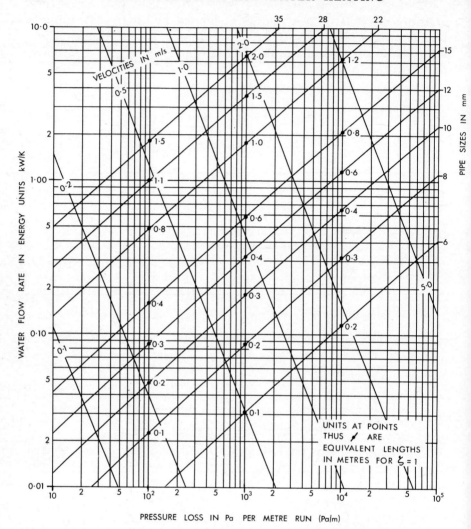

Fig. 8.15.—Pipe sizing chart for small copper pipes to B.S. 2871, Table Y.

Pipe sizes being so small, heat losses therefrom can be ignored. A temperature drop of 10 K may be assumed.

Where the piping is to be of copper, Fig. 8.15 gives the necessary data. Pressure loss in the header or manifold may be 2·5 to 5·0 kPa: if of proprietary make, the makers' information should be available.

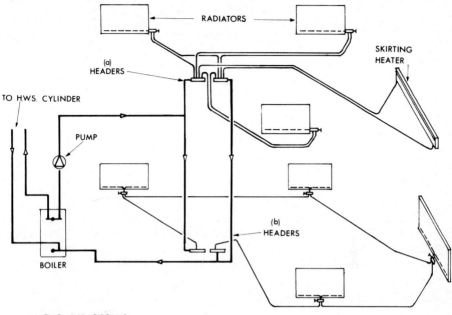

(a) TWO PIPE CIRCUITS
(b) SINGLE PIPE CIRCUIT

Fɪɢ. 8.16.—Diagrams of mini-bore systems: note that cold water feed, expansion pipe and hot water supply indirect cylinder are not shown.

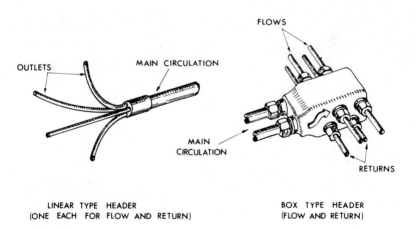

LINEAR TYPE HEADER
(ONE EACH FOR FLOW AND RETURN)

BOX TYPE HEADER
(FLOW AND RETURN)

Fɪɢ. 8.17.—Alternative types of mini-bore connection header.

CHAPTER 9

Hot-Water Heating Auxiliaries

IN THE PRESENT CHAPTER, the various accessories associated with a hot-water heating system will be described.

The general use of pumped circulation has already been mentioned, and this is the first item to be considered. *The pump* has now become a vital and essential part of the installation and with small-bore and micro-bore heating systems for houses, this applies even to the smallest installations.

It is then necessary to consider how the system is filled with water and what happens to the water when it expands due to heating.

Next, it will be appreciated that the system itself becomes larger as its temperature is raised, and provision for expansion of piping has to be carefully arranged and designed for.

The control of the heating system is of the greatest importance both from the point of view of comfort of the users and economy of fuel consumption.

Another matter to be considered is the safety of the system under a variety of fault conditions, and the various fittings which are mounted on boilers for checking their operation.

For small installations, the pump, boiler, burner and various accessories are now to be obtained in one combined packaged unit. The tendency in larger installations is also in the same direction, but whether or not this is the case, the designer will always want to know the basis on which the capacities and sizes have been determined.

This chapter attempts to deal with these problems.

PUMP CIRCULATION—CAPACITY OF PUMP

The capacity of the pump is determined from the total kW emission of the system. Pump duties are measured by volume per unit time which, in SI units, is litres per second. The mass flow of water per second is thus energy in kW divided by specific-heat capacity × system-temperature drop. Taking specific-heat capacity of water as 4·2 kJ/kg K and 1 kg of water = 1 litre,

$$\text{litre/s} = \frac{\text{kW}}{4\cdot2 \times \text{K}}$$

To limit the pump to this duty would mean that every circuit and radiator is expected to take exactly its correct volume of water and no more. This is obviously impracticable, and a margin to allow for the hazard of short circuiting is desirable of the order of 20 to 30 per cent.

In the case of single-pipe or ladder systems where the water is not sub-

Plate XII. Continuous convector forming barrier rail in front of tall glazing (see p. 141)

Plate XIII. Metal plate acoustic ceiling showing grid pipes and insulation (p. 147)

Plate XIV. Floor panel heating. Copper pipes are continuous without buried joints (see p. 140

Plate XV. Micro-bore piping before floor screed is laid. Note headers arranged for access vi *floor trap (see p. 179)*

divided, or is split into some three or four circuits only, 10 per cent margin would be sufficient.

Taking as an example:

Total emission of radiators and mains - 1000 kW
Temperature drop - - - - 15 K

$$\text{Litre/s} = \frac{1000}{4 \cdot 2 \times 15} = 16$$

A pump of, say, 20 litre/s would be suitable. The pressure drop would be that previously determined for the pipe sizing plus pressure loss in boiler and in heat emitter on the index run.

Taking a total pressure loss of 30 kPa, the power delivered by the pump to the water will be

$$20 \times 30 = 600 \text{ W}$$

Pump efficiencies vary between 50 per cent and 75 per cent. If the former, the power absorbed at the pump shaft would be

$$600 \times \frac{100}{50} = 1200 \text{ W} = 1 \cdot 2 \text{ kW}$$

The electric motor for such a drive would require to have a margin over this to allow for the possibility of error in the estimation of the head, insofar as this would affect the flow quantity in practice. Probably a motor of 1·5 kW output would be provided.

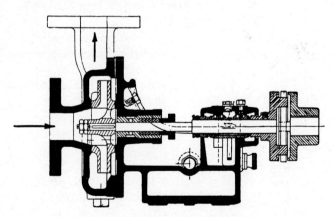

Fig. 9.1.—End-suction type centrifugal pump.

CIRCULATING PUMPS

Centrifugal Pumps—The centrifugal type of pump is most suitable for the purpose. This consists of an impeller rotating in a fixed casing of volute shape. The impeller may be of the end suction type, as in Fig. 9.1, although a split casing type is sometimes preferred, in which the top half of the casing is removable for inspection.

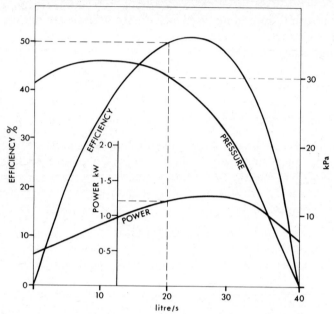

FIG. 9.2.—Typical characteristic curves for centrifugal pump.

Centrifugal pumps of either type are more suitable for heating circulations than other types of pump since they are very conveniently and simply driven by electric motors, are low in cost, require little maintenance and may be made silent in running.

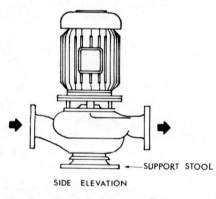

SIDE ELEVATION

FIG. 9.3.—Vertical spindle centrifugal pump.

The characteristic curves of a pump such as that in the example are shown in Fig. 9.2. It will be noted that the selected duty is near the point of maximum efficiency, and that if the pressure varies from that calculated so does the performance in terms of volume. As the pressure reduces, power increases to a limiting point, after which it falls away to a minimum on free discharge. Motors are usually selected to cover the power limit, thus being 'non-overloading'.

Pump Arrangement—The simplest form of drive is direct where the horizontal pump shaft is connected to that of an electric motor via a flexible coupling, both pump and motor being mounted on a common bedplate. The coupling is incorporated to allow for possible misalignment and to facilitate dismantling for maintenance. Another arrangement is with a vertical spindle, the pump casing being specially designed to incorporate the base mounting and to carry the motor above it. This has the merit of saving floor space, as shown in Fig. 9.3.

An alternative method of drive is one in which the motor is separate and coupled to the pump shaft via pulleys and V-belts. This permits the motor speed and pump speed to be independent of one another, which is often an advantage. In this case, when commissioning a system, a simple change of pulleys and of V-belts can be made and the pump capacity adjusted quickly without recourse to draining down, fitting an alternative impellor and refilling. A pump of this type is as Fig. 9.4.

For larger installations it is common practice to fit pumps in pairs, each with valves on suction and delivery, one pump acting as standby to the other.

Automatic By-pass—This fitting was common in the days when systems were designed for gravity circulation, a pump being fitted as an

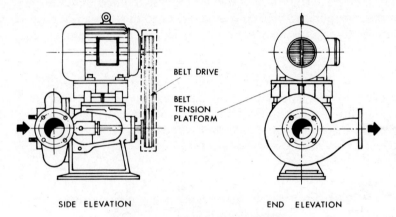

SIDE ELEVATION   END ELEVATION

FIG. 9.4.—Belt driven centrifugal pump.

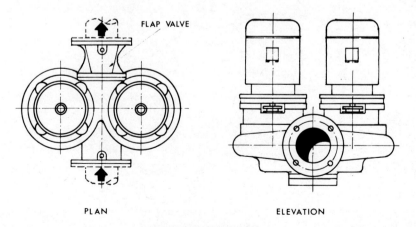

PLAN   ELEVATION

FIG. 9.5.—Twin pump set.

'accelerator'. With the pump off during the night, for instance, or in emergency, gravity circulation would take over via the by-pass which was, in effect, a non-return valve. With systems designed wholly for pump circulation and with automatic firing of boilers, present practice is to omit the by-pass.

Self-Contained Pumps—This is a class of equipment incorporating in one unit duplicate pumps and non-return valves and in some cases including isolating valves to the pumps. A vertical spindle type (Fig. 9.5) is so designed as to have only one suction and one delivery connection, a non-return valve being incorporated in the casing to prevent short-circuiting. The merit of these self-contained pumps is compactness and convenience in operation.

Submerged Rotor Pumps—Pumps specially designed for heating systems are produced in which the rotor of the driving motor is contained within the circulating water. The rotor is surrounded by a non-magnetic water barrier and the field coils are exterior to this. Rotating parts require no lubrication, and there is no gland, so that this type of pump is put forward as requiring no maintenance of any kind, and is generally silent. One make of this type is illustrated in Fig. 9.6. It will be noted that it is mounted in the pipeline, and requires no base or other fixing. It is suitable only for alternating current and the driving spindle should always be fitted horizontally. In domestic ranges such pumps are produced with crude integral means for control of capacity.

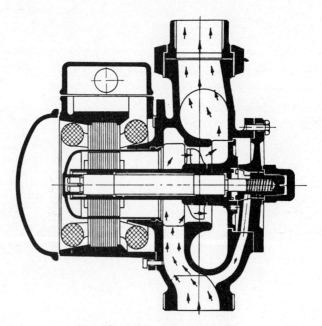

FIG. 9.6.—Submerged rotor pump.

Pump mountings—Each pump, or pair of pumps where in duplicate, should be provided with pressure gauges, or at the least with pressure tapping points, such that suction and delivery readings may be observed. Such readings, in conjunction with characteristic curves (Fig. 9.2) enable pump performance to be checked.

Self-Accelerated Systems—Mention should perhaps here be made of accelerated systems of the past in which a pump was not employed. Most of these depended for their operation on steam generated by the heating boiler, or independently, and have fallen into disuse because of the simplicity and cheapness of the centrifugal pump, and the wide availability of electric current.

DUAL TEMPERATURE SYSTEMS

It is sometimes necessary to supply water at two different temperatures from the same boiler plant, such as when a constant temperature circuit is

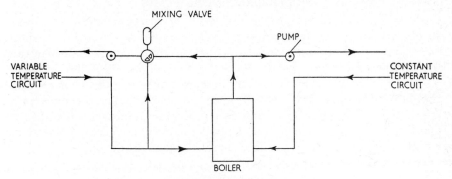

FIG. 9.7.—Diagram of dual temperature circuits.

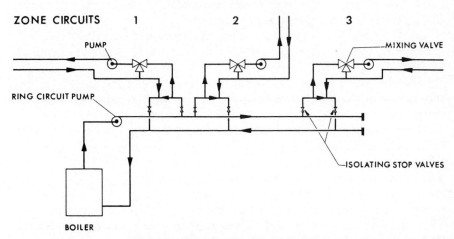

FIG. 9.8.—Ring circuit and zone controls.

desired for fan convectors, unit heaters and possibly hot water supply and, in addition, a radiator heating system is also to be served but with a variable water temperature to suit the weather.

This can conveniently be achieved by a mixing arrangement whereby part of the return water from the variable-temperature circuit is re-circulated direct into the flow without passing through the boiler. The amount of water re-circulated may be controlled by hand, or by thermostatically-operated mixing valve. See (Fig. 9.7.)

The same method of variable temperature flow may be extended to a number of circuits by using a ring circuit for boiler circulation, maintained by a separate pump or pumps, and each circuit with its own pump and mixing arrangements connected by means of 'accept-reject' connections (see Fig. 9.8). The capacity of the ring-circuit pump must of course exceed the total of the zone-circuit pumps.

BOILER-CIRCULATING PUMPS

In order to avoid the corrosive effects of oil firing, referred to on page 76 due to low water temperatures, boiler-circulating pumps are being increasingly used and are standard fittings on some makes of boiler in the larger range of sizes.

A boiler-circulating pump acts independently of other pumps and circulations. It draws hot water from the top of the boiler and delivers it into the cooler bottom, so raising its temperature. Another advantage is that it ensures some circulation in the boiler in the case where mixing valves serving the heating system are liable to shut-off against the boiler.

GENERAL

Feed and Expansion Tanks—The water in a heating system expands on being heated, and the purpose of the feed and expansion tank is to receive this water when the system is hot and return it when it cools down. In so doing the water in the system is not changed, and encrustation or cor-rosion which might otherwise occur with a constantly changing supply is avoided.

For the same reason it is inadvisable to empty the system in the summer or to change the water at all, except when repairs or alterations call for it.

The expansion of water from 7° C to 100° C is one twenty-third of its volume at the initial temperature. In order to determine the size of expan-sion tank for a certain system it is therefore necessary to estimate the total water contents. For boilers and radiators, makers' catalogues may be con-sulted. As an approximation is all that is necessary, however, Table 9.1 will be found to give a fair average. This table also gives the contents of piping.

In addition, about 100 mm of water is necessary permanently in the tank to float the ball valve, and a fair margin of space above, before the overflow is reached. This will call for a tank capacity about double that estimated.

The method of connection of the tank by means of a 'feed and expansion pipe' to the system requires some consideration. In a gravity system the feed and expansion pipe simply connects into the boiler return; a vent off the boiler or from the system is turned over the tank and there is no problem.

With pump circulation, which is now usual, there are three possible alternatives (see Fig. 9.9):

A pump in return, feed and expansion pipe into suction;
B pump in return, feed and expansion pipe into delivery;
C pump in flow, feed and expansion pipe into boiler return.

In case A, the vent pipe must be carried above the tank water line to a height exceeding the pump head to prevent water discharging. With high pump heads this is often impossible.

In case B, the vent pipe is not subject to the pump head and no extra height is necessary, but there is a tendency to draw air in at the air cocks on top-floor radiators so that, in order to release air, the pump must be stopped.

In case C, with pump in flow, the feed and vent are in balance and, the

TABLE 9.1

CONTENTS OF HEATING APPARATUS

(a) Boilers, cast-iron sectional type - - - - 1·5 to 2 litre/kW
(b) Boiler, steel types—vary greatly - - - - 0·5 to 3 litre/kW
(c) Radiators, steel panel types - - - - 5 litre/kW
(d) Radiators, steel column types - - - - 10 to 15 litre/kW
(e) Piping, steel to B.S. 1387 (medium grade) and copper to B.S. 2871 (average of Tables X, Y and Z):

Nom. Bore mm	Litre per m run	
	Steel	Copper
10	0·12	—
12	—	0·09
15	0·21	0·14
20	0·37	0·32
25	0·59	0·54
32	1·02	0·84
40	1·38	1·24
50	2·21	2·07
65	3·70	3·22
80	5·12	4·20
100	8·68	8·61
125	13·25	13·35
150	18·95	12·72

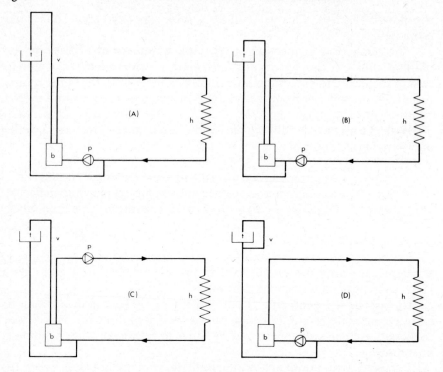

FIG. 9.9.—Alternative feed and vent arrangements: b, boiler; p, pump; h, heating system; t, feed and expansion tank; v, open vent.

heating system being under pressure from the pump, air release presents no difficulty.

In no circumstances should the feed and expansion pipe be combined with the system vent pipe as case D. This practice was shown to be dangerous more than 50 years ago.

The feed and expansion tank is usually made of galvanized iron or, in small systems, of plastic, and is provided with a ball-cock connected to the cold-water supply. The lever of the ball-cock is bent so as to keep the ball near to the bottom, and the valve itself is above overflow level. The overflow should be of large dimensions such as 32 mm minimum for small systems and 100 mm or more for large ones. If in an exposed position, or in a cold roof-space, the tank and its connections should be protected against frost.

Expansion Taken in Pressure Cylinder—The open expansion tank suffers, particularly in small installations, from being out of sight and out of mind, and not infrequently is found to have run dry due to the ball valve having stuck. Also it is frequently placed in a position liable to frost, and so constitutes a danger should it become frozen with the boiler being fired. The use of a pressure cylinder removes this distant object from the scene altogether; the pressure cylinder may be placed alongside the boiler, and thus be under constant observation.

Fig. 9.10, shows the arrangement of such a system and it comprises a closed cylinder containing nitrogen above a neoprene diaphragm. As the water expands the gas is compressed and pressure rises. A safety valve prevents undue pressure being reached. Proprietary forms of this equipment of continental origin exist which include a special form of safety valve, pressure gauge and non-return valve for filling. The size of vessel is dictated by the amount of expansion water to be accommodated within the limits of rise and fall of the diaphragm. (Plate XVI facing p. 214.)

In small domestic systems the filling may be checked annually using a removable hose from the water main attached to the non-return valve.

For larger systems a filling unit is provided comprising a small tank with ball-cock and an electrically driven pump as in Fig. 9.10. A pressure switch connected to the system brings on the pump to maintain a minimum water pressure sufficient to ensure that the highest point is filled. A second pressure switch may be arranged to cut off the firing unit on over-pressure before the safety valve operates.

This kind of system is suitable only for automatically-fired boilers, not for hand-fired solid-fuel boilers where undue temperature swings may occur.

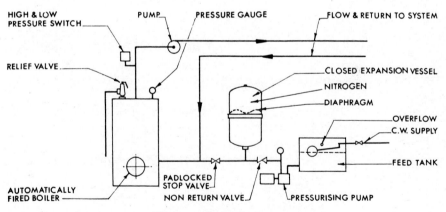

FIG. 9.10.—Pressurized system.

Insulation of Pipes—Piping is insulated to reduce heat loss to a minimum, where such heat would be wasted or a nuisance. This applies, for example, particularly to main piping run in trenches or ducts, and through basements or other unheated spaces. Table 7.3 (p. 151) gives heat losses from pipes insulated with glass fibre, expressed as a percentage of bare-pipe loss. It will be seen that the larger the pipe, the less the percentage loss for a given thickness of insulation—due to the increase in outer radiating surface being proportionately greater for small pipes than large.

Conductivities of insulating materials may be compared in Table 9.2, the values being applicable to the range of temperature used in low-pressure systems.

As to the materials themselves, dry or sectional laggings avoid wet trades on site. All materials may be finished in a variety of ways with enamels, fabric, bituminous paint, metallic sheathing, etc., according to position and cost.

TABLE 9.2
THERMAL CONDUCTIVITIES OF PIPE INSULATING MATERIALS
At Pipe-Face Temperature 60° to 90° C

Material	W/m K
Calcium Silicate	0·068
Glass Fibre	
(rigid sections)	0·042
Mineral Wool	0·041

Expansion of Pipes—From Table 1.2 (page 9), it will be noted that the coefficient of linear expansion of steel is 11·3 × 10⁻⁶ per kelvin. Thus for a steel tube 100 metres long raised from 0° C to 100° C, the expansion will be roughly 0·11 m or 110 mm.

In the practical installation, flexibility in pipework is achieved by changes of direction and intentional offsets, but where long straight runs are involved flexible bellows of the type shown in Fig. 9.11 are introduced. For larger mains, *horseshoe loops* are sometimes used—for which, see makers' data.

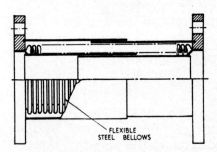

FIG. 9.11.—Bellows expansion joint.

HOT-WATER BOILER MOUNTINGS

Boiler mountings for a hot-water heating boiler comprise:

(a) Safety valve.   (c) Pressure gauge.
(b) Thermometer.   (d) Open vent.

(*a*) **Safety Valve**—The safety valve or relief valve is a device to relieve pressure should this rise above normal for any reason. Thus, if boiler valves have been shut and feed and vent perhaps frozen, the safety valve is the last resort to prevent an explosion should the boilers have been fired through an act of stupidity. Safety valves are either of dead-weight or spring type. Fig. 9.12 shows one form of the latter approved by insurance companies: it will be noted that the spring is enclosed and the setting cannot be altered. Sizes are given in Table 9.3 according to boiler rating (based on BS 779.1961).

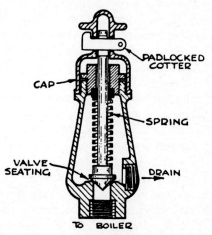

FIG. 9.12.—Spring loaded safety valve.

TABLE 9.3

RELIEF VALVE SIZES

	mm
Up to 270 kW	20
270 to 360 ,,	25
360 to 450 ,,	32
450 to 540 ,,	40
540 to 750 ,,	50

(*b*) **Thermometer**—The thermometer is a most important accessory to the boiler, and may be a plain mercury type in a brass case or one of dial pattern. The latter is desirable as being more easily read and, in many cases, more reliable.

In order that the thermometer (of whatever type) may be removed for repair, an immersion well is always provided (see Fig. 1.2, page 4). This is screwed into the boiler and forms a water-retaining joint. The well may be of steel or of brass; a steel well may be filled with mercury, but a brass well should be filled with oil for the reason that mercury might form an amalgam with the parent metal.

Thermometers are also useful on other parts of the system, such as after a mixing valve, and on individual circuit returns in large systems; they should assist in regulation—the main object of which is to ensure that all returns are at the same temperature.

(*c*) **Altitude Gauge**—An altitude gauge is simply a pressure gauge, and is calibrated so as to read in kPa (or bars), or alternatively in metres head of water. One bar equals nearly 10 m head of water. The gauge is often fitted with a red index hand which is set to the normal pressure in the system, and any variation from this up or down will at once indicate trouble which requires immediate investigation.

(*d*) **Draw-off Cock**—This item needs little comment and is often now fitted as a standard component of the boiler.

(*e*) **Open Vent Pipe**—This is simply a pipe taken from the top of the boiler (as in Fig. 9.9), being turned over the feed tank. In the event of valves being closed, preventing flow, or the boiler running on uncontrolled and causing steam to be generated, the worst that can happen is for steam and water to be discharged over the tank, the water returning to the boiler via the feed pipe.

Where more than one boiler occurs and to save the cost of a separate vent pipe from each, use is made of three-way 'safety' cocks as shown in Fig. 9.13. In the position shown, each boiler is connected to the common vent. If, however, the boiler valves are shut for draining down the boiler, the cock is turned to the opposite position, isolating the common vent pipe and opening the boiler connection to atmosphere.

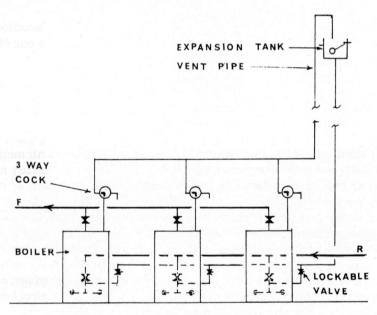

FIG. 9.13.—Open vent pipes for multi-boiler system.

Sizes of open vent pipes are given in Table 9.4 (from BS 779.1961 converted to SI). Open vent pipes are not of course required with closed pressure cylinder systems.

TABLE 9.4

SIZES OF OPEN VENT PIPES

kW	mm
45 to 150	25
150 to 300	32
300 to 600	40
over 600	65

WATER TREATMENT

In the newer forms of heating system, often using thin steel radiators, small-bore copper pipes and other metals, there is a danger of corrosion internally which can be overcome, however, by dosage of the initial filling of water with a suitable corrosion inhibitor.

With larger systems it is piously hoped that the water will not be changed; but for one reason or another, such as leakage from pump glands and alterations necessitating draining down, water is subject to change. The problem is probably one more of preventing scale formation than of

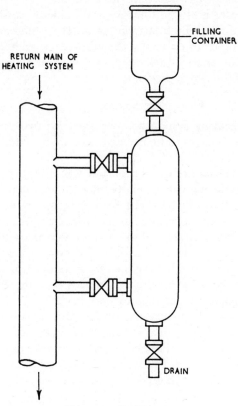

FILLING
CONTAINER

RETURN MAIN OF
HEATING SYSTEM

DRAIN

FIG. 9.14.—Dosing pot.

preventing corrosion, or it may be both. It is necessary, then, for the treatment to be repeated periodically and use may be made of a dosing pot for this purpose, as shown in Fig. 9.14. The usual addition is sodium metaphosphate, in order to raise the alkalinity, sometimes combined with tannin. The effect of such treatment is to destroy the scale-forming properties of calcium carbonate and other products of a hard water district by raising the pH value to about 10 or 11. One such chemical goes under the trade name of Fernox.*

The use of the dosing pot is as follows:

The chemical dose is placed in the top vessel. The valves connecting the pot to the system are closed, and the small cylinder is drained off. The chemical dose is then admitted to the small cylinder by the opening of the top valve. This is then closed, and the two valves connecting the cylinder to the system, usually in the return main, are opened, so that the liquid can pass into the general circulation. Such a method of treatment requires careful and periodical control in order to check the pH value, this being usually done by a simple indicator test which any ordinary boilerman can carry out. It is desirable to obtain advice in regard to such treatment and certain proprietary firms undertake this function.

Where the water is of a soft nature and liable to cause corrosion, there may equally be a need for treatment in order to remove the acidity and raise the pH value to 10 or 11, by the addition of an alkali in some form, so as to avoid corrosive attack of steel piping and particularly of steel radiators, should they form part of the system.

VALVES

Valves in a heating system are provided for two purposes: one is to enable parts of the system to be isolated, and the second, to enable regulation to be carried out.

These two purposes are quite separate and distinct, and it is indeed a fact that a good isolating valve usually makes a very poor regulating valve.

Isolating valves will be required on boilers, pumps and throughout the system on all main branches, and, in the case of a multi-storey building, at the base of each pair of risers and so on. Such valves are to enable a section of the installation to be isolated, drained down, repaired and altered as may be required without affecting the rest of the system. In order to enable the section or sections to be drained down, it is therefore necessary to provide emptying cocks at each pair of valves.

Valves for isolation on heating systems are usually of full-way gate type, which, when open, cause virtually no resistance to flow. Valves of 80 mm and over are usually flanged; smaller sizes are usually screwed.

Emitting equipment also requires to be provided with valves, partly for isolation and partly for regulation, since, if regulation is necessary, it is

* Industrial (Anti Corrosion) Services, Waltham Cross.

much more easily carried out at the terminal ends than in a general way on mains.

Radiators, convectors, ceiling panels, unit heaters and the like on any extensive installation are usually provided with two valves, one in the flow and one in the return, so that each piece of apparatus may be disconnected and removed, if required, without affecting other parts of the system. At the same time, for purposes of regulation, usually the valve in the return is fitted with a lockshield, whereby the valve can be adjusted by a key, and this valve is then called upon to serve the double purpose of isolation and regulation. The ordinary screw-down valve is virtually useless for regulation, as it is far too critical. A curve of flow against amount of opening for this type of valve is of the form indicated in Fig. 9.15. Various attempts have been made to devise a valve form which will give nearer to a straight line, and the acorn disc nearly achieves this object. The amount of heat emitted, however, is dependent on the overall temperature drop of the water and will be greater the smaller the flow. The ideal valve curve for straight line heat-output is shown dotted.

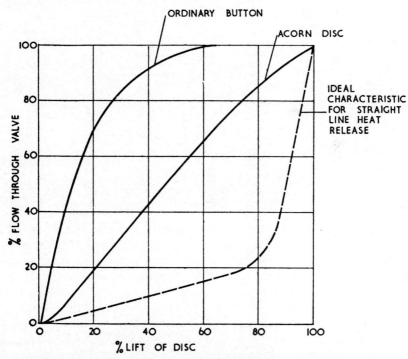

FIG. 9.15.—Valve characteristics.

Where such a valve is used for regulation, there is always the danger, when it is used for isolation purposes, that it will not be reset back to its original setting after it has been shut off for some reason. To overcome this disadvantage, various designs of double regulating valve have been devised which can be used for isolation, but which, on being opened up, can do so no further than by a given amount.

In order that the process of regulation, referred to previously on p. 174, may be effected with precision it is becoming more usual in good practice to fit each principal sub-circuit with a special type of balancing valve as shown in Fig. 9.16. Such valves have known pressure loss/water flow characteristics, and details are published by the manufacturers in chart or table form. As the heating system is commissioned, a test instrument is connected across the two special tappings on the valve and from the readings taken and the published data, a conventional double regulating valve fitted in series is adjusted such that the required water flow rate is obtained.

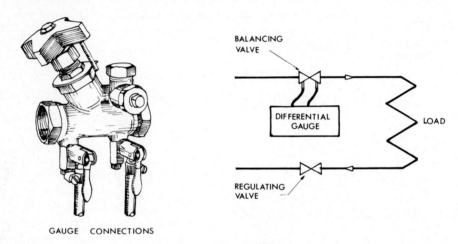

GAUGE   CONNECTIONS

Fig. 9.16.—Special balancing valve.

What has been said above refers mainly to two-pipe installations. In the case of a single-pipe system such as a small bore installation for a house, a single valve on a radiator may be all that is necessary, since, with all radiators virtually in series, no question of regulation as between one and another arises. The only purpose of a valve on a radiator in such case is so that the heat can be shut off or turned on as required, and there are now a number of neat forms of such valves incorporated in the radiator unit as a standard. If in such an installation there is need to remove a radiator for painting or repair, it becomes a matter of emptying down the circuit complete, which is of no great disadvantage seeing that the water content is so small.

In the micro-bore system there is much to be said for the single connection valve to radiators, etc., for the two-pipe entry as referred to on p. 179.

CONTROL OF BOILERS AND HEATING SYSTEMS

A separate book could be written about control systems, as they have tended to become more and more sophisticated as time goes by.

The elements of control are simple enough; first, there is control of the heat producing equipment, such as a boiler, in order that it may deliver just so much heat as is required by the heating system and no more; second, there is control of the heat emitters in the various rooms of a building or in the different parts of a factory, so as to ensure that the required temperature is achieved without overheating, which might cause discomfort and in any event would waste fuel.

Whereas in the past, with a hand-fired coke boiler, a crude form of damper-regulator sufficed to control the air admitted below the grate for combustion, this meant that the boiler often ambled along at much below its rated output. This is bad for the boiler, particularly with oil-firing or gas-firing, due to the likelihood of condensation being caused under such conditions with corrosive effect on the boiler itself.

It is now considered better practice to control the boiler in such a manner that it runs at a constant high temperature, so as to reduce the tendency for condensation to occur, and then to mix the return water with the flow, by means of a mixing valve, for the purpose of serving the heating system. Where heat at a constant temperature is required from the boiler, such as hot-water supply, or for heat for ventilation air, or for forced convectors, unit heaters and the like, this would be served direct from the boiler with a separate pump, as indicated in Fig. 9.7 (p. 187).

The control of the boiler may also be under the influence of a time switch, by means of which the firing is started up at some predetermined time in the morning and shut off at night. Alternatively, it may be considered preferable simply to cut the temperature down at night; which involves a second thermostat, and the clock switch, in effect, changes over from one stat to the other. Time switches may have special features to cut out heating at weekends, or for a certain period in the middle of each day.

Control by Weather—A general control of temperature of water supplied to the heating system may be by means of a compensator system, one version of which is shown diagrammatically in Fig. 9.17. This comprises an external temperature-sensing element, and a similar temperature-sensing element in the flow main to the heating system. By means of a control box, in cold weather a balance of the circuit is achieved only by the element in the water flow being at a high temperature; similarly, in warm weather a balance is achieved with the water at a lower temperature. At any point between the two extremes the water temperature is adjusted proportionately to the outdoor temperature. Refinements have been introduced into this system by way of a heating element in the external unit, so that the effect of wind is taken into account. This type of compensator control may be used to control directly an oil burner or a gas-fired boiler;

but if, as stated above, and as is preferable, the boiler is run at a constant temperature, the compensator system will then control the mixing valve. If there are a number of main circuits to various parts of the building, it is possible for each one to have its own compensator and circulating pump with mixing valve.

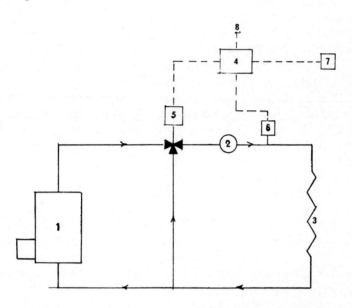

1. Boiler-burner unit.	5. Modulating motor to mixing valve.
2. Pump.	6. Water-temperature detector.
3. Heating system.	7. Outside temperature detector.
4. Compensator control box.	8. Electric supply.

FIG. 9.17.—Diagram of compensator control system.

Zone Control—Control of the heating system for a building or series of buildings by zone control, is based, in effect, on the assumption that certain areas are thought to have common characteristics and hence similar heat requirements. For instance, in a rectangular building with north and south aspects on the long face, it is probable that rooms on the south aspect will often require less heat than on the north, and therefore might well be served off a different zone with a different mixing valve, as referred to above.

In a tall building, likewise, due to the greater exposure of the upper floors, it might well be desirable to zone the upper half of the building separately from the lower half. Again, if there are two or more main elevations, this might involve four or more zones.

Individual Room Control—The greatest refinement is to be achieved by control of the heat emitter in each room. In buildings where some rooms are sparsely occupied and others crowded or with their own internal heat gains, it is probably only by individual room control that uniform temperatures can be maintained. The alternative is, of course, to assume that the occupants themselves shut off the heat when it is not wanted and turn it on again later. In practice, however, this seldom seems to happen, but, rather, the window is opened if the room becomes too hot, with obvious wastage of heat.

Individual room control appears to be particularly necessary in ceiling-heated systems, where the ceiling is one of the forms of suspended metal trays.

A combination of compensator control and individual room control has considerable merit, as the thermostatic equipment in the various rooms then has less work to do and there would be much less tendency for fluctuations of temperature to occur.

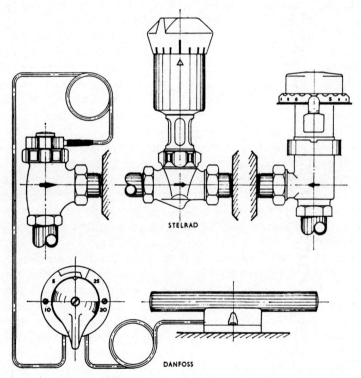

FIG. 9.18.—Thermostatic control valves for radiators etc.

Control Systems—The simplest form of controller is direct-acting, comprising a sensing element in the room or in the water flow, and, by liquid expansion or vapour pressure through a capillary, power is transmitted to a bellows or diaphragm operating a valve spindle. The valve spindle may be a mixing valve, mixing water from the boiler and return water on the suction or delivery of a pump, as referred to previously, or the valve may be in the supply pipe to a radiator, convector or other heat emitter (as in Fig. 9.18), so serving the purpose of an individual room controller.

Direct-acting thermostats have little power, and their control band is sometimes somewhat wide, but they have been considerably improved of recent years and are applied extensively, particularly to the smaller systems. Direct-acting thermostatic equipment may be said to be modulating or gradual-acting. An application to a domestic small-bore system is shown in Fig. 8.14 on page 179.

For larger installations, electrical systems of control are by far the most common, and can be devised to perform any of the functions desired in a great variety of ways. The simplest electrical system comprises an on-off thermostat connected to an electrically-operated valve or damper, such that, when the thermostat calls for heat, the valve or damper is opened and, likewise having become satisfied, the valve or damper is closed. An on-off thermostat may also be used to control the stopping and starting of a motor driving a fan, as in the case of forced convectors and unit heaters. This type of control also appears to be satisfactory for controlling ceiling heating. Indeed, it is probably true to say that most heat emitters in rooms may be satisfactorily controlled on an on-off method, particularly if the heat supplied through the water has been adjusted according to weather.

Electrical modulating controls enable, for instance, a mixing valve to be so adjusted automatically that it finds some mid-position and so supplies the desired flow-temperature to the heating system. For example, for a quick heat-up in the morning, the control system may be so arranged as to call for full heat for the first hour or so according to external weather, after which the temperature will be dropped proportionately to the external temperature.

A recent re-development, the 'optimum start' controller, has been installed as an energy conserving device in many Government and other buildings. In brief, this controller is a refinement of a compensator system which, in conjunction with an external thermostat and a timing mechanism, selects both the time prior to occupation and the appropriate system water temperature for a pre-heating period after night time shut-down. The maximum advantage, commensurate with daytime comfort conditions, is thus taken of fuel savings due to intermittent plant use. It has been estimated that a reduction of 30 per cent. of annual boiler fuel consumption results from use of this system in a suitable building but it seems more than probable that much of this theoretical saving has resulted from belated overhaul of the previously installed control systems.

Furthermore, buildings vary in their response to heating according to their construction and exposure: at medium external temperatures, some may require higher water temperatures than others. This kind of adjustment or compensation can be made in the control box.

Electrical control systems now make use of transistors and thermistors which, with electronic circuits, eliminate mechanical moving parts to a large extent, and hence tend to make for greater reliability.

Controls using compressed air as a means of actuating valves and dampers are also available, but are more used in air-conditioning and industrial applications, so need not be considered here.

Control Panels—Where a comprehensive system of controls is to be provided, it is advantageous to locate the whole of the equipment at one point on a control panel, which can be pre-wired, and greatly facilitates installation. On the same panel would be mounted controls to boilers and starters for pumps, etc., together with any time-switching equipment, temperature-indicating dials, pressure indicators, if required, and combustion control equipment such as CO_2 instruments, flue-gas thermometers, and so on. Plate XVII, facing page 214, illustrates a control panel of this type.

CHAPTER 10

Heating by Steam

STEAM AS A MEDIUM FOR HEATING in radiators and the like is a thing of the past. Hot water, with its flexibility to meet variable weather conditions and its simplicity, has supplanted it in all new residential, commercial and public buildings.

Steam is, however, often used for the heating of industrial buildings where steam-raising plant occurs for process or other purposes. It is also used as a primary conveyor of heat to calorifiers such as in hospitals, where again steam-boiler plant may be required for sundry duties such as in kitchens, laundry and for sterilising. Heating is then by hot water served from the calorifiers, further sets providing hot-water supply.

The present Chapter will mainly be confined to the principles involved in the generation of steam and to its utilisation for the purposes mentioned.

Generation of Steam—It will be as well to recapitulate here what was stated briefly in Chapter 1: namely that steam is produced when heat is applied to water in a partially filled closed vessel; and that, when boiling point is reached (100° C at atmospheric pressure), the addition of further heat causes a change of state to occur from water to steam. The quantity of heat involved in the process—the *latent heat of evaporation* (see page 14)— is considerable: 2258 kJ/kg at atmospheric pressure, compared with 420 to raise it to boiling point from 0° C.

In a closed vessel, the steam has no means of escape and the addition of further heat causes the pressure to rise. Means for preventing the rise from being above the strength of the vessel are provided by safety valves. As the pressure rises so does the temperature. Thus, this further heat entering the water is in the form of sensible heat in the liquid and steam, the latent heat falling as the pressure rises.

Steam in contact with water is termed *saturated*. If it carries some water droplets in suspension, it is referred to as *wet*.

If steam is removed from the vessel in which it is generated and further heat is added out of contact with water, the steam is said to be *superheated*. In this condition it tends to behave as a dry gas and is of little use for heating until it has cooled and become saturated.

The utilisation of steam for heating involves the process of condensation, in which the latent heat is removed by the heat-emitting surfaces of the heating system and reverts to water at the same temperature as the steam. This hot water or *condense* must be removed as soon as it is formed, or the heating apparatus will become water-logged and useless. The condense is, however, under pressure and, as it is released to atmospheric pressure, it suffers a reduction in temperature. In effect, part of the heat in the condense above atmospheric boiling point goes to re-evaporate a proportion of the liquid into steam at the lower pressure, this being termed *flash-steam*. Use can be made of this in a variety of ways, but if it is not usefully employed it constitutes a loss, as also does the remaining heat in the condense if this is run to waste. Thus, in practice, condense is collected and returned to the boiler for re-use, which at the same time affords a supply of distilled water and saves on water consumption.

Unfortunately, condense also may carry with it uncondensible gases such as chlorine, CO_2 and O_2 which are liable to cause severe corrosion of condense lines.

STEAM TABLES

Tables of properties of saturated steam in SI units are to be found in the *Guide, Section C2*. Tables 10.1 and 10.2 at the end of this chapter give extracts from this data. Fig. 10.1 gives a graphical representation thereof. From this data, the following may be noted:

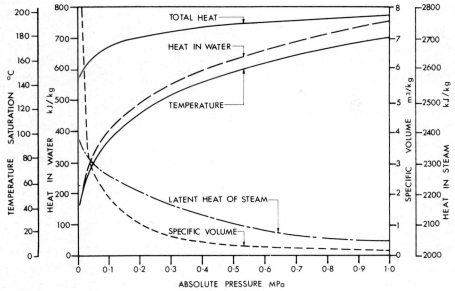

FIG. 10.1.—Graphical representation of properties of saturated steam.

Pressure. This is stated in absolute terms, the units being kilopascals (kPa). Atmospheric pressure very nearly equals 100 kPa (101·3 kPa = 1 bar); and thus, as pressure-gauge readings are relative to atmosphere, gauge pressures are equivalent to absolute pressures minus 100. Subatmospheric pressures are given in the same units, though it is well to recall that, in the past, vacuum has been referred to in *inches of mercury column*—a descendant of the mercury barometer (1 in. mercury = 3·386 kPa = 0·034 bar). Pressures over 1 MPa (10 bar) need not concern us for heating purposes.

Temperature. Atmospheric boiling point is taken as 100° C. The 'standard' atmosphere is 1·013 bar, and thus the boiling point at 100 kPa (1 bar) is less (99·63° C). Steam temperatures at sub-atmospheric pressures fall to levels comparable with hot water, which feature was made use of in the vacuum and vapour systems now obsolescent in this country. Steam at temperatures above atmospheric will be seen from Table 10.2 to advance to 180° C at 1 MPa (10 bar). The higher the temperature the greater the output from a given heating surface. On the other hand, the higher the pressure, the greater the potential heat losses from mains and from flash steam and condense. Furthermore, the higher the pressure the greater the cost of pipework and apparatus.

Enthalpy (or heat content). The next three columns give the heat content or *specific enthalpy* in terms of kJ/kg for the water, for the latent heat, and for the sum of the two (total heat of saturated vapour). The enthalpy of water is calculated from nought at 0° C. The decline in the latent heat with the increase of pressure will be noted; thus, per unit mass of steam, the lower the pressure the greater the ratio of latent heat to total heat, and consequently the less the loss from flash steam.

Volume. The last column quoted in Tables 10.1 and 10.2 give volumes in m³/kg, these being of use in pipe sizing where velocity of flow is used as a criterion. As would be expected, volume varies inversely with pressure from 1·69 m³/kg at an absolute pressure of 100 kPa (1 bar) to 0·19 m³/kg at 1 MPa absolute (10 bar).

Other values. The steam tables in the *Guide* include in addition the Prandtl number, which is the ratio of viscosity to thermal diffusivity and is a factor involved in heat-transfer problems under forced convection. It need not concern us here.

Another column in the *Guide* Tables gives values of specific-heat capacity for steam at each pressure. The value varies from 2·01 kJ/kg K at an absolute pressure of 100 kPa (roughly atmospheric pressure) to 2·62 at 1 MPa absolute.

Use of Steam Tables—By way of example, assume that a heat load of 1000 kW (= 1000 kJ/s) is to be served from a steam boiler at a distance.

The mass flow of steam involved is obtained by dividing the heat energy content by the latent heat, which in turn is dependent on pressure. Thus,

at 100 kPa (1 bar)   1000 ÷ 2258 = 0·44 kg/s
at 500 kPa (5 bar)   1000 ÷ 2109 = 0·47 kg/s
at   1 MPa (10 bar)   1000 ÷ 2015 = 0·50 kg/s

In terms of volume, these mass flows represent

at 100 kPa (1 bar) $0.44 \times 1.69 = 0.744 \ m^3/s$
at 500 kPa (5 bar) $0.47 \times 0.38 = 0.179 \ m^3/s$
at   1 MPa (10 bar) $0.50 \times 0.19 = 0.095 \ m^3/s$

Thus it is clear that the higher the pressure the smaller the pipe size to carry the load.

When the steam is condensed at the distant point to which it has been conveyed, assuming no losses on the way, we have as heat in the liquids

at 100 kPa (1 bar) $= 417 \ kJ/kg$
at 500 kPa (5 bar) $= 640 \ kJ/kg$
at 1 MPa (10 bar) $= 763 \ kJ/kg$

At an absolute pressure of 100 kPa (1 bar), being almost atmospheric, the condense issues as boiling water and may be conveyed back to the boiler as such.

At 500 kPa (5 bar), when released to atmosphere, $640 - 417 = 223 \ kJ/kg$ of heat re-evaporates part of the parent water (as explained previously);
   i.e. $223 \div 2109 = 0.106 \ kg/kg$ as flash steam.
Similarly, at 1 MPa (10 bar), $763 - 417 = 346 \ kJ/kg$ of heat;
   i.e. $346 \div 2015 = 0.172 \ kg/kg$ as flash steam.

Thus it is apparent that the higher the pressure the greater the potential loss due to flash steam: at atmospheric pressure, nil; at 500 kPa, 10·6 per cent; at 1 MPa, 17·2 per cent. In practice, the re-condensing of flash steam may be achieved in equipment for the purpose, or, if it is passed into return-condense lines within the building the loss can be turned into useful heating. Otherwise, the overall efficiency of the system in terms of heat input to heat usefully employed is reduced to the extent of the percentages above.

Entropy—This is a conception which is much used when considering steam used in a heat engine. For ordinary heating problems it is not important, and need only be considered very briefly.

Entropy does not correspond to any physical property of heat of which we can have direct knowledge. It is simply a mathematical ratio for expressing the availability of heat, and the entropy of any *closed* system involving heat-transfer increases as the transfer takes place, e.g. from steam at 150° C to secondary water at 10° in a calorifier, where finally we are left with condensate at, say, 95° and water at 80°. By no known method could the heat be taken out of the water and put back into the condensate to make steam (without adding external energy as in a heat pump), and this irreversible process takes place in the direction of an increase of entropy. Similarly, the expansion of steam through a throttling valve is an irreversible process which causes an increase in entropy of the steam.

Entropy may be depicted graphically by drawing a curve of the heat content of a closed system, the curve having ordinates of absolute temperature. If the area under the curve represents heat, then the abscissae represent entropy.

For the development of this subject the reader is referred to any of the standard works on Heat Engines.

Steam Boiler Evaporation Ratings—are commonly listed *from and at* *100° C.*

This denotes that if the water in the boiler is at atmospheric pressure the steam generated at that pressure will be the quantity given, i.e. it is assumed that only latent heat is added.

Such a rating, however, is of little practical use as it stands. The feed water is seldom exactly 100° C and the pressure is usually above atmospheric.

For any other set of conditions the *actual evaporation* can be derived as follows:

Actual evaporation

$$= \frac{\text{Evaporation from and at } 100° \text{ C} \times 2258}{\left\{\begin{array}{l}\text{Total heat at pressure} \\ \text{required}\end{array}\right\} - \left\{\begin{array}{l}\text{Heat in water at} \\ \text{feed-water temp.}\end{array}\right.} .$$

It will generally be found that the actual evaporation is less than the 'from and at' rating.

STEAM GENERATION IN PRACTICE

A whole bibliography exists on this subject but, within the limits of the future use of steam in heating projects, the scope of this chapter may

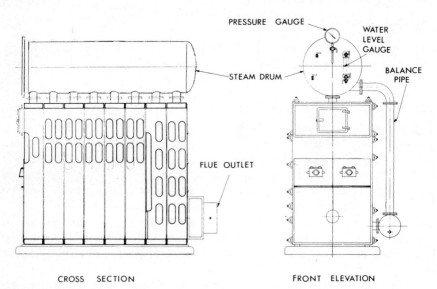

CROSS   SECTION   FRONT   ELEVATION

Fig. 10.2.—Steam boiler for low pressures.

be limited to considering firstly the question of pressure and secondly the
types of equipment available for this purpose.

It will be clear from the foregoing that high pressures are not necessarily desirable. Absolute pressures of 150 to 200 kPa (1·5 to 2 bar) may
suffice for a canteen with a steam-cooking load and for some heating in
the canteen itself. In such a case, the simplest form of steam boiler, as in
Fig. 10.2, would be possible. It will be noted that the steam separating
space is contrived by means of a drum external to the boiler. The limitation of this type of boiler as to absolute pressure is about 250 kPa (2·5 bar).

Higher pressures are dealt with by means of steel boilers, now generally
of packaged welded construction, automatically fired by solid fuel, oil or
gas. A boiler of this type is shown in Fig. 10.3 covering a range from
1 MW to 10 MW. Beyond this output, water-tube boilers are employed,
but these are not relevant in the present context.

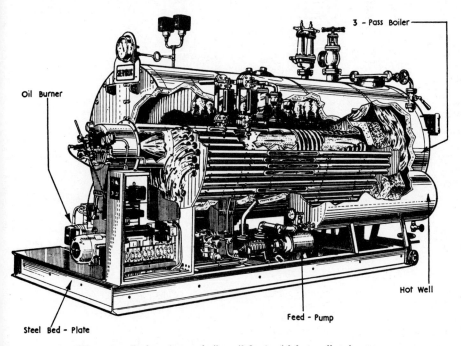

FIG. 10.3.—Packaged steam boiler, oil-fired, with hot-well and pump.

Boiler Pressure—The selection of the most economical boiler pressure
depends on a number of conflicting factors.

The lower the pressure the lower the temperature and, hence, losses
from piping surfaces, even if insulated, are less. But the volume of steam
is greater and larger mains are involved. Heat-emitting apparatus
requires larger surfaces than when using higher temperatures. Pressure-

retaining equipment (boilers, flanges, valves, pipework etc.) is less costly for low pressures than for high ones; but low-pressure steam, when conveyed over any considerable distance, becomes progressively wetter in losing temperature until a limit is reached where it is no more applicable.

Thus—except for small, compact systems—absolute steam pressures of the order of 400 kPa (4 bar) may be suitable, whilst for extensive runs of main absolute pressures of 700 kPa to 1 MPa (7 to 10 bar) are often used.

Pressure Reduction—It is often necessary to generate steam at one pressure but desirable to serve equipment at a lower pressure. For instance, sterilizing and laundry plant may require an absolute pressure of 700 kPa (7 bar), but, for reasons stated above, the rest of the system can be run more economically at some decreased pressure, which at the same time reduces losses.

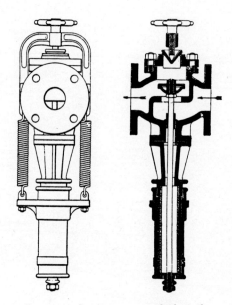

Fig. 10.4.—Steam-pressure reducing valve.

This is done by means of a pressure-reducing valve as shown in Fig. 10.4. In fact, if the steam entering such a valve is dry-saturated, it will be apparent that it becomes superheated when pressure is reduced. Some initial wetness in the steam and some losses following pressure reduction may, however, be enough to restore its condition to saturation.

The installation of a pressure reducing valve involves a number of auxiliary components, as shown in Fig. 10.5, the complete array being known as a 'pressure reducing set'.

Flash Steam Recovery—This has been mentioned already and it only remains to refer to means for taking advantage of it. In a calorifier system, a separate coil may be used to receive the flash steam. In a unit heater system, one or more of the units may be designed to receive flash steam.

Condense Return—Having been condensed in the heating equipment it is grossly wasteful to discharge condense to drain. It may represent as much as 20 per cent. on the fuel bill. Thus it is normally returned for re-use in the boiler, as explained already.

At the drain point of the heat emitting apparatus or calorifier, some means is required to allow water to pass but not steam. This device is termed a steam trap. Various types exist, and these fall into three broad classes, identified by the means adopted to distinguish and separate condensate from steam, as follows:

1. Mechanical, incorporating
   (*a*) Open top bucket
   (*b*) Inverted bucket
   (*c*) Ball float

2. Thermostatic, with various elements
   (*a*) Balanced pressure
   (*b*) Liquid expansion
   (*c*) Bi-metallic

3. Miscellaneous, including
   (*a*) Labyrinth
   (*b*) Thermodynamic
   (*c*) Impulse

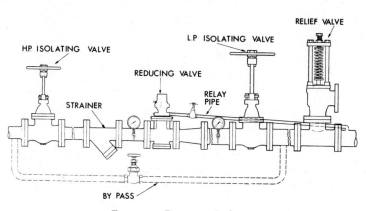

FIG. 10.5.—Pressure reducing set.

Some of these types are illustrated in Fig. 10.6 but manufacturers' literature should be consulted for specific details. As with pressure reducing valves, it is usual to add auxiliary components and a 'trap set'

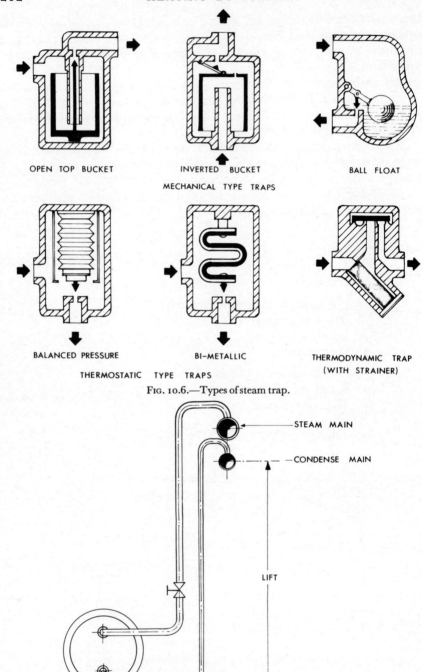

OPEN TOP BUCKET   INVERTED BUCKET   BALL FLOAT

MECHANICAL TYPE TRAPS

BALANCED PRESSURE   BI-METALLIC   THERMODYNAMIC TRAP (WITH STRAINER)

THERMOSTATIC TYPE TRAPS

FIG. 10.6.—Types of steam trap.

STEAM MAIN

CONDENSE MAIN

LIFT

TRAP SET

FIG. 10.7.—Lifting condensate to a higher level.

will commonly comprise a strainer in addition to the trap itself plus an outlet check valve, a sight glass and a drain branch valve as may be appropriate to the circumstances.

Choice of trap type depends upon a number of factors; load characteristics (constant or fluctuating); inlet and outlet pressures; associated thermostatic or other controls on the steam equipment; and the relative levels of trap and condense main piping to name but a few. Whilst it is not possible to generalise for all varieties of application, the following trap types are generally suitable for space heating equipment:

Radiators and natural convectors	Thermostatic balanced pressure type
Fan convectors	Mechanical ball float or inverted bucket type
Unit heaters (small)	Thermostatic balanced pressure type
Unit heaters (large)	Mechanical ball float or inverted bucket type
Plenum or air conditioning heater batteries	Mechanical ball float or inverted bucket type: may be multiple for large units
Storage and non storage steam/water calorifiers	Mechanical ball float or open bucket type
Drain points on steam mains	Mechanical open top or inverted bucket type or thermodynamic type
Oil storage tank coils or outflow heaters	Mechanical open top or inverted bucket types
Oil tracing lines	Thermostatic bi-metallic type or thermodynamic type

It is always preferable that a trap discharges into a condensate main run below it and that the main then falls in level back to the boiler plant. Often, however, a trap fitted at low level must discharge into a main above it and this is quite practicable *provided that* the steam pressure at the trap inlet is adequate to overcome the back pressure imposed by water in the vertical discharge pipe; in simple terms, an available pressure of 10 kPa is required for each metre height of vertical 'lift', as Fig. 10.7.

In cases where the steam-using equipment is fitted with thermostatic or other automatic control in the form of a throttling valve, it must be remembered that such a valve acts by causing a reduction of steam pressure to the equipment (and hence to the trap). Thus, although the initial steam supply to the control valve may be adequate to provide the necessary lift, at low loads this will not be the case.

It is good practice to avoid lifting condensate from any trap where this is possible and *always* to avoid a lift from a trap serving equipment fitted with automatic steam control.

For further information on steam traps and application techniques, the reader should consult various specialist texts.*

Air—One of the bugbears of steam is air. Its presence in heating equipment acts as a blanket preventing steam from condensing on the heating surfaces; this is particularly so at low pressures. Means for air removal are therefore essential. Types of steam trap with a thermostatic element allow air to pass through, due to the air causing a lowering of temperature. Certain types of float traps contain a separate thermostatic element for this purpose. Air is more dense than steam and collects at the lower portions of equipment. Thus, another fitment is an air eliminator which is connected near the bottom of the equipment and is also thermostatically controlled.

At higher pressures, air is usually purged by hand-opening of the blow-through cocks on mains and equipment, after which the steam traps may be expected to cope with it mixed with the condense.

By 'air', in this connection, is meant the mixture of uncondensable gases already alluded to.

Boiler Feeding—The simplest method now common is to provide, near the boiler, a collecting tank or 'hot-well' into which condense return is delivered. In addition, the tank receives cold water make-up through a ball valve from the mains, which water may in turn have passed through a water-treatment plant.

From the tank, water is pumped to the boiler through a non-return valve by means of an electrically-driven or steam-driven feed pump under the control of a water-level regulator attached to the boiler. This comprises a float in a vessel connected to the boiler at the water line, such that if the level drops the pump is started. When the level is restored the pump is stopped.

The condense return generally serves to keep the feed-water hot; but where this is not the case, a steam coil in the tank is used to pre-heat the feed so as to avoid the injection of cold water into the boiler, which is liable to cause thermal shock.

In the packaged boiler (see Fig. 10.3, page 209) the feeding equipment forms part and parcel of the unit: i.e; feed tank, pump and level control.

---

* Lyle, Sir Oliver, *The Efficient use of Steam*, HMSO, 1947.
Northcroft, L. G., *Steam Trapping and Air Venting*, Hutchinson, 1945.
Spirax-Sarco, Various excellent instructional texts.

Plate XVI. Pressure cylinders in boiler house serving a large country mansion (see p. 191)

Plate XVII. Composite control panel, with mimic diagrams, at the Imperial Cancer Research Fund Laboratories (see p. 203)

Plate XVIII (bottom). Heating and H.W.S. calorifiers at the new Royal Free Hospital (see p. 224)

Plate XIX. Steam and condense mains within gantry showing anchor points and expansion loops (see p. 228)

Plate XX. Circulating pumps, headers and mixing valves in H.P.H.W. boiler house (see p. 248)

Plate XXI. Radiant panels in two rows around heavy factory bay (see p. 254)

Steam Boiler Mountings—The essential fittings of any steam boiler are:

water gauges: to give a visual indication of water level in the boiler;
pressure gauge: to give indication of steam pressure;
safety valves: to limit the pressure to that for which the boiler is designed;
feed valve: for admittance of feed water, and incorporating a non-return valve;
steam stop-valve: to shut off the outlet;
blow-down valve: for draining-off and clearing sludge.

In addition, it is common practice to add:

high and low-water alarms: these give an audible alarm on the levels rising or falling below safe limits; at the same time, with automatic firing, the burner may be cut-off. They are usually electrical;
pressure control: this acts in the same way as a thermostat on a hot-water boiler by simply opening contacts when the set pressure is reached, so shutting off the firing equipment. On fall of pressure below a minimum setting, the reverse occurs. For duties over about 300 kW, high-low or modulating control is used (as explained in Chapter 5).

EMISSION FROM STEAM PIPING

Piping Systems—It is unlikely that exposed steam piping will be used as a heating surface. It was a common and cheap method in the older type of factory. It may, however, occur for drying or other special purposes and emissions are given in Table 10.3 (see page 233) for various steam pressures.

Mains—Piping used for mains will be insulated and the losses must be taken into account in determining the steam flow to be carried. Table 10.4 (see page 233) gives the emission where insulated by means of a commonly used material—glass fibre. This Table has been compiled in a form convenient for reading emissions direct according to the steam pressure.

Emission from condense piping may be taken as at 100 K temperature difference and according to whether bare or insulated. Whilst this is on the high side it is usually of little account and ignored.

UTILISATION OF STEAM

Probably the most general method of utilising steam for the heating of industrial and similar buildings is through the medium of warm air.

Plenum System—The Plenum System accepts steam at a central point in a set of heating coils or 'battery', and a fan is arranged to blow air over the coils, delivering it warmed into a system of ducts for distribution throughout the space to be heated. This system, however, lacks flexibility

—the ducts are often cumbersome, and distribution of heat is liable to be upset by wind pressure and the opening of doors.

Unit Heaters—The unit-heater system makes use of warm air as the final distribution medium but avoids the use of ducts. Thus, steam is conveyed to a series of units throughout the space to be heated, each unit being equipped with a fan discharging the heated air into a limited zone such that, with a suitable lay-out of units, fairly uniform distribution of temperature is achieved. Local cold spots such as at doors can be dealt with by units directed to such areas, and local control of temperature is achieved by on-off switching thermostatically of individual units or groups of units.

Various types of unit heater have been devised: the horizontal type (see Fig. 3.14, page 61), the vertical-downward type (Fig. 10.8) and the projector type (Fig. 10.9). Ratings vary as follows:

horizontal type:	10 kW to 30 kW
vertical type:	30 kW to 200 kW
projector type:	100 kW to 300 kW

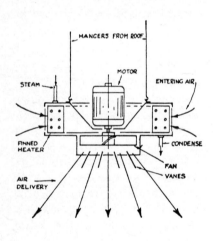

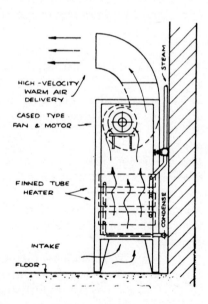

Fig. 10.8.—Downward discharge unit heater.   Fig. 10.9.—Projector or blast type unit heater.

For desirable air volumes, discharge temperatures and mounting heights, makers' lists should be consulted. Lower air-delivery temperatures such as 40° C to 50° C, with correspondingly larger air volumes, are preferable to small air quantities at higher temperatures.

Typical lay-outs of horizontal unit heaters are shown in Fig. 10.10.

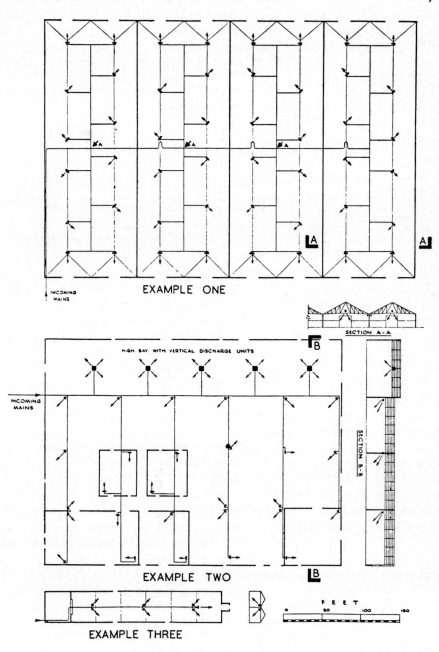

EXAMPLE ONE

EXAMPLE TWO

EXAMPLE THREE

FIG. 10.10.—Typical unit heater layout:
example 1, store;
example 2, workshop;
example 3, canteen.

The projector type relies on an unimpeded path for the discharge into the heated space, the air-flow pattern being as in Fig. 10.11.

Where quiet running is essential, the selection of type and speed should take this into account. Some noise in industrial applications is often unimportant.

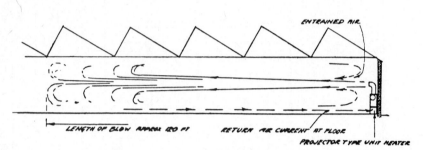

FIG. 10.11.—Air circulation with blast type unit heater.

Fresh Air Inlets—It is often necessary to provide artificial inlet ventilation to a building where unit heaters are used. This may conveniently be done by connecting the suction side of the unit by means of a duct to external air, terminating either with a roof cowl or with a louvred opening in the wall as in Fig. 10.12. The basis of design must take into account the volume of fresh air introduced, and that the unit inlet will be at 0° C or below in cold weather. This, of course, affects the rating of the unit. When some units in a building are provided with fresh air inlets and some are recirculating, it is desirable that their final air temperatures should be as nearly the same as possible.

It is sometimes arranged that the fresh air inlet is controlled by a damper, so that in cold weather the unit may be made to recirculate.

It should be noted that the battery of a unit connected by vertical ducting to a roof cowl will induce a strong reverse air current when the fan is stopped, with risk of overheating of the motor. It is generally arranged that these units should be not under thermostatic control by switching off the fan. If they are switched off, the steam should be shut off also.

Motors—The sizes of motors for unit heaters vary from about 60 W to 1000 W for horizontal and vertical types, and up to 4 kW for the projector type. Single-phase motors are usually of the 'Capacitor start-run' type which have no brush gear to give trouble, but sometimes the condensers break down, due to overheating from the radiant heat. Where it is possible, it is preferable to use three-phase current, though protection against single-phasing seems to be desirable. Motors are usually best if totally enclosed, and the bearings should be suitable for long running without attention.

Margin—The required output of unit heaters is sometimes not achieved

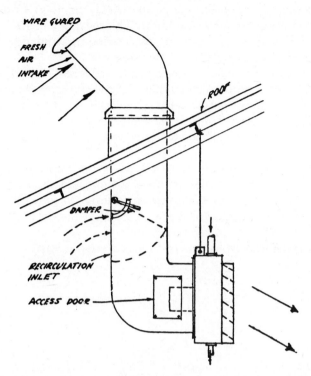

FIG. 10.12.—Unit heater with fresh air intake from roof; unit heater with fresh air intake through wall.

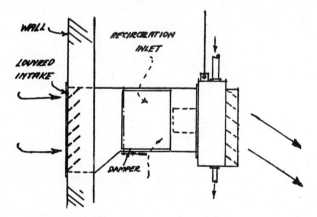

in practice, due to steam pressure drop, dirt in the fins of the heater, direct escape of warm air through doors before mixing, unsatisfactory distribution and other causes. It is customary to make the installed load somewhat larger than the calculated heat losses. A margin will furthermore allow the thermostatic control to take charge even during coldest weather when the units would otherwise be running continuously.

Such a margin must be purely empirical and is generally taken at from 15 to 20 per cent of the heat losses. This margin will also be available for sub-design temperature and quick heating-up, and the piping and boiler power should be installed to include for it.

Piping Connections—The large concentrated steam loads of unit heaters call for care in the piping connections. The steam inlet requires a valve, preferably of the fullway gate type. The condense outlet requires a trap capable of passing large volumes of condense rapidly as, on heating up a cold building, the condensation rate will be much higher than when warm.

For small units, up to about 30 kW on low pressure steam, a thermostatic trap will suffice; for larger units a float and thermostatic trap is necessary. For medium and high-pressure units an inverted bucket and certain types of float trap are suitable. Unions or flanges are necessary for disconnection of steam and condense.

Fig. 10.13 shows a typical arrangement of connections.

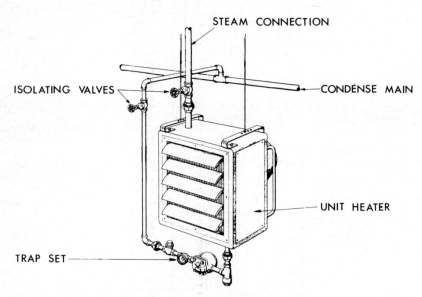

FIG. 10.13.—Typical unit heater connections.

CONVECTORS

Convectors of the same general type as those already described (see page 139) may be used with steam as the heating medium, and while it is physically possible to operate them with the same range of pressures as, for instance, unit heaters, it is preferable to avoid the use of absolute pressures over about 200 kPa (2 bar).

Steam convectors are an economical way of heating factory offices particularly where these occur singly or in small groups over a wide area. For larger office layouts it is preferable to install a steam-water calorifier serving hot-water radiators, either by gravity or pump circulation.

Fan Convectors—Fan convectors illustrated in Fig. 3.13 (page 141) may satisfactorily be used with steam, particularly where large outputs are required. Fan convectors may be used with fresh air inlets behind and a change-over recirculating damper to meet cases such as a canteen where inlet ventilation is required to replace kitchen exhaust.

Radiant Metallic Panels using Steam—These are an alternative method of heating factories and similar large spaces, and are dealt with under *High-Pressure Hot Water* in Chapter 11. When used with steam each will be fitted with a valve and steam trap.

HEATING CALORIFIERS

A calorifier is, in effect, a heat exchanger; that is to say, a device whereby heat from a medium at a high temperature is transmitted to a second medium at a lower temperature. The high-temperature medium we are concerned with here is steam, and the low-temperature medium hot water to be used for heating.

Types of Calorifier—A horizontal steam calorifier for heating (non-storage type) is shown in Fig. 10.14 and consists of:

(a) An outer shell.
(b) An internal battery of piping.
(c) A chest in which the ends of the tubes terminate so as to admit the heating medium.

The steam is generally passed through the tubes, the water to be heated being outside. This gives a lower temperature on the outer casing and consequently less heat loss than if steam is in the shell.

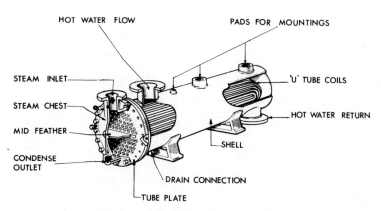

FIG. 10.14.—Horizontal U tube calorifier.

The outer shell is either of cast-iron or steel. Inlet and outlet connections or flanges are necessary on the casing. The thickness of shell depends on the head pressure and on the diameter of the vessel: a minimum test pressure should be not less than one and a half times the working pressure.

The tube battery may be of steel, brass or copper. Usually the latter is to be preferred, as it may be thinner than steel, allowing a more compact arrangement of the tubes, apart from which it has a slight advantage in conductivity.

The tube battery may be made removable, as shown in Fig. 10.14, with the tubes of U (or hairpin) form, or a 'floating header' may be adopted, the tubes then being straight. A further type is vertical (Fig. 10.15), being economical in floor space but requiring height for withdrawal of the shell.

The chamber for the admission of the heating medium is called *the steam chest*, and it is generally of cast-iron with inlet and outlet connections for the steam and condense.

Calorifier/Condensate Cooler Units— Reference has been made on page 210 to the recovery of flash steam from calorifiers. Where a calorifier is supplied by other than low pressure steam, discharge of the resulting condensate may give rise to problems due to flash steam: in any event a potential for recovery will exist. Equipment appropriate to this application is available in the form of two heat exchangers close-coupled as shown in Fig. 10.16. The upper unit is a normal steam to water calorifier equipped with a steam trap the discharge from which passes, as flash steam mixed with condensate, to coils in the lower unit. The heating water circuit is directed through the two heat exchangers in series.

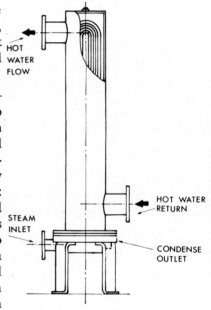

FIG. 10.15.—Vertical U tube calorifier.

Rating of Calorifiers—The subject of heat transfer is a separate study. Within the limited scope of a steam calorifier, the parameters include:

(*a*) difference of temperature between the steam and water, inlet and outlet;

(*b*) the velocity of water over the tubes;

(*c*) the material of the tubes and thickness;

(*d*) the transfer coefficient which is itself dependent on temperature;

(*e*) water quality: i.e., whether clean or liable to foul the tubes.

The *Guide, Section C3,* gives data under the above headings from which calorifiers may be designed. Table 10.5 (see page 234) has been prepared from these data and gives in a simplified form the essential information for a conventional set of conditions.

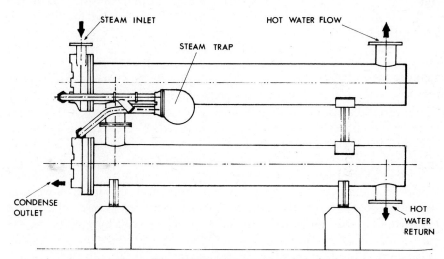

Fig. 10.16.—Calorifier–condensate cooler unit.

The user will, however, as with a boiler, refer to makers' lists for actual outputs and sizes, particularly so as by the use of controlled flow paths and high velocities within the shell of a calorifier a much enhanced performance is possible; but this can only be related to a particular design.

Example of Calorifier Sizing—

Output required	1000 kW
Steam pressure (absolute)	300 kPa (3 bar)
Steam temperature	133° C
Water temperature mean	60° C

Difference ..   ..   ..   73 K

From Table 10.5 (page 234) for
       300 kPa steam   7·25 kW/m
For 73 K mean temp.
       difference
       7·25 × 73 ÷ (133 − 75) = 9·13 kW/m
Hence, length of tube
       required = 1000 ÷ 9·13 = 109·5 m

If the tubes are U form and the average length of one U tube ($2\frac{1}{2}$ × 2) equals 5 m, then the number of U tubes = 109·5 ÷ 5 = 22.

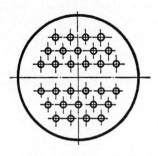

FIG. 10.17.—'Staggered' tubes.

The tubes would be arranged in staggered formation, as in Fig. 10.17. Allowing a reasonable pitch, the shell might be about 0·7 m diameter and 3·2 m long, plus 0·3 m for the steam chest.

Mountings for Calorifiers are much the same as for boilers. Safety valve, thermometer, and drain cock are essential on the water side, and a steam pressure gauge and trap on the steam side. An altitude gauge connected to the shell is desirable. Open vents direct from calorifiers are as for boilers.

Thermostatic Control to calorifiers is a necessary provision. If control at a constant temperature is required, as when the heating system is divided into a number of zones each with its own pump and mixing valve, a simple direct-acting thermostatic valve may be adequate. Where the calorifier is controlled by compensator, an electrical system will be required using a modulating steam valve on the steam inlet. A word of warning here, as, on shut-off, a vacuum may be formed preventing escape of condensate which must in any event go out by gravity, as has previously been emphasised. The addition of a vacuum breaker on the steam side may be necessary. A trap giving ready air release is then a corollary.

General—Other items required in connection with calorifiers are supporting cradles, runways or other means for tube withdrawal, and lagging. These do not call for comment. (An installation of heating and h.w.s. calorifiers is shown in Plate XVIII, facing p. 214.)

PIPE SIZING FOR STEAM SYSTEMS

For the sizing of steam mains and piping it is necessary to determine:

(a) The mass flow of steam in grams per second. This may be derived in the case of a heating system directly from the kW transmitted, by dividing by the latent heat at the appropriate pressure. The value at atmospheric pressure is 2258 kJ/kg (2·258 kJ/g).

In the case of kitchen or laundry apparatus the steam condensed in g/s may be arrived at by calculation if the duty is known, or from manufacturers' data for each item.

Where the mains are lagged and comparatively short their heat loss may be ignored in estimating the steam flow. If such is not the case due allowance should be made from Tables 10.3 or 10.4 (page 233).

(b) The initial pressure of the steam.

(c) The pressure drop permissible between the two ends of the system. This may depend upon the known or required end

conditions e.g. at boiler and/or equipment served. Otherwise, it is normal to allow for a pressure drop equivalent
to between 5 and 10 per cent. of the initial pressure, depending upon the size of the distribution system, subject to a
limiting velocity of about 50 m/s.

(d) The resistances in the pipe line due to bends, tees, valves, etc.

In order that flow rates may be represented by whole numbers, it has
been necessary to adopt the unit of *grams of steam per second*. This may seem
strange to engineers used to pounds per hour, but if it be remembered that
1000 grams equals 2·2 pounds and one hour equals 3600 seconds, we have

$$1 \text{ lb/h} = \frac{1000}{2 \cdot 2 \times 3600} \text{ g/s}$$
$$= 0 \cdot 126 \text{ g/s}$$

i.e: roughly 8 lb/h = 1 g/s.

Pipe sizing for steam may be based on a pressure-drop method or on a
velocity method.

Sizing by Pressure Drop—This method, first developed by Rietschel
and followed in earlier editions of this book and by the *I.H.V.E.* in the
Guide, Section C4, makes use of a pressure factor Z, where $Z = P^{1 \cdot 929}$ and
where P is pressure in bar absolute. Table 10.6 (p. 234) gives Z values
up to 10 bar, multiplied by 100 in each case to avoid decimal places.

Fig. 10.18 gives the flow of saturated steam in steel pipes in terms of
difference between initial and final Z values per metre length of pipe,
multiplied by 100 throughout to be compatible with Table 10.6. By way
of example, assume

Heat to be transmitted	= 1000 kW
Steam pressure (absolute), initial	= 500 kPa (5 bar)
Steam pressure (absolute), final	= 450 kPa (4·5 bar)
Latent heat, final	= 2121 kJ/kg
Steam flow $\dfrac{1000}{2121}$	= 0·472 kg/s
	= 472 g/s
Length of main (including single resistances)	= 50 m
Z_1 initial	2230
Z_2 final	1821
	————
$Z_1 - Z_2$	= 409
$\therefore \dfrac{Z_1 - Z_2}{L}$	= 8·18

Reading this value on the base scale of Fig. 10.18, it will be seen that a
65 mm pipe carries 460 g/s and, thus, this would be the appropriate

size of pipe. The pressure drop, 50 kPa (0·5 bar) over a 50 m run of pipe equals 1 kPa/m.

Sizing by Velocity—Piping in boilerhouses such as main headers, offtakes, short runs of mains, connections to calorifiers etc. is more conveniently sized on the basis of velocity. Table 10.7 (page 235) gives some conventional steam velocities and also pipe areas in square metres. For a given pressure, the volume of steam in m³/kg can be read from Tables 10.1 or 10.2. This multiplied by the mass flow per second will give the volume to be handled, which divided by the velocity selected from Table 10.7 then gives the pipe area necessary, and hence the pipe size.

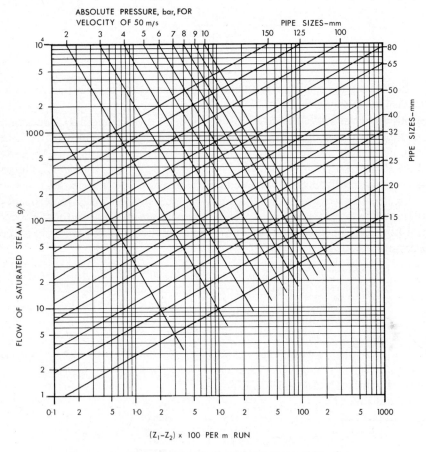

Fig. 10.18.—Flow of saturated steam in steel pipes to B.S. 1387.

Equivalent length of single resistances in metres for $\zeta = 1$ given in the *Guide* Table may be approximated as:

15 mm: EL=0·5	32 mm: EL=1·5	65 mm: EL=4	125 mm: EL=8
20 mm: EL=0·8	40 mm: EL=2	80 mm: EL=4·5	150 mm: EL=10
25 mm: EL=0·1	50 mm: EL=3	100 mm: EL=6	

The value of 50 m/s velocity has no quantitative significance. Quantities corresponding to other velocities may be obtained by simple proportion and, to other initial pressures, by interpolation.

Alternatively, the velocity lines on Fig. 10.18 may be used for the smaller sizes, though probably with less facility.

Whilst reasonable care should be taken in sizing steam mains, it must be remembered that imponderables exist in that heat losses from the pipework will lead to condensate formation and, in consequence, any steam main will be carrying a mixture of vapour and liquid. The proportions of the two components may vary from time to time with changes in ambient air temperature and in load imposed upon the steam main in terms of quantity carried.

Sizing of Condense Returns—Similarly, the presence of air and flash steam, as pressure is relieved on passing the steam trap, has been mentioned. Any precise calculation of pipe sizes is consequently impossible and various rule-of-thumb methods are commonly employed. One is to make the condense pipe one size smaller than the steam pipe.

The *Guide, Section C4,* postulates two conditions: one where air is released by other means and the trap discharge is at low pressure, when the condense lines may be sized using the hot-water-flow data. The pressure in pascals can be derived from the head causing flow (1 metre head = 9·8 kPa). The second condition is that referred to above, where gases and re-evaporated steam add greatly to the resistance to flow and it is recommended that a pressure drop ten times that of ordinary hot water be used. Table 10.8 provides data representative of these two conditions based upon a pressure drop of 100 Pa/m run of condense main.

Where condense is collected in a receiver and pumped back to the boilerhouse, the pumping main can be assumed to run full and hence may be sized as for normal hot water. In general terms, a well designed system of condense mains will arrange for flow by gravity from steam trap outlets either back to the boiler plant or, where this is impracticable, to one or more points of collection and pumping. In no circumstances should a steam trap be arranged to discharge into a pumping main.

STEAM ACCESSORIES

Provision for Expansion—Reference has been made in Chapter 9 to the need to make provision for expansion, but, as the temperatures employed with steam are much greater, the problem is more severe.

As before, the amount of expansion to be accommodated in metres is simply length in metres × coefficient of expansion per kelvin × temperature rise. By way of example:

Initial temperature	10° C
Steam pressure (absolute)	500 kPa (5 bar)
Steam temperature	152° C, thus the rise = 142 K
Coefficient expansion of steel	11·3 × 10⁻⁶
Length of main	50 m
Expansion	= 50 × 11·3 × 10⁻⁶ × 142
	= 0·08 m = 80 mm

Methods of providing for expansion are:

1. By changes of direction of the piping in the manner indicated in Fig. 10.19 (and see also Plate XIX, facing page 215). It will be noted that certain points are fixed or anchored. This is to prevent successive creep which might otherwise eventually rupture the pipe. The U expansion type or the Lyre type are standard fittings for which data on movement allow able per unit is available.

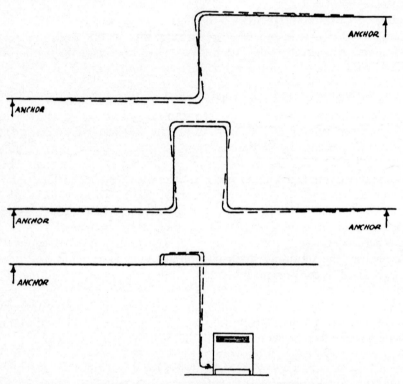

FIG. 10.19.—Use of changes of direction for expansion.

2. By flexible bellows of the type shown in Fig. 9.11 (page 192). These are best used in an off-set to rack sideways in various forms, being restrained from extension due to pressure by external pivoted links.

3. By sliding expansion joints. These are, however, prone to leakage if not well maintained and have largely gone out of favour.

The provision for expansion in whatever form needs to be considered in conjunction with the form of supports: hangers, rollers, guides, spring suspensions and the like. But this is beyond the scope of the present book.

Steam Main Drainage (Fig.10.20)—Where possible, steam mains should fall in the direction of steam flow at about 1 in 200. Reducers should be eccentric to allow no pocket for condense to collect. At the end of a run the main should be drained with a steam trap discharging to a condense main. Mains longer than about 30 m should be drained intermediately.

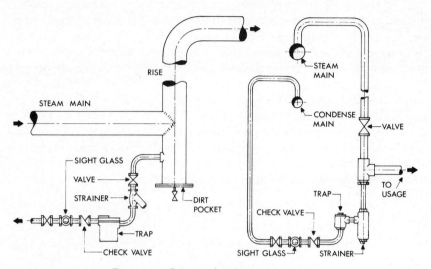

Fig. 10.20.—Steam main drip and relay point.

Where it is necessary to step the steam main up to a higher level due to the fall having brought the pipe too low, this may be done by arranging a

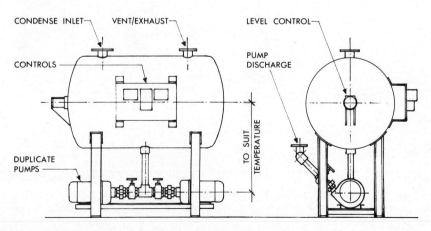

Fig. 10.21.—Condense return unit.

'relay' suitably drained as shown. This may be combined with provision for expansion at the same point by taking the vertical pipe up in the form of a loop.

Condense Return Pumping Unit (Fig. 10.21)—This is used to collect and return water of condensation when, due to level or distance, it cannot be returned to the boiler plant by gravity. It consists of a receiver containing a float switch, and a motor driven centrifugal pump or pumps, stopped and started by the float switch. The rating of the pump is usually about three times the condense return rate. The receiver requires a large vent to deal with the vapour from re-evaporation. It is necessary for the pump to be lower than the receiver to provide a 'head' on the pump suction.

Super-Lifting Traps (Fig. 10.22)—As an alternative to the pumping unit, and where the condense is from steam at too low a pressure to lift the condense the required height, use may be made of the 'super-lifting' trap. The steam supply thereto should be provided from a source having an absolute pressure of at least 300 kPa (3 bar). A small vented receiver is necessary to accommodate the incoming condensate during the period of discharge.

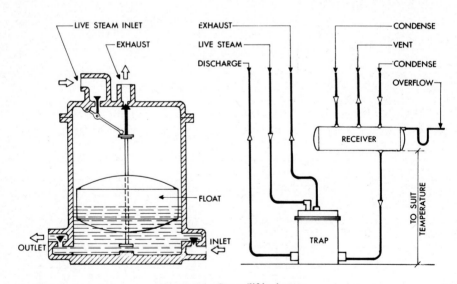

FIG. 10.22.—Super 'lifting' trap.

Steam Equipment Generally—In addition to the above brief list of steam apparatus particularly concerned with heating systems, there are numerous other fittings and equipment to which it is not possible to refer in detail.

These include:

Feed water-treatment plant.
Feed pumps and injectors.
Hot-well arrangements and fittings.
Boiler mountings for high pressures.
Steam valves of various types.
Condense valves.
Steam separators.
Flanged joints.
Welding and screwing.
Cast-iron and cast-steel fittings.
Steam and feed-water meters.
Instruments and Controls.

For further information on these matters, reference should be made to a textbook on Steam Engineering, and to the various makers' catalogues and handbooks.

TABLES: CHAPTER 10

TABLE 10.1*

PROPERTIES OF SATURATED STEAM
(SUB-ATMOSPHERIC PRESSURES)

Absolute Pressure		Temper-ature °C	Specific Enthalpy kJ/kg			Specific Volume m³/kg
kPa	bar		Heat in Water	Latent Heat	Total Heat	
20	0·2	60	252	2358	2610	7·65
40	0·4	76	318	2319	2637	3·99
60	0·6	86	360	2293	2653	2·73
80	0·8	94	392	2274	2666	2·09
100	1·0	100	417	2258	2675	1·69

* See p. 205.

TABLE 10.2*

PROPERTIES OF SATURATED STEAM
(100 kPa to 1 MPa)

Absolute	Pressure	Temperature °C	Specific Enthalpy kJ/kg			Specific Volume m³/kg
kPa	bar		Heat in Water	Latent Heat	Total Heat	
100	1·0	100**	417	2258	2675	1·69
120	2	105	439	2244	2683	1·43
140	4	109	458	2232	2690	1·24
160	6	113	475	2221	2696	1·09
180	8	117	491	2211	2702	0·98
200	2·0	120	505	2201	2706	0·89
220	2	123	518	2193	2711	0·81
240	4	126	530	2185	2715	0·75
260	6	129	541	2178	2719	0·69
280	8	131	552	2170	2722	0·65
300	3·0	133	561	2164	2725	0·61
320	2	135	571	2157	2728	0·57
340	4	138	580	2151	2731	0·54
360	6	140	589	2145	2734	0·51
380	8	142	597	2140	2737	0·49
400	4·0	144	605	2134	2739	0·46
420	2	146	612	2129	2741	0·44
440	4	147	620	2123	2743	0·42
460	6	149	627	2118	2745	0·41
480	8	150	634	2113	2747	0·39
500	5·0	152	640	2109	2749	0·38
520	2	153	647	2104	2751	0·37
540	4	155	653	2099	2752	0·35
560	6	156	659	2095	2754	0·34
580	8	157	665	2090	2755	0·33
600	6·0	159	670	2087	2757	0·32
620	2	160	676	2082	2758	0·31
640	4	161	682	2078	2760	0·30
660	6	163	687	2074	2761	0·29
680	8	164	692	2070	2762	0·28
700	7·0	165	697	2066	2763	0·27
720	2	166	702	2063	2765	0·27
740	4	167	707	2059	2766	0·26
760	6	168	712	2058	2767	0·25
780	8	169	716	2052	2768	0·25
800	8·0	170	721	2048	2769	0·24
820	2	171	725	2045	2770	0·24
840	4	172	730	2041	2771	0·23
860	6	173	734	2038	2772	0·22
880	8	174	738	2034	2773	0·22
900	9·0	175	743	2031	2774	0·21
920	2	176	747	2028	2775	0·21
940	4	177	751	2025	2776	0·21
960	6	178	755	2021	2776	0·20
980	8	179	759	2018	2777	0·20
1000	10·0	180	763	2015	2778	0·19

* See page 205.
** Accurately, 99·63°C. 100°C occurs at standard atmospheric pressure of 101·3 kPa (1·013 bar).

TABLE 10.3*

THEORETICAL HEAT EMISSION FROM BARE
SINGLE HORIZONTAL STEEL PIPES IN AMBIENT AIR AT 15° C
Pipes to B.S. 1387

Nom. Bore mm	Heat Emission W/m run For stated absolute steam pressures kPa to ambient air at 15° C									
	100	200	300	400	500	600	700	800	900	1000
15	110	140	170	190	200	220	230	250	260	270
20	130	180	210	230	250	270	290	300	320	330
25	160	210	250	280	310	330	350	370	390	410
32	200	260	310	350	400	420	430	450	470	490
40	220	290	350	390	420	450	460	490	530	560
50	270	360	420	470	510	560	590	610	640	680
65	330	440	520	580	630	680	720	760	790	830
80	380	500	600	670	730	790	830	870	910	960
100	470	620	750	840	910	980	1030	1080	1140	1190
125	560	750	890	1010	1090	1180	1240	1300	1370	1440
150	650	870	1040	1170	1270	1380	1450	1520	1600	1680
200	840	1110	1340	1510	1630	1750	1820	1940	2060	2160

Note: 1. The above Table is derived from the *Guide, Section C3*.
2. For 2 pipes in bank, factor = 0·95
   For 4 pipes in bank, factor = 0·85
* See page 215.

TABLE 10.4*

HEAT EMISSION TO AMBIENT AIR AT 15° C FROM PIPE INSULATED WITH
MATERIAL OF THERMAL CONDUCTIVITY 0·045 W/m K—e.g. GLASS FIBRE

Nom. Bore mm	Heat Emission W/m run For stated absolute steam pressures kPa to ambient air at 15° C											
	Insulation 25mm thick						Insulation 50mm thick					
	100	300	500	700	900	1100	100	300	500	700	900	1100
15	18	23	27	31	34	35	14	17	20	24	26	27
20	20	26	31	36	38	41	15	19	23	27	29	30
25	24	30	36	42	45	47	17	22	25	30	32	34
32	27	35	41	48	51	54	20	25	29	35	37	39
40	30	38	44	52	56	59	21	26	30	36	38	41
50	35	44	52	61	66	69	23	29	34	41	43	46
65	41	52	61	72	77	81	26	33	39	46	50	52
80	47	59	70	83	88	93	31	39	46	54	58	61
100	57	72	85	102	107	113	38	45	53	63	67	71
125	68	87	102	120	128	135	43	52	63	75	80	84
150	77	98	116	136	156	154	48	60	71	84	90	95
200	100	127	150	177	189	199	60	76	89	105	112	118

Note: 1. Values are derived from *Guide* factors.
2. Other thicknesses between 25 mm and 50 mm may be interpolated with slight inaccuracy.
3. For other insulating materials and conductivities, the above values may be multiplied in ratio of the factors in the *Guide* to 0·045.
* See page 215.

TABLE 10.5*

STEAM HEATING CALORIFIERS

Heat Transfer with Free Convection at 0·4 m/s Water Velocity

Absolute Pressure		Steam Temperature °C	Heat Transfer to water at 75° C by 25 mm plain copper tubes kW/m run
kPa	bar		
100	1	100	3·15
200	2	120	5·60
300	3	133	7·25
400	4	144	8·65
500	5	152	9·65
600	6	159	10·50
700	7	165	11·30
800	8	170	11·90
900	9	175	12·50
1000	10	180	13·15

Note: 1. Factor for water velocity 0·2 m/s = 0·6
Factor for water velocity 0·6 m/s = 1·26
2. For tubes of other internal diameters, heat transfer is in direct ratio of diameter.
3. Transfer rates are based on tube wall thickness of 1·6 mm.
4. Tubes are assumed to be suitably pitched.
* See page 222.

TABLE 10.6*

VALUES OF 100 Z FOR USE WITH FIG. 10.18

P = Absolute pressure in kPa

P	Z	P	Z	P	Z	P	Z
100	100	240	541	460	1899	740	4751
110	120	250	586	480	2061	760	5001
120	142	260	632	500	2230	780	5258
130	166	270	679	520	2405	800	5522
140	191	280	729	540	2587	820	5791
150	219	290	780	560	2775	840	6066
160	248	300	832	580	2969	860	6348
170	278	320	943	600	3170	880	6636
180	311	340	1060	620	3377	900	6930
190	345	360	1183	640	3590	920	7230
200	381	380	1313	660	3810	940	7536
210	418	400	1450	680	4036	960	7849
220	458	420	1593	700	4268	980	8167
230	499	440	1743	720	4506	1000	8492

* See p. 226.

Note: For convenience in using whole numbers, the factors quoted are actually (100 Z). Fig. 10.18 has been similarly adjusted.

TABLE 10.7*

CONVENTIONAL STEAM VELOCITIES (SATURATED STEAM)

	m/s
Boiler outlet connections (LP) - - - - - -	5 to 10
Boiler outlet connections (HP) - - - - - -	15 to 20
Headers (LP and HP) - - - - - -	20 to 30
Steam mains (LP and HP) - - - - - -	30 to 50
Vacuum steam mains - - - - - - -	100
Calorifier connections - - - - - - -	25 to 30

CROSS-SECTIONAL AREA OF PIPING

Steel to B.S. grade 1387 heavy grade and B.S. 806 (over 150 mm)

Nom. Bore mm	Area m²	Nom. Bore mm	Area m²	Nom. Bore mm	Area m²
15	0·000175	40	0·001272	100	0·00838
20	0·000326	50	0·00207	125	0·01305
25	0·000518	65	0·00353	150	0·0187
32	0·000927	80	0·004905	200	0·0344

TABLE 10.8

APPROXIMATE DATA FOR FLOW OF CONDENSATE IN
STEEL AND COPPER PIPES

Nom. Bore mm	Flow of condensate in g/s for listed conditions and pressure loss of 1 m head/100 m (100 Pa/m)			
	Liquid condensate without air or vapour		Mixture of condensate air and steam	
	Steel pipe*	Copper pipe**	Steel pipe*	Copper pipe**
10	25	20	5	5
15	50	40	15	10
20	120	125	30	30
25	220	250	60	70
32	480	460	135	125
40	730	770	210	210
50	1390	1560	400	430
65	2820	2800	825	775
80	4380	3970	1290	1100
100	8870	10400	1900	2900

* Heavy grade to B.S. 1387.
** Light gauge to B.S. 2871, Table X.
† See page 227.

CHAPTER 11

Heating by High-Pressure Hot Water

THE FATHER OF ALL HIGH-PRESSURE hot-water systems was Perkins, whose patent was filed in 1831. In his system, shown in Fig. 11.1, the piping was extremely strong, about 22mm bore, and formed one continuous coil, part of it passing through the boiler and the remainder forming the heating surface in the rooms to be heated. Expansion of the water was allowed for in a closed vessel at the top. Two or more coils could be used

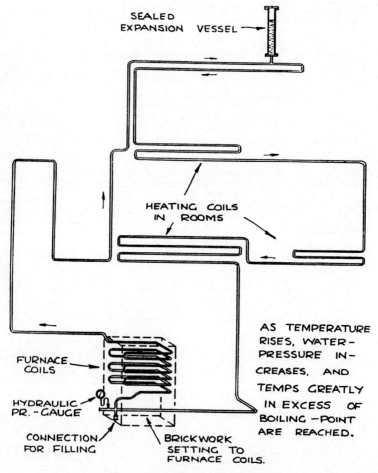

SEALED EXPANSION VESSEL →

HEATING COILS IN ROOMS

FURNACE COILS →

HYDRAULIC PR.-GAUGE

CONNECTION FOR FILLING

BRICKWORK SETTING TO FURNACE COILS.

AS TEMPERATURE RISES, WATER-PRESSURE IN-CREASES, AND TEMPS GREATLY IN EXCESS OF BOILING –POINT ARE REACHED.

FIG. 11.1.—Diagram of Perkins system.

236

where one was insufficient. As the water was heated, its expansion compressed the air in the expansion vessel and considerable pressures were reached. The principle was that the formation of steam was prevented by the pressure to which the water in the system was subjected. The system is now obsolete, though examples may still be found at work chiefly in churches and chapels.

Present-day Systems—The same principle is the basis of all modern systems of this type. Hot water is the medium in a closed system and is subjected to pressure either by steam or by a gas such that its temperature is raised well above atmospheric boiling point. The essential differences between this and the method used by Perkins is that circulation is by pump, and the working pressure is controlled so that boilers and heat-emitting equipment of various types suitable for moderate pressures may be used. There are, in fact, two ranges of pressure in common use for space heating:

high pressure   -   -   -   500 kPa to 1 MPa (absolute)
(approx. 150° C to 180° C)

medium pressure -   -   -   200 kPa to 300 kPa (absolute)
(approx. 120° C to 133° C)

Temperature Ranges—The higher the initial temperature, obviously the greater the temperature drop which can be allowed between flow and return, and, correspondingly, the larger the heat content per unit mass flow of water, requiring smaller pipe sizes for a given load. On the other hand, the higher the pressure the more costly the boilers and other equipment. Temperature drops are usually of the following order:

high-pressure system   -   -   45 K to 65 K
medium-pressure system   -   -   25 K to 30 K

Thus, a high-pressure system, with 50 K drop and a flow temperature of 160° C, has a temperature difference at the emitting apparatus, mean water to air at 20°C, as follows:

$$\left(\frac{160 + 110}{2}\right) - 20 = 115 \text{ K}$$

A medium-pressure system, with 30 K drop and flow at 120° C, has a difference of

$$\left(\frac{120 + 90}{2}\right) - 20 = 85 \text{ K}$$

The lower pressure system requires more heating surface for a given output—in this case about one-third more. For smaller installations, however, there are many compensating advantages.

For this and for other reasons, it has come to be accepted that there is no point in designing for high pressures for systems under say 2000 to 3000 kW, below which ratings medium-pressures are most suitable. The high-pressure system comes into its own with increasing size and extensive runs of mains. Furthermore, where process equipment has to be supplied, as in many industrial applications, the temperature of operation is controlled by the needs of the process and may be up to 200° C at an absolute pressure of 1·5 MPa.

Comparison with steam—Hot water in a closed system under pressure may be run at any temperature up to its design maximum. Where serving space-heating apparatus, the temperature of the water can be varied according to the weather, so saving on mains heat losses and by better control generally. Variability of temperature is not possible with steam, which must be either on or off and any attempt at throttling is liable to cause water logging at the remote ends. (This assumes that the vacuum steam system with its further complications is ruled out anyway.)

Hot water requires no steam traps. The potential loss of heat through flash steam and condense return has been referred to in Chapter 10 and may amount to ten per cent or more of the fuel bill. Traps also require maintenance, as do other steam accessories.

Hot-water mains may be run with complete freedom as to levels, whereas steam mains require careful grading and draining. Also corrosion of condense lines is avoided.

In terms of pipe sizes and cost it can be shown that, taken overall, there is little difference in the two systems. By way of example, consider a heating load of 4 MW (4 MJ/s):

Steam, at an absolute pressure of 500 kPa (5 bar)
> Latent heat, Table 10.2 (page 232) = 2109 kJ/kg
> Steam quantity = 4000 ÷ 2109
> $\qquad$ = 1·897 kg/s
> Volume = 1·897 × 0·38 = 0·72 m³/s
> Velocity in a 125 mm pipe = 55 m/s
> Velocity in a 150 mm pipe = 38 m/s
> ∴ Select 150 mm for steam main

The condense main, from Table 10.8 (page 235), for a flow rate of 1897 g/s in copper pipe would fall between 80 mm and 100 mm. The latter would be chosen unless the run was very short.

Hot Water (50 K drop)
> The flow rate would be 4000 ÷ 50
> $\qquad$ = 80 kW/K

From Fig. 8.6 (page 164), at a unit pressure loss of, say, 130 Pa/m, the flow and return mains would each be 125 mm.

The saving on the smaller condense main will, however, be eaten up by traps and drainage fittings, plus the cost of the copper pipe.

PROPERTIES OF WATER AND PIPE SIZING AT HIGH TEMPERATURES

Table 11.1 gives the specific heat capacity and density of water from 75° C (low-pressure hot water) to 200° C. It will be noted that, whilst the specific heat capacity increases, density decreases; but, when the two are multiplied together giving heat per unit volume, there is only a small correction needed where velocity of flow is the criterion, as in pipe sizing.

It can also be shown that as temperature increases viscosity decreases and a correction in pressure loss for a given flow rate is required on this account. Table 11.2, abstracted from the *Guide*, shows the order of these corrections, which are quite small. Corrections to equivalent length factors are also required, but these are still smaller and may be ignored.

Thus, there is little error if Fig. 8.6 (page 164) is used as it stands for

TABLE 11.1

SELECTED PROPERTIES OF WATER AT HIGH TEMPERATURES

Temperature °C	Absolute Pressure kPa	Specific Heat Capacity kJ/kg	Density kg/m³	Heat Content per litre	Correction Factor
75	—	4·19	974·9	4·094	1
100	100	4·22	958·3	4·043	0·99
110	140	4·23	951·0	4·023	0·98
120	200	4·25	943·1	4·007	0·98
125	230	4·26	939·1	4·000	0·98
130	270	4·27	934·8	3·987	0·97
140	360	4·29	926·1	3·973	0·97
150	480	4·32	916·9	3·961	0·97
160	620	4·35	907·4	3·948	0·96
170	790	4·38	897·3	3·930	0·96
175	900	4·40	892·2	3·926	0·96
180	1000	4·42	886·9	3·919	0·96
190	1250	4·46	876·0	3·906	0·95
200	1550	4·51	864·7	3·900	0·95

TABLE 11.2

CORRECTION FACTORS FOR USE WITH FIG. 8.6
FOR HIGHER TEMPERATURE WATER (150° C)

Pressure Loss Pa/m from Fig. 8.6	Pipe Size mm				
	15	25	50	100	150
2	0·81	0·84	0·87	0·90	0·91
5	0·83	0·86	0·89	0·92	0·93
10	0·85	0·88	0·91	0·94	0·95
20	0·87	0·90	0·93	0·96	0·97
50	0·90	0·92	0·95	0·98	0·99
100	0·92	0·94	0·97	0·99	1·00
200	0·95	0·96	0·98	1·00	1·01
500	0·97	0·98	0·99	1·02	1·02
1000	0·98	1·00	1·01	1·02	1·02

high-pressure and medium-pressure pipe sizing. For accurate use, the corrections may be made in accordance with Tables 11.1 and 11.2.

PRESSURIZATION BY HEAD TANK AND BY STEAM

Head Tank—An elevated tank at sufficient height above the highest point of the system is rarely possible, though the simplest way of obtaining pressure. 1 metre head of water is approximately equal to 10 kPa (9·84). For a medium-pressure system at an absolute pressure of 200 kPa, some 20 metres height would be required plus an excess to prevent ebullition at the highest point of the system.

Where the medium-pressure system is confined to low level, possibly using calorifiers for low-pressure heating to the buildings, an open expansion tank may be practicable, as referred to in Chapter 9.

Pressurization by steam—Steam is generated above the water in a steam boiler, as shown diagrammatically in Fig. 11.2. The water for circulation to the heating system is taken from below the water-line of the

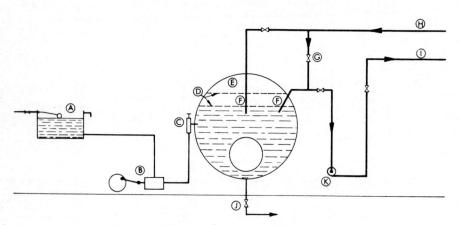

FIG. 11.2.—Diagram of steam-pressurizing system.

A. Feed tank	E. Steam space	H. Return
B. Feed pump	F. Dip pipes	I. Flow
C. Check valve	G. Cooling water by-pass	J. Blow down
D. Water line, upper and lower		K. Circulating pump.

boiler and returned thereto. The danger of draining the boiler, should a serious leak occur in the system, is overcome by using dip-pipes as shown.

It will be clear that the steam and water are at the temperature of saturation and that any reduction of pressure on the water side will thus cause water to flash into steam. To overcome this, a cooling supply from the return is injected into the flow outlet as it leaves the boiler, so bringing the temperature below ebullition point. Also, the flow outlet is taken down to low level, so increasing static pressure.

Where more than one boiler occurs, their respective water levels are

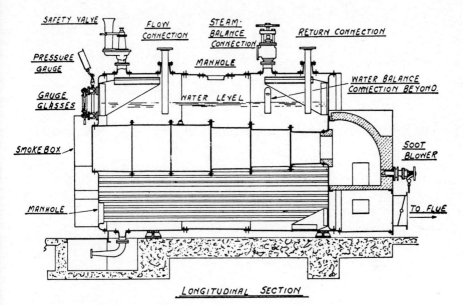

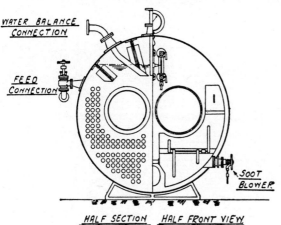

FIG. 11.3.—Super Economic boiler for H.P.H.W.

kept uniform by steam and water balance pipes, the former from the steam space and the latter by dip pipe. Fig. 11.3 shows the various connections to a boiler of shell type.

Alternatively, the high-velocity tubular type of boiler has been much used with these systems, as shown in Fig. 11.4. The boiler tubes connect to a drum in which the steam space is formed. Where a number of such boilers occur, a common drum or drums are used, as in Fig. 11.5.

In order to feed the steam-pressurized system, a normal feed pump is

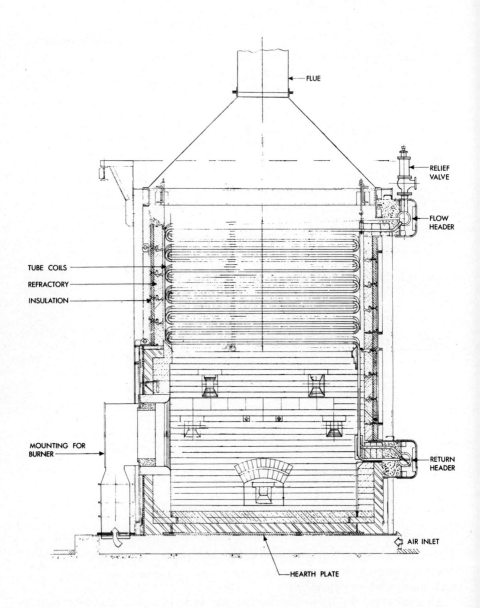

FIG. 11.4.—Section of LaMont boiler.

used, drawing its water from an open feed tank. It is usually hand-controlled.

Water of expansion is to some extent contained in the rise in water level in the boiler or drum—sufficient to contain diurnal variations; but the greater volume of expansion from cold is discharged to drain.

The disadvantages of steam pressurization are that the system tends to be unstable and requires skilled operation and maintenance. Varying rates of output from boilers causing pressure changes and close interdependence of pressure and temperature tend to fickleness in behaviour not conducive to unskilled attention.

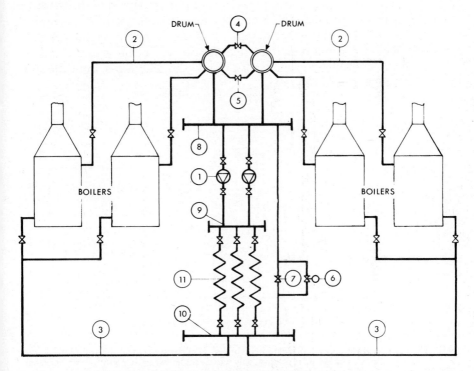

FIG. 11.5.—Diagram of multi-boiler plant with steam drums.

1. Circulating pumps. 2. Boiler flow mains (each run separately to drum). 3. Boiler return mains. 4. Steam balance-pipe. 5. Water balance-pipe. 6. Thermostatic mixing-valve. 7. Hand-operated by-pass. 8. Mixed-water header (pump suction). 9. Flow header (pump discharge). 10. Return header. 11. Heating circuits to buildings. Air bottles where shown.

PRESSURIZATION BY GAS*

Instead of using the steam generated by heating the water in the boiler of a high-pressure hot-water system for creating the pressure, an alternative method is the use of an air or gas cushion, applied to the

* See also paper by J. R. Kell, *Methods of Pressurization. I.H.V.E. Journal*, Vol. 26, p. 1.

system in a cylinder.

With this system the boiler or boilers are completely filled with water, as in the case of a low-pressure hot-water system, and the circulation is by pump in the normal manner. A pressure cylinder is connected to some part of the system, generally the return near the boiler where the water is coolest. The pressure cylinder is maintained partly filled with water and partly with air, or gas, derived either from an air compressor or from bottled gas. In this way, pressure may be applied to any amount desired, within the pressure limits of the system and well above boiling point as determined by the working temperature of the system.

The whole of the water of expansion could be contained in the pressure cylinder if this were made large enough; but in one system the arrangement is made for surplus water to spill out of the system through a pressure control valve into a spill tank open to atmosphere. For the return of water, as the system cools down, a pump is provided to take water from the spill tank and return it to the system.

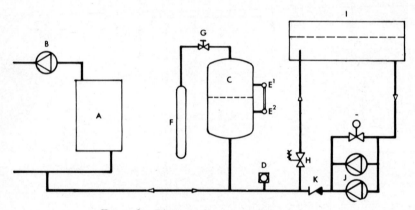

Fig. 11.6.—Diagram of gas-pressurizing system.

A: Boiler. B: Circulating pump C: Pressure cylinder. D: Pressure control. E^1 and E^2 High and Low level cut outs, F: Gas cylinder or air supply. G: Gas inlet regulating valve, hand operated. H: Water pressure spill valve. I: Expansion water spill tank. J: Water pumps. K: Non-return valve, L: Pump pressure relief valve.

The elements of the system are shown in Fig. 11.6. An essential part of the apparatus is the pressure controller D, which regulates the admission of water from the pump J or its expulsion via the regulating valve H, level controllers EE and safety-limit controllers.

The advantages of this arrangement are that, apart from temperature being independent of pressure, no balance pipes are needed to a multi-boiler installation, boiler mountings are simplified, and the system is more stable in operation due to the absence of steam.

The reason for using gas such as nitrogen as the cushion, instead of air, is that it is less soluble in water than air. Furthermore, it is inert, so that córrosive tendencies from this source are eliminated.

In order to remove the tendency for atmospheric air to be dissolved into the spill tank water, in an alternative system this is arranged as a closed cylinder lightly pressurized with nitrogen. In another system the water of expansion is taken up by a flexible bellows. The expansion water is cooled in a heat exchanger (not shown in the diagram) before passing to the spill tank.

Gas Pressurization: Working Pressures—The calculation of the working pressure on the boiler and on the cylinder in this system starts from a decision as to the flow temperature required at the highest point in the system, such as in some high level main. Having settled this point, and the temperature, the rest follows as in the example given below.

Example of Pressure Differentials

Assume 170° C at flow to system, absolute pressure   = 800 kPa
    Plus 15 K anti-flash margin = 185°C $= 1121.3$ kPa   = 325 kPa
    Highest point of system 10 m head
        above W.L. in cylinder   = 100 kPa

    Minimum pressure at cylinder   = 1225 kPa
    Pressure differential on control switch   50 kPa

    Normal operating pressure absolute of system   = 1275 kPa

Head pump operates between 1225 and 1275 kPa
Spill valve starts to open at +25 kPa   = 1300 kPa
Safety valve on boiler set at   = 1400 kPa

It will be noted that an anti-flash margin of 15 K has been selected, but this might be varied according to circumstances. The object of this is to provide pressurization at the highest point of the system such that at no time can ebullition take place. The head of the circulating pump greatly augments this anti-flash margin during working conditions, but it has to be assumed that at some time the circulating pump may stop when the system is up to full temperature, and it would be unfortunate if ebullition should occur under such conditions.

Gas Pressurization: Sizing of Pressure Cylinder—It obviously simplifies control if the movement of water line for a given pressure differential is at a maximum. If no loss of heat occurs from the cylinder and associated piping during expansion of the water and compression of the gas in the cylinder, the law of compression will approach the adiabatic $PV^{1.405} = C$. If, however, the vessel and piping are left unlagged, the change becomes more nearly isothermal and it has been found by experiment that compression follows the law approximately $PV^{1.26} = C$.

It follows from the above that the movement of the water level can be calculated from the formula:

$$x = h\left[1 - \left(\frac{p_1}{p_2}\right)^{\frac{1}{1.26}} \right]$$

where x = movement of water line in metres
      h = height of air space in metres before compression starts
      p_1 = initial absolute pressure in air space, kPa
      p_2 = final absolute pressure in air space, kPa

It should be noted that the movement of the water line is independent of the diameter of the vessel, but the latter must be considered in conjunction with the quantity to be dealt with by the head pump, which should run for about two minutes between low and high water levels.

Sizing of Head Pressure Pump—The maximum duty which this pump could be called upon to deal with at any time would be in the case of rapid contraction throughout the system due to a sudden shut down of the boilers. This is not a matter of accurate calculation, but some approximation can be made to it and an ample margin allowed. Pressure against which the pump must work will be that of the highest pressure to which the cylinder is subjected before the spill valve starts to open with the differential selected, as in the example previously given. In this case the head pump would operate between absolute pressures of 1225 and 1275 kPa, plus a margin for friction loss in piping etc.

With regard to the various items of control equipment and fittings associated with this method of pressurization, it is not possible here to consider details, but there are a number of proprietary systems which have been developed, and reference to makers' data is suggested for those seeking further information.

PRESSURIZATION BY PUMP

For medium pressure hot water systems operating at temperatures up to about 110° C and working at absolute pressures of 400 kPa, an alternative type of pressurization unit has recently been introduced. In principle this consists, as shown in Fig. 11.7, of a pressure pump and a spill valve connected to an expansion tank. The tank is fitted at boiler house floor level thus avoiding long pipelines up to roof level.

On start-up from cold, the spill valve allows the water of expansion to escape into the expansion tank after a pre-set pressure is reached. Whilst the system remains at working temperature (and thus constant pressure) the valve remains closed and the pump is out of action. When system temperature and pressure fall, owing to cycling of boiler firing or as a result of thermostatic control, the pump is started automatically and the pre-set pressure in the system restored. Since the rate of fall in pressure may vary considerably, the pump does not run continuously but is arranged to cycle on/off by means of a timed interruption of the electrical

Plate XXII (above). Radiant strip heaters integrated with roof structure in vehicle repair depot (see p. 255)

Plate XXIII (right). Electric floor heating cables laid ready for screening (see p. 273)

Plate XXIV (below). Electrode boilers for a thermal storage system (see p. 274)

Plate XXV (above)
Open roof ventilation i[n]
factory (see p. 360)

Plate XXVI (left)
Recirculating fan-co[il]
units integrated wit[h]
structure (see p. 380)

supply; this arrangement provides very simply what is, in effect, a variable pump output.

Such units are factory assembled; they are small in size and may be fitted with duplicate pumps and controls. A safety cut-out to shut down the boiler plant in emergency is incorporated in the control circuit.

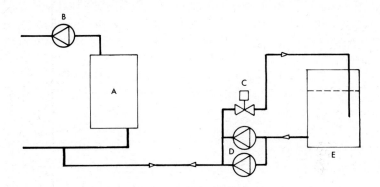

Fig. 11.7.—Diagram of pump-pressurizing system: A, boiler; B, circulating pump; C, water pressure spill valve; D, pressure pump; E, expansion tank.

Expansion of Water—The amount of expansion may be calculated from the water contents as in the following example:

Water capacity of system $\qquad = 100\ 000$ kg

Volume at 15° C (Table 1.2): $\quad \dfrac{100\ 000}{998 \cdot 5} = 100 \cdot 15$ m³

„   „   mean temp. 125° C (Table 11.1):

$$\dfrac{100\ 000}{939 \cdot 1} = 106 \cdot 48 \text{ m}^3$$

$$\text{increase} \quad = \quad 6 \cdot 33 \text{ m}^3$$

Volume at mean temp. 150° C: $\dfrac{100\ 000}{916 \cdot 9} = 109 \cdot 06$ m³

Increase from 125° C to 150° C $\qquad = \quad 2 \cdot 58$ m³

Increase overall 15° to 150° C $\qquad = \quad 8 \cdot 91$ m³

As already stated, in a steam pressurized system 2·58 m³ may be accommodated by variation in water level in boiler or drum: the initial warming up expansion (6·33 m³) is blown down, and on cooling restored by feed pump.

In a gas pressurized system the capacity in the cylinder is small and the whole of the water of expansion is therefore to be accommodated in the spill tank (i.e.; 8·91 m³). For a pump-pressurized system, this same principle applies.

I   K.H.A.C.

CIRCULATING PUMPS

In a steam pressurized system it is preferable to place the circulating pumps in the flow main. Fig. 11.8 illustrates this problem. It will be noted that, whatever the pressure in the boiler, the additional head set up by the pump is added so that the flow main is at a pressure well above that in the boiler. Water travels through the system until, at the point of return to the boiler, the lowest pressure is reached, and therefore the whole of the rest of the system is above this pressure, excepting for the short length of main between the boiler and the pump suction.

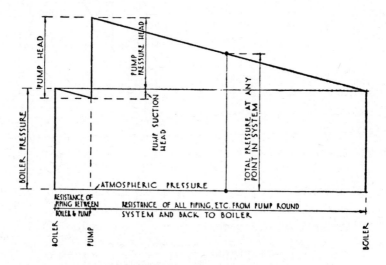

FIG. 11.8.—Pressure diagram for typical H.P.H.W. system.

In gas or pump pressurized systems, it may be better for the pump to be in the return to the boiler with the pressure cylinder connected in the suction line. Virtually all parts of the system are then subject to pump pressure in addition to static pressure.

Control of Temperature by Mixing—Where a boiler plant serves process heating at a constant temperature, also in the case of space heating where a variable temperature is required according to weather, use may be made of a mixing connection (controlled either by hand, or thermo-statically, in such a manner that return water is by-passed into the flow from the boiler) for the heating system. Separate pumps would be provided for the process work and for the heating system. Such an arrangement is illustrated in Plate XX, facing p. 215.

In a similar manner, where it is desired to apply zoning, i.e. to arrange for different areas to be supplied at dissimilar temperatures, a series of mixing valves and pumps may be used, each delivering to its own circuit. Fig. 11.9 shows diagrammatically a gas-pressurized system which has a boiler circulating pump and three zones, together with process heating.

The advantages of including a boiler circulating pump are that boiler flow is uninfluenced by the variations in demand of the mixing valves. Also, where tubular boilers having high resistance are used, the necessary pressure is disposed of by the boiler circulating pump, leaving the circuit pumps only their own circuit head to deal with. Where large load variations are expected, there is a case for providing one pump per boiler.

It will be noted in Fig. 11.9 that, whilst the boiler circulating pump is in the return for the reason stated earlier, the sub-circuit pumps are in the flow—again for the reason that their pressure is added to the sub-circuit.

Fig. 11.10 shows one of the types of pump available for this duty. It is of the end-suction centrifugal type direct-coupled to an electric motor: for large duties, split casing pumps having both connections at right angles to the driving shaft are often used.

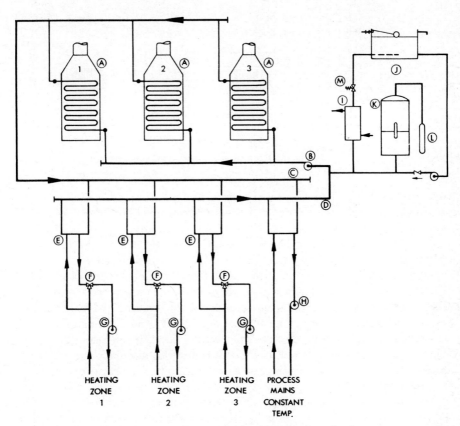

FIG. 11.9.—Example of zoned system with process circuit.

A. Tubular boilers
B. Boiler circulating pump
C. Flow header
D. Return header

E. Accept and reject connection
F. Mixing valve
G. Zone pump
H. Process pump
I. Cooler

J. Nitrogen pressurising unit
K. Pressure cylinder
L. N₂ bottle
M. Spill valve.

Points usually taken care of in the design of circulating pumps for this purpose are:

Suction connection designed to avoid sudden changes of velocity, and consequent cavitation.

Casing generally of cast steel to withstand high pressures.

Gland cooled by water fed under pressure and with visible discharge into a tundish.

Bearings suitable for high temperature.

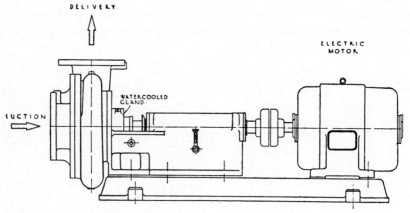

FIG. 11.10—Circulating pump for H.P.H.W.

Pump Duty—In the type of system shown in Fig. 11.9, the boiler circulating-pump duty will be determined by the total boiler loading and design-temperature drop. The pressure of this pump will be the sum of the loss in the boiler, headers and connections, the units being kPa.

The zone pumps will be sized to the loading of each zone circuit, and to the overall design-temperature drop or some lesser drop—assuming there is always some return water entering the mixing valve. The total quantity handled by the zone pumps and process pump must always be exceeded by the boiler circulating pump. The pressure in kPa of the zone and process pumps will be that of their own circuit, piping, emitter and mixing-valve loss in each case, unrelated to boiler pressure loss.

In practice, standby pumps would no doubt be provided, particularly to the boiler circulating pump. The zone pumps might use a common standby with valving arrangement permitting it to be used on any circuit. **Treatment of Feed Water**—High-pressure and medium-pressure systems are prone to give trouble from scale deposit on valves, sludge deposits in apparatus such as unit heaters, and corrosion of feed-connecting pipework. Treatment of the feed water is thus highly desirable and one of the manufacturers of pressurizing equipment offers a combined pressurizing and demineralising plant. On smaller systems some form of dosing pot may be used as described in Chapter 9 (see Fig. 9.14, page 195). For feed-connecting pipework it is probably worth using cupro-nickel or a similar material.

BOILERHOUSE INSTRUMENTS

The number and type of instruments to be provided in the boilerhouse depend on the size of the installation, the degree of control required and cost. The following instruments, however, are generally considered essential:

(a) Boiler or drum pressure gauge.
(b) Boiler thermometer.
(c) Flow-main thermometer.
(d) Return-main thermometer.
(e) Pressure gauges on pump suction and delivery.
(f) Water gauges on boiler, or drum, with steam-pressurized system; water gauge on pressure cylinder with gas-pressurized system.

Other instruments usual in large systems, roughly in order of usefulness:

(g) CO_2 indicator.
(h) Boiler or drum pressure recorder.
(i) Flue-gas temperature indicator.
(k) Draught gauge.
(l) Flow and return thermometers on each outgoing circuit.
(m) Differential pressure gauge across each boiler.
(n) Outside temperature remote-thermometer.
(o) Fuel meters.

Most of the above may be combined with recorders to give a continuous chart of the conditions obtaining.

Another instrument which is sometimes provided, and which is in any case useful for keeping an accurate check on operating costs, is a *Heat Meter*.

SAFETY DEVICES

If the pressure in a h.p.h.w. system were suddenly reduced, as for instance by the bursting of a pipe, large quantities of steam would be generated by the boiling of the superheated water. Automatic isolating valves were at one time installed in such systems to shut off in the event of excessive flow, but these have fallen into disfavour due to the considerable hydraulic forces set up on closure. Failures of piping systems have fortunately been rare and accidents do not appear to have had the serious consequences anticipated.

Low-flow devices are sometimes provided, particularly with tubular type boilers, with the object of sounding an alarm in the event of pump failure or cessation of flow from one cause or another.

PIPING FOR H.P.H.W. SYSTEMS

Piping is invariably of mild steel, suitable for the pressure. In this connection it is to be noted that at certain points of the system the pressure is in excess of the boiler working-pressure by an amount equal to the pump head.

Joints—Screwed joints are not suitable for H.P.H.W. piping, and should not be used, as they are liable to leak. All valves should be flanged, and elsewhere the piping should be butt-welded, either by oxy-acetylene or by electric arc to an approved specification.

All bends should be of large radius (at least 3 pipe diams.), and tees should be 'swept' so as to avoid sudden drops of pressure in the piping.

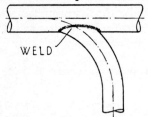

WELD

Connections—Connections to heating apparatus are best made with flanges, but to save expense the piping is sometimes welded direct to a 'tail' left on the heater battery, etc.

Headers and branches in the boiler-house may also be built up from standard tubing and bends welded together, but it is desirable to provide a certain number of flanges in the boilerhouse to facilitate erection and dismantling. Connections to pumps and boilers will be flanged.

Expansion of the Piping is a matter of the first importance, and should be carefully considered at every stage of the design. Short rigid connections between boilers, drums, headers, pumps, etc., should be avoided, and it is worth pointing out that whereas such a short connection might be practically unstrained if all the piping adjacent to it is full of hot water, cases will arise when part of the piping is out of use, and consequently full of cold water, with resulting uneven contraction, and strain.

The mains outside the boilerhouse will behave as regards expansion in exactly the same way as steam mains at the same temperature, and the basic considerations are the same as those which have already been discussed in Chapter 10.

Air Bottles—All high points of piping should be provided with vent pipes taken off the top of the main and run to an air bottle, as shown in Fig. 11.11 (a). This is made up from a short length of pipe about 150 mm bore, and has an air-pipe taken from the top to a convenient level and provided with a stop-cock, which should always be locked in some way to prevent unauthorized use.

Alternatively, a pocket may be formed on top of the pipe as in Fig. 11.11 (b), and an air-pipe taken off the top as before. This has the advantage that air bubbles carried along the top of the main by the water have a greater chance of being trapped in the large diameter pocket than of entering the relatively small vent pipe in (*a*).

Venting—Mains should be fixed to a slight fall to assist in venting, and ideally all mains should rise in the direction of flow so that the entrained air does not have to collect against the flow of water. This, however, would entail fixing the mains 'with falls in different directions, which is inconvenient, and is not generally necessary. The main which would be expected to contain the greater quantity of air (normally the flow main) should rise in the direction of flow, with the return main lying parallel to it.

Once the initial air has been removed there is generally little further accumulation. Due to the considerable pressure in the system, air is compressed to a much smaller volume than in the normal low-pressure system.

Valves—Trouble is often experienced in H.P.H.W. systems due to leakage past the glands of the valves. As soon as it is at atmospheric pressure, the water leaking out immediately evaporates and leaves behind any salts which it may contain, and this deposit may in time, if not regularly removed, be sufficient to prevent the valve from being turned. Trouble is particularly noticeable with small motorised valves where the operating torque is limited by the power of the motor. To overcome this difficulty many attempts have been made to design special glands for H.P.H.W., and this point should be borne in mind when specifying makes and types.

Valves used for H.P.H.W. should always be flanged, and many engineers prefer a screw-down oblique type valve as giving more satisfactory service than other types.

Regulating Valves—The amount of water passing through a H.P.H.W. system

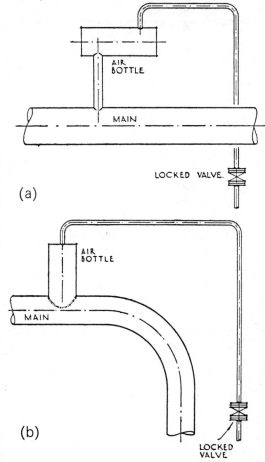

Fig. 11.11.—Air bottles on mains.

is relatively small, and the pump head available at the nearer branches may be large, so regulation is often critical. Hence some designers provide, in addition to the isolating valves on the flow and return to the branch, a third valve in the return which is used for regulating only, and once set this valve need never be disturbed. Alternatively, combined regulating and shut-off valves may be used, in which case regulation is effected by means of an auxiliary seating, which is so arranged that it cannot be tampered with after adjustment.

PIPING FOR MEDIUM-PRESSURE SYSTEMS

Medium-pressure hot water does not demand the high standards of high-pressure. Screwed joints at valves and equipment may be used and flanges are not necessary. Welding of pipe runs is nevertheless preferable to screwing, in view of the expansion forces involved, these being more severe than those associated with low pressure systems.

HIGH-TEMPERATURE AND MEDIUM-TEMPERATURE RADIANT HEATING, CONVECTION HEATING AND OTHER APPLICATIONS

Radiant Heating—The use of this type of heating is specially suitable for large spaces such as factories. The heating panels are of steel, having a coil welded to the plate, (see Fig. 11.12). The coil is supplied with high-

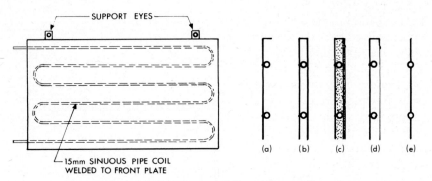

FIG. 11.12.—High temperature radiant panels.

Various types of panel:
(*a*) exposed coil; (*b*) double-sided; (*c*) insulated back; (*d*) shielded back; (*e*) plate welded between coil. The sizes of the panels are approximately 2·5 m by 1·25 m and 2 m by 1 m. The pipe coil is welded to the steel plate at 150 mm centres.

pressure or medium-pressure hot water. Temperatures of the surface are of the order of 90° to 135° C depending on the temperature and pressure of the water. At these temperatures, flat metal plates emit powerful radiation, felt at a long distance. (See Plate XXI, facing p. 215.)

It is usual to suspend the panels vertically along the walls or between columns of a large shed about three to four metres above the floor. They may alternatively be fixed horizontally overhead or inclined at an angle.

Another method of achieving the same object is by the use of strip heating (see Fig. 11.13) which in one form comprises aluminium plates, clipped to heating coils, for running continuously down a workshop, overhead. (See Plates III and XXII, facing pages 86 and 246, the latter showing an example integrated with the structure.)

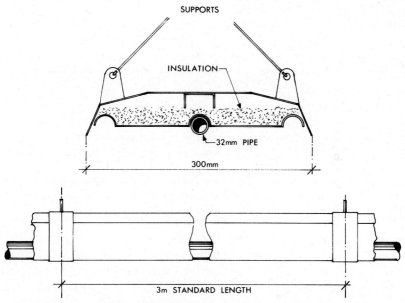

Fig. 11.13.—Strip heating (foil insulation is omitted for the convective type).

A further development, in the case of strip heating, is to combine it with provision for lighting by means of fluorescent tubes; the heating surface then also forms the lighting reflector.

The advantages of high-temperature radiant heating are:

(1) Uniform distribution of heat.
(2) Absence of temperature gradient.
(3) Objects are warmed rather than air.
(4) Reduced air change losses due to the above.
(5) No moving parts or maintenance.

The calculation of heat requirements for radiant systems in buildings is referred to in Chapter 2 (see p. 40).

Table 11.3 gives emissions for one make of panel. It will be noted that the convection and radiation components are given separately. In designing such a system it will be borne in mind that the convection component, together with convection losses from piping, serves to counteract the cooling effect of the roof. Convection currents rising to above may be regarded as supplying the heat necessary to counteract the cooling effect of the roof glass, and it is desirable to achieve a rough balance between these two quantities. It will, for instance, generally be found desirable to insulate large main piping, otherwise the convection component in the building may greatly exceed what is necessary to deal with roof losses, and such would then lead to temperature gradients which are to be avoided.

TABLE 11.3

EMISSION OF HIGH-TEMPERATURE RADIANT PANELS
kW PER PANEL

Panel size 2·5 m × 1·25 m. Freely exposed and uninsulated. Ambient air, 15° C. Water temperature drop across panel, 30 K.

Hot Water mean Temp. ° C.	Convection	Radiation	Total
100	1·9	3·1	5·0
125	3·5	4·5	8·0
150	4·7	5·9	10.6

Values are approximate for one make with plate welded between coils. For accurate emissions, consult makers' data.

Radiant panels may be freely exposed back and front where hung between columns down the centre of a shop but, where against an outside wall, it is usual to protect the back against back radiation either by insulation or by a backing plate with air space. Where fixed horizontally overhead it is also probably desirable, in most cases, to insulate the back to avoid undue wastage of heat to above.

The advantages of strip heating are that fewer connections are involved and therefore the overall cost should be less. Also in certain cases there should be less obstruction than with hot surfaces supported near walls and columns etc.

Outputs range from about 0·5 kW/m run with a single pipe at 100° C (air at 15° C) to 3kW/m run with four pipes at 150° C. Mounting heights as given by the makers are a minimum of three metres, using low-pressure hot water, and five metres using medium-pressure or high-pressure hot water. There is no upward limit to the mounting height.

High-Temperature Heating by Convection—High-pressure hot water may also be used for the heating of large spaces by means of unit heaters, convectors, and pipe coils.

Unit heaters for high-pressure hot water follow closely the designs already referred to for steam in Chapter 10, except that connections flow and return are the same size and no steam trap is required. In the design of equipment for high-pressure hot water heating an attempt is usually made to keep the velocity of water high, by arranging circuits in series rather than in parallel.

Convectors used in conjunction with H.P.H.W. also follow the designs already referred to for steam and low-pressure hot water. Their use is usually confined to the heating of small offices. Where it is a matter of serving an extensive office block adjacent to factory buildings, more advantageous results can usually be obtained by adopting low-pressure hot water.

Hot-Water Heating—For a building or group of buildings where H.P.H.W. heating is not suitable, a low-temperature radiator system may be installed, and served from a water-water calorifier using H.P.H.W. for the primary heating.

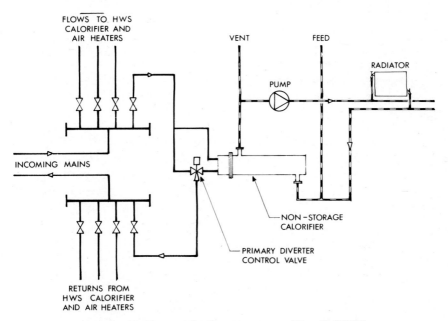

FIG. 11.14.—Diagram of radiator system served from H.P.H.W.

Fig. 11.14 shows the piping and the apparatus for the heating and ventilation of the office-block in a factory heated by H.P.H.W. The incoming H.P.H.W. mains are connected to headers which in turn serve the heating calorifier, with its own low-pressure radiator system, pump-circulated. In addition, connections will be seen from the H.P.H.W. headers to H.W.S. units and plenum air-heaters.

The design of the low-pressure system will be exactly as given in Chapter 8. In sizing the calorifier a suitable margin should be added to the net heat losses to allow for time-lag and heating-up.

Other Applications—H.P.H.W. may be used as a source of heat for many sorts of apparatus, among which the following may be mentioned.

Steam Generators. (which are in effect water-steam calorifiers) comprise a primary coil served from the H.P.H.W. system which heats water in a cylinder above which is connected a second cylinder partly filled with water, the space above forming a steam space. Steam so produced indirectly is usually at a low absolute pressure of up to about 200 kPa (2 bar).

Vats and Boiling Vessels of all types. In these a coil is placed in the bottom of the vessel, or in a separate chamber at the side with mechanical circulation.

Baking and Enamelling Ovens, and similar apparatus requiring an intense radiant heat, and hot air for drying.

Vulcanizing and Plastic Presses where hot surfaces are required to promote a physical or chemical process.

Laundry Dryers, etc., in which hot dry air is used. The air is circulated over the H.P.H.W. coil by means of a fan.

The details of the design of such apparatus are beyond the scope of the present book, but it will be seen that all the processes normally served from a steam-heating system may equally well be provided by H.P.H.W. with the added advantage of greater range of control by valve operation.

PIPE SIZING FOR HIGH-PRESSURE AND MEDIUM-PRESSURE SYSTEMS

Emissions from bare and insulated piping may be estimated from Tables 11.4 and 11.5 derived from data in the *Guide*, Section C3.

The process of pipe sizing follows that for a low-pressure two-pipe system. Loadings in kW are summated, mains losses are apportioned and

TABLE 11.4

THEORETICAL HEAT EMISSION FROM BARE SINGLE HORIZONTAL
STEEL PIPES IN AMBIENT AIR BETWEEN 10° C AND 20° C

Nom. bore mm	Heat Emission W/m run for stated temperature difference pipe surface to air, K							
	100	110	120	130	140	150	160	170
15	140	150	170	190	220	240	260	290
20	160	190	210	240	260	290	320	350
25	200	230	260	290	320	350	390	430
32	240	280	310	350	390	430	480	520
40	270	310	350	400	440	490	530	590
50	330	380	430	480	530	590	650	710
65	410	470	520	590	650	730	800	880
80	470	530	600	680	750	830	920	1010
100	580	670	750	840	940	1040	1140	1260
125	700	800	900	1010	1120	1250	1370	1510
150	810	920	1040	1180	1310	1460	1600	1760
200	1040	1190	1340	1520	1690	1880	2060	2270

Note: 1. The above table is derived from the *Guide*, Section C3.
2. For 2 pipes in a bank, factor = 0·95
For 4 pipes in a bank, factor = 0·85

provisional sizes determined on the basis of the design temperature drop, using Fig. 8.6 as already explained. Accurate sizing may then follow circuit by circuit, be duly balanced and result in the final pressure-loss calculation from which the pump head is established

A word of caution should be added here with regard to the accuracy necessary in the sizing and balancing of piping for H.P.H.W. systems. For a given heat load, the water quantity circulated in this type of system is much less than that discussed in Chapter 8, the temperature drop, flow to return, being greater. In consequence of this fact, balancing is much more critical to the achievement of successful overall distribution and operation.

TABLE 11.5

HEAT EMISSION FROM PIPE INSULATED WITH
MATERIAL OF THERMAL CONDUCTIVITY 0·045 W/mK e.g. GLASS FIBRE

Nom. bore mm	Heat Emission W/m run For stated temperature difference pipe surface to air, K							
	Insulation 25 mm thick				Insulation 50 mm thick			
	110	130	150	170	110	130	150	170
15	23	27	31	35	18	21	25	28
20	28	33	38	43	19	22	26	29
25	31	37	43	49	22	26	30	34
32	35	41	47	53	24	28	33	37
40	39	46	53	60	26	31	35	40
50	45	53	61	69	30	35	41	46
65	54	64	74	84	34	40	46	53
80	61	72	83	94	39	46	53	60
100	74	87	100	113	46	54	63	71
125	88	104	120	136	54	64	74	83
150	101	119	137	155	62	73	85	96
200	130	154	178	202	77	91	105	119

Note: 1. Values are derived from *Guide* factors.
      2. Other thicknesses between 25 mm and 50 mm may be interpolated with slight inaccuracy.
      3. For other insulating materials and conductivities, the above valuers may be multiplied in ratio of the factors in the *Guide* to 0·045.

CHAPTER 12

Heating by Electricity

ELECTRICITY IS THE MOST REFINED of all forms of energy supply and the most versatile. This is not the place to enlarge on its manifold benefits to mankind whether for lighting, power, traction, electronics or communications. To make use of such a sophisticated supply for the common run of space heating applications is however to degrade the refinement and misapply the versatility.

As a source of energy for space heating, nevertheless, electricity offers many advantages to the user.

(a) Transmissibility to any point regardless of physical levels or limitations found with other media.
(b) Absence of fumes or products of combustion.
(c) Avoidance of on-site labour in operation.
(d) Availability at any temperature at point of use and hence great freedom in design of apparatus.
(e) Shortness of time lag in availability.
(f) Ease of thermostatic control.

To produce electrical energy from a base fuel however involves much wastage since more than three quarters of the calorific value of that fuel, as supplied to the generating station, is lost in cooling towers, rivers or canals or in the distribution system. The contra argument usually advanced is that the grade of base fuel used at generating stations is so low that it would be difficult if not impossible to burn it other than in large power plant boilers. Whilst this argument is valid as advanced, natural gas is burnt in the boilers of some generating stations and the value of the low grade fuels as chemical feedstock cannot be ignored. No argument, however, can excuse the heat wastage implicit in present generation policies and methods.

Practical Economics—In the day to day sense, the question of whether electricity is used for a particular application revolves around comparative assets. In making any comparison not only must the cost of the alternative forms of energy be considered but also the capital cost to the user of the heating system. Furthermore, the total cost must be set against the amount of heat available to the user and not the energy input into the system; electrical systems are often more efficient convertors of energy input into useful heat output than other systems.

Electrical Supplies—As with any product, the cost of supplying electricity is a mixture of capital and day-to-day running costs: in the case of electricity however the capital component is a higher proportion of the total than is usual.

In the U.K., most of the CEGB output capacity (over 60 per cent) is accounted for by conventional coal-fired steam stations. Oil-fired stations account for about 20 per cent and some 14 per cent is provided by natural gas, gas turbine, diesel and hydro electric plant. The remainder, 6 per cent, originates from nuclear power stations. These main types each have a different capital/running cost relationship and as such each is more suitable in terms of cost for different patterns of loading. Nuclear stations are more economical for meeting the base load and gas turbine stations for the peak load of the supply system.

Capital costs are the reflection of the production assets required to meet the maximum simultaneous demand of all consumers for electricity and thus, in order to satisfy the statutory obligation of equitable pricing, it is necessary to determine consumers' responsibility for capital costs. Running costs are incurred predominantly for fuel to produce the electrical energy requirements and can thus be equitably recovered per unit (kWh) of electricity consumed. Generating stations will themselves have different costs of production depending on age, design, etc., but the National Grid allows the most efficient stations to be run at any given time to match the load at that time.

Variations in Demand—Because of climatic conditions, the demand for electricity in the U.K. is greatest in winter, when casual heating requirements are superimposed upon those for lighting and the normal industrial loads. Because of normal living and working habits, this demand varies as between weekdays and weekends and between one hour of each day and another.

A typical load curve has a morning peak which usually extends from 8 a.m. onwards to about noon and a lesser peak in the late afternoon. Such changes in demand, in-so-far as they relate to lighting and motive power etc., are inevitable but nevertheless create problems for the supply industry. It has been therefore to the heating component of the demand curve that the supply industry has looked in search of some solution.

Whilst electricity, uniquely among energy supplies, cannot be stored as such, it may be converted without loss into heat which can be stored using both old techniques and new approaches developed over the last 30 years. Hence, the supply industry has done much to encourage the development and use of heating equipment of storage type since this presents the dual advantages of not only reducing peak demand but also of increasing off-peak demand.

Tariff Structures—Because plant of overall higher efficiency operates at night, generation costs are lower then than during the day. If the costs of metering were of no consequence either to the supply industry or to the consumer, then metering and pricing of both demand and energy might be the universal and preferable tariff practice.

For the larger consumer, the method of charge usually takes the form of a maximum demand tariff with separate components for demand and for energy. The demand charge may be presented as a single annual rate related to the highest demand taken in the year; as separate but equal demand charges related each month to the highest demand of that month or as separate demand charges each month varying from high prices in the mid-winter months to low prices in summer months. In each alternative case, the charge for energy used may be at a single rate or in day/night form.

Such metering methods would, however, be disproportionately costly for domestic consumers. For them, the tariff used takes the form of a standing charge which reflects the cost of making supply available, plus unit rates set at the appropriate level to recover both demand-related and energy-related costs. This may be presented as a single rate for all units of energy no matter when consumed; as a separate and lower rate for energy consumed for specific purposes only during the night, or in day/night form with one rate for all use in daytime and a lower rate for all use at night. For domestic consumers the 'off-peak' rate is between 50 and 60 per cent of the price for normal tariff consumption.

One effect of the alternative forms of tariff is that electrical energy used at night is available at the lower end of the price range to all consumers. Hence, the storage techniques previously mentioned are commonly applied, to use electricity by night to store heat in some form for use during the following day.

Electricity Boards in the U.K. are independent organisations and their tariffs are related to the particular supply and distribution circumstances arising in the geographical areas which they serve. These factors vary from area to area and are reflected in the hours during which night time 'off-peak' tariffs apply. In general terms, however, roughly eight hours of such supply are available each day, commonly from 11 p.m. (2300 GMT) until 7 a.m. (0700 GMT). It is important to note, of course, that during the period when British Summer Time is in force the supply availability period remains as for the winter months and may thus be out of phase with storage requirements.

Distribution Capacity—If a building is to be heated entirely by electricity, the installed load will be many times that of the supply required for normal lighting, power and other services.

It is also the case that, whereas the other services have some diversity as between maximum demand and connected load, it is not safe to assume a diversity where heating load is concerned, seeing that in extremely cold weather every available heater is more than likely to be full-on at the same time.

By transferring the heating load to the off-peak period, the distribution capacity for normal supplies can, in effect, be available for the heating load, but even so in a given area the total of a large number of buildings electrically heated may bring about a condition where the available capacity is fully used up. In such a case, additional capital charges may become involved in order to meet the electric heating load, and the viability of the off-peak tariff is in jeopardy. The three hour afternoon boost period, which some Electricity Boards provided at off-peak rates up to about 1970, is thus no longer available for new consumers and, instead, new 'White Meter' tariffs are being offered designed for an eight to ten hour night-charging period. In consequence, all new storage heaters are uprated as 'high-capacity' heaters.

Units—It is as well to restate here that the watt is a *rate of energy flow*. One watt equals one joule per second. Thus the kilowatt-hour (kWh), or unit of electricity, is equivalent to

$$\frac{kJ \times 3600\,s}{s} = 3600\,kJ = 3 \cdot 6\,MJ.$$

In passing, however, it should be noted that the kilowatt-hour is not pure SI since the *hour* is not a preferred unit for time.

In British units, the *therm* of 100 000 Btu equals $1 \cdot 055 \times 10^5$ kJ. The equivalent of the therm is then

$$\frac{1 \cdot 055 \times 10^5}{3600} = 29 \cdot 2\,kWh.$$

The therm, in fact, is roughly equal to $MJ \times 10^2$, which would be a convenient unit which might possibly be called the *new therm*. But, again, a multiple of 10^2 is not preferred in SI.

Comparative costs of heat from electricity and from various fuels were given in Table 3.2 (page 68) in terms of 100 MJ but these relate only to the market at the time of going to press. Fig. 3.16 (page 68) may be used to prepare other comparisons.

If on-peak supply is taken, the use of electricity is an expensive way of heating; but in spite of this, in view of its flexibility and almost universal availability, it is still used to a considerable extent. This is particularly so where heating is required intermittently and energy can be taken just when it is required with no no-load losses: heating systems which rely by design upon on-peak electricity as the sole source of energy input are wasteful with respect to dwindling raw fuel supplies. Some of the various methods of using electricity direct for heating in rooms were referred to in Chapter 3 and it is not proposed to consider them further.

Given some guarantee of a reasonably acceptable tariff such as may apply to an off-peak supply, and if the building is designed with a view to economy of heat-requirement by general weathertightness and exceptionally good use of insulation (U $= 0 \cdot 4$ W/m² K average),* the kind of

---

* The Electricity Council recommend $0 \cdot 3$, $0 \cdot 4$ and $0 \cdot 5$ W/m²K respectively for roofs, walls and floors.

running cost involved may in many cases be acceptable, particularly bearing in mind the saving of labour, maintenance, storage space, flue, etc. It is in view of the importance of this method of heating that further attention will be devoted to it in this Chapter.

OFF-PEAK ELECTRIC HEATING

The various methods in common use for storing heat taken electrically off-peak at night for use during the following day are:

Storage Radiators, in which the heat is stored in a solid material contained in an insulated casing and given out again continuously and uncontrolled;

Storage heaters, in which the majority of heat is retained and given out when required, known as 'storage fan heaters'—a type which is the basis of the 'Electricaire' warm-air system;

Warmed floors in which the heat is stored in the screed and floor slab;

Warmed walls in which the heat is stored in a 50 to 100 mm masonry covering to conducting cables or sheets. These are, in effect, in-built storage radiators;

Warmed walls in which the heat is stored in purpose made cores incorporated in the structure and provided with air-ways and a fan for circulation. The wall surfaces are insulated. These are, in effect, in-built 'Electricaire' units;

'Centralec' systems in which heat is stored in solid material but used to supply an air/water heat exchanger connected to normal radiators or where storage is in water in small cylinders for similar connection;

The traditional Thermal Storage system, in which the heat is stored in water in large cylinders for use in industrial or commercial buildings.

Amount of Heat Required—The amount of heat required over a 24-hour period will depend on which of the above methods is adopted. Where the heat is emitted uncontrolled, as in the case of floor- or wall-warming and *input-controlled* block storage radiators, heat is liberated from the surfaces of floor or storage casing during the whole period of warming-up at night, as well as during the period of use during the day. In effect, this amounts to continuous heating throughout the 24 hours, though with some dropping off of temperature towards the end of the discharge period, depending on the capacity of the storage medium in relation to the heat losses.

In the case of *output-controlled* storage fan-heaters, 'Electricaire',

'Centralec' and the water thermal storage system, although there will be some losses at night, the bulk of the heat to be taken during the off-peak period will be that used for heating during the day.

The calculated heat losses per 24 hours may take into account the fact that at night-time a lower air-change rate is to be expected (see Table 12.1).

TABLE 12.1

RECOMMENDED DESIGN VALUES OF RATES FOR VENTILATION LOSS BY NATURAL INFILTRATION IN MULTI-STOREY OFFICE BUILDINGS*

Type of Building	Ventilation Loss (W/m³ K)			
	Height of Building			
	15 m and under		Over 15 m	
	Day	Night	Day	Night
Buildings with little or no internal partitioning (i.e. 'open plan' buildings), or with partitions not of full height, or buildings with poorly-fitting windows and internal doors.	0·5	0·3	0·7	0·3
Buildings with internal partitions of full height, without cross ventilation, and having self-closing doors to staircases, lift-lobbies, etc.	0·3	0·2	0·3	0·2

Note: The basic design temperature should take into account the fact that there is no inherent reserve and a basic temperature of − 3·3° C for single-storey buildings, and − 1·7° to − 1·1° C for multi-storey buildings is proposed.*

In the case of a floor-heated building, it is necessary to add to the net heat requirement the expected downward loss to the ground or to a room beneath the heated floor. If the total heat requirement so found is denoted by QkW, the installed load for an *input-controlled* system will be $24Q/n$ where n is the available hours of charge.

Where the *heat-output* is controlled, for an occupied period of 12 hours, and losses assessed at the equivalent of 4 hours, the installed loading required would be $16Q/n$.

It is apparent from experience that the successful application of off-peak heating whether by floor or block storage depends on adequate thermal storage in the building structure to prevent too great a swing in temperature during the day when the power is off. The temperature swing should not exceed about 3·5 K; which means that if it is desired to maintain 18° C average, the temperature in an office block will be up to 19·75° C in the morning and down to 16·25° C in the afternoon. For domestic usage this might not be satisfactory, since it is in the evening that heat is usually most required, and hence the importance of a mid-day boost in such cases. Some of the heater output during the day will,

* B.S.R.I.A. (late H.V.R.A.) Report, May 1961, The Heating of Buildings by Off-Peak Electricity Supplies.

however, have been stored in other room surfaces, walls, partitions and furniture and release of this during the evening may meet the apparent deficit.

In the case of thermal storage systems using water, these considerations do not arise, as the available storage can be made as great as required simply by designing the capacity of the storage vessels accordingly.

Block Storage Radiators—Block storage radiators enable heat produced electrically to be stored during off-peak hours actually within the rooms to be heated. They are therefore 100 per cent efficient, and incur no losses in piping etc.

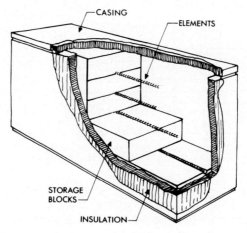

FIG. 12.1.—Block storage radiator (controls not shown).

The radiator (as in Fig. 12.1) comprises a series of electric resistance elements enclosed in a ceramic or concrete based material contained in an insulated steel casing. Loadings vary with different makes but are commonly 2, 2·6 and 3 kW. The maximum temperature achieved on the surface of the casing after the charging period is about 85 K above ambient (BS3456, Part A11) and this declines to about 40 K at the end of the day. The storage material reaches a temperature of about 200° C and the maximum rate of output is approximately 1 kW/m² of casing. Table 12.2 gives leading particulars of typical units.

External (room) thermostatic control is rarely used—the 'charge controller' with its somewhat limited ability to control energy input is usually found adequate. A high temperature cut-out is included to prevent overheating. A typical curve of output from a storage radiator is shown in Fig. 12.2. The 'half-life' indicated is the point in time when output has fallen to 50 per cent of the maximum. It will be noted that about 30 per cent of the 24 hour output (the area under the curve) is released during the charging period but some of this, as has previously been mentioned, will pass to storage in other room surfaces.

Storage Fan Heaters—This development of the storage radiator is one in which the lagging is increased so as to reduce the amount of standing (uncontrolled) heat discharge, the unit being equipped with a fan arranged

TABLE 12.2
BLOCK STORAGE RADIATORS.
TYPICAL RATINGS AND DIMENSIONS (DIMPLEX)

Item	Input Rating (kW)		
	2·0	2·6	3·3
Energy (MJ)			
Acceptance (8 hour)	58	75	95
Active store	39	50	64
Output (W)			
Maximum	1050	1345	1685
Minimum	400	540	670
Temperature (K)			
Front panel, above room	80	80	80
Dimensions (mm)			
Height	614	699	699
Length	735	735	960
Depth (overall)	292	292	292
Weight (kg)			
Total	123	150	199

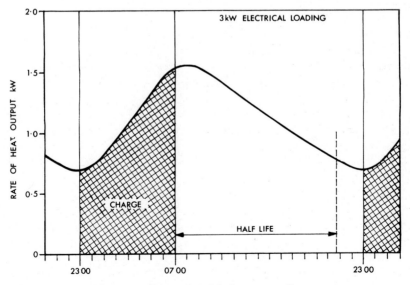

FIG. 12.2.—Output from block storage radiator.

to drive air through passages in the heat storage blocks and out into the room to be heated. Standing losses from the casing over 24 hours are less than 20 per cent of the heat input.

So as to provide a uniform air-outlet temperature, a certain degree of mixing of air from the room with the heated air from the block takes place by means of a thermostatically controlled damper. A unit of this type can be controlled by means of a thermostat in a room, and, if desired, by a

TABLE 12.3

STORAGE FAN HEATERS.
TYPICAL RATINGS AND DIMENSIONS (CONSTOR)

Details	Input Rating (kW)			
	3	4·5	6	7·5
Energy (MJ)				
Acceptance (8 hour)	86	130	173	216
Active Store*	77	118	152	204
Output (W)				
Uncontrolled	300	400	400	400
Controlled†	1480	2330	3350	4320
Dimensions (mm)				
Height	762	762	762	762
Length	1118	1118	1118	1118
Depth (overall)	331	432·	432	432
Weight (kg)				
Total	185	230	278	355

* Assessed as 90% of 8 hour acceptance.
† Balance of active store, 12 hour constant output rate.

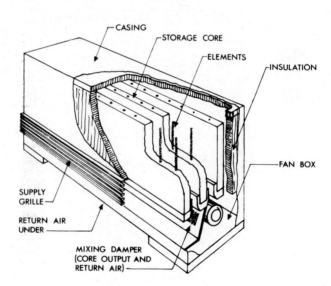

FIG. 12.3.—Storage fan heater.

time switch, so that although the heat supply is taken only during the night, the bulk of the heat stored is released according to requirements during the day. Such units are commonly provided with a two speed fan so that capacity for rapid room temperature rise is available.

If the day is mild, room temperature will be quickly achieved and a small amount of heat only will be withdrawn from the unit. If the day is cold, a larger amount will be withdrawn. Provided the heat storage is adequate to last throughout the day, the result should be one in which a uniform temperature is maintained internally without any problem of anticipatory control or swings of temperature, such as are inherent in other systems so far described. A unit of the controlled type is seen in Fig. 12.3 and dimensional details etc. are listed in Table 12.3.

'Electricaire' Heaters—An extension of the principle of the fan storage heater led to the development of a larger unit designed to heat a complete dwelling. Whereas the normal run of fan heaters provides for service to a single room and, not withstanding weight, could be thought of as being semi-portable, the 'Electricaire' type of unit has a larger capacity and is a permanent fixture within the dwelling.

Ratings are in the range of 6 to 12 kW, to the dimensions given in Table 12.4, but much larger units of similar type have been made, for application to commercial and industrial premises, up to about 100 kW. Construction is as illustrated in Fig. 12.4. In common with domestic warm

TABLE 12.4

'ELECTRICAIRE' HEATERS.
TYPICAL RATINGS AND DIMENSIONS. (CREDA)

| Details | Input Rating (kW) | | | | | | | | |
| | Concealed or free standing | | | | | | | Free standing | |
	5	6·25	7·5	8·75	10	12	14·4	6·75	9
Energy (MJ)									
Acceptance (8 hour)	144	176	209	241	274	317	382	184	248
Active Store	133	162	194	220	252	292	356	162	216
Output (W)									
Uncontrolled	560	600	650	700	730	750	860	700	800
Controlled*	2330	2950	3620	4160	4860	5760	7090	2820	3930
Dimensions									
Height†	1191	1191	1275	1438	1438	1605	1605	2245	2245
Width	610	610	610	610	610	610	610	832	832
Depth	610	610	610	610	610	610	610	350	350
Weight (kg)									
Total	263	318	375	431	488	579	627	436	499
Air flow (litre/s)									
At 60° C	104	118	132	146	160	188	222	86	118

* Balance of active store, 12 hour constant output rate.
† Above plenum for concealed/free standing type. Overall for others.

air units fired by gas or oil, air distribution to the various spaces within a dwelling may either be direct via stub-ducts from a plenum box fitted below the heater or by insulated below-ground air ducting from the heater to chosen discharge positions.

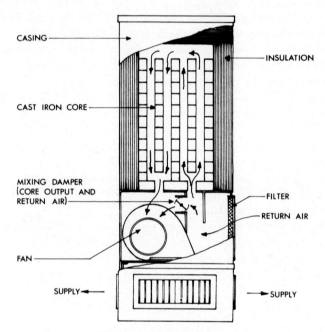

FIG. 12.4.—'Electricaire' unit.

As in the case of fan storage heaters, control of the night-time input of energy is arranged so as to reflect seasonal output requirements and speed control of the fan provides a heating boost facility. Ideally, the unit should be selected such that all energy input is during off-peak periods but limitations in available space (and hence in storage capacity) may require that a day-time boost charge is taken via a White Meter. Units are commonly arranged such that half of the supply elements are activated when the output temperature from the store falls to, say, 50° C.

EMBEDDED FLOOR HEATING

The withdrawal of facilities for a mid-day boost at off-peak rates has reduced the application of floor heating to those circumstances where a decline in temperature during the afternoon and evening is unimportant, or to cases where the extra cost of an on-peak boost is acceptable. The 'White Meter' tariff takes care of such a means of operation.

In concept, an electrically heated floor is not dissimilar to that described in Chapter 3. A large proportion of the heat output is radiant, the temperature gradient in the space served is small. A floor screed at least

75 mm thick is required in either case to prevent cracking and to avoid local 'hot spots' above the source of heat supply. In the present context, the heating elements are resistance wires, the total length being such as to produce the loading required, calculated as described hereafter. This length of element is laid out in grid-iron formation in one, two, three or more circuits according to size of cable, its rating and the length required. Spacing of the coils will vary according to the loading required and to type of floor finish, bearing in mind that a hard finish emits heat more readily than say a carpet. Centres are often 150 mm, 230 mm or 300 mm. The surface temperature should not exceed 24° C, though it may be assumed to run up to 30° C first thing in the morning at the end of the charge with the result that the mean 24 hour temperature will be about 25° C.

If it is not possible, in the area of the floor available, to accommodate sufficient loading to meet the heat losses over the 24 hours, supplementary heating will be required, either direct on-peak or by night storage radiators.

Various forms of resistance elements are available; proprietary systems make use of:

Copper sheathed mineral insulated cable laid solid;
PVC sheathed/EPR (Ethylene propylene rubber) insulated;
Silicone rubber.

Methods of laying include:

Burying solid in the screed;
Drawing through steel tubes running between troughs, as for instance on the two sides of a room;
D-shaped ducts connecting to troughs overlaid with expanded metal mesh.

Construction—The screed requires to be not less than 75 mm in thickness, of a strong mix, such as one part cement to $3\frac{1}{2}$ sand and aggregate. Where the floor is at ground level, edge insulation as shown in Fig. 12.5 is the minimum necessary but overall coverage is to be preferred. Where the heating is applied to a multi-storey structure and the screed is on a suspended slab, the insulation is preferably placed below the slab so that the heat-storage effect thereof assists in reducing temperature swing, though it is undesirable that the slab mass should be too great or the system will become unduly sluggish. If the insulation is placed immediately beneath the cables, the swing will be much greater and the safe operating temperature of the cables might be exceeded.

Most floor finishes have been used on floor-warming installations, but precautions are necessary to ensure that if adhesives are used, such as for linoleum, rubber and cork, they will be suitable under heating conditions. Residual moisture should be driven off by operating the system before final finishes are laid, particularly where a wet finish is to be applied.

Various hardwoods and softwoods are suitable. Where carpets are laid with an underfelt, if too thick, the temperature below may rise to too high a level. Underfloor insulation is essential if a high resistance floor finish is used; otherwise there will be considerable loss of heat to the ground if on a ground floor, or to the room below in the case of an upper floor.

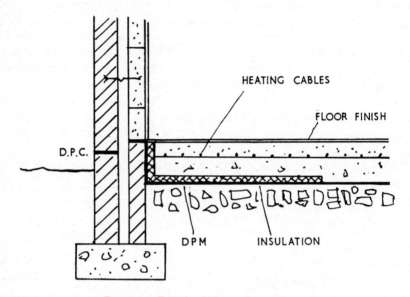

FIG. 12.5.—Edge insulation to heated floor.

Heat Transmission to Below—The loss of heat from a heated floor at the ground level, depending on size and shape, may amount to about one-third of the total input in the case of small floors, or to about one-tenth for larger floors. The major part of the downward loss occurs at the edges of the floor, hence insulation at the centre might have an insignificant effect although application beneath the whole area is to be preferred. The downward heat loss through intermediate floors may be 25 per cent of the heat supplied if insulation is not provided, but would be reduced to perhaps 5 per cent with insulation.

The downward losses are susceptible of calculation, but it is clear that they should be taken into account when estimating loadings and consumptions.

Design Calculations—Because the operational characteristics of a floor heated off-peak are relatively critical, the design calculations necessary are somewhat tedious. The unit emission of the heated area is a function of the difference between the 24 hour mean surface temperature of the floor and the Resultant Temperature of the room. This latter is derived as the mean between the required design air temperature and a weighted average of all room surface temperatures 'seen' by the heated floor.

Taking a 24 hour mean floor temperature of 25° C and assuming that a Resultant Temperature value of 17° C has been calculated, the unit emission from the floor would be

$$q = 9 \cdot 65 \ (25 - 17) = 77 \ W/m^2 \ (average)$$

Knowing this value, the proportion of the floor which must be heated may be calculated from a 24 hour heat loss which has been determined in the conventional way, allowing for reduced night-time ventilation loss as proposed in Table 12.1.

To the upward emission must be added the unwanted downward loss mentioned in a previous paragraph, the sum being the total output. This loss, in practice, is calculated from knowledge of floor size, insulation values and temperature of the plane of the heating cable. For the purpose of illustration, assume a value of 15 W/m² (20 per cent of upward emission) and hence the required installed loading may be calculated, for an 8 hour charge, as

$$q = 24/8 \ (77 + 15) = 276 \ W/m^2$$

From these results, further calculations are necessary, principally to ensure that the temperature at the heating element is not excessive and that the maximum daily swing in room air temperature will not exceed 3·5 K.

Reference may be made again to the B.S.R.I.A. (late H.V.R.A.) publication, dated May 1961, on the *Heating of Buildings by Off-peak Electric Supplies* and also to *Recommendations for the Design of Floor Warming Installations* issued by the Electricity Council.

Plate XXIII, facing page 246, illustrates the installation of an electric floor-warming system.

Control of Floor-Warming Systems—The simplest method of controlling a floor-warming installation is by a room air-thermostat which switches the power off when the desired temperature has been reached. The time switch installed will bring the supply on once again at the commencement of the charging period and, when the desired indoor temperature has been reached during the night, the off-peak supply will be cut off by the thermostat. Should the temperature drop, the thermostat will bring the supply on again, and so on until the off-peak period of availability is over. Such a system of control may be wasteful of energy in that unnecessarily high night-time temperatures may be maintained.

Greater economy may be achieved by a control system which delays the commencement of the night-charging period as long as possible, such that in the remaining hours of supply the temperature of the floor will have been brought up to that desired by about the time that the supply is terminated. This involves a time-temperature controller which is responsive to external weather. What is required is some form of anticipatory control, seeing that the heat input during the night will have to last for the next twelve hours or so and, should there be a sudden drop of temperature, internal warmth may suffer. In a building of substantial construction with good thermal mass, such sudden changes may not be serious. On the other hand, a warm day following a cold one might result in discomfort in the rooms through overheating. There is no means of cutting down the amount of heat liberated from a warm floor.

In the case of a block of flats, time-temperature controllers for each user may not be practicable, and a single master controller may be used.

A thermostat may be placed in the floor itself, where it acts as a high-temperature limit to prevent the floor temperature rising beyond an acceptable level. It will not serve for controlling the room temperature.

An application of electric floor warming not concerned with buildings may be mentioned here: namely, the heating of road-way ramps to prevent frost. For this application, a steel-mesh element operating at a low voltage has become popular in recent years.

THERMAL STORAGE—HOT-WATER SYSTEM

This system is indirect. Heat is stored in the form of hot water raised to as high a temperature as possible, and contained in large cylinders generally placed in the basement. The heat is generated either in an electrode heater utilizing the resistance of the water as the element, or in smaller systems by immersion heaters. The apparatus is automatically controlled by thermostatic and time switches. (See Plate XXIV, facing page 246).

Operation—Fig. 12.6 shows the general arrangement of a typical thermal storage system. The principle of operation may be described as follows:

(a) The electrode water heater (or separate immersion heater) warms the water in the storage cylinder, the pump accelerating the circulation. If the immersion heaters are contained in the main storage no pumping is necessary.

(b) This heating takes place at off-peak hours, and in order to economize in capacity of storage the temperature is raised as high as possible without generating steam. This may be as high as 185° C, and is determined solely by the pressure which can be applied to the storage vessels. This, in the case of a tall building, may be the static pressure available from a tank fixed at roof level. Alternatively, an artificial head can be produced for pressurizing by means of a separate air or gas cushion cylinder, as described in Chapter 11. The storage temperature is generally kept about 10 K below that at which steam would be produced.

(c) At a given time the current to the heater is cut off automatically

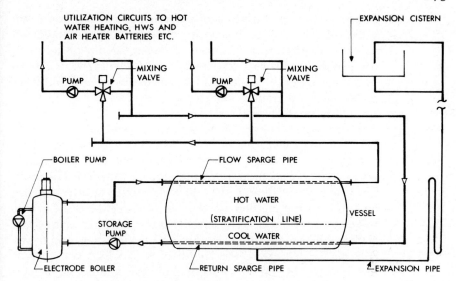

FIG. 12.6.—Diagram of hot water thermal storage system.

by time switch, or from the supply authority's sub-station. After a short delay, the storage pump is stopped.

(d) When heating in the building is called for, the heating pump is started; high-temperature water is drawn off from the cylinder and returned cooler at the bottom. A proportion of return water is mixed with the flow through a thermostatically controlled mixing valve. Thus, with possibly 140° C in the cylinder, 80° C may suffice for radiators or other heating equipment.

(e) This process continues all day, the level of cool water at the bottom of the cylinder gradually rising, but not mixing with the hot, and by night time in cold weather probably little hot water is left in reserve.

(f) When heating is no longer required the secondary pump is stopped.

(g) At a given time, when off-peak current is available, the supply is turned on to the boiler and the process repeated. The primary pump is simultaneously started.

The storage capacity is controlled by the heat output of the plant, its period of use and the temperature at which it can be run. The electrical loading of the heaters is likewise determined by the daily total heat load and the number of hours during which current is available.

Types of Heater—The heater may be of two types according to the voltage and size of the installation:

(a) *Immersion heaters*, consisting of resistance elements inside metallic tubes or blades. These are suitable for low voltages up to 250, or, when in a balanced arrangement, up to 500, and are generally of 3 to 4 kW loading each, arranged in groups or banks of 50 kW or more. This type of heater is often placed direct in the bottom of the storage vessel as in Fig. 12.7. Installations rarely exceed 300 kW.

(b) *Electrode heaters*, connected to a medium voltage (up to 650 volts) or high voltage (3·3, 6·6 or 11 kV) three-phase alternating current supply. They are suitable for installations up to 5000 kW each, and one such type is shown in Fig. 12.8. They are vertical and the chief difference in the various designs is in the method adopted for load regulation. Current passes from electrode to electrode, using the resistance of the water itself as the heating element, and the load is varied by increasing or decreasing the length of path which the current has to take by the interposition of non-conducting shields between or around the electrodes. The conductivity of the water may have to be adjusted by the addition of soda or other salts if need be, to render it suitable for the purpose. This type of heater is kept separate from the main storage because means of load regulation involving raising and lowering or rotating gear to the sheaths complicates a direct application; and, further, one heater may be connected to a series of storage vessels to provide the capacity required.

Loading of Heaters—The total heat requirements per 24 hours must include all radiation and other losses over the period.

The hourly load in kW depends on the duration of the off-peak supply, which is often restricted to eight hours but may be longer if additional charges are acceptable. The hourly rating of heaters in kW is thus:

$$\frac{\text{total heat requirement (kW) over twenty-four hours}}{\text{duration of supply in hours}}.$$

Thermal Storage Cylinders—Storage vessels are usually cylindrical, either horizontal or vertical, the size and shape being largely determined by space conditions. Diameters up to 4 metres are usual, with lengths up to 10 metres.

The storage capacity to be provided depends on the total heat requirements per day and on the temperature range.

The capacity in kg of water is thus:

$$\frac{\text{kJ required during the time current is off}}{\text{storage temp. °C minus minumum heating return temp. °C} \times 4·2}$$

then

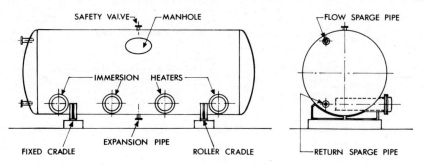

FIG. 12.7.—Immersion type hot water storage heater.

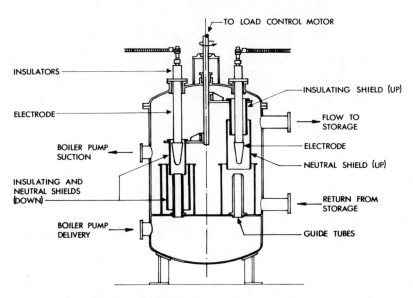

FIG. 12.8.—High voltage electrode water heater.

$$\frac{\text{kg}}{\text{density kg/m}^3 \text{ at storage temp.}} = \text{storage capacity m}^3$$

A selected range of densities will be found in Table 11.1 (page 239).

The head required to prevent ebullition, allowing 10 K excess over saturation temperature, is given in Table 12.5. Where tankage at these heights is impracticable, a pressurizing plant may be used, as referred to in Chapter 11.

Capacities of cylindrical storage vessels are given in Table 12.6.

Expansion of the heating water is much larger on thermal storage systems than with heating boilers, on account of the greater volume and the high temperature of the water stored. For instance, when water is

TABLE 12.5

Head Pressures for Hot-Water Thermal Storage

Storage Temperature °C	Pressure, kPa		Head of Water m
	Absolute	Above atmospheric (10 K margin)	
125	232	213	22
150	476	518	53
175	893	1023	104
200	1550	1808	184

TABLE 12.6

Capacities of Cylindrical Vessels with Flat Ends
m³ per m length of cylinder

Diameter m	m³ per m Length
1·0	0·785
1·25	1·227
1·5	1·767
1·75	2·405
2·0	3·142
2·25	3·976
2·5	4·908
2·75	5·940
3·0	7·069
3·25	8·296
3·5	9·621
4·0	12·566

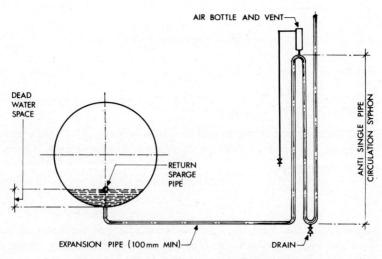

Fig. 12.9.—Expansion connection to storage vessel.

raised from 50° to 120° C the increase in volume is about 4·5 per cent, and when raised to 150° C the increase is 8 per cent.

If the water of expansion is allowed to pass heated from the water heater, and then is allowed to cool off in the expansion pipe and tank before returning gradually during the day, a certain loss of heat is incurred. For this reason a sufficient space at the bottom of the cylinders should be arranged below the level of the return connection to contain the diurnal water of expansion. From the bottom of this space the feed and expansion pipe is connected as in Fig. 12.9. When the water is heated it depresses that at the bottom up into the tank and in theory only cold water should be forced up the expansion pipe. In practice, there is probably some mixing, in addition to the small heat transmission by conduction, and a small loss from this source is unavoidable.

In any case, this expansion space is only considered to accommodate the daily expansion, i.e. from about 50° C to the storage temperature, since the expansion from 10° to 50° only occurs once during the heating season and it may be accommodated in the expansion tank additionally to the daily expansion.

The height of the return pipe from the bottom of a horizontal cylinder should therefore be about one-eighth of the diameter, corresponding to an expansion allowance of 8 per cent.

Insulation—Loss of heat from the cylinders and heater is reduced as far as possible by efficient insulation, such as 125 mm of glass fibre with metal or other cladding. Cradles are also insulated with hard material, such as compressed cork or hard wood, to minimize heat loss by conduction.

Mixing Valve—The mixing valve, which is an essential part of every thermal storage plant, consists of three ports. One is the high-temperature water inlet, one the cool return water inlet, and one the mixed water outlet. The proportions of the two former are controlled by a valve or valves operated by means of water pressure, electrical solenoids, or motor, from a thermostat in the mixed outlet pipe.

Expansion Tanks—As has been said, relatively large expansion tanks are required for thermal storage systems capable of containing the water of expansion resulting from

(a) heater and storage capacity raised from 10° to maximum storage temperature;

(b) heating system capacity raised from 10° to maximum working temperature : e.g., 80° C with radiator heating.

Control—As the temperature in an electrode heater rises the resistance of the water becomes less and the load correspondingly increases. Thus a 100 kW heater at 40° C would at 150° C have an output of about 220 kW. Means to provide a constant outlet temperature, and hence constant load, are included by some installers, making use of a further thermostatically controlled mixing valve in the boiler-cylinder circulation.

Protective gear is necessary in the case of electrode heaters to prevent operation on two out of three phases, or with out-of-balance currents flowing. Such faults might cause heavy earth leakage currents since the latter are only avoided when the current in all three phases is equal.

The protective and control gear for thermal storage systems is perhaps outside the scope of the present book. It is a great advantage to have all the control instruments, relays, contactors and switchgear mounted on a common switchboard so that faults can be more easily located and proper supervision given.

'Centralec'—A number of attempts have been made in the past to apply the principle of hot-water thermal storage to domestic premises. Since, in such circumstances, the pressure available from a roof tank is limited to perhaps 6 m, storage temperature cannot be much more than 105° C and the quantity of water stored for full service must therefore be considerable.

Some ten years ago,* the Electricity Council sponsored a novel development in this area and although storage of energy was not in water, it deserves mention under this heading. Taking advantage of the fact that heat may be stored in a cast iron core over a temperature range of 750 to 150° C as against 100 to 50° C in water, to provide a storage capacity of about 2·4 GJ/m³ as against 0·2 GJ/m³, a 14 kW Electricaire type unit was modified as shown in Fig. 12.10. The principle, as may be seen, was that the air circulation over the heated core was not released as

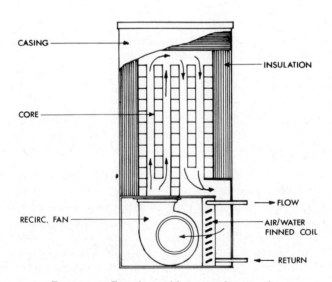

FIG. 12.10.—Experimental hot water heater unit.

* Mitchell, H. G., Parker L. C., and Haslett G. *Storage Heating Systems.* IEE/IHVE Conference, April 1971.

is normal but used to transfer heat to water for use in conventional radiators, via a simple heat exchanger. To provide for dwelling heat requirements during the charging period, a 4 kW 'flow heater', as described later, was fitted in series with the heat exchanger.

Although field trials indicated satisfactory results and some production models were made, it is understood that problems arose in connection with the heat exchangers and such units are no longer made.

Quite recently, under the same type name, more conventional hot water thermal storage equipment has reached the market in a range of sizes as listed in Table 12.7. Each unit consists of a copper cylinder, factory insulated with sprayed polyurethene foam and provided with two banks of immersion heaters. One bank is fitted low in the cylinder and one near the top outlet. The electrical wiring is arranged, via relays for 'White Meter' connection, such that all heaters are available for off-peak supply but only the top bank for on-peak connection should the water storage temperature drop below some pre-determined level.

The capacity of the cylinder as a thermal store will depend upon the temperature of storage which is, of course, a function of the head pressure available. Cylinders are tested to suit a maximum of 10 m (100 kPa gauge pressure). As with the larger type of system previously described, an

TABLE 12.7

HOT WATER THERMAL STORAGE UNITS.
RATINGS AND DIMENSIONS (CENTRALEC)

Item	Unit Size		
	100	125	150
Manufacturer's data			
Water store (litre)	450	560	700
Loading (kW)			
Input available	8	8	8
Output to radiators	3–4	5–6	7–8
Dimensions			
Height	1900	1900	1900
Width	800	800	800
Depth	700	700	700
Calculated data			
Energy (MJ)			
Acceptance (8 hour)	113	140	166
Active store			
(105° C to 50° C)	104	131	157
Output (W)			
Uncontrolled*	300	300	300
Controlled†	2020	2620	3220
Weight (kg)			
Estimated	500	610	750

* Estimated pro rata to surface area.
† Balance of active store, 12 hour constant output rate.

outflow mixing valve is an essential: recommended temperatures for the connected radiator circuit are 70° C flow and 50° C return. Calculation suggests that, for continuous discharge at full load, supply at on-peak rates would be required after about ten hours running, depending upon thermostat settings.

For domestic water supply, a non storage type heat exchanger may be added, immersed in the water store, operating on a similar principle to that described for Combination Boilers in Chapter 13 (page 296).

Electrical Flow Heaters—Although this chapter deals predominantly with off-peak systems, reference should be made to the direct electric water-heater generally known as a 'flow heater', which may be used to replace a boiler in a conventional hot-water system. Cases exist where it is impracticable to arrange for a flue of any sort, and yet where heating by hot water is preferred for various reasons. In such cases the use of a flow heater might be an acceptable expedient, provided that use was not continuous. This type of unit, for instance could fill a need in churches where the labour for dealing with any sort of boiler is non-existent, and therefore a means of automatic heating is necessary. Many Electricity Boards offer special tariffs for church-heating during periods restricted to week-end and evening use.

THE HEAT PUMP

The heat pump is a means of using energy such as is produced from electric supply to better advantage than by merely degrading it into heat in the normal way. It operates on the same principle as a domestic refrigerator by extracting heat from a low temperature source such as air, river water or the ground and upgrading it to a higher and more useful temperature. A heat pump thus delivers more energy to the point of use than that which has been expended in driving the compressor and any necessary fans or pumps.

The following example of a practical electrically driven unit type heat pump is given by way of introduction to this subject about which a considerable literature has been built up. Further reference to such equipment is made in more detail in Chapter 24 (page 567).

This heat pump illustrated in Fig. 12.11 is for room heating and derives its low grade heat supply from air outside the building by means of a fan drawing air over the external coil. The condenser forms the internal coil over which room air is drawn and delivered into the space to be heated. The compressor is self-contained in the casing, the whole unit fitting into a window or into a hole in the external wall.

Using a mass produced unit, with an outside air temperature of −4° C and an electrical input of 1400 W, a heat output of about 4·9 kW is obtained, i.e. an advantage of about 3·5. At an outside air temperature of 7° C however, typically average of winter in the U.K., this advantage would rise to about 4·1, the output being 7·4 kW.

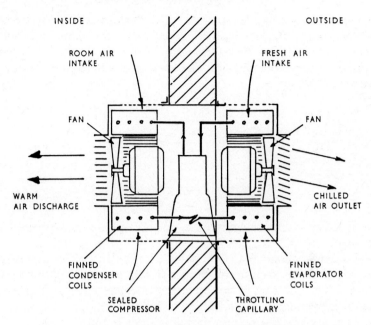

INSIDE   OUTSIDE

ROOM AIR   FRESH AIR
INTAKE   INTAKE

FAN   FAN

WARM   CHILLED
AIR DISCHARGE   AIR OUTLET

FINNED   FINNED
CONDENSER   EVAPORATOR
COILS   COILS

SEALED   THROTTLING
COMPRESSOR   CAPILLARY

FIG. 12.11.—Self contained electric heat pump for single room heating.
(Change over valve for summer cooling is not shown)

ADDITIONAL NOTES AND CALCULATIONS

In conclusion, by way of an appendix to this chapter, some notes concerning electricity generally and its application to heating systems are appended here.

Units—Current is measured in *amperes*. Pressure, or potential difference, in *volts*. Resistance in *ohms*. The current I passed through any resistance R, when the potential difference is E, is given by Ohm's law:

$$I = \frac{E}{R},$$

so that

$$E = IR \text{ and } R = \frac{E}{I}.$$

Power supply P is measured in *watts*.

$$1 \text{ watt} = 1 \text{ volt} \times 1 \text{ amp.}$$

or $P = EI$. Thus power P, which is equivalent to the heating effect, is given by substitution:

$$P = I^2 R.$$

Types of Supply—Electricity is almost universally supplied as alternating current: direct current is obsolete as a public supply.

With *alternating current* the polarity of the supply is reversed in regular cycles. The unit of frequency in SI being the hertz (Hz), 1 Hz = 1 cycle per second. Thus in a 50 Hz supply (standard in this country) the reversal takes place 50 times per second. The cycle of a single-phase supply is of sine wave from, thus:

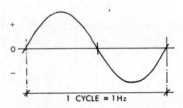

A.C. is usually either single-phase, or three-phase (two-phase supplies are obsolete).

Single-phase current may be treated as D.C. supply for the purpose of heating loads which are all non-inductive and for which Ohm's law still holds good. The question of '*power-factor*'* does not enter on this account, i.e. power-factor is taken as 'unity'.

Three-phase three-wire may be visualized as three single-phase supplies 120° out of phase, represented thus:

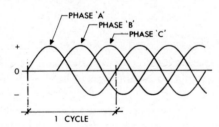

The three wires may be connected to apparatus in star or delta formation:

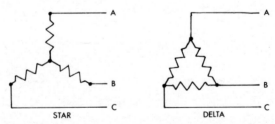

The voltage between points *A* and *B*, *B* and *C*, *C* and *A* is the declared voltage; 415 volts is now standardized in the U.K.

* Where the load is inductive, as in a motor, the voltage and current get out of step with one another. The apparent watts as shown by a voltmeter and ammeter are greater than the true watts as shown by a watt meter. The power factor is the ratio true watts ÷ apparent watts.

Three-phase four-wire is the same as the three-wire, but a neutral line is introduced at the centre of the star formation, and this is earthed, thus:

The voltage between A and N, B and N, C and N, is then the declared three-phase voltage divided by $\sqrt{3}$.

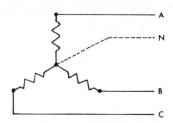

With the standard voltage of 415, it is $\dfrac{415}{\sqrt{3}} = 240$ volts.

Thus this supply is often referred to as 415/240 V. 50 Hz.

Each of the phases A, B, and C then give a single-phase supply, and a two-wire system is possible for each phase, using the neutral for the return of each.

When a three-phase four-wire system is balanced, the load on each phase is the same and no current flows in the neutral. If any one of the loads is varied, current flows in the neutral wire.

This system is the standard now adopted in Great Britain.

Application to Heating Systems—Direct electric heating elements up to 3 kW (13 amp. at 240 volts) are served by a two-wire circuit. Heavier duty elements are usually arranged in banks and the load divided over three phases of a three-phase supply if available.

For thermal storage with immersion heaters the usual arrangement is balanced on a three-phase three- or four-wire supply.

Thermal storage with electrode heaters is invariably served from three-phase three-wire supply at medium or high voltage.

The current passing in the conductors may be calculated from the load and voltage. Taking as an example a heating load of 500 kW, the current passing through the circuit will be as follows:

for two-wire D.C. *or single-phase* A.C. at 240 volts,

$$\frac{500 \times 1000}{240} = 2080 \text{ amperes};$$

for three-wire A.C., *three-phase* at 415 volts between phases, load balanced,

$$\frac{500 \times 1000}{\sqrt{3} \times 415} = 700 \text{ amperes per phase};$$

for four-wire A.C. *three-phase* at 415 volts between phases (240 volts phases to neutral), load balanced, the current will be the same as with the three-wire system, no current flowing in the neutral.

In the case of a number of small heaters, each will be served from a separate way on the distribution board. Having established the current flowing in the circuit supplying one heater, these are totalled up to give the current supplied by the main to the distribution board. From these currents the electrical distribution system is then designed. For a more complete study, reference should be made to one of the many text-books on electrical engineering.

kW and kVA—The difference between kilowatts (kW) and kilo-volt-amperes (kVA) should be understood. The latter term is used in connection with alternating currents and refers to 1000 volt-amperes, or 1000 'apparent' watts. The kilowatt refers to 1000 'true' watts. At unity power-factor (see page 284) the kVA and the kW are the same, but, if the load is inductive, then kVA × power factor = kW. Normal heating elements are non-inductive and the terms are then synonymous, but such is not the case where motors or power transmission lines are concerned.

CHAPTER 13

Hot-Water Supply

LOCAL AND CENTRAL SYSTEMS

THE HEATING OF HOT WATER for baths, basins, showers and sinks, may be by a local or a central system.

Local hot-water supply systems derive their heat either from gas or electricity, and the heater is placed near to the point of consumption. A central system is one in which the hot water is heated at some point remote from the fittings to be served, and is conveyed thereto through a system of piping. The heat supply for a central system may be produced from the burning of solid fuel, oil or gas, or by the consumption of electricity.

Choice of System—The type of system will depend on the type of building. Small houses and flats may very often be conveniently served by local systems which require little attendance, cleaning or maintenance. A small domestic boiler system in a house, provides some general warmth to the kitchen and, where fired with solid fuel, a means of disposal of a certain amount of refuse. In addition, one or two small radiators may be added to heat the hall, etc.

Large Block of Flats—The tendency of late has been to provide each flat with its own local hot-water heater, heated either by gas or electricity. The tenant then pays for just what he or she consumes through the gas or electricity meter. The alternative of a large central system for the complete block of flats or a number of blocks is favoured by certain Authorities and in the better class of property, in which case the cost of hot water may be included in the rent or charged on a basis of consumption via some form of meter (see Chapter 25).

Office Blocks, Factories, etc.—If the lavatories are few and widely separated, the advantage will be with the small separate local hot-water supply units, electrically or gas heated. If, however, the washing facilities are arranged in a compact manner the central system will be more economical and will give greater reserve for sudden peak load draw-off.

Hospitals, Hotels, Institutions—In this class of building the central system is almost exclusively adopted for the reason that it alone can give a great reserve for heavy demands, and the dispersed maintenance of a large number of small units is avoided. In building complexes of this type the heat is often derived from a central boiler plant which provides all the heat requirements, thus centralising maintenance work and isolating it from occupied areas.

Losses—The determining factor of local versus central system may, in a

287

doubtful case, be considered in terms of losses versus heat usefully supplied. In a central system there are mains losses from circulating piping continuing during the hours of running per day whether or not any water is drawn off. These losses may continue both day and night unless some means be incorporated in the system design to stop circulation during the hours when the building is unoccupied, or not in full use. Where large quantities of hot water are drawn off daily, these losses may bear a small ratio to the total heat supplied. If, however, draw-off is small in quantity and spasmodic, it may well be found that the mains losses far exceed the actual amount of heat in the water drawn off. In the former case a central system would be well justified; in the latter there would be much in favour of a local system.

LOCAL SYSTEMS

There are two basic types of local hot water system, fundamentally different in concept. The first makes use of some form of instantaneous heater and has no hot water storage capacity. Except for fuel consumption by a pilot flame in the case of gas firing, this type has no associated heat loss when not in use. The second type incorporates hot water storage, adequate in capacity to meet the local demand, and however well insulated the storage vessel may be, a standing heat loss will occur.

Apart from any question of heat losses, the two types of system will for a given hot water production consume equal amounts of energy. In terms of rate of energy supply however, an instantaneous heater will impose a greater load since the water outflow must be brought up to the temperature of use, from cold, during the short time of actual demand. With a storage unit, no such limitation exists and the rate of energy supply required will be a function of vessel size and demand pattern.

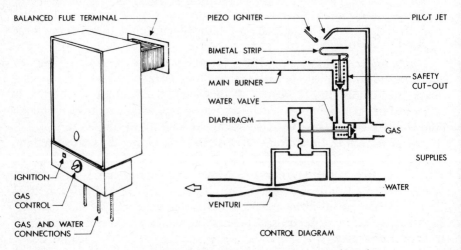

FIG. 13.1.—Instantaneous multipoint gas water heater (Ascot).

Instantaneous Heaters—The earliest type of heater in this category was the old fashioned bathroom geyser, gas fired. The much refined modern equivalent is made in single and multi-point form and Fig. 13.1 shows a well known example of the latter type. A number of draw-off points may be served provided that these are within the limit of length permitted for a dead-leg (see page 329). A gas pilot burns continuously and when a hot-water tap is opened water begins to flow and, by the pressure difference across the venturi, the main gas valve is opened, whereby the gas burners are ignited. A safety device cuts off the gas should the pilot light be extinguished. A hot water output of between 0·03 and 0·1 litres per second may be expected.

In the case of a small single point heater, used intermittently, the products of combustion may be discharged into a well ventilated room and no flue is necessary. For larger heaters or small heaters which are either used continuously or fitted in inadequately ventilated rooms, a flue with a suitable outside terminal is necessary. Models having balanced flues, with inlet and outlet ducts communicating with the outside, are available.

A comparatively recent innovation is the single-point electric instantaneous heater, designed to serve a wash basin or a shower as shown in Fig. 13.2. Various models are available which will provide an outflow of be-

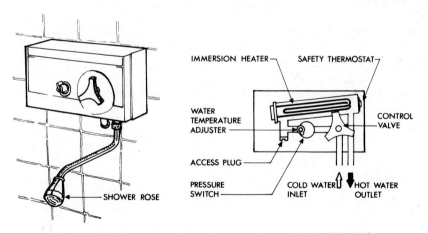

FIG. 13.2.—Instantaneous single point electric water heater (Heatrae).

tween 0·02 and 0·05 litres per second. Such units are provided with thermostatic controls and a diaphragm pressure switch which permits current to be available only when water is flowing. Preset cut-out and other safety devices are incorporated. Physically, the casing of a unit of this type is small being only 200 to 300 mm square with a depth of perhaps only 60 to 80 mm. Electrical loadings are high, at from 3 kW to 6 kW, and since usage of water is immediately adjacent it follows that particular care must be taken in providing electrical protection and good earthing facilities.

In most applications instantaneous heaters, gas or electric, may be connected directly to the cold water main supply but the appropriate water authority must be consulted. Where supply via a cold water storage tank is envisaged, a static head of at least 3 m, equivalent to a pressure of about 30 kPa, is necessary. The rate of water flow from all such heaters is restricted and, in the case of multi-point units it should not be assumed that a good supply may be obtained from more than one tap at a time. In the case of single point units, water control is normally on the inlet connection and outlet is via an inconvenient swivel spout, permanently open to atmosphere.

Storage Heaters—For single point draw-off, local storage heaters are available in the capacity range of about 7 to 70 litres and are normally provided with 3 kW electrical heating elements. If the capacity is small, this type may be connected directly to the cold water main supply. Strictly speaking, present regulations permit only units of 13·64 litres (3 gallons) storage to be so connected but a waiver can usually be obtained to allow the use of metric 15 litre models. In this application, the simplest form of heater is the free-outlet or non-pressure type illustrated in Fig. 13·3.

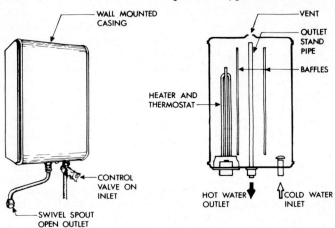

FIG. 13.3.—Non-pressure type electric water heater (Sadia).

It is the cold water inlet which is the point of control, discharge of hot water being by displacement as the cold supply is admitted. The storage is always open to atmosphere via a swivel spout and as the contents expand on heating, a drip may occur at the outlet.

To overcome the inconvenience of a swivel arm, the outlet may be piped to a spout fixed to the point of draw-off as shown in Fig. 13.4(a): an alternative introduced by one manufacturer combines this spout with the cold inlet water control as Fig. 13.4(b).

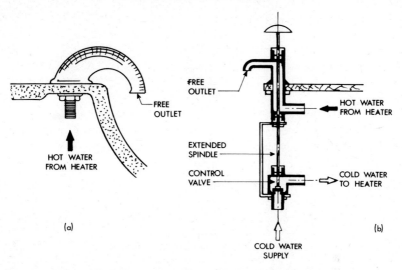

FIG. 13.4.—(a) Spout outlet (b) combination tap outlet (Santon).

For multi-point local use, as might arise in each of a number of dispersed toilets in an office building, both gas fired and electrical storage heaters are available. Packaged gas fired units are, in effect, cylinders incorporating a stainless steel heat exchanger direct fired as shown in Fig. 13.5, cased in sheet metal. Such heaters have a water capacity of about 80 litres and a heat input of 8 kW: the products of combustion must be discharged to outside via a conventional or balanced flue arrangement.

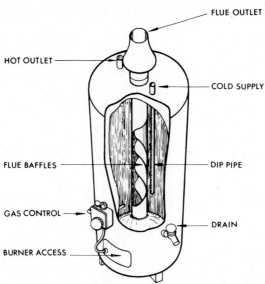

FIG. 13.5.—Storage heater—gas fired (Lennox).

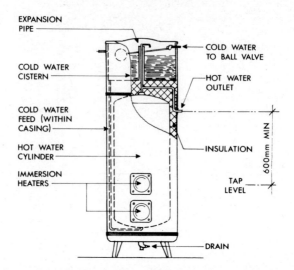

FIG. 13.6.—Combination heater with integral cistern (Sadia).

A cold water supply from a remote storage cistern must be provided as must arrangements for venting. For applications where larger storage or heating capacities are required gas firing becomes less tidy to apply since the storage cylinder is separated from the fired heat exchanger and the equivalent of a small central system thus arises.

With electrical heating, local multi-point supply is most conveniently arranged using a so-called combination type unit where heater and cold water cistern are fitted in one casing as shown in Fig. 13.6. The heater may be cylindrical or rectangular, for easy wall fixing, and capacities in the range of 25 to 140 litres are available, normally with a 3 kW heating element. The pressure available at the draw-off points will be meagre if the combined unit cannot be fixed well above the level of the taps.

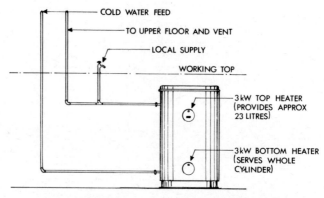

FIG. 13.7.—Electric pressure type water heater (Santon).

For applications where larger storage or heating capacities are necessary, or where a combination unit cannot be fitted high enough above the draw-off, pressure type units are used as illustrated in Fig. 13.7. Siting of the storage in these circumstances becomes unimportant and thus the UDB (under-draining-board) heater which is less than a metre high has many applications. Pressure type units however require a cold water supply from a remote storage cistern and arrangements for venting. Capacities are in the range of 50 to 450 litres with heating elements to suit: the larger of such units, however, do not really apply to this present context.

Local storage heaters are well insulated and neat in appearance as suitable for exposed mounting. They are however inevitably bulky, the equivalent to the electrical instantaneous unit previously mentioned occupying more than ten times the space. Storage units can however be arranged to take advantage of electrical off-peak supplies under the 'white-meter' tariff (see page 263) and the ratio between volumetric and heating capacities adjusted accordingly. Special cylinder shapes are available, tall and smaller than usual in diameter, in order to preserve stratification of the hot and cold water. The Authors are familiar with installations where families of up to five persons have been served satisfactorily by a 200 litre storage, heated only by off-peak supply.

CENTRAL SYSTEMS

A central system comprises a boiler or water heater of some form, coupled by circulating piping to a storage vessel, the combination of the two being so proportioned as to permit of all the normal demands for hot water being met upon the opening of the taps according to the type of

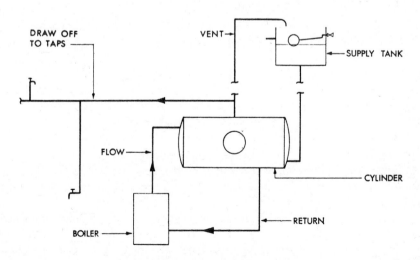

FIG. 13.8.—Direct hot-water supply system.

usage. For instance, in a hospital there may be a continuous demand for hot water all day, and a comparatively small storage but a high heating-up rate may be suitable. In a sports pavilion, however, there may be only one sudden demand at the end of games, in which case a large storage and a small heating-up rate may be adequate.

The system may be *direct*, in which case the water drawn off is the same water as circulates through the boiler as shown in Fig. 13.8. If the water is hard, scale deposit may occur in the boiler and hence boilers for direct hot-water supply have to be of a simple type with large waterways and big clean-out openings. Fig. 13.9 shows a cast-iron sectional boiler so arranged. If, alternatively, the water is soft, direct boilers of iron or steel may cause discoloration of the water through rusting, and hence in the past such boilers have sometimes been made of copper or treated with an anti-corrosion treatment such as Bower-barffing, or one of the vitreous coatings.

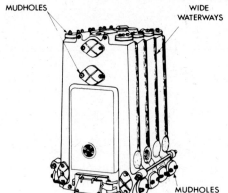

FIG. 13.9.—Cast-iron sectional boiler for direct hot-water supply.

In the *indirect system* all this is avoided. Fig. 13.10 illustrates an indirect system, in which any type of heating boiler may be used. The storage cylinder becomes indirect containing some form of heating surface, a common form being of annular type as shown, and within which the water from the boiler circulates. The water to be heated is contained in the vessel outside the indirect surface, and hence never comes in contact with the boiler. An indirect cylinder is alternatively termed a *Calorifier*.

In the indirect system scale formation is generally negligible as the surface temperature is not usually high enough to cause scale deposit. Although the use of copper for indirect cylinders was at one time confined to soft-water areas, this material is now frequently used in hard-water areas also, the piping system being in copper throughout.

The boiler in such a system may be of much more efficient type than is possible with a direct type of boiler. It will be noted that the boiler, or primary system, requires a separate feed tank or cistern as in the case of any ordinary heating system (but see p. 311 also).

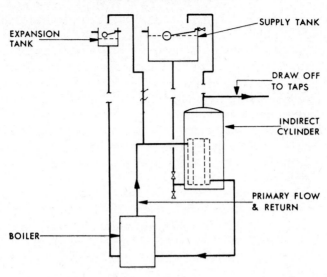

FIG. 13.10.—Indirect hot-water supply system.

Combined Systems—It has become common practice to combine the
duties of heating and hot-water supply served from one boiler or group of
boilers. This can only be done by using the indirect system, the boiler or
boilers being run at a constant temperature with the heating served
through a mixing valve at varying temperature, as discussed on page 187
(and see Fig. 9.7).

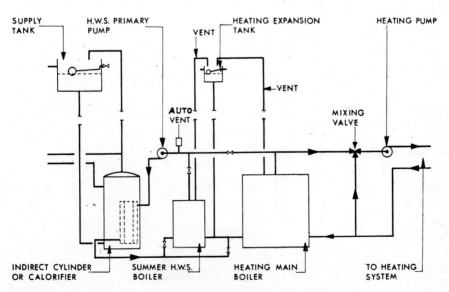

FIG. 13.11.—Combined heating and hot-water supply system.

During summer when the heating is off, the boiler is of course over-sized—which, in the days when hand-fired solid fuel was usual, was considered unsatisfactory due to the likelihood of occasional overheating or boiling. With automatically-fired boilers burning gas or oil, this dis-advantage does not arise and it is quite common to find a wide disparity of load successfully handled from one boiler, though there is clearly a limit. A large boiler running intermittently on a light load may involve corrosion troubles if oil-fired, and hence this practice is best confined to plant burning class D oil and to duties where the combined load is under 300 kW. With gas firing the question of corrosion does not arise, but a large gas boiler on light load tends to low efficiency.

Thus, where combined system loadings are above the limit suggested above, it is preferable to provide a small boiler for summer hot-water sup-ply duty but so arranged that, when the main boiler or boilers are in use during the heating season, they carry the hot-water supply load and the small boiler is out of commission. Such an arrangement is shown in Fig. 13.11.

The circulation from boiler to indirect cylinder may be by gravity, but it is worth considering the use of a pump, as shown in the Figure. Not only does this enable the size of piping connections to be reduced but, by thermostatic switching of the pump on and off, a simple means of temper-ature regulation of the hot-water supply is achieved.

The Combination Boiler—Another form of combined system is one in which the boiler and hot water supply calorifier form a self-contained unit. One type is illustrated in Fig. 13.12 and the diagram of connections for another in Fig. 13.13. Advantages are that space and time in erection is saved and that losses are reduced. In addition, of course, the combined boiler/calorifier unit takes full advantage of the thermal capacity of the heating and the hot water supply systems. It is unlikely that both will

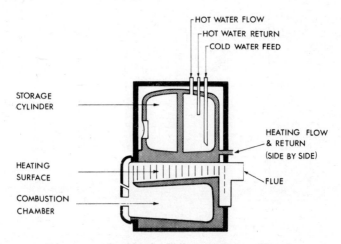

FIG. 13.12.—Combination boiler arrangement (Hoval).

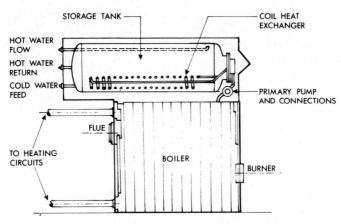

FIG. 13.13.—Piping within combination boiler (Ideal Falcon).

require full boiler output simultaneously, bearing in mind the time lag in demand of the two diverse systems and the mass of the hot water and/or building structure.

As will be seen from Fig. 13.13 the primary circulation from the boiler to the coil in the cylinder may be by a separate pump. This is thermostatically switched so as to keep a constant storage temperature. The heating circuit also is pump circulated independently and may include a mixing valve as in Fig. 13.11. In the domestic field, a new approach has recently been introduced which incorporates some of the features of a Combination

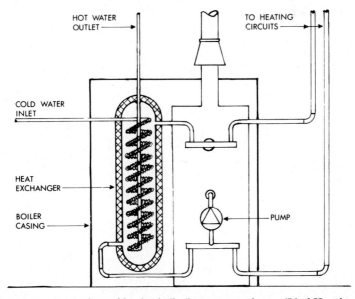

FIG. 13.14.—Domestic combination boiler/instantaneous heater (Ideal Heatslave).

boiler and some of those of an instantaneous heater. Fig. 13.14 shows the arrangement in question where a small vertical vessel is close coupled to a boiler with pumped circulation to what, in the case of a normal indirect cylinder, would be the 'secondary' contents. Through this vessel passes a high efficiency finned tube coil which conveys water to hot water taps. The unit has been designed to provide hot water at mains pressure by direct connection but can operate from a normal cistern supply provided that an adequate pressure is available. It remains to be seen whether the principles of the design will be acceptable to water supply authorities.*

Heat Supply—The heat supply to a central system may therefore be derived from either independent, combined or conventional central boiler plant fired by any of the various known fuels.

In the case of gas, the supply unit may be quite independent ranging in size from a circulator developed to fit within a domestic warm air heater (however fired), as shown in Fig. 13.15 and rated at about 4 kW, to any of the standard range of purpose made boilers. In each case, one or more indirect cylinders or calorifiers will be required having capacity to cater for the load imposed.

Electricity may be used on-peak or off-peak year round or, perhaps, in either way for summer supply only. In either case, use will be made of an immersion heater fitted to a storage vessel which, in the case of off-peak supply, must be large enough to cater for demand during the periods when current is not available. The special case of domestic hot-water supply provided from a thermal storage system has been referred to in Chapter 12.

Hot-Water Supply Draw-off—The draw-off piping from a central

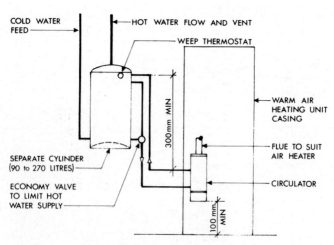

FIG. 13.15.—Gas fired heater for use with warm air system (Ascot).

* See D.O.E. consultative document *The consequences for the introduction of domestic unvented hot water systems in England and Wales*. National Water Council. 1976.

system may comprise a series of dead legs as shown in Fig. 13.16, but the runs are strictly limited by virtue of the fact than an undue time would have to pass before hot water came from the tap, in the case of a very long run. This causes wastage of water and Water Authorities commonly limit the length of dead-leg draw-off to about 8 metres. On anything other than small systems, therefore, it is necessary to form a secondary circulating system as illustrated in Fig. 13.17, such that the length of branches to any draw-off points is kept to a metre or less and hot water is available almost instantly. Naturally, such circulations are insulated as there is a continuous loss of heat from them summer and winter. The secondary circulation may also serve towel rails, as indicated, and linen cupboard coils.

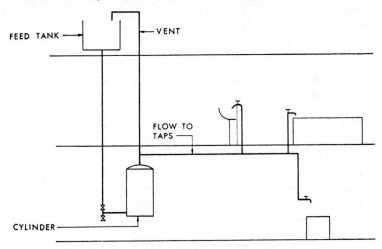

FIG. 13.16.—The dead-leg system.

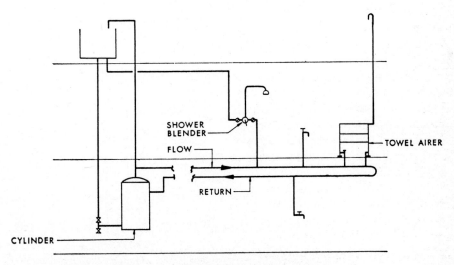

FIG. 13.17.—The secondary-circulation system.

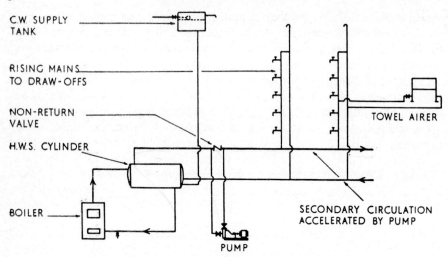

C.W. SUPPLY
TANK

RISING MAINS
TO DRAW-OFFS

NON-RETURN
VALVE

H.W.S. CYLINDER

BOILER

TOWEL AIRER

SECONDARY CIRCULATION
ACCELERATED BY PUMP

PUMP

FIG. 13.18.—Pumped secondary circulation.

In more extensive systems still, where a gravity secondary circulation is not practicable, a circulating pump is introduced as shown in Fig. 13.18. Alternatively, the pump may be in the secondary return but such an arrangement will not allow any advantage to be taken of the reversal of flow at time of draw-off which is described in Chapter 14, page 326.

Head Tank System—Fig. 13.19 illustrates an old type of hot-water supply system, which is mentioned only with the object of drawing attention to its many defects, because it is still in existence.

In this system, heat from the boiler is applied at the bottom and the hot-water tank is at a higher level, perhaps in a roof space, and when no draw-off occurs the water gradually rises from the heat source to the head-tank, which then acts in the same way as a cylinder, the level between the hot and cold water gradually falling in the normal manner.

The circulating pipes between the two may be, however, extremely long, so that the circulation is sluggish and it is frequently necessary for the heat source to provide a very high temperature before there is much circulation.

When a draw-off occurs, it is largely a matter of chance whether the water will be taken from the heat source or from the top tank and, in consequence, it may happen that a mixture of hot and cold water

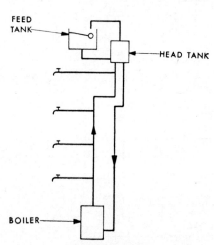

FEED
TANK

HEAD TANK

BOILER

FIG. 13.19.—Old-type head tank system
(not recommended).

is delivered to any individual tap, since the cold water runs down the return pipe and passes the heat source so quickly as to benefit very little by such passage before reaching the draw-off point. It should be regarded as obsolete.

Capacity of Storage Cylinders—The decision as to the size of cylinder and boiler necessary depends on considerations quite different from those which apply in a heating system where, as a rule, the quantity of heat which is required under design conditions may be calculated with a fair degree of accuracy and reasonable assumptions made, based upon meteorological data, of demand over any given period of 24 hours.

A hot-water supply system, on the other hand, as a rule functions intermittently. For example, in a domestic installation, there is very little hot water required except when hot baths are drawn off in the morning and in the evening, with a certain amount of water taken intermittently by basins, kitchen wash-up, etc. In a hospital or an office block however, the pattern of use will be entirely different.

A calculation of the total quantity of heat required in twenty-four hours will therefore give no criterion of the capacity of cylinder required, unless the cylinder is designed to give a twenty-four hour storage—which would usually result in its being grossly uneconomical.

In general, the more generous the cylinder capacity, the smaller the boiler power that may be used, as it has a long time in which to catch up the draw-off at peak load. On the other hand, the more sluggish will be the raising of the temperature when starting from cold, or when the cylinder temperature has for any reason been allowed to fall below normal. Between these two extremes a compromise is to be effected and this, apart from the capacities of storage and boiler plant, must also take into account the reaction characteristics of each.

In many installations it will be found reasonable to give the cylinder a capacity equal to the maximum draw-off of hot water in any one hour at peak-load conditions, and the boiler may generally then be sized on a basis of heating this quantity of water up to the desired temperature in some longer time, depending on the installation. In many cases it will be adequate if this heating takes 2 to 3 hours, but where there is little draw-off between the peak-load conditions, this period may be further extended, and, on the contrary, where the supply approximates more to a continuous one, it may need to be shortened.

In most installations hot baths constitute the peak load for the hot-water supply system, especially in hotels and similar buildings. In any case, it is necessary to consider whether during such peak loads a supply is also required for kitchen, basins, etc.

Table 13.1 gives the capacities of various standard fittings, and Table 13.2 the approximate figures of consumption for various types of building, as a guide to the cylinder capacity.

As an example, in a hotel with 100 rooms, each having its own bath-

TABLE 13.1

CAPACITY OF VARIOUS STANDARD FITTINGS

	Capacity in Litres	Temperature usually required, ° C.
Lavatory basin, normal filling - -	5–10	43–60
Sink - - - - - - -	12–18	60
Bath, average - - - - -	80–120	48–60
Shower Bath, spray type - - -	0·15 litre/s	43
,, ,, 150 mm rose type - -	0·6 litre/s	43

TABLE 13.2

HOT-WATER CONSUMPTION IN VARIOUS TYPES OF BUILDING

Type of Building	Consumption per Day per Occupant in Litres (Water at 65° C.)	Peak Consumption per Hour per Occupant in Litres
Schools		
Boarding	115	25
Day	15	5
Dwellings		
Max.	140	45
Min.	70	25
Hotels		
Max.	135	45
Min.	115	35
Factory		
Toilets	20	5
Offices		
Total	15	5
Hospital		
Infectious	225	45
General	135	30
Geriatric	70	25

room, taking the peak demand at 45 litres per hour from Table 13.2 it would be desirable to allow

$$100 \times 45 = 4500 \text{ litres at } 65° \text{ C.}$$

This is equivalent, when mixed with 2000 litres of cold water at 10° C, to 6500 litres at 48° C or the equivalent of 80 baths taken during one hour.

Here it should be noted that the 4500 litres is the hot water at 65° actually required, but as there is bound to be some mixing with the entering cold supply, something should be added to arrive at a satisfactory storage capacity. It is usual to take the effective storage at 90 per cent of the actual to allow for this incomplete stratification. Thus in the example, the actual storage to be provided would be:

$$4500 \times 100 \div 90 = 5000 \text{ litres}$$

Public bathrooms not attached to a particular bedroom may, of course be used much more frequently in one hour, particularly when the number of bathrooms is small compared with the number of bedrooms. Special consideration is necessary in such cases.

This sort of assessment of storage capacity probably overstates the case. Records of hot-water demands taken in university halls of residence* show that the actual demands are much lower than those forecast by conventional methods. Thus a considerable diversity factor on storage capacity appears justifiable, and can best be determined by a careful study of possible load curves and records of comparable installations where they exist. There must, however, always remain some element of intelligent guesswork in arriving at storage capacity.

Table 13.3 gives the capacities for flat-ended cylinders of various dimensions.

Boiler Power Required—The boiler power is arrived at, as already stated, by assessing the allowable re-heating period. In addition to the heat required for raising the hot water to the required temperature, the losses due to radiation from boiler, mains, cylinder and fittings must be properly allowed for.

Taking the previous example, and assuming a re-heating period of two hours, with radiation losses from the system of 10 kW, the boiler power is estimated as follows:

Heat to raise water temperature in two hours

$$4500 \text{ litre} = 4500 \text{ kg (approx.)}$$

$$= \frac{4500 \times (65 - 10^\circ \text{ C.}) \times 4 \cdot 2}{2 \times 3600} = 144 \text{ kW}$$

Radiation	10 kW
Net boiler power	154 kW

HOT-WATER SUPPLY CALORIFIERS

Calorifiers for hot-water supply differ from those for heating in having a much larger shell so as to provide storage. The capacity of the storage is determined as in the case of cylinders already discussed, making due allowance for the space occupied by the heating surface. With any type of heat exchanging surface the element is best kept near the bottom of the shell so as to promote convection over as great a volume as possible. A vertical arrangement gives better stratification than a horizontal one.

Water-to-Water—For low pressure hot water, where the storage capacity is below about 900 litres, an annular type element is commonly

* Harper A., Hot Water Consumption at Leeds University *JIHVE.* 1964. 32. 353.

TABLE 13.3

CAPACITIES OF HOT-WATER SUPPLY CYLINDERS IN LITRES (ASSUMING FLAT ENDS)

Diameter metres	Length: metres					
	1	1·5	2	2·5	3	4
0·4	125	188				
0·5	196	296				
0·6	282	423	564			
0·7	385	580	770			
0·8	503	755	1006	1258		
0·9	636	956	1272	1592		
1·0	785	1180	1570	1965	2355	
1·2		2240	2262	3371	3393	
1·4		2300	3080	3850	4620	
1·6			4022	5040	6033	8044
1·8			5090	6360	7635	10 180
2·0			6286	7850	9429	12 572

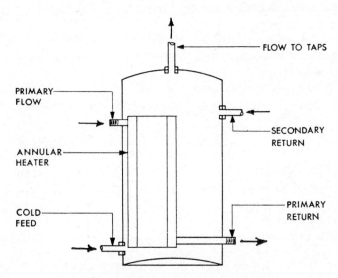

FIG. 13.20.—Indirect cylinder for hot-water supply.

used, as shown in Fig. 13.20, being economical in cost. Calorifiers of this type are generally referred to as 'indirect cylinders'.

For larger capacities a tubular battery is used, as shown in Fig. 13.21. Means for withdrawal are provided for cleaning purposes and pipe connections should be arranged so that this may be achieved with minimum disturbance.

For medium-pressure and high-pressure hot water, the calorifier frequently takes the form shown in Fig. 13.22, in which the primary water follows a single path in order to maintain a high velocity which is neces-

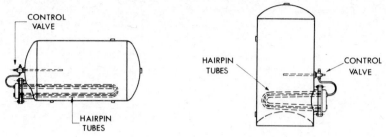

FIG. 13.21.—Tubular type calorifier for large systems.

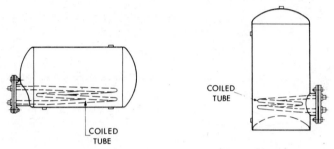

FIG. 13.22.—Tubular type calorifier for HPHW.

sary due to the relatively small amount of water in circulation. Tube withdrawal for cleaning purposes is provided for, as before, the cylinder 'neck' being suitably enlarged to suit the alternative configuration.

A new form of indirect heater for water-heater heat transfer is shown in Fig. 13.23 and comprises an immersion heater having waterways so arranged as to give a high rate of transmission. This is specifically designed for domestic use in conjunction with a micro-bore system.

Control of water-water calorifiers can be simply achieved by a direct-acting valve actuated by an expansion element located in the shell about one-third up from the bottom, or by electrical on-off control. Or control may be by on-off switching of the primary pump if such is included, as referred to earlier. In very small systems the primary circulation is often left uncontrolled.

Steam-to-Water—The heating surface generally adopted is a tubular battery, similar in most respects to that shown in Fig. 13.21. The steam, on condensing, collects in the lower part of the steam chest and passes thence to the steam trap. The battery is withdrawable for cleaning—hence the horizontal arrangement is preferred. Control may be by direct-acting 'Horne's' valve, as shown, or by various other means.

Centralised Distribution of Hot-Water Supply—Where a number of buildings are to be served from a central boilerhouse, or where a number of flats are to be served there are two methods of distribution which may be employed: *Centralised Secondary* and *Centralised Primary*. Centralised Secondary means that water is heated at the central point and is distri-

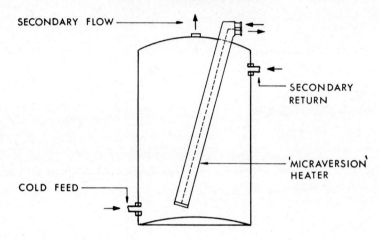

SECONDARY FLOW

SECONDARY
RETURN

'MICRAVERSION'
HEATER

COLD FEED

FIG. 13.23.—Above. Hot water immersion heater: Below, system diagram.

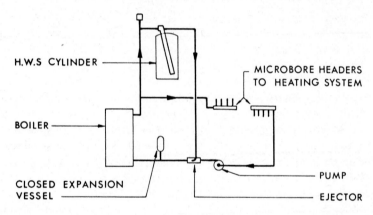

H.W.S CYLINDER

MICROBORE HEADERS
TO HEATING SYSTEM

BOILER

CLOSED EXPANSION
VESSEL

PUMP

EJECTOR

buted through a system of circulating pipework serving all points directly—i.e., the water circulating is the water drawn off. This is simply an enlargement of the system shown in Fig. 13.18. Centralised Primary means that the water circulated or the steam supplied from the central heat station is passed into a number of calorifiers from which the hot-water supply is derived locally. Fig. 13.24 shows diagrammatically, such a system served from high-pressure hot water.

The choice as to which is the best system in a certain set of circumstances depends on many factors such as size, water characteristics, space, maintenance and cost. Further consideration to this matter will be given in Chapter 25 on *District Heating*.

Rating of Calorifiers—The heating surface required depends on the temperature difference between the primary water or steam and the secondary water, and on the transmission rate per degree kelvin.

Water-to-Water Calorifiers—Considering Fig. 13.20, for simplicity in

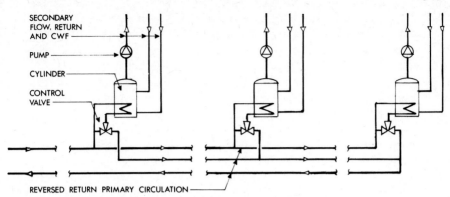

SECONDARY
FLOW. RETURN
AND CWF

PUMP

CYLINDER

CONTROL
VALVE

REVERSED RETURN PRIMARY CIRCULATION

FIG. 13.24.—Centralised primary hot water system.

calculation, it is generally assumed that the primary temperatures of flow and return are suitably fixed at say 80° C flow and 70° C return. It is then also assumed that cold feed is entering continuously at some temperature such as 10° C and hot water is leaving also continuously at say 70° C. The mean temperatures are then: primary 75° C, secondary 40° C, which equals 35 K difference.

In practice, conditions of stability such as this may never exist. In periods of heavy draw-off, the whole cylinder may be emptied of hot water when the temperature difference will then be nearer to 65 K. As the cylinder warms up the difference will become less, until when fully heated the difference may be only about 10 K. In the former case considerable boiler power would be necessary to maintain the full rate of transmission at the increased temperature difference, and in the second case when the cylinder is warm, the heat requirement is much less. This shows that some means of control of the primary heat source is needed.

Steam-Water Calorifier—Where steam is used as the heating medium, the temperature of the steam will depend on its pressure (see Chapter 10 on *Steam Heating*, p. 206). The temperature of the water may be taken at the arithmetic mean of the secondary. The temperature difference is then multiplied by the coefficient per unit area of heating surface. This transfer rate, divided into the estimated duty, gives the area of surface required. Certain data for heating calorifiers were given in Table 10.5, page 234, based on the *Guide*, but these are not directly applicable to storage calorifiers where scaling has to be allowed for; also, water velocities on the secondary side are those due to natural convection only.

The *Guide* coefficients for heat transfer in storage calorifiers using 25 mm copper tube, range from 400 to 1170 W/m²K for water-to-water and 640 to 1180 W/m²K for steam-to-water, depending on whether clean or fouled, on water velocity and temperature differences. But, due to variables involved, the size and spacing of tubes and types of design, it is best to consult the makers for ratings and not rely on theory only.

The Angelery Hot-Water Generator—This equipment (illustrated in Fig. 13.25), suitable for steam or hot-water as the primary medium, tackles the hot-water supply problem in a different manner from the conventional. In certain cases it is possible to dispense with storage completely. It will be seen that the heat-exchange surface consists of a series of interlacing coils giving a high capacity in a small space and being largely self-de-scaling. It is not an instantaneous heater, but, by means of a 'fluid operated computer' and a special flow-regulator valve in the primary medium, the heater relates heat input to 'periodic' demand, so avoiding peaks. To apply this principle successfully, the form of load must be known and the size of unit determined from maker's data. Storage (termed an 'accumulator') is required: for instance where batch or surge loads occur.

For normal load variations without storage, the Angelery heater in effect draws on the surplus capacity of the boiler plant. Similar features are inherent in the 'Load Leveller' and other specialities introduced over

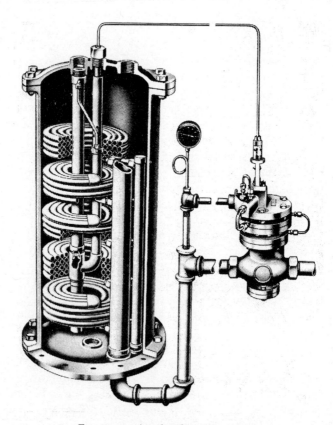

Fig. 13.25.—Angelery hot-water generator.

recent years as the investigations* undertaken by the Hospital Research Unit at Glasgow University have been absorbed into practice.

MATERIALS FOR CYLINDERS, CALORIFIERS AND PIPING

Galvanized steel being cheaper than other materials was at one time most in use; however, copper piping is now preferred for secondary systems regardless of water characteristics. As mixed metals are undesirable in such a system due to probable electrolytic corrosion, the storage vessel must in this case also be of copper.

A galvanized cylinder with galvanized pipework is suitable with waters having a pH value as follows:

pH value	temporary hardness
7·3 - - - -	greater than 210 ppm
7·4 - - - -	150
7·5 - - - -	140
7·6 - - - -	110
7·7 - - - -	90
7·8 - - - -	80
7·9 and over - -	70

Copper for cylinders and pipework is necessary with waters having a pH value of 7·2 and under.

It has been common practice for many years in Europe to use storage vessels of steel, copper lined. In addition, 'thermo-glazing' has been similarly employed, this taking the form of a ceramic enamel coating applied internally to a steel shell. A range of thermo-glazed calorifiers in factory made casings is now available in the U.K., as shown in Fig. 13.26.

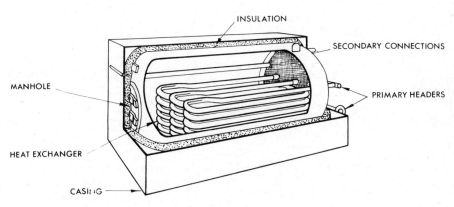

FIG. 13.26.—Packaged thermo-glazed calorifiers (Combustions Ltd.).

* Maver, T. The design and performance of hot water storage calorifiers. *JIHVE*. 1964. 32. 330.

In hard water districts there is often a good case for water softening for the feed to the hot-water supply system in order to reduce soap consumption, remove calcium scale in baths, etc., avoid furring of heating surfaces and improve the washing characteristics of the water generally. This is particularly the case with laundries. Materials for the hot-water supply system will then be selected according to the resultant water condition.

Insulation—With a hot-water supply system it is most desirable that every metre of circulating pipe should be insulated, as heat losses therefrom will continue not only throughout the period of heat supply but also during the hours when the system is cooling down unless means are incorporated to prevent circulation, as referred to on page 288. The heat emitted from such pipes if not insulated is also objectionable from the point of view of the temperature in the building in summer.

Such insulation may take the form of any of the insulating coverings already referred to for heating systems. It is needless to say, of course, that the boiler and cylinder should also be adequately lagged, unless the latter is used for the warming of a linen cupboard in a small domestic system.

Valves—Valves on hot-water supply systems often give trouble due to fur or scale collecting on the faces, which in time renders them useless for shutting off the water. For this reason plug cocks are much to be preferred, especially if of the lubricated type. These can be relied upon to shut off tight even with the most severe internal encrustation, though approval may be required by certain Water Authorities.

Showers—Blending valves for showers are either hand-operated or thermostatic. It is a great advantage in both cases for the pressures of cold and hot to be roughly equal. This is achieved by arranging for the cold feed to the blender to be fed from the same tank as the hot-water supply, preferably from a separate down-feed so that the pressure is not affected by other draw-off points.

FEED CISTERNS

Feed cisterns (often referred to as tanks) are required on hot-water supply systems, to supply the water drawn off from the taps, and in addition to allow for expansion as in the case of a heating system.

The sizing of the cistern depends to some extent on the supply of water available. If the pressure and flow are good, a one-hour storage under peak load conditions will probably suffice, but if poor, sufficient for two or three hours would be advisable.

The feed pipe from the cistern has to be connected to the cylinder so as to prevent mixing with hot water, thus preserving stratification. This may be achieved by finishing with a tee or bend inside the cylinder when entering through the bottom of the shell.

Indirect systems require a feed cistern to both boiler and calorifier circuits, and these have been shown in the diagrams already illustrated. In order to avoid the need for two cisterns, special indirect cylinders have been devised for domestic use in which the primary is fed from the second-ary side as shown in Fig. 13.27. Obviously, with the inevitable expansion and contraction, this would lead to continual change of water in the boiler and heating system with great risk of furring-up or other hazards. Hence these units contain means to prevent such mixing—as, for instance, by an air seal. Within the limits they are intended to cover, they appear to be successful.

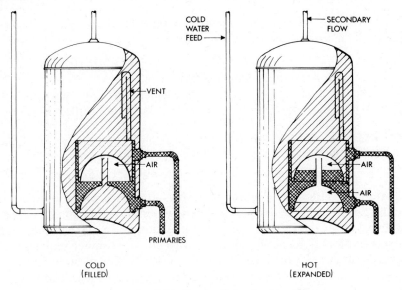

FIG. 13.27.—Special indirect cylinder for domestic use.

CHAPTER 14

Piping for Hot-Water Supply Systems

PIPE RUNS FOR HOT-WATER SUPPLY systems fall into three categories, which call for separate consideration. These are:

(a) Primary pipes, which have to circulate water steadily from the boiler to the cylinder or calorifier.

(b) Secondary pipes and feed pipe which are required to pass water spasmodically and in relatively large quantities when a tap is opened.

(c) Secondary circulating pipes, which in addition to (b), should circulate enough to make good their own heat-loss, and so keep hot water constantly available at the draw-off points.

Primary Circulation (Direct System)—The sizing of the pipes connecting boiler and cylinder on the direct system should be ample in order, first to allow for the furring which will inevitably take place in them if the water is hard, and, secondly, so that the frictional resistance may be so low that a brisk circulation takes place. Even for the smallest systems, pipe connections should be 25 mm minimum in order to make allowance for furring.

The sizes may be calculated in exactly the same manner as already described for a heating system operating by gravity circulation, taking the circulating head as the difference of pressure between the hot rising and cold falling columns from centre of boiler to centre of cylinder, with a temperature drop of, say, 30 K. This head, divided by the length of travel of flow and return, including bends, etc., will give the circulating pressure per metre run, from which the pipe size necessary may be determined from Fig. 8.6 (page 164). Alternatively, the sizes listed in Table 14.1 may be used, these being based upon a flow temperature of 40° C, with return at 10° C, the height from centre of boiler to centre of cylinder being taken

TABLE 14.1

SIZE OF PRIMARY FLOW AND RETURN MAINS FOR 'DIRECT' SYSTEM

Pipe size (mm)				Energy flow (kW)
32	-	-	-	10
40	-	-	-	15
50	-	-	-	30
65	-	-	-	50
80	-	-	-	80

at 1·5 m and travel assumed as 6 m plus 4 bends. An allowance for a 20 per cent decrease in flow rate due to furring has been made.

The maximum output of a hot-water boiler is required at periods of heavy draw-off—i.e., when the bottom of the cylinder is certain to be cold. As the heating of water in the cylinder proceeds and the return to the boiler rises in temperature, the flow temperature will also rise, though not by the same amount since the circulating head at higher temperatures will be increased and more water will be passed. Thus, when the water is returning to the boiler at 60° C, an equal heat transmission will be taking place with the same pipe sizes with a flow at 80° C—that is to say, a 20 K rise as against 30 K. At this point, however, the output from the boiler should be reduced so as to meet the radiation losses only, otherwise the temperature will continue to mount, which, in most cases, would be unnecessary and wasteful.

It will be seen from the above that the primary circulation is a constantly varying one, and as the maximum duty is required at the poorest circulating temperatures, great care should be exercised in seeing that the piping is kept as short as possible with easy radius bends to avoid undue resistance.

Primary Circulation (Indirect System)—With the indirect system, no allowance need be made in the sizing of the primary circulation for furring since the same water is constantly re-used as in a heating system.

A rapid circulation is essential, as the transmission of heat from the primary to the secondary water through the walls of the heat exchanger depends on a difference of temperature being maintained right through to the outlet.

The flow and return temperatures may be taken at higher figures than with the direct system; for example, with a combined heating and hot-water supply apparatus the primary water circulation will be the same as that supplied to the radiators in cold weather at, say, 80° C flow and 60° C return.

Where the indirect cylinder or calorifier is served from a boiler whose sole duty is to provide hot-water supply, it is of advantage to run this at as high a temperature as possible so as to economize in heat-exchange surface.

Based on a flow and return temperature of 80° C and 60° C respec-

TABLE 14.2

SIZE OF PRIMARY FLOW AND RETURN MAINS FOR 'INDIRECT' SYSTEM
Flow 80° C, Return 60° C

Pipe size (mm)	Energy flow (kW)
25	7
32	13
40	20
50	35
65	60
80	85

tively, a height from centre of boiler to centre of calorifier heat exchanger of 1·5 m, and a travel of 6 m, plus 4 bends, plus the equivalent of 3 m of pipe for the resistance of the heating element, the capacity appropriate to various sizes of pipes is shown in Table 14.2.

SECONDARY SUPPLY

Outflow from Taps—The sizes of the pipes from the cold water storage cistern to the cylinder and from the cylinder to the taps depend on the outflow from the latter, and not in any way on the boiler load. The pressure available for delivering the water is the gravity head or distance between the lowest water level in the supply cistern and the topmost tap. Obviously the topmost must be taken as this is bound to be the worst case.

This pressure must deliver the calculated quantity of water through the pipes, and as the length of the latter can be measured from the plans, making due allowance for bends, etc., as for a heating system, the available pressure per metre run of pipe may be calculated.

Sizing Secondary Flows—The rate of flow from any tap will depend on its size and on the pressure or head available at that point and thus, in a building with various floors fed from a cistern on the roof, the flow rate will progressively increase going down the building. It would be an added complication to allow for this in calculating pipe sizes; furthermore, at high discharge rates there is liable to be considerable splashing and throttling-down may be expected.

The *Guide* gives recommended rates of flow from fittings from which Table 14.3 gives adapted values. Rates for cold water supply are, for practical purposes, the same as those quoted for hot water fittings.

TABLE 14.3

APPROXIMATE DISCHARGE RATES FROM HOT AND COLD-WATER FITTINGS, AND SIZES

Fitting	Rate of flow (litre/s)	Size of connection (mm)
Bath (private) - -	0·3	20
(public) - -	0·6	20 or 25
Basin - - - -	0·15	15
Basin with spray tap -	0·05	10
Shower nozzle - -	0·15	15
Shower 100 mm rose -	0·4	15
150 mm rose -	0·6	15
Sink - - - -	0·3	20

Where there are only a few taps to be served, as in a private house, it may be assumed that at some time all the taps will be open at once, although it is unlikely.

With any sizeable installation there will be increasing diversity of demand the greater the number of taps. For instance, it may take three minutes to fill a bath, but at least twenty minutes before the next filling may occur, during which time other baths may be filling. A hot tap to a basin may be open for thirty seconds and at least another sixty seconds will elapse before the tap is opened again. In a row of ten basins this would mean a maximum of about three taps open simultaneously even if people were queuing up to use them.

The type of usage is important. A sports pavilion when play ends may have nearly all showers running simultaneously but, in a hospital, it is probable that only a small proportion of baths would be filling at precisely the same time.

Various attempts have been made to establish a basis for evaluating simultaneous demand.* The *Guide* postulates a method based on the theory of probability and adopts the conception of a 'demand unit'. This unit is taken as unity for a lavatory basin tap. The type of application is taken into account by weighting the units according to an assumed interval of use varying from five minutes to twenty minutes for a basin and from twenty minutes to eighty minutes for a bath. From this theory, a simplified comparative table (Table 14.4) is derived under three categories of application:

'congested' (where times of draw off are regulated or co-ordinated)
'public' (normal random usage)
'private' (infrequent or spasmodic)

TABLE 14.4
COMPARATIVE DEMAND UNITS (APPLIED)

Fitting	Category of application		
	Congested	Public	Private
Basin	10	5	3
Bath	47	25	12
Sink	43	22	11

Having established progressive totals of 'demand units' throughout the system to be sized, the flow required may be read from Fig. 14.1 which is based on probability of simultaneous use. For showers and basin spray taps, no diversity of use can be taken and the whole flow rate, fitting by fitting should be added to the sectional and total flow rates read from Fig. 14.1. Similarly, fittings such as wash fountains and washing-up machines requiring a continuous flow may be included additionally acccording to type and maker's flow requirement.

* e.g. Griffiths T. and Burberry P. Demand and Discharge sizing for Sanitary Fittings. *The Architects Journal*, Nov. 1962. p. 1185.

No method such as the above can be more than a guide—a certain amount of common-sense must be exercised in applying it. For instance, in an hotel, cleaner's sinks on various floors are unlikely to be in use simultaneously with baths, but sinks in the kitchen may well be.

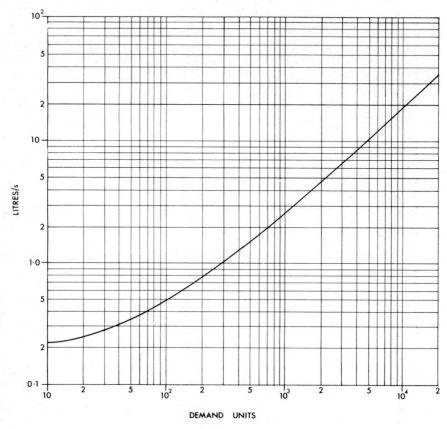

FIG. 14.1.—Flow probability and demand units.

Example of Flow Pipe Sizing

To illustrate the method of sizing, consider Fig. 14.2 which is a diagram of a typical riser system serving four floors. From this the number of fittings and hence the demand units may be set down in tabular form, as shown below the Figure, finally arriving at the flow required in litre/s for each section of pipework.

To proceed with the sizing: use may be made of the pipe-sizing charts (Fig. 8.6, page 164, or Fig. 14.4, page 324), but for greater convenience pressure drops should be based on the head of water, as it is the head from cistern to tap which causes flow. Also, flows expressed in volume per unit time, not kW/K, are required.

Table 14.5 has been compiled in this form from *Guide* data for water flowing at 75° C in galvanised heavy-weight steel pipes, and Table 14.6 for copper pipes to B.S. 2871 (Table X).* The differences with medium-weight steel are slight, but the accurate data are in the *Guide* if required.

Next, consider the pressure due to the head of water at the taps. The worst case is the top-floor tap on riser *A* which is subject to only 3 m head and has the longest travel.

Travel T = cold feed = 24

    allow for

    fittings   5   29 m

    main   9

    fittings   2   11 m

    riser   11

    fittings   3   14 m

                   54 m

$$\frac{H}{T} = \frac{3}{54} = 0.055 \text{ m/m run}$$

Thus, taking the litre/s flow rates from the tabulated example, it is possible to set down the nearest pipe size for each section, reading from Tables 14.5 or 14.6 according to the material used for piping.

	Pipe size (mm)	
Riser A	Galvd.	Copper
Branch to 4	25	28
Riser OA	25	28
,, OB	32	35
,, OC	40	42
,, OD	50	42
Main OE	50	42
,, OF	65	54
,, OG	65	67
C.W. Feed	65	67

It will be noted there is a slight saving in pipe size in this example by using copper.

The branches at each floor next require attention: floors 3, 2 and 1 on riser *A* each have a greater pressure head available. The loss on the index run to the point of branch may be taken out in each case and deducted from the head available. The surplus may then be used with a new H/T to size the branch. The same applies to risers *B* and *C* which have a shorter travel, even to the top floor. However, if the plant is of modest size, the

---

* Since the density of water at 10° C is 999·7 kg/m³ and standard gravity is 9·808 m/s², the force exerted by a column of water one metre high (at that temperature) is 999·7 × 9·807 = 9801 kg/ms² (Pa). By taking this as being approximately 10 000 (as in Tables 14.5 and 14.6), an error of only five per cent arises, which is insignificant in the context of the other variables existing.

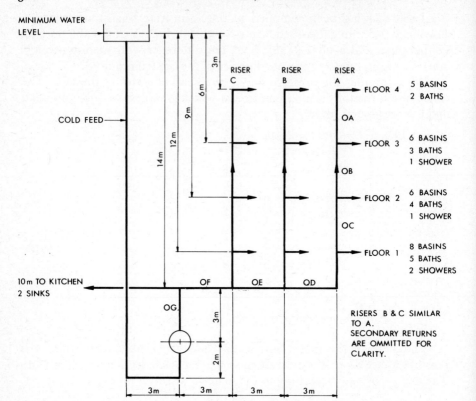

FIG. 14.2.—Pipe sizing example, up feed system flows.

Individual branches from risers (taken as 'public' category).

Floor	Fittings	Demand Units			Litre/s	
					From Fig. 14.1	Total
4	5 basins 2 baths	5 25	25 50	75	0·4	0·4
3	6 basins 3 baths	5 25	30 75	105	0·51	
	1 shower	× 0·15 litre/s		=	0·15	0·66
2	6 basins 4 baths	5 25	30 100	130	0·56	
	1 shower	× 0·15 litre/s		=	0·15	0·71
1	8 basins 5 baths	5 25	40 125	165	0·66	
	2 showers	× 0·15 litre/s		=	0·30	0·96

Risers and Mains

Item	Section	Demand Units	Litre/s		
			From Fig. 14.1	Showers	Total
Riser	OA	75	0·4		0·4
A	OB	180	0·74 +	(1 × 0·15)	0·89
	OC	310	1·0 +	(2 × 0·15)	1·3
	OD	475	1·45 +	(4 × 0·15)	2·05
Main	OD	475	1·45 +	(4 × 0·15)	2·05
	OE	950	2·5 +	(8 × 0·15)	3·7
	OF	1425	3·45 +	(12 × 0·15)	5·25
Add for kitchen sinks (assumed to run simultaneously)				2 × 0·3	0·6
Main OG and cold water feed					5·85

TABLE 14.5

FLOW OF WATER AT 75° C, LITRE/S, IN GALVANISED STEEL PIPES TO B.S. 1387 (HEAVYWEIGHT)

Head loss (m water per m run)	Water flow (litre/s) in pipes of stated nominal bore (mm)										
	15	20	25	32	40	50	65	80	100	125	150
0·01			0·1	0·4	0·6	1·2	2·4	3·8	7·8	14	22
0·015		0·1	·2	·5	0·8	1·5	3·0	4·7	9·6	17	27
0·02		·1	·2	·6	0·9	1·7	3·5	5·4	11·1	20	32
0·025		·1	·3	·6	1·0	1·9	3·9	6·1	12·5	22	36
0·03		·2	·3	·7	1·1	2·1	4·3	6·7	13·7	24	39
0·035		·2	·3	·8	1·2	2·3	4·6	7·2	14·8	26	42
0·04		·2	·3	·8	1·3	2·4	5·0	7·8	15·8	28	45
0·045		·2	·4	·9	1·3	2·6	5·3	8·2	16·8	30	48
0·05		·2	·4	·9	1·4	2·7	5·6	8·7	17·7	32	51·
0·06	0·1	·2	·4	1·0	1·5	3·0	6·1	9·5	19·4	34	56
0·07	·1	·3	·5	1·1	1·7	3·2	6·6	10·3	21·0	37	60
0·08	·1	·3	·5	1·2	1·8	3·5	7·1	11·1	22·4	40	64
0·09	·1	·3	·5	1·2	1·9	3·7	7·5	11·7	23·8	42	68
0·10	·1	·3	·6	1·3	2·0	3·9	7·9	12·4	25·1	45	72
0·15	·2	·4	·7	1·6	2·5	4·8	9·7	15·2	30·8	55	89
0·20	·2	·4	·8	1·9	2·9	5·5	11·3	17·5	35·6	63	103
0·25	·2	·5	·9	2·1	3·2	6·2	12·6	19·6	39·8	71	
0·3	·2	·5	1·0	2·3	3·5	6·8	13·8	21·5	43·6		
0·4	·3	·6	1·2	2·6	4·1	7·8	16·0	24·8			
0·5	·3	·7	1·3	3·0	4·6	8·8	17·9	27·8			
0·6	·3	·8	1·5	3·2	5·0	9·6	19·6				
0·7	·3	·8	1·6	3·5	5·4	10·4					
0·8	·4	·9	1·7	3·8	5·8	11·1					
0·9	·4	·9	1·8	4·0	6·1	11·8					
1·0	·4	1·0	1·9	4·2	6·5						
Equivalent lengths for k = 1 m (average)	0·4	0·6	0·8	1·1	1·4	1·9	2·7	3·4	4·7	6·2	7·8

Pipe fittings, K

Bends, tees, reducers, enlargements—allow = 1·0
Screw down valve or tap   ,,   10·0
Connections to cylinder   ,,   1·0

TABLE 14.6

FLOW OF WATER AT 75° C, LITRE/S, IN COPPER PIPES TO B.S. 2871 (Table X)

Head loss (m water per m run)	Water flow (litre/s) in pipes of stated outside diameter (mm)										
	15	22	28	35	42	54	67	76	108	133	159
0·01		0·1	0·3	0·5	0·8	1·6	2·8	4·0	10	18	29
0·02		·2	·4	·6	1·1	2·3	4·1	5·8	15	27	43
0·03		·2	·5	·8	1·4	2·9	5·1	7·3	19	33	53
0·04		·3	·6	1·0	1·7	3·4	6·0	8·5	22	39	62
0·05	0·1	·3	·6	1·1	1·9	3·8	6·8	9·6	25	44	70
0·06	·1	·3	·7	1·2	2·1	4·2	7·5	11	28	49	77
0·07	·1	·4	·8	1·4	2·3	4·6	8·2	12	30	53	84
0·08	·1	·4	·8	1·5	2·4	4·9	8·8	12	32	57	90
0·09	·1	·4	·9	1·6	2·6	5·2	9·4	13	34	61	100
0·10	·2	·5	·9	1·6	2·8	5·6	9·9	14	36	64	102
0·12	·2	·5	1·0	1·8	3·0	6·1	11	16	40	71	
0·14	·2	·6	1·1	2·0	3·3	6·7	12	17	44	77	
0·16	·2	·6	1·2	2·1	3·6	7·2	13	18	47		
0·18	·2	·6	1·3	2·3	3·7	7·7	14	19	50		
0·20	·2	·7	1·3	2·4	4·0	8·1	14	20			
Equivalent length for $K = 1$ m (average)	0·6	1·0	1·5	2·0	2·5	3·6	4·7	5·6	8·9	12	15

Pipe Fittings
Values of K may be taken from Table 14.5

time spent in further refinement of calculation may not result in much saving in cost and the designer may content himself with using one value of H/T throughout. On an extensive installation this could not be regarded as economical design.

Nevertheless, in this example the kitchen would be too glaringly oversized if taken at the index H/T. It will in fact be as follows:

Available head $= 14$ m
Head absorbed, at design flow rate, from cistern through cylinder to junction of OG with OF,

Cold Feed, $T = 29$
$OG = 3 + 1 = 4$
_____
33 m

From Table 14.5, unit head loss is 0·055 m/m run

$\therefore 33 \times 0.055 = 1.82$ m

Thus residual head available for kitchen connection
$= 14 - 1.82 = 12.18$ m

$$\frac{H}{T} = \frac{12.18}{(10+2)} = 1 \text{ m/m run}$$

At this rate of head loss, a flow of 0·6 litre/s requires 20 mm galvanised pipe or 22 mm copper pipe. Had it been sized on the index pressure loss, the size would have been 32 mm galvanised or 28 mm copper.

With regard to showers, it should be borne in mind that a head of about 1·5 m is required for the spray to function. This should be allowed for by deduction from the total available.

Sizing Secondary Returns—Having established the sizes of the secondary flow pipes, it is a simple matter to calculate the heat emission from the circulating portions of these from Tables 7.2 and 7.3 (page 151), but for convenience Table 14.7 may be used. In this Table it is assumed that mean water temperature is 60° C and air is at 20° C—a difference of 40 K. Insulation has been taken as 25 mm thick glass fibre.

TABLE 14.7

EMISSION FROM HOT-WATER SUPPLY SECONDARY
CIRCULATING PIPING

Based on 40 K difference mean water to air W/m run

Galvanized Steel			Copper B.S. 2871 (Table X)		
Nom. Bore mm	Emission W/m		O.D. mm	Emission W/m	
	Bare	Insulated		Bare	Insulated
15	42	8	15	25	9
20	51	9	22	34	10
25	62	10	28	42	12
32	75	12	35	50	13
40	84	13	42	58	16
50	102	15	54	72	18
65	125	18	67	85	19
80	143	20	76	95	22
100	179	24	108	128	28
125	214	29	133	153	31
150	248	33	159	178	37

Notes
Insulation is taken as 25 mm thick glass fibre, or equivalent.
Emission from towel airer: average = 250 W.
Emission from linen cupboard coil: average = 250 to 500 W.

To this must be added the emission from linen cupboard coils and towel airers, average values for some examples being also given in the Table.

The secondary returns must also carry sufficient water for their own heat emission so that the temperature drop back to the cylinder does not exceed a predetermined maximum such as 10 K. Emissions from these returns must therefore be added to each section, and since their size is not yet known they may be assumed at, say, one or two pipe sizes less than the flow in each case, or the emission may be taken at a rough approximation of two-thirds of the associated flow pipe.

With the above totals marked on the plans for each branch or section of main it is possible to arrive at totals working back to the cylinder exactly as for a heating system.

The circulating pressure if by gravity may be obtained as described on p. 174, by determining the average height from the centre of cylinder to average point of heat emission. The appropriate circulating head per metre of height may be taken from Table 8.1 (page 175), allowing flow at say 65° C and return at say 55° C: the value in this case will be found to be 47·36 Pa per metre height. Alternatively if a pump is necessary owing to long runs or mains below the cylinder level, the pump pressure may be taken at 20 kPa overall for most systems, and 40 kPa to 60 kPa for large buildings or institutions with considerable distances between the blocks. The pump pressure necessary may be estimated by allowing a maximum of 10 Pa per metre of travel, allowing the total run of return pipe, plus one-third of the run of flow pipe (on account of its larger size) plus 25 per cent allowance for bends and resistances.

Having determined the pressure, whether for gravity or pump circulation, the resistance of the flow mains (the sizes of which have already been determined) to the furthest point should be calculated as for a heating circuit. This figure, deducted from the total circulating pressure, will give the available pressure for the return mains. The latter divided by the metres run of return main, including allowance for bends and single resistances, will give the available pressure per metre.

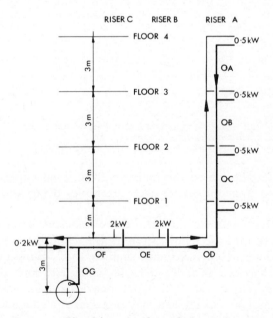

FIG. 14.3.—Pipe sizing example, up feed system returns.

Example of Sizing of Secondary Returns

Consider Fig. 14.3, which is the same system as Fig. 14.2 but with returns added.

The emission of heat from the flow piping, which has already been sized and is assumed as insulated copper, is as follows:

Section	Pipe Size	Emission (W/m)	Length (m)	Emission (W)	
OA	28	12	3	36	
OB	35	13	3	39	
OC	42	16	3	48	
OD	42	16	2	32	155
Riser B					155
Riser C					155
OD	42	16	3	48	
OE	42	16	3	48	
OF	54	18	3	54	
OG	67	19	3	57	207
					672
Taking emission of returns as 2/3 of that of flows					448
Total pipe emission					1120 W

The branches serve one coil and one towel airer per floor and amount to 2 kW per riser, i.e. a total of 6 kW plus 200 W in the kitchen, 6·2 kW in all. Mains losses thus represent an addition of 18 per cent to the useful heat output and it will be sufficiently near to apportion these losses equally to each pipe section for this example but, in the case of an extensive circulation, it would be necessary to make a full calculation as described in Chapter 8.

Assuming gravity circulation, the circulating pressure to floor 1 from the centre of the cylinder

$$= 5 \text{ m} \times 47 \cdot 36 \text{ Pa/m} = 236 \cdot 80 \text{ Pa}$$

and to floor 4

$$= 20 \text{ m} \times 47 \cdot 36 \text{ Pa/m} = 947 \cdot 20 \text{ Pa}$$

Taking the nett heat loads for each pipe section and adding for mains losses progressively, the pressure loss in the flow pipework to floor 1 may then be set out as below, dividing kW by the temperature drop of 10 K to give kW/K, and reading the required unit pressure loss values from Fig.

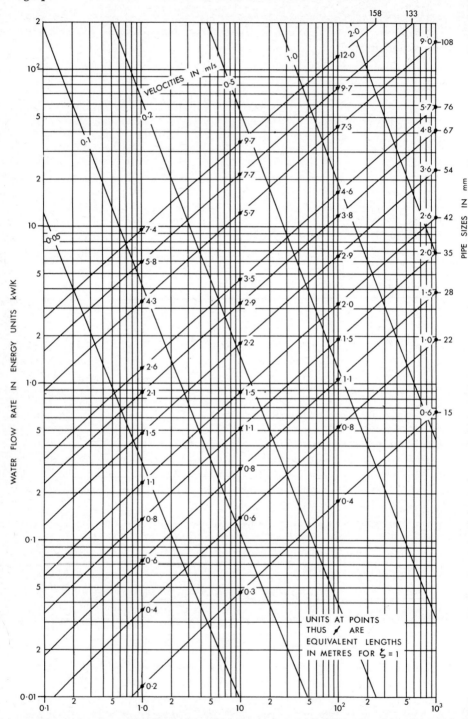

Fig. 14.4.—Pipe sizing chart for water at 75° C (Copper pipes to B.S. 2871, Table X).

14.4. This figure is a pipe-sizing chart for copper tube which has been prepared by computer, like Fig. 8.6 (page 164). For galvanised steel, Fig. 8.6 may be used even though drawn for black steel, there being little practical difference.

Section	Size (mm)	Length including fittings (m)		kW per K	Pressure loss (Pa/m) from Fig. 14.4	Pressure loss (Pa)	
OG	67	5		0·73	0·75	3·8	
OF	54	5		0·71	2·0	10	
OE	42	5		0·47	3·2	16	
OD	42	7		0·24	1·2	8·4	
at Floor 1	28	5	27	0·06	0·7	3·5	41·7
		—				—	
OC	42	4		0·18	0·6	2·4	
OB	35	4		0·12	0·8	3·2	
OA	28	4		0·06	0·7	2·8	
at Floor 4	28	5	17	0·06	0·7	3·5	11·9
		—				—	

Note:
In arriving at the equivalent length for fittings, an arbitrary basis has been used for this example. In a practical case, equivalent lengths would be assessed as explained in Chapter 8.

Since these flow pipes have been sized for outflow from taps and the water quantity required for circulation is small in comparison, the corresponding pressure loss is low also. The difference between the available circulating pressure (CP) and the pressure loss in the flow pipes is available to overcome friction loss in the return pipes, as yet unsized. Omitting decimal quantities:

At floor 1.

$$\text{Available } CP \quad = 236 \text{ Pa}$$
$$\text{Loss in flow} \quad = \ 42 \text{ Pa}$$

$$\text{Residual} \quad = 194 \text{ Pa}$$

Taking return travel as 27 m as for the equivalent flowpipe

$$\frac{CP}{T} \quad = \frac{194}{27} = 7 \text{ Pa/m}$$

At floor 4.

$$\text{Available } CP \quad = 947 \text{ Pa}$$
$$\text{Loss in flow}$$
$$(41\cdot7 + 11\cdot9 - 3\cdot5) = \ 50 \text{ Pa}$$

$$\text{Residual} \quad = 897 \text{ Pa}$$

Taking return travel as 39 m as for the equivalent flow pipe

$$\frac{CP}{T} \quad = \frac{897}{39} = 23 \text{ Pa/m}$$

It will be seen that the index circuit is that to floor 1. Reference to Fig. 14.4, using the same values of kW/K as for the parallel sections of flow piping, now enables return sizes to be entered, taking $CP/T = 7$, Pa/m. Fig. 8.6 would be used for galvanised steel. For the remainder of the riser a separate calculation might be made for each floor, but, taking them as for floor 4 for purposes of this example, we use $CP/T = 23$ Pa/m. The resultant sizes would be as follows:

OG – 42 mm OD – 28 mm
OF – 42 mm OC – 22 mm
OE – 25 mm OB – 22 mm

The return pipe at each floor would be 15 mm.

The branch to the kitchen is a special case, the circulating water quantity being small and the travel short. This would result in a pipe size less than 15 mm which is, however, the minimum usable on account of possible deposits. In fact, this circuit will run at a lesser temperature drop, lower CP and a greater flow to balance. Some designers prefer 22 mm as a minimum size for reasons of mechanical strength and to allow for possible furring-up of the pipe bore.

Were this a pumped system, regulating valves in the returns would be desirable to enable any short branches such as this to be checked down.

Draw-off through returns—No account has been taken in the example of water flowing to the taps from the return pipe. In Fig. 14.5, when a tap is opened at X, water will obviously flow to it in both directions from the cylinder, the water in the return pipe flowing in the opposite direction to the usual owing to the circulation pressure being completely overcome by the much greater head from the cistern to the tap.

The reduced frictional resistance brought about in the out-flowing

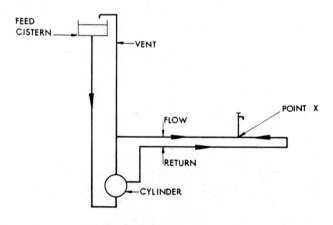

Fig. 14.5.—Direction of water flow to an open tap.

system, due to the double pipe, will vary inversely as the distance of any tap from the cylinder. This effect is usually ignored when sizing pipes for small systems, but on a large layout it may be taken into account.

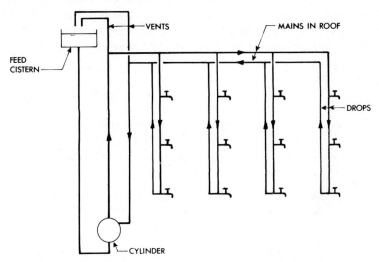

FIG. 14.6.—Hot water supply, drop system (not satisfactory).

DROP SYSTEM

There are occasions where building configurations may lead to consideration being given to running circulation piping in a roof void, or other top-level route, not far below the cold water feed cistern as shown in Fig. 14.6. Such arrangements will not provide satisfactory service since,

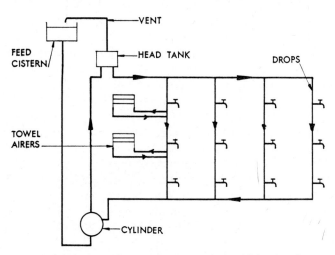

FIG. 14.7.—Hot water supply, drop system with head tank.

at times of heavy draw-off, the flow main will be emptied of water and air admitted through the vent.

One method which has been used to overcome this problem is as shown in Fig. 14.7. In this case the flow riser delivers direct to a hot-water head tank above the level of the topmost tap, and the supplies to the taps are taken from a system of drops connected from the bottom of the tank.

The advantage of this method over the up-feed in the case of high buildings, is that the pipe sizes are generally less. The upper floors are fed downwards from the top tank or partly downwards and partly upwards.

As a result, the head of water (H) available at the upper taps may be assumed to derive from the hot head tank and the effective travel (T) will be only the pipe length between that tank and the taps. An assumption that the top one-third of the drops are fed downwards is generally reasonable. The remaining two-thirds of the height is fed from the returns working in the reverse direction

The disadvantages of the drop system are:

(i) That it involves ranges of large piping at or near roof level, often difficult to accommodate.

(ii) That towel airers and linen cupboard coils connected to the drops operate by local gravity circulation only, so that there is not the same possibility of connecting them when at a distance from the drop.

UP-FEED HEAD TANK SYSTEM

In order to avoid mains at or near roof level necessary with the drop system, the up-feed method may be used with a separate hot-water head tank to each riser as shown in Fig. 14.8.

At periods of heavy draw-off on the lower floors of any particular riser, when the supply to the top floor might tend to be reduced, this tank will supply the riser downwards by drawing on its storage, so ensuring a good supply at all levels. As it is unlikely that the heavy draw-off continues for more than a minute or so, the tank will not have emptied before the demand ceases, when the level in the system will balance up again by water flowing up the riser in the normal way.

The sizing of this system may be simplified by ignoring the top one or two floors and sizing flow mains and risers for discharge to the lower floors only. The effective height to these lower floors is thus greater and smaller pipes may be used throughout the system.

Dead Legs—These are the non-circulating branches to taps and it is desirable to keep them as short as possible. *Code of Practice* No. 342, Part 2 lists the lengths given in Table 14.8 as the maximum permissible. Spray taps are a special case where the circulation should be taken almost up to the tap—a dead leg of no more than 1 m if possible—on account of the small flow.

Sizing of Head Tanks—The size of the head tank can be determined by

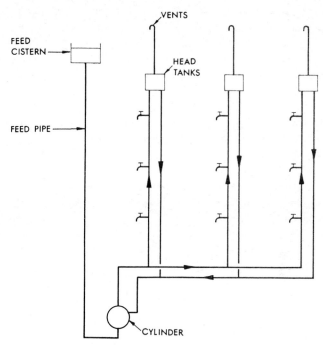

FIG. 14.8.—Hot water supply, up feed to multiple head tanks.

TABLE 14.8

MAXIMUM PERMISSIBLE LENGTH OF DEAD LEGS

Pipe size (mm)		Length (m)
Steel Nom. bore	Copper O.D.	
up to 15	up to 15	12
20	22	8
over 25	over 28	3

allowing two or three minutes' capacity for the discharge which is assumed to come from it. Thus if 1 litre per second comes downwards in a drop system, and 2 litres upwards, a head tank of $1 \times 3 \times 60 = 180$ litres would be adequate. This storage should not be taken as reducing the main cylinder capacity.

Pump Sizing—In the case of an accelerated secondary circulation, the size of the pump may be determined from the emission in kW as for a heating system.

For example, if the total emission is 50 kW, and a 10 K drop has been assumed, the capacity of the pump will be

$$\frac{50}{10 \times 4 \cdot 2} = 1 \cdot 2 \text{ kg/s} = 1 \cdot 2 \text{ litre/s}$$

Allowing a margin, a suitable size pump would be 1·5 litre/s. The pressure of the pump will be arrived at as discussed on page 183.

As for a heating system, pumps of centrifugal type are most suited for the purpose. The pump body should be of a type readily opened for inspection and for removal of scale, and is preferably of gun-metal or bronze.

GENERAL

Vent and Feed Pipes—Vent pipes from the top of each riser carried above the feed cistern are necessary unless it can be so arranged that the topmost tap acts as a vent when it is opened. Frequently it is possible to accommodate only one open vent, other risers being vented by the taps. In the case of the head tank system, particular care must be taken to ensure that each tank is vented; otherwise a collapse could occur when water is drawn off.

It should be noted that the cold feed pipe has to furnish the whole supply, and an adequate size is essential. Even in the smallest installation 25 mm should be the minimum size for this pipe, and is so called for in *Code of Practice* No. 342, Part 1.

CHAPTER 15

Running Costs of Heating Systems

THE COST OF OPERATION of any system providing space heating and/or hot water supply will depend upon a number of variables including:

(1) Fuel consumption
(2) Power consumption of auxiliaries
(3) Labour
(4) Maintenance
(5) Insurance and similar on-costs
(6) Interest on capital and depreciation.

The amenity value of various fuels may enter into these considerations. In some cases, such as an industrial application, convenience and cleanliness might be unimportant whereas in a bank or office block they would merit first priority. Similarly, the character and anticipated life-span of the building may bear upon choice and a fuel which would be suitable for a well-equipped solid structure would not match the expendibility of, say, a system-built school.

Where a full comparison must be made in order to select the most economical method of operation, detailed calculations must be made for each alternative. The present instability in energy costs, as illustrated by Fig. 15.1, does not allow interpretation of the results to be presented with the confidence customary in the past.

FUEL CONSUMPTION

In order to estimate fuel consumption, it is first necessary to arrive at an approximation of the annual consumption of energy for the heating and/or hot water supply system. This will in principle be a function of the maximum heat loss, as calculated; the hours of use; the likely ratio of normal to peak loading which will apply over the heating season; the manner in which the building structure responds to heat input and the routine of daily and weekly operation proposed for the system. If hot water supply is to be included there will be a significant standing no-load loss to be taken into account, additional to energy supplied to heat the water actually drawn off throughout the year.

The next step is to derive from this approximation of annual energy consumption the equivalent in terms of energy supply. This introduces the question of efficiency of conversion over the whole year at varying rates of demand.

Heat Losses—The totals calculated for design purposes will be in excess of those to be used as a basis for an estimate of energy consumption. This is due to the fact that if one considers any building, heating design must be such that on *each* external aspect, sufficient warmth can be provided to maintain a satisfactory temperature. In practice, air infiltration resulting from wind will occur only on the windward side; other aspects, i.e. the leeward side will exfiltrate.

Furthermore, adventitious heat gains from occupants, lighting, solar heat and sundry other sources all contribute to the sum total of heat supply and some allowance for the effect of these must be taken into account.

Based on a variety of evidence, a fair assessment of requirements for energy input will take account of:

(1) Heat losses as normally calculated in kW but deducting between 10 and 15 per cent for air change not applying in all parts simultaneously.

(2) An allowance, by adjustment of the design inside/outside temperature difference, for incidental heat gains. This aspect is taken into account by use of the 'Degree-Day' concept described later in this Chapter.

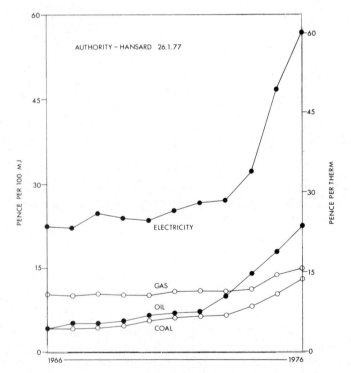

Fig. 15.1.—Changes in domestic fuel costs over the decade 1966/76
(Hansard 26 Jan. 1977).

Any losses from piping in a central system which do not contribute to the building heat requirements must be calculated as a separate exercise and added to the nett heat loss figure. It should be remembered that with certain types of system mains losses may be constant throughout the heating season and thus disproportionately high. An example would be a fan-convector system, controlled by room thermostat switching of the fan motor, fed from constant temperature heating mains.

Proportion of Full-Load Operation—This will be a variable factor depending on the weather. Obviously no system will be called upon to operate at 100 per cent output (based on $-1°$ C or $-3°$ C outside) during the whole season. What proportion of this full load can be assumed for the purpose of calculation, and how does it vary for different parts of the country?

A basis of comparison is afforded by the Degree-Day method. The British degree-day assumes that in a building maintained at $18.3°$ C, no heating is required when the external temperature is $15.5°$ C or over. The difference between the external daily mean temperature and $15.5°$ C is then taken as the number of degree-days for the day in question. Thus, for an external daily mean temperature of $5.5°$ C there are 10 degree days and if this mean were constant for 7 days then 70 degree days would have accumulated. The total degree days for the year forms a basis of com-

TABLE 15.1

DEGREE DAYS FOR THE BRITISH ISLES, 20 YEAR AVERAGE (1957/76),*
FOR A BASE TEMPERATURE OF $15.5°$ C.

| Area | June July Aug. | Monthly Totals | | | | | | | | | Seasonal Totals | | |
		Sept.	Oct.	Nov.	Dec.	Jan.	Feb.	Mar.	April	May	Annual	Sept. to May	Oct. to April
Thames Valley	98	56	132	256	333	346	304	282	197	113	2117	2019	1850
South East	166	84	163	280	356	370	329	310	224	145	2427	2261	2032
South West	120	55	114	215	276	293	272	267	197	131	1940	1820	1634
Severn Valley	121	69	143	259	328	344	311	292	209	129	2205	2084	1886
South	151	76	145	258	328	339	307	294	214	141	2253	2102	1885
Midlands	173	92	172	290	358	371	335	318	233	152	2494	2321	2077
East Anglia	160	74	154	283	360	378	334	315	232	143	2433	2273	2056
Wales	172	72	136	239	301	323	301	292	228	156	2220	2048	1820
East Pennines	148	77	157	281	350	362	323	304	217	139	2358	2210	1994
West Pennines	150	79	155	280	346	359	323	304	222	139	2357	2207	1989
North East	174	87	171	295	360	374	334	317	234	154	2500	2326	2085
North West	193	94	169	296	357	366	333	319	239	163	2529	2336	2079
Borders	251	108	184	300	361	376	343	332	259	193	2707	2155	2456
East Scotland	240	106	185	308	368	379	343	326	252	189	2696	2456	2161
West Scotland	207	106	179	303	354	368	335	316	235	163	2566	2359	2090
N. E. Scotland	282	124	199	322	381	396	359	345	270	206	2884	2602	2272
N. Ireland	210	100	170	288	343	359	325	311	237	167	2510	2300	2033
Dublin	187	83	148	263	324	335	309	292	226	155	2322	2135	1897
Cork	163	68	122	223	273	293	272	263	204	146	2027	1864	1650
West Coast	209	83	134	240	286	313	286	266	214	157	2188	1979	1738
Possible Maximum for constant outside temperature of $-1°$ C	1519	495	512	495	512	512	462	512	495	512	6023	4505	3498

* 9 year average for values in Republic of Ireland (1961/70).

parison between the heat requirements of one place and another and, for a given building, of energy consumption from one year to another.

It should not be regarded as an absolute unit, nor should it be pressed beyond its usefulness for comparative purposes. For instance, the effect of low night temperatures may give exaggerated degree-day readings even though the buildings may not be heated at night.

If it be assumed that full load on a heating system takes place when the outside is at $-1°$ C, the maximum degree-days may be taken as 16.5 in any one day. For a heating season of 30 weeks (October/April), there are a possible 3498 degree-days and, for 39 weeks (September/May), a possible 4505 degree-days. Traditionally, the 30 week season has been used but over recent years with rising standards, the 39 (or nominally 40) week season is often considered to be more appropriate, particularly for the case of residential heating.

The British Gas Corporation have for many years published monthly and annual degree-day records for various parts of the country based on Meteorological Office records, but this task has now been taken over by the Department of Energy.* From these data fractions may be deduced, taking 3498 or 4505 as unity, for 30 and 40 week heating seasons respectively.

Such fractions we herein term as *weather factors* since each illustrates, in simple terms, the proportion of full load imposed by the external weather and provides a comparison from year to year and from one part of the country to another. Table 15.1 lists 20 year average degree-days and it may be seen that the weather factors for, say, Thames Valley are 0.53 and 0.45 over 30 and 40 week heating seasons respectively.

Meteorological Office records may be obtained in respect of any particular town, if desired, and the degree-days calculated therefrom.

Period of Use—Here we have to assess the occupancy of the building and the length of time the heating system must be at work. It is necessary that a clear distinction be made between the period of use of the building by the occupants and the period of use of the heating plant. The former may be defined with ease but the latter will depend upon the particular characteristics of each individual building in-so-far as the response of the structure to heat input and the ability of alternative types of heating system to match that response are concerned. For instance, with a light construction, some drop in temperature of the fabric may occur at night-time, but correspondingly less heat is required to restore the internal temperature in the morning. Similarly a radiator system will respond more slowly to energy input at the associated boiler than will a direct-fired warm-air heater.

For a heating system, the hours of use may be resolved into periods per year, per week or per day. For the yearly use, as has previously been

* *Fuel Efficiency Booklet No. 7.* Department of Energy, 1977.

explained, a figure of 30 weeks is most commonly assumed for commercial and industrial premises. Since, as illustrated in Fig. 15.2, the diurnal external temperature range in May and September is such that evening and early morning heating is required, a 40 week season is more appropriate to houses, flats and hostels etc.

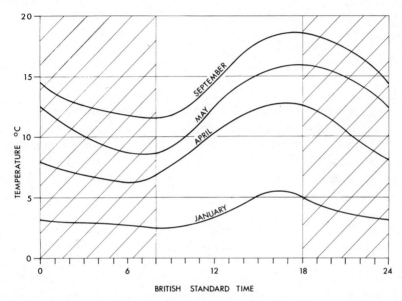

FIG. 15.2.—Diurnal variations in outside temperature for Manchester. September and May curves displaced to suit time base.

The weekly use depends again upon the kind of building considered. Domestic premises, hotels and the like require service over seven days; commercial and industrial buildings need not be heated for more than a five-day week and one- or two-day plant use per week may be appropriate to village halls, churches and the like.

For daily use, similar conditions apply. It will be appreciated that few buildings are heated continuously for twenty-four hours each day. The shorter the period of heating per day, the greater will be the no-load losses or, where plant operation is intermittent, the greater the requirement for pre-heating.

Hence, if energy supply is to be by oil or gas firing or by direct on-peak electrical input, no night-time plant operation will be necessary except in the most severe weather: pre-heating of the building prior to occupancy will however be required. On the other hand, with solid fuel firing (by magazine boiler or automatic stoker) no-load losses will occur even with little heat input to the building. Nevertheless, some advantage will have accrued and pre-heating time will decrease in consequence.

It is therefore convenient to classify energy input under two groupings:

Semi-automatic, using solid fuel via magazine boilers or automatic stokers or electricity via an off-peak supply.

Automatic, using gas or oil fuel or electricity via an on-peak supply.

As to the use of the building, as distinct from the heating system, it is suggested that this may be divided into about six categories, as follows:

(1) *Continuous Heating.* Hospitals, three-shift factories, police stations.

(2) *Continuous with Reduction at Night-Time.* Houses, flats, hotels, boarding schools, university hostels.

(3) *Daytime Heating.* Offices, public buildings, shops, factories (one shift).

(4) *Part Daytime Heating.* Day schools, cinemas, clubs.

(5) *Occasional*—two days per week. Churches with mid-week meetings. Public halls used two days per week.

(6) *Occasional*—one day per week. Churches, Sunday schools, sports accommodation.

Many more intermediate divisions are no doubt possible, but these cover the majority of cases.

There are a number of methods available by which the various aspects outlined in preceeding paragraphs may be taken into account in a calculation for assessment of fuel consumption. The more common of these is analytical in that it takes account of the various known aspects and, by simple deduction, derives an estimate of equivalent running hours at full load. Such a calculation does not, however, take full account of the effect of the character of the building structure as this may bear upon economies arising from intermittent plant operation.

An alternative method, as described in the *Guide, Section B18,* results from Research by B.S.R.I.A. and goes some way towards making allowance for all the variables. We present a simplified version of this approach in the form of a series of multiplying factors.

Hours of Use per Annum (Analytical Method)—Tables 15.2 to 15.4 summarise an attempt to work through the various considerations in a logical manner, using the six categories of building use previously listed. For each category a weather factor of 0·53 has been used as being representative of the annual proportion of full load use. This fraction is, of course, that applicable to the Thames Valley and other values would need to be used for other areas.

Table 15.3 works through the various hours of operation appropriate to typical plants and indicates the order of the allowance which should be made for pre-heating prior to occupation. Further data are included to show how account may be taken of energy use by semi-automatic plant burning solid fuel during hours of 'shut down'.

Table 15.4 gives the hours of firing per annum (30 weeks heating season.) It will be noted that the weather factor has not been applied to the figures for night-time use since this will be constant at the 15 per cent

TABLE 15.2
PREHEATING AND NO-LOAD ENERGY USE

Building Use Category	Daily Preheating, Hours	Semi-automatic Firing			Automatic Firing. Additional Preheating in Morning, Hours
		No-load Energy use at Night (15%)	Extra pre-heating after week-end etc. Hours.	No-load Energy use at Week-end (15%)	
1	—	—	—	—	—
2	1	24 hours, less occupation period, less preheating	—	Total hours unoccupied, less preheating	1 to 3 hours per day of use
3	2		2		
4	3		2		
5	6		—		
6	8		—		

TABLE 15.3
EQUIVALENT HOURS PER ANNUM FULL USE FOR VARIOUS HEAT REQUIREMENTS

Type of Heat Requirement	Hours of Use	Number of Hours per Day	Daily Preheating in Hours	Days per Week	Total Hours per Week — A	No-load hours at night (15%)	Days per Week	Equivalent Total Hours per Week — B	Extra Preheating after Week-end etc. Hours — C	No-Load Hours at Week-end (15%)	Equivalent Total Hours per Week — D
1. Continuous	—	24	—	7	168	—	—	—	—	—	—
2. Continuous except night-time	7 a.m.–11 p.m.	16	1	7	119	7	7	7½	—	—	—
3. Daytime, excluding week-ends	9 a.m.–6 p.m.	9	2	5	55	13	4	8	2	59	9
4. Part daytime	10 a.m.–4 p.m.	6	3	5	45	15	4	9	3	60	9
5. Occasional, 2 days per week	—	10	6	2	32*	14	2	4¼	—	—	—
6. Occasional, 1 day per week	—	10	8	1	18†	14	1	2	—	—	—

* For use at 7 a.m. on Sunday, preheating would commence at 2 p.m. on Saturday and would go on until 6 p.m., followed by 14 hours no-load use, with a further 2 hours preheating before occupation commenced. This complete cycle occurs twice a week.

† A similar cycle to No. 5, but starting at noon on Saturday and occurring once a week.

given, no matter what the outside temperature may be.

It is admitted that there is, in Table 15·4, plenty of room for argument as to what allowances should properly be made, and it is put forward solely for what is believed to be average conditions. There can be nothing final or dogmatic about it, especially when divorced from any particular set of circumstances.

Hours of Use per Annum (Factorial Method)—The theory underlying this alternative approach* takes into account variations in the level

---

* Billington. N. S., *Estimation of annual fuel consumption*. J.I.H.V.E. 1966. 34. 253.

TABLE 15.4

Type of Heat Requirement	Semi automatic Firing							Automatic Firing			
	Hours per Week / Useful Heat / Columns A + C	Weather Factor	Column E × Weather Factor E	Equivalent Hours Banking per Week —Columns B + D	Total Equivalent Hours per Week F	Number of Weeks' Use per Annum	Equivalent Hours per Annum	Total as Column E	Add for Extra Pre-heating	Total Hours per Week G	Column G × Weather Factor (0.53) × 30 Weeks = Equivalent Hours per Annum
1. Continuous - -	168	0·53	89	—	89	30	2670	168	—	168	2670
2. Continuous except night-time	119	0·53	63	7½	70½	30	2115	119	7	126	2003
3. Daytime, excluding week-ends	57	0·53	30¼	17	47¼	30	1418	57	10	67	1065
4. Part daytime -	48	0·53	25½	18	43½	30	1305	48	10	58	922
5. Occasional, 2 days per week	32	0·53	17	4¼	21¼	30	638	32	6	38	604
6. Occasional, 1 day per week	18	0·53	9½	2	11½	30	345	18	3	21	334

of adventitious heat gains within the building, the thermal mass of the structure and the ability of the heating system to respond to demand. Allowance is made also for the fact that, as has previously been explained, standard British degree-days relate to the maintenance of an internal temperature of 18·3° C: For other internal temperatures the degree-day totals must be adjusted to suit.

As far as internal gains are concerned, it is suggested that buildings fall into five classes in this respect:

High—Large areas of window: dense levels of occupancy: much heat producing equipment.

Moderate—Any one or two of the above characteristics.

Traditional—Normal window areas, level of occupation and equipment.

Low—Small windows: sparse occupancy: minimum heat producing equipment.

Dwellings.

The effects arising from heat gains, relative to design internal and external temperatures are brought together and represented by the factors listed in Table 15.5. Similarly, further sets of factors, as Tables 15.6 and 15.7, have been produced for use in evaluating the interaction of plant response rate, structural mass and hours of occupation. The influences of these aspects, as they bear upon plant operation, were discussed in Chapter 2 (page 42).

Application of the factors towards estimation of equivalent running hours at full load is extremely simple. The appropriate degree-day total is read from Table 15.1 and multiplied in turn by a value chosen from each of Tables 15.5 to 15.7 to represent the various aspects of the problem.

TABLE 15.5

FACTORS TAKING ACCOUNT OF BUILDING TYPE AND USE RELATIVE TO EXTERNAL
AND INTERNAL DESIGN TEMPERATURES

Building Characteristic	Outside design temp. −3° C			Outside design temp. −1° C		
	Inside design temp. °C			Inside design temp. °C		
	16	18	20	16	18	20
High	0·52	0·67	0·85	0·58	0·75	0·93
Moderate	0·63	0·79	0·99	0·71	0·87	1·07
Traditional	0·75	0·93	1·12	0·83	1·02	1·22
Low	0·87	1·07	1·26	0·97	1·19	1·38
Dwelling	0·47	0·62	0·78	0·52	0·68	0·86

TABLE 15.6

FACTORS TAKING ACCOUNT OF EFFECT OF MASS OF BUILDING STRUCTURE,
RESPONSE RATE OF PLANT AND METHOD OF OPERATION

Building Structure	Continuous Plant Operation (24 hours)		Intermittent Plant Operation†			
			Slow Response Plant		Quick Response Plant	
	7 day week	5 day* week	7 day week	5 day* week	7 day week	5 day* week
Heavy	1·0	0·85	0·95	0·81	0·85	0·72
Light	1·0	0·75	0·70	0·53	0·55	0·41

* With week-end shut down.　　　　† With night-time shut down.

TABLE 15.7

FACTORS TAKING ACCOUNT OF EFFECT OF BUILDING STRUCTURE AND PERIOD OF
OCCUPATION

Building Structure	Period of actual occupation			
	4 hours	8 hours	12 hours	16 hours
Heavy	0·96	1·0	1·03	1·05
Light	0·68	1·0	1·23	1·40

For example, take a traditional building in Wales having a heavy structure and a fast response heating system such as warm air units. The outside and inside design temperatures are −3° C and 20° C respectively. The building is to be occupied for 12 hours per day over a 5 day week and the plant will be operated intermittently for 40 weeks per annum.

Degree days (Table 15.1)   = 2048

Factor for building type,
use and temperatures
(Table 15.5)   = 1·12

Factor for intermittent
operation of a quick response
system in a heavy building
(Table 15.6)   = 0·72

Factor for 12 hours occupancy
in a heavy building (Table 15.7) = 1·03

Hence,

Equivalent running at full load = 2048 × 1·2 × 0·72 × 1·03
                                   = *1700 hours/annum*

This total would apply of course only if automatic firing were to be used since allowance for no-load use at night time is not made in any of the factors. Between 400 and 500 hours would have to be added to the calculated total if firing were to be semi-automatic.

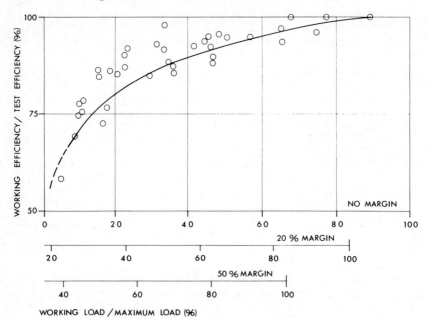

FIG. 15.3.—Part load efficiencies for gas and oil-fired boilers. Note the effect of boiler margin as shown by alternative base scales.

Comparison of Methods—For continuous or near continuous operation, either method will produce approximately the same estimate of annual hours. Where plant is to be operated intermittently, the factorial method will produce a range of results on average rather higher than those calculated from analysis. This situation is not unexpected since the factorial method makes allowance for aspects not otherwise taken into account.

It must of course be realised that such estimates cannot be expected to be accurate within better than plus or minus 20 per cent, in absolute terms, since so many imponderables exist. Their principal value is as a basis for comparisons in circumstances where assumptions are either constant or may be adjusted pro-rata with confidence.

Average Working Efficiency of Plant—This is a most controversial matter. Test figures can be produced which, while probably true for the conditions under which the test was made, are misleading for ordinary working—due to boiler heating surfaces not being always clean, excess air owing to lack of adjustment, periods of light load working, intermittent operation and so on. A recent analysis carried out by B.S.R.I.A., see Fig. 15.3, relates annual working efficiency to test efficiency and average figures derived therefrom are listed in Table 15.8 for oil and gas fired boilers. As may be seen, the seasonal average efficiency could be expected to be between 62 and 66 per cent for the normal run of such boilers.

TABLE 15.8

ANNUAL EFFICIENCIES OF OIL AND GAS FIRED BOILERS
RELATIVE TO TEST EFFICIENCY AND MARGIN PROVIDED

Seasonal temperature	Test efficiency 80%			Test efficiency 75%		
	Boiler margin provided over peak load					
	Nil	20%	50%	Nil	20%	50%
Above 9° C	62	55	53	59	52	49
4 to 9° C	73	71	66	68	67	62
Below 4° C	78	77	72	73	72	67
Average over season	69	66	63	65	62	59

In the case of solid fuel firing by semi-automatic means, the size of plant concerned will affect seasonal efficiency since it is to be expected that large boilers will have attributes, and be better attended and maintained leading to better performance, than small boilers. For sizes over, say, 750 kW an annual efficiency of 60 per cent could be expected and 5 per cent less for smaller units.

Electricity is provided to an off-peak hot water thermal storage system at 100 per cent efficiency. Heat losses from the plant are disproportionately large and a seasonal efficiency of 95 per cent is sometimes considered to be appropriate in consequence. A full calculation must however be made if these losses are to be quantified with accuracy.

Table 15.9 lists derived heat quantities provided by various fuels at the seasonal efficiencies here postulated.

TABLE 15.9

ENERGY AVAILABLE FOR USE FROM ALTERNATIVE SOURCES

Fuel	Heat content per unit of measurement	Derived heat sent out per unit of measurement	
		Large	Small
(a) Solid fuel, automatic feed -	30 GJ/tonne	18 GJ/tonne	16·5 GJ/tonne
(b) Gas boiler - - - -	105·5 MJ/therm	69·6 MJ/therm	65·4 MJ/therm
(c) Oil, Class D - - - -	38 MJ/litre	25·1 MJ/litre	23·6 MJ/litre
(d) Electric thermal storage - -	3·6 MJ/kWh	3·42 MJ/kWh	—

Annual Energy Consumption—Knowing the hours of use, the seasonal efficiency of the boiler plant and the peak energy demand of the heating system, calculation of the annual fuel consumption becomes a matter of arithmetic.

For example, consider an office block having a calculated heat load of 600 kW, after adjustment (see page 89). Equivalent full load running hours per annum are 1418 and 1065 for semi-automatic and automatic firing, respectively. Mains losses arise over the whole period of plant operation, including pre-heating hours, at the rate of 30 kW.

Hence, annual loads are:

Semi-automatic
$$0{\cdot}0036\,[(600\times1418)+(30\times1710)]=3247\ \mathrm{GJ}$$
Automatic
$$0{\cdot}0036\,[(600\times1065)+(30\times2010)]=2517\ \mathrm{GJ}$$

Thus, consumptions would be:

(a) Solid fuel
$$\frac{3247}{18}=180\ \mathrm{tonnes/annum}$$

(b) Natural Gas
$$\frac{2517\times1000}{69{\cdot}6}=36\ 160\ \mathrm{therms/annum}$$

(c) Oil Fuel (Class D)
$$\frac{2517\times1000}{25{\cdot}1}=100\ 300\ \mathrm{litres/annum}$$

(d) Electric thermal storage—hot water
$$\frac{3247\times1000}{3{\cdot}42}=950\ 000\ \mathrm{kWh}$$

Electrical Heating Systems—For the case of electrical off-peak or 'White Meter' systems where heat output is controlled and night time losses are not too great, annual energy consumptions may be expected to be not dissimilar to those of equivalent centrally supplied plant, with semi-automatic firing. 'Electricaire', 'Centralec' and storage fan heater systems fall into this category.

The Electricity Council, in their Electric Floor Warming Design Manual referred to previously in Chapter 11 (page 273), suggest that degree-days may be used as a basis for estimates of consumption as follows.

Annual energy (kWh) = Design heat loss × degree-days

Applying this approach to the previous example but adjusting the heat load figure back to that representing design heat loss, suggests that a floor-heated building would have an annual consumption of about 1 250 000 kWh.

As far as storage radiators are concerned, the heat output from these is controllable only by guesswork and no real systemised method can be proposed for estimation of annual energy consumption.

It is common practice for annual energy consumptions to be quoted by electrical supply authorities in terms of kWh/kW of calculated heat loss. Such figures have a statistical base from records of many installations but wide variations are known to occur. Table 15.10 lists such data and for comparison the consumptions for other fuels, as previously calculated here, transposed into similar units.

TABLE 15.10

COMPARATIVE ANNUAL FUEL CONSUMPTIONS FOR VARIOUS FUELS AND SYSTEMS, kWh/kW

Type of System	Annual Consumption kWh/kW	Electrical Supply (%)	
		On peak	Off peak
Electric off-peak systems*			
Storage radiators	2100	10	90
Floor warming	1800–2000	—	100
Fan storage heaters	1900	—	100
'Electricaire'	1800–2000	10	90
'Centralec'	1800–2200	40	60
Electric on-peak systems*			
Wall mounted panels	1600–2000	50	50
Ceiling heating	1600–2000	50	50
Conventional systems†			
Solid fuel	2300–2600	—	—
Natural gas	1600–1800	—	—
Fuel oil (Class D)	1600–1800	—	—
Thermal storage-water	1400–1700	—	100

* Data from Electricity Council publications.
† As calculated, with + margin of 15%.

M

K.H.A.C.

Fuel Consumption for Hot-Water Supply—The consumption of heat for a central system may best be assessed by estimating:

(a) heat losses from storage vessel(s) (Table 15.11), circulating piping towel airers, linen cupboard coils, drying coils, etc.,

(b) actual hot water drawn off per day by the occupants.

It is worthy of mention that components (a) and (b) are often equal. A knowledge of the type of building will decide whether (a) is continuous for 24 hours per day, as in a hospital, or for some lesser period such as 8 hours per day in a school. Similarly the days per annum will vary.

The heat consumption of (b) taken from cold at say 10° C to hot at say 60° C, will be derived from data of water consumptions of comparable buildings. Table 13.2 (page 302) provides some guidance in this respect.

TABLE 15.11

HEAT LOSS FROM STORAGE CYLINDERS.INSULATED WITH 75 mm OF GLASS FIBRE
TO AMBIENT TEMPERATURE OF 25° C

Capacity of cylinder (litre)	Heat loss from cylinder	
	Watts	MJ/annum
150	40	1300
250	55	1800
300	60	2000
450	80	2500
650	100	3300
1000	135	4300
1500	180	5600
2500	250	7600
3000	280	8800
4500	360	11300

The total of (a) and (b) will then form the basis of a sum in which calorific value of fuel and efficiency are taken into account as for heating.

Domestic Fuel Consumption—A useful summary of annual energy consumption for heating systems in dwellings can be derived from data included in the *Guide, Section B18*. This suggests that input energy may be related to the total floor area and the fuel type, as follows:

	MJ per m²
Solid fuel, semi-automatic	920
Oil fuel, automatic	830
Natural gas, automatic	790
Off-peak electricity	770
On-peak electricity	570

For a family of four, other authorities suggest that energy input for hot water supply would average about 15 GJ per annum.

RUNNING OF AUXILIARIES

A further item in running costs is the electric current consumed by motors driving blowers, and other auxiliaries, for automatically fired boilers. This applies to oil firing, automatic coal stokers and to gas boilers with forced air.

As a rule, for the lower range of boiler sizes up to say 1000 kW, the cost of the current is not a significant item and is ignored.

For an oil-fired boiler of 1000 kW burning oil requiring preheating,

the consumption is about 0.03 kg/s of oil;

the temperature rise (say 10° to 80° C) is 70° K;

the specific heat capacity of oil class E or F is approximately 2kJ/kg.

Thus, approximately:

the heat required $= 0.03 \times 70 \times 2$	$= 4.2$ kW
the current needed for the boiler	$= 2.0$ kW, approx.
the current for the induced draught fan (if any)	$= 3.0$ kW ,,
the current for the controls	$= 1.0$ kW
Total per hour	$= 10.2$ kW

If the running hours are 1500 per annum, the annual consumption of electricity for boiler auxiliaries could be expressed as 15 kWh per kW of boiler power. Such an approximation is invalid over a wide range of boiler size but useful as a quick guide.

A forced-air gas-fired boiler of the same size will involve only the burner power and the controls, i.e. about one third of the above.

Similarly, the cost of current for the circulating pumps is relatively of small magnitude and indeed most of the energy paid for comes out as heat somewhere in the system and so is not lost. A system of 1000 kW boiler load, as above, might involve a pumping power of 2 kW, the pump running continuously for the heating season (say 5000 hours). Annual electrical consumption might then be stated to be 10 kWh per kW of boiler power, with similar qualifications to those applied above.

The above examples may serve to illustrate a method of estimating these costs; but, for any particular case, the makers lists will provide actual power requirements from which consumptions may be calculated.

LABOUR

The cost and scarcity of labour for manual work being such as it is, all new heating plant is assumed to be fully automatic. Thus the old estimates based on hand-stoking, or even the filling of hoppers and the clearing of ash, are no longer relevant.

Below a certain size, the occasional attention required for oil-fired, gas-fired or electric systems is so small as to be no more than a very part-time occupation, and it cannot be given a value. What that size is it is difficult to say, as there comes a point, such as in a hospital or a factory, where boilermen must be employed anyway, even though their duties are more in the nature of shift engineers or technicians than involving actual labour. The cleaning of boilers, keeping the log, turning pumps off and on and minor maintenance jobs, such as oiling and greasing, are still required. This may be worked on a three-shift basis which with weekends and holidays involves four men with a possible fifth as stand-in.

Costs for this scale of staff are then bound up with the whole staff structure of the enterprise and become but a tiny fraction of the overall running expenses. In other words, the old methods of evaluating the labour required to run such and such a system as against another one are no longer the criteria that matters. Other broader aspects of management enter in which are beyond our present scope.

However, if costing must be applied to heating apparatus, an assessment can be made at current rates of remuneration for the kind of attention mentioned above.

MAINTENANCE

Average maintenance of small installations is often covered by a maintenance contract with the installer or fuel supplier, involving two or three visits a year and costing a nominal sum per visit.

In the medium range of installations it is now common practice to employ one of the firms specialising in the maintenance of heating plant— which covers the cleaning of boilers, adjustments of controls, lubrication, checking burner equipment and operation, the cleaning of calorifiers and fan convectors (if any) and general overhaul. Annual charges will vary according to the size of the installation. In the larger scale plants such as are dealt with by hospital authorities, estate companies and industrial concerns, maintenance staff is part of the organisation, probably operating on a planned maintenance basis.

Taken by and large, maintenance of heating plant may be expected to be covered by an allowance of the order of ten per cent of the fuel cost per annum.

Interest and Depreciation—In making comparisons between one system and another, it is often necessary to take depreciation into account. In certain plant, such as boilers, the combustion equipment may be given a life of fifteen to twenty years, rotating equipment such as pumps twenty to thirty years, and static equipment such as pipes and radiators fifty to sixty years. A sinking fund may be established over these periods to cover replacement.

Interest on capital expended on the basis of a diminishing principle should take into account builders' work in addition to engineering work, such as for a flue where an alternative system would require no flue. The same applies to space for plant and fuel storage. The rate of interest in the comparison will of course depend on the current market rate or Government loan sanction rates where these apply.

Insurance—Insurance and annual inspection of steam boilers and other pressure vessels is mandatory, but this does not apply to low-pressure hot-water apparatus. Nevertheless, in any sizeable system it is customary for the owner to insure as a matter of self-protection, and also to cover against fracture, burn-outs and accidents of all kinds.

DIRECT HEATING SYSTEMS

If these are of convective type, the same method of calculation of the fuel consumption may be followed as for central systems—inserting the appropriate efficiencies, running hours and weather factor, assuming thermostatic control. There will, of course, be no mains loss for inclusion.

If the system is to be of radiant type, a true comparison with a convection system must be based on equation of demand related to the maintenance of environmental temperature. This criterion, as explained in Chapter 2, takes the mean radiant temperature of the enclosure into account and is thus particularly suitable for such an exercise in comparison. An assessment of running cost can best be made by estimating the time when the heaters are likely to be on and multiplying by their rated consumption. There is often no true control of temperature with such systems, and data from comparable installations are probably the best guide.

CHAPTER 16

Ventilation

THE WORD *Ventilation* MEANS LITERALLY the causing of air movement or wind (L. *Ventus*), but has acquired the meaning of a system which gives a regulated supply of air to an enclosed space so as to make the conditions better for human habitation.

The purpose of ventilation is fundamentally to supply the untainted air necessary for human existence, because life depends on a constant supply of oxygen. The supply of this air involves the removal of a corresponding volume of expired or vitiated air: with the smells and noxious gases which are associated with concentrations of people.

The normal adult at rest inhales about 0.5 m³ of air per hour (0.14 litre/s). Of this about five per cent is absorbed as oxygen by the lungs. The exhaled breath gains from three to four per cent of carbon dioxide (CO_2), equal to about 0.02 m³ per hour (0.005 litre/s). Thus, if it is desired to keep the concentration of CO_2 down to say 10 parts per 10 000, with external air containing say 3.5 parts per 10 000, it will be necessary to supply

$$0.02 \times \frac{10\ 000}{10 - 3.5} = 30 \text{ m}^3 \text{ air per hour (8.3 litre/s) per occupant.}$$

In addition, the human body at normal temperatures and at rest gives out about 50 grams of water vapour per hour and 100 W (0.1 kW)* of sensible heat.

The 30 m³ per occupant, as above, will become warmed and humidified as follows:

$$\text{Temperature rise} = \frac{0.1 \times 3600}{30 \times 1.22\dagger} = 10 \text{ K assuming no heat loss from room.}$$

$$\text{Increase in humidity} = \frac{50}{30} = 1.7 \text{ grams moisture per m}^3.$$

It has long been accepted that the CO_2 content is not a satisfactory criterion of good ventilation. The CO_2 content may be as much as 50 parts per 10 000 and yet the air feel fresh and pleasant, or it may be under 10 and yet feel stuffy owing to other more important effects.

To state what is a satisfactory criterion is more difficult. Freshness of air in a room may be judged by the sense of smell, an analytical method

* This varies greatly with temperature. See Fig. 17.17.
† Volumetric specific heat capacity of air at 20° C. See Table 1.3.

more delicate than any laboratory one. Freshness or stagnation are in addition judged by the rate of loss of heat and moisture from the skin, on air-speed, and by dust content.

The interplay of air temperature, air movement, and radiation have been referred to under Equivalent Temperature (p. 37). Where heating to a room is from a radiant source, the air temperature requires to be lower and the air movement greater than in a room with the same equivalent temperature, but with air and surfaces all at a uniform temperature, for the same degree of comfort.

The inter-relation of all these factors, as depicted in Fig. 16.1, is very complicated, and nothing short of an exhaustive investigation of each one

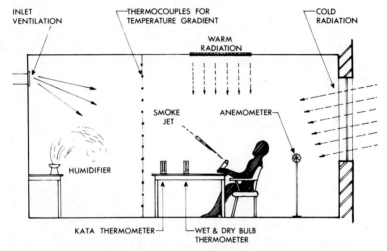

Fig. 16.1.—Means for determining a state of comfort.

will suffice to determine the precise conditions, and even with such knowledge it may be difficult to say whether a room is well ventilated or not. Yet, as is well known, a person (particularly of the female sex) entering the same room will be able to say at once, 'This room is stuffy' or 'This room is draughty'. After a time, due to acclimatization, the same room may appear to have become more tolerable.

The practice of ventilation is made somewhat more indefinite by the lack of an easy method of detecting slight air currents. A speed of between 0·2 and 0·5 m per second is generally regarded as the maximum which can be allowed without a feeling of draught but a lower level will be necessary if the air stream impinges on the back of the neck or if the temperature of the moving air stream is below that in the room.

Such relatively small velocities can be measured either by the Kata Thermometer or by the Hot Wire Anemometer, in which air flow is measured by means of the cooling effect on a heated wire. These do not, however, indicate direction, for which some visual detector is necessary, such as a cold smoke produced by the mixture of ammonia and hydrochloric acid fumes, or titanium tetrachloride. Smoke, however, tends to diffuse rapidly and is useful only as a local index.

Good ventilation cannot, therefore, be defined in simple terms, and reference has to be made to conditions which have been found in practice to give reasonably satisfactory results. These may be discussed under the following headings:

Volume of air necessary.

Distribution and air movement.

Temperature.

Humidity.

Purity.

VOLUME OF AIR NECESSARY

(a) **Temperature Rise Basis**—Where the occupancy is known the temperature rise can be estimated as with *Air-Conditioning* (see Chapter 17). If a rise of, say, 8 K is accepted as reasonable and 0·1 kW is assumed to be emitted per occupant, the volume of air required will be

$$\frac{0 \cdot 1 \times 3600}{8 \times 1 \cdot 22} = 36 \text{ m}^3 \text{ per hour (10 litre/s) per occupant.}$$

This assumes no building heat losses or gains other than from occupants.

It frequently happens that in crowded rooms much heat is also released from electric lamps, motors, cooking appliances, plant, and other apparatus. The heat from these can generally be calculated, and an increased air-change is needed to carry this extra heat away (see p. 40).

In such cases the air per person per hour may greatly exceed 36 m³ per hour (10 litre/s).

For summer a higher rate may be necessary, as there may be heat gains from the sun and a higher inlet air temperature; but it will be seen from a consideration of Chapter 17 that the removal of this heat gain without cooled air calls for very large ventilation volumes, which may be undesirable from other points of view and unnecessarily high for winter. In the case of a heavy structure, it can be shown that better day-time conditions may be obtained during the summer if a high air change rate is provided during the night when the available outside air is cooler and a reduced rate during the day when the building is occupied.

(b) **Legislation Basis**—For theatres and music halls, including all places licensed for music and dancing, boxing, etc., a minimum of 28 m³ of outdoor air per hour (7·8 litre/s) per occupant is required by most licensing authorities.

Factories where manual labour is employed for gain are governed by the current Factories Act, 1961. This does not, however, lay down

numerical standards. H.M. Factories Inspectors interpret these requirements according to each separate case. Six changes per hour may be considered an average for normal conditions, but there are many exceptions
such as where noxious vapours, steam, dust, heat and fumes occur. Orders
made under the Act lay down requirements for special industries. In the
case of textile factories special requirements cover the humidity permissible.
The Health and Safety at Work Act, 1974, confers wide powers with
respect to Building Regulations etc., including standards of mechanical
ventilation. The Act, however, includes no detail, this remaining to follow
by publication of Orders. Offices, Shops and Railway Premises are subject
to the Act of this name of 1963. Technical Data Note 19, published in 1970
by the Department of Employment, HM Factory Inspectorate (now
Health and Safety Executive, Factory Inspectorate), recommends that a
minimum of 4·7 litre/s of outside air be provided, per person, to meet the
requirements of section 7 of this Act.

The London Building (Constructional) By-laws 1972 contain a
requirement in Part XIV that where a habitable room cannot be provided
with natural ventilation and mechanical means are hence required,
outside air is to be supplied at a rate of not less than 22 m³ per hour
(6·1 litre/s) per person or 5 m³ per hour (1·39 litre/s) per square metre of
floor area, whichever is the greater.

(c) **Air-change Basis**—Where the occupancy is unknown or variable, an
arbitrary basis must be taken. Table 16.1 may be used as a guide.

TABLE 16.1

VENTILATION ON AIR-CHANGE BASIS

Type of Room	Air-changes per Hour
Offices above ground - - - - - -	2–6
Offices below ground - - - - - -	10–20
Factories, large open type - - - - -	1–4
Factories and workrooms closely occupied - ＼ -	6–8
Workshops with unhealthy fumes - - - -	20–30
Laundries, dye-houses, spinning mills - - -	10–20
Kitchens above ground - - - - -	20–40
Kitchens below ground - - - - -	40–60
Lavatories - - - - - - -	6–12
Boilerhouses and engine rooms - - - -	10–15
Foundries, with exhaust plant; rolling mills - -	8–10
Foundries, without separate exhaust plant - -	10–20
Laboratories - - - - - - -	10–12
Hospital operating rooms - - - - -	20
Hospital treatment rooms - - - - -	10
Restaurants - - - - - - -	10–15
Smoking rooms - - - - - - -	10–15
Stores, strong-rooms - - - - - -	1–2
Assembly halls - - - - - - -	3–6
School class rooms - - - - - -	3–4
Living rooms - - - - - - -	1–2
Sleeping rooms - - - - - - -	1
Entrance halls and corridors - - - -	3–4
Libraries - - - - - - - -	2–4

(*d*) **Effect of Cubic Content on Ventilation Rate**—The cubic content per occupant and length of time occupied obviously have an important effect on the rate of air supply necessary. Thus, in a large hall of 15 000 m³ capacity, seating 500 persons, each will have 30 m³ of air already stored in the building. If occupied for three hours and the desired rate of supply is say 30 m³ per hour per occupant, the actual outside air needed is only

$$500 \times 30 - \frac{15\ 000}{3} = 10\ 000 \text{ m}^3 \text{ per hour.}$$

$$= 20 \text{ m}^3 \text{ per hour per occupant instead of 30.}$$

(*e*) **Odours**—One of the essentials of good ventilation is the removal of odours arising from human occupation. The problem only becomes serious in crowded places. A supply of outside air at the rate of 17 m³ per hour (5 litre/s) per person is found to be the minimum to obviate trouble from this source. With workpeople in their dirty working-clothes, the figure may need to be 40 to 50 m³ per hour (10 to 15 litre/s).

Much recent research* has been directed to assessment of ventilation requirements arising from smoking in offices and work rooms. The results are not easy to quantify in simple terms since the statistical relationship between office size and smokers must be taken into account. In a large open office, it may be assumed that only half the occupants will smoke whereas for a small private office it is probable that the ratio of smokers to abstainers will be much higher. Over twice the outside air necessary to disperse body odours is required to dilute cigarette smoke.

(*f*) **Condensation**—In certain special circumstances the need to combat condensation can be the criterion for determination of the ventilation rate. One such instance is that of the swimming-pool hall where, of the requirements to supply air to occupants, to dissipate the characteristic 'chemical smell' and to combat condensation on glazing, the last is by far the most important factor. It has been shown† that, with double glazing, the outside air supply rate should be not less than 0·02 m³/s (20 litre/s) per m² of wetted surface. In this context 'wetted surface' is defined as 1·2 times the water surface of the pool or pools in order to allow for splash-over etc. onto the pool surrounds.

In housing, it has been shown that a mean ventilation rate of one air change per hour is required to prevent condensation on single glazed windows in a uniformly heated dwelling. In domestic kitchens, the ventilation requirement to avoid condensation is about 100 litre/s for electric cooking and half as much again for gas cooking. For un-heated bedrooms, an air change rate of between 3 and 4 per hour is required in the coldest

---

* Brundrett, G. W., *Ventilation requirements in rooms occupied by smokers: a review*. Electricity Council Research, 1975.

† Doe, L. N., Gura, J. H. and Martin, P. L., *Building Services for Swimming Pools*, JIHVE. 1967.35.261.

weather for this purpose* but it is questionable whether such an air flow would be tolerable.

DISTRIBUTION AND AIR MOVEMENT

The admission of outside air to a room should be such that:

(a) It is evenly diffused over the whole area at the breathing level.
(b) It should not strike directly on the occupants.
(c) It should give a feeling of air movement and prevent stagnant pockets.

Various methods of approaching this end are discussed in Chapter 18.

The extraction of vitiated air from the room has little directional effect apart from a general upward, downward or crosswise motion.

Distribution often determines the air volume to be circulated. Though a small quantity of outside air may suffice in the case of a large room sparsely populated, it may be impossible to diffuse this evenly throughout and avoid stagnation. The volume must then be increased, either all as outside air, or by re-circulating room air mixed with the small quantity of outside air.

Similarly, in a small crowded room, distribution difficulties may prevent admission of the necessary air-change rate without draught, and the volume then has to be reduced with unavoidable rise of temperature unless full air-conditioning with cooling is employed.

Distribution limits, consistent with the maintenance of comfort, appear to be about 3 air-changes per hour minimum, and 20 air-changes per hour maximum where mechanical means are employed. With natural inlet, lower rates than 3 will, of course, be possible and often sufficient, but distribution can then be only a matter of chance.

TEMPERATURE

The temperature of the air admitted must be not too much below that of the room, or the air will fall to the floor without proper mixing and cause cold draughts. The air must not on the other hand be too warm, or it will rise to the ceiling without adequate distribution.

In a 'straight' ventilation system, which is here under discussion, it is assumed that the warming of the building is accomplished by a separate radiator or other direct heating system to supply the heat necessary to balance the heat losses in winter, say to 20° C. Under these conditions the air supply will preferably be about 3 K lower than the desired room temperature and, when occupied, the leaving air will be of room temperature.

Under summer conditions without provision for air-cooling, no control can be exercised over the inlet temperature except by the small reduction possible by the process of evaporation, referred to later.

* Loudon, A. G. and Hendry, I. W. L., *Ventilation and Condensation Control*. RIBA. IHVE. IOB. Conference, 1972.

HUMIDITY

The ventilation air admitted from outside will have the same moisture content as outside. In cold weather this may result, as has been shown earlier, in an unduly· low relative humidity after heating. This may be adjusted, where mechanical inlet is provided, by humidification by means of water sprays or other device for adding water vapour, and the amount of moisture added may be controlled to give the desirable relative humidity of about 40 to 70 per cent at room temperature.

In normal mild weather the outside humidity in Great Britain will generally be satisfactory without alteration. The human body is not critical of humidity variations at normal temperatures.

There is, further, a reservoir effect caused by the hygroscopic contents of buildings, such as fabric, wood, paper, etc., which steadies up humidity changes, tending to equalize them over long periods.

In summer, when the outside humidity is high, there can be no control of the inside humidity without full air-conditioning which includes means for de-humidification. An air washer or other humidifying device, unless supplied with chilled water, can only *add* moisture, but due to evaporation may cool the air to a temperature approaching the wet-bulb.

Thus, with outside air at, say, $24°$ C dry bulb and $18°$ C wet bulb (relative humidity 55 per cent), a single bank air washer may be able to reduce the temperature to $19°$ C, at the same time increasing the relative humidity to 90 per cent. This final condition may be more oppressive than the initial, and it is indeed often found that in hot weather, air washers, where provided, are shut off for this reason. Reference to Fig. 2.8 will show that the above initial condition is just inside the summer comfort zone, but the latter condition is well outside it.

AIR PURITY

Most buildings requiring special ventilation are in the centre of towns where atmospheric pollution is highest. Such pollution is produced partly by the burning of coal and other fuels, and consists of soot, tar, ash and sulphur dioxide and partly by road traffic causing dust and fumes.

When air is introduced into a building for ventilation a proportion of solids will be brought in also. If the inlet is natural, nothing satisfactory can be done to prevent this, but where mechanical inlet is provided, filters and washers may be employed to remove the greater part. The various devices available are discussed in Chapter 18.

Vitiated air may be re-vitalized by removal of deleterious products in exhaled breaths by using activated carbon filters, the same also serving to reduce the content of sulphur dioxide and other noxious gases. Such filters are costly and in the past have been used only in special cases: in the interests of energy conservation there are some circumstances where the use of such equipment should be examined in more detail.

METHODS OF VENTILATION

Ventilation may be:

Natural, in which air movement is induced by effect of temperature difference or wind.

Mechanical, in which air movement is caused by power-driven fan or fans.

Inlet and extract have to be considered separately.

There are thus four possibilities:

	Inlet	Extract
(1)	Natural	Natural
(2)	Natural	Mechanical
(3)	Mechanical	Natural
(4)	Mechanical	Mechanical.

(1) **Natural Inlet and Extract**—This applies to most rooms and buildings with low occupancy, such as living rooms, bedrooms, small offices, schools, hospitals, shops, etc., and even then is dependent on clean outside air and conditions which permit open windows.

If an open fireplace or gas fire with flue is provided, the flue serves the double purpose of carrying away the products of combustion and of exhausting air from the room (Fig. 16.2). This air is replaced by cold air

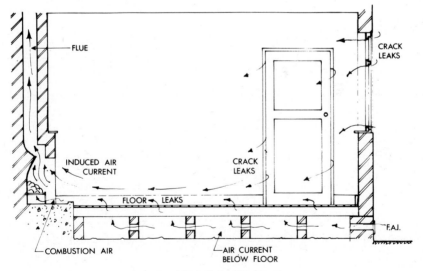

Fig. 16.2.—Ventilation by open fire.

drawn through cracks around doors and windows, between floor-boards, etc., thus causing ventilation of the room, though often at an unnecessarily high rate causing draughts. A restriction of flue throat is advised in order to reduce this loss.

Various methods of achieving natural ventilation without relying on opening of windows have been devised, one of which is shown in Fig. 16.3.

For larger rooms, assembly halls, factories, etc., roof ventilators combined with fresh-air inlets behind radiators (Fig. 16.4) provide a cheap solution, but the operation is spasmodic and unreliable, depending partly on temperature difference and partly on wind. Thus, in hot still weather

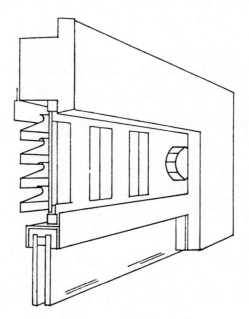

Fig. 16.3.—Natural window ventilator.

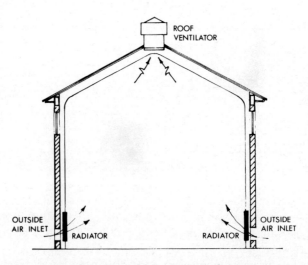

Fig. 16.4.—Natural inlet and natural extract.

when ventilation is perhaps needed most, no appreciable air flow occurs. In freezing weather with high winds the fresh-air inlets cause draughts, and are frequently found permanently closed.

Consider the warm air inside a flue with cold air outside as forming two legs of a U-tube.

Outside temperature, kelvins T_0
Inside temperature, kelvins T_1

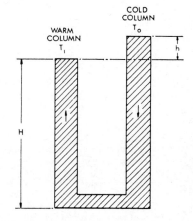

The density of air at the two temperatures is proportional to the absolute temperatures.

If the height of column T_1 is H, then the cold column, being denser, will in terms of the density of T_1 be equivalent to H plus some increased height h such that

$$h = H \cdot \frac{T_1 - T_0}{T_1}.$$

This difference in the height of the two columns causing flow may be substituted in the usual formula

$$P = \tfrac{1}{2}\rho v^2$$

where P is a pressure difference in Pa$(=N/m^2)$ and ρ is the specific mass in kg/m^3. By transposition,

$$\frac{P}{\rho} = \frac{N}{m^2} \times \frac{m^3}{kg} = \frac{kg\,m}{m^2\,s^2} \times \frac{m^3}{kg} = \frac{m^2}{s^2}$$

and thus

$$h = \frac{v^2}{2g}$$

where g is acceleration due to gravity ($9\cdot81\,m/s^2$), v is in m/s and h is in metres.

Then
$$v = \sqrt{2 \cdot g \cdot H \cdot \frac{T_1 - T_0}{T_0}}.$$

This applies to any flue in which it is assumed that atmospheric pressure is exerted at the top. In the case of a building with inlet at floor and outlet at ceiling, it is generally considered that a neutral zone at atmospheric pressure exists about halfway between floor and ceiling; the lower half is at an increasingly negative pressure from the neutral zone downwards, thus causing inward flow, and the upper half at an increasingly positive pressure from the neutral zone upwards to the ceiling, causing outward flow.

The top of the imaginary U-tube must therefore in this case be at the zone of atmospheric pressure, which is at a level of $\dfrac{H}{2}$.

Then
$$v = \sqrt{2g \cdot \frac{H}{2} \cdot \frac{T_1 - T_0}{T_0}}$$
$$= \sqrt{g \cdot H \cdot \frac{T_1 - T_0}{T_0}} \, .$$

Table 16.2, calculated for an outside temperature of 5° C., gives the theoretical velocity in metres per second for various heights between inlet and outlet and inside temperatures.

TABLE 16.2

Indoor Temp. °C	Theoretical velocity in m/s for following heights between inlet and outlet								
	2 m	4 m	6 m	8 m	10 m	15 m	20 m	25 m	30 m
10·0	0·60	0·84	1·03	1·19	1·33	1·63	1·88	2·10	2·30
12·5	0·73	1·03	1·26	1·46	1·63	1·99	2·30	2·57	2·82
15·0	0·84	1·19	1·46	1·68	1·88	2·30	2·66	2·97	3·26
17·5	0·94	1·33	1·63	1·88	2·10	2·57	2·97	3·32	3·63
20·0	1·03	1·46	1·78	2·06	2·30	2·82	3·26	3·63	3·99
22·5	1·11	1·57	1·93	2·22	2·49	3·04	3·52	3·92	4·30
25·0	1·18	1·68	2·06	2·38	2·66	3·26	3·76	4·21	4·60
27·5	1·26	1·78	2·18	2·52	2·82	3·45	3·99	4·46	4·88
30·0	1·32	1·88	2·30	2·66	2·97	3·64	4·20	4·70	5·15

The theoretical velocity is not obtained in practice, due to resistance to air flow. The practical flow is commonly taken at $\frac{1}{2}$ to $\frac{2}{3}$ of the theoretical. The effect of wind is to increase the velocity, but this depends on the type of roof outlet. Most makers give factors of performance for the various forms of roof-ventilating appliances.

Natural Ventilation (Industrial)—There are a great many heavy industries where natural ventilation can be supplied successfully to single-storey shed-type constructions. These industries are those in which there is a considerable release of heat or steam, such as foundries, rolling mills, welding shops, plastics processing shops and dye-houses. The quantity of heat to be removed in such cases can usually be derived from the known input of energy per unit of time in the form of gas, electricity, oil, coal or steam as well as that from the sun. A temperature rise of the order of 2 K per metre of height of building may be assumed (taking as a maximum 25 K rise).

On this basis, the air quantity required to be moved may be assessed, which, divided by velocity, will give the area of opening necessary. The velocity in this case will be the practical velocity—i.e. the theoretical velocity from Table 16.2 reduced by the factor of performance of the ventilating unit. It will be noted that this will hold good for one outside temperature only (namely 5° C) and that for other outside temperatures the velocity should be recalculated. It will be obvious that the higher the outside temperature the lower the velocity obtainable by thermal motive force, unless the inside temperature is allowed to rise in proportion.

One form of roof ventilator for industrial use is the well-known Robertson's, as shown in Fig. 16.5.

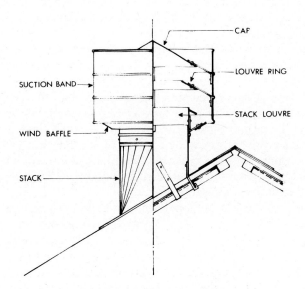

FIG. 16.5.—Robertson's ventilator.

A more versatile form is the Colt M.F. (multi-function) shown in Fig. 16.6. This is rectangular, giving an opening from about 350 to 600 mm in width and from about 1250 to 2200 mm in length. The side dampers may be moveable so that they may be closed when extraction is not required. The weather cap also may be openable, giving a completely unobstructed outlet as shown in the right-hand diagram.

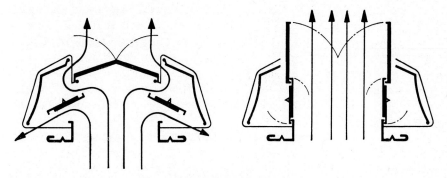

FIG. 16.6.—Colt type MF ventilator.

Another form of Colt roof ventilator consists of a series of openable louvres (as shown in Plate XXV, facing p. 247) giving a very large area of aperture and a view of the sky which may have a psychological benefit. Both this and the M.F. type, when open, may of course admit rain or snow but, where there is a considerable up-draught of hot air, such may not reach the shop floor or it may be evaporated before it does so. Controls are available, however, to close the louvres almost completely, acting through a rain-sensing device.

The layout of roof vents is largely a matter of common sense—for instance, where a high concentration of heat or fumes is likely to occur, the vents should be closely grouped over this area.

Fire Venting—A development of free-opening roof vents of the types just described is in their use as fire vents. These may be made to open automatically in the event of fire, so allowing smoke to be cleared rapidly and thus preventing it from spreading to other areas. This is a valuable aid to fire-fighting where smoke is generally the principal hazard. The principles underlying this application are covered in research papers prepared by the Joint Fire Research Organisation.*

FIG. 16.7.—Inlet louvre panel.

Air Inlets—Air inlets behind radiators have been referred to earlier (page 356). The free area of inlet gratings should be the same as the extract, or based on a velocity of 1·5 m/s. Convectors may similarly be fitted with fresh-air inlets, but the heater elements are liable to become quickly clogged with airborne dirt. Fan convectors are better for the purpose as they may be fitted with air filters.

Natural air inlets on an industrial scale often present a serious problem. In heavy industries, particularly where there is much radiant heat, wall openings at low level fitted with louvres may be allowable. These may be controllable and will match in size the area of roof outlet. One form of panel louvre is shown in Fig. 16.7.

Where some means of warming the incoming air is required to prevent cold air impinging on the working population, mechanical inlet becomes necessary to overcome the resistance of heating elements and is discussed later (page 367).

(2) **Natural Inlet and Mechanical Extract**—A mechanical-extract

---

* *Investigations into the Flow of Hot Gases in Roof Venting* (Fire Research Technical Paper No. 7); also *Design of Roof Venting Systems* (Fire Research Technical Paper No. 10). Both are published by H.M.S.O.

system will function irrespective of wind and temperature differences, and is positive in action. Owing to the negative pressure set up in the space there is a tendency to inward leakage rather than outward, thus preventing escape of steam and smells, etc., to other parts of the building, and it should be particularly suitable for laundries, kitchens, lavatories, laboratories, and rooms where fumes or noxious vapours are given off from process work and where natural extract would be too unreliable.

However, the difficulty arises, as in the previous section, in achieving a satisfactory means of admitting air to balance the extract and warming it in winter. Fresh-air inlets behind natural convectors (Fig. 16.8) have only a

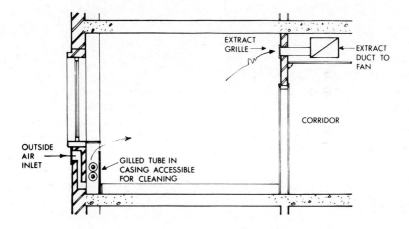

FIG. 16.8.—Natural inlet and mechanical extract.

limited application. On a small-scale building, leakage may suffice. Sometimes replacement air may be drawn from another part of the building, as in a kitchen where air may come from the restaurant or the canteen. In general, however, mechanical extract goes with mechanical inlet. Nevertheless, in this section we will deal with the means for mechanical extraction alone.

A typical industrial application is as in Fig. 16.9 where a series of

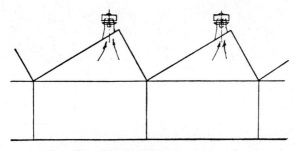

FIG. 16.9.—Mechanical extract to factory.

extractor fans are mounted on the roof. In Fig. 16.10(a) the fan is centri-fugal, and in Fig. 16.10(b) the fan is of the propellor type.

The air discharged from such units tends to hug the roof; so, where fumes or other pollutants are carried in the air, it is better to use a vertical-

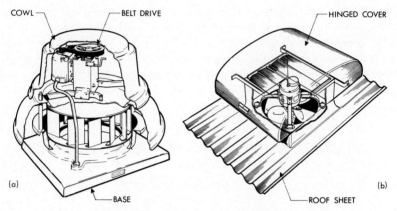

FIG. 16.10.—Roof extract units (a) centrifugal or (b) propellor fan.

discharge unit, as in Fig. 16.11. This unit contains vanes to exclude rain when not running but, when in operation, these open and the velocity is such that rain cannot enter.

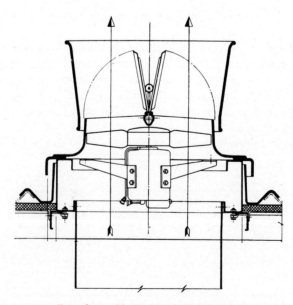

FIG. 16.11.—Vertical jet discharge unit.

For use in a vertical or horizontal duct, the arrangement may also be as in Fig. 16.11.

Where exhausting through a wall, the fan would be fixed as in Fig. 16.12.

The effect of wind blowing in opposition to a fan discharging through a wall opening is to hold up the delivery of air, or to make it hunt, producing a surging gusty action, often with some noise, and with a considerable reduction in volume exhausted. If it is possible to turn the discharge vertically upwards this is preferable, as the wind then aids the extraction. If this is not possible a baffle placed in front of the fan discharge will improve the operation.

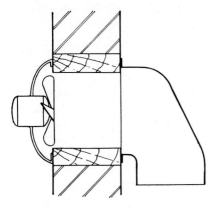

FIG. 16.12.—Wall extract fan.

To prevent the escape of warm air when the fan is shut off, self-closing dampers may be fitted (Fig. 16.13).

For small rooms one of the moulded plastic type window or wall fans shown in Fig. 16.14 will frequently give a useful solution to a ventilating problem. These are also made to provide inlet as well as extract or to be reversible.

In the case of a larger hall, or where a number of rooms has to be dealt with, an extract duct system

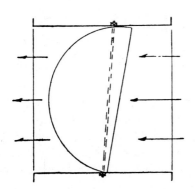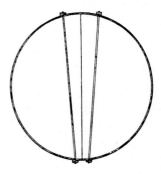

FIG. 16.13.—Butterfly self-closing fan dampers.

becomes necessary (Fig. 16.15). The fan must then be capable of overcoming the resistance of the ducts, and the ordinary propellor fan then tends to become noisy. Use may then be made of a cased fan (p. 442) or of an 'axial flow' fan (p. 443).

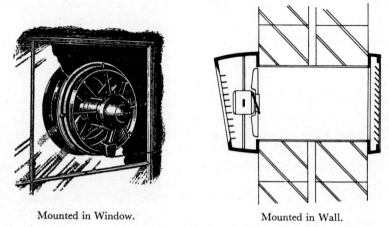

Mounted in Window.   Mounted in Wall.

FIG. 16.14.—Domestic extract fans.

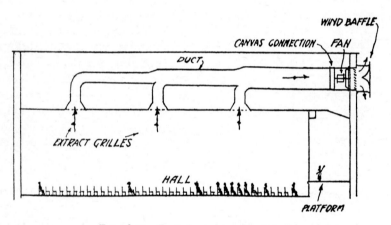

FIG. 16.15.—Ducted extract system to hall.

Sizes of fans can only be selected from makers' data. It is not possible to give a summary here as the range for each size and type is so wide.

The arrangement of ducting may be as simple or as extensive as is necessary to meet the case. A typical arrangement for a kitchen is shown in Fig. 16.16. Here it will be noted that hoods are provided over the main fume-producing items of equipment.

Internal Lavatories—A review of previous practice and existing Local Authorities' requirements carried out by the BRE has revealed many inconsistencies. As a result a recommended minimum basis and a method of ensuring absence of interference of one floor with another has been evolved.*

* Wise, A. F. E. and Curtis, M., *Ventilation of internal bathrooms and water closets in multi-storey flats.* JIHVE. 1964. 32. 180

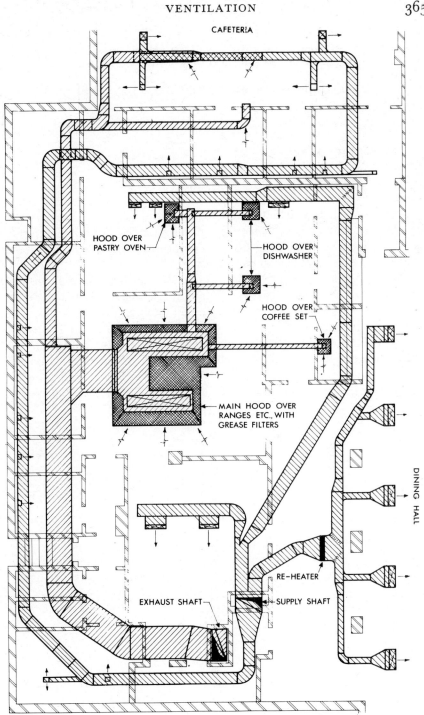

CAFETERIA

HOOD OVER
PASTRY OVEN

HOOD OVER
DISHWASHER

HOOD OVER
COFFEE SET

MAIN HOOD OVER
RANGES ETC., WITH
GREASE FILTERS

DINING HALL

RE-HEATER

EXHAUST SHAFT

SUPPLY SHAFT

FIG. 16.16.—Typical kitchen ventilation system.

This enquiry and research was primarily directed to blocks of flats, but the principles apply generally.

Briefly the recommendations may be summarised as follows:

Per W.C. compartment for housing a minimum of 20 m³/hour (5·5 litre/s)
Per Bathroom without W.C.   ,,   ,,   ,, 20 m³/hour (5·5 litre/s)
Per Bathroom with W.C.   ,,   ,,   ,, 40 m³/hour (11 litre/s)

In order to overcome stack effect, and other chance effects such as doors open on one floor and not on others, a design of duct system as in Fig. 16.17 is advised. It will be noted that there is to each floor a leg with sizeable resistance, this tending to provide inherent regulation. A simplified system, using standard PVC circular drain piping for air ducts has been developed by BRE* but, whilst this promises much advantage in terms of basic capital cost, the complications imposed by the necessity for the inclusion of fire dampers can be expensive.

Local Authority regulations require that exhaust fans must be in duplicate with arrangements for automatic change-over in the event of fan failure. This involves an arrangement such as shown in Fig. 16.18, using a purpose made twin unit, or two conventional fans fitted in parallel.

Provision must be made for inlet air by suitably placed fresh-air inlets. In a block of flats, for instance, where lavatories are vertically above one another, these inlets may connect to a rising shaft communicating with outside air at the bottom, where a heater may be placed. The shaft may

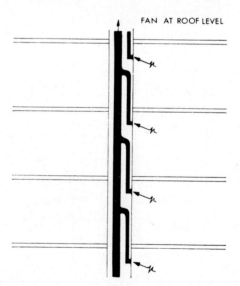

FAN AT ROOF LEVEL

Fig. 16.17.—Extract duct to multi-storey flats.

* Wise, A.F.E. and Curtis, M., *Development of ventilation systems for multi-storey housing.* JIHVE. 1968. 36. 35

form the duct for the various pipes serving the lavatory fittings. Alternatively air may be drawn from a hallway or corridor, subject to approval under local by-laws. In office blocks and other large buildings an extract rate of 10–12 changes per hour is desirable. Some authorities require a separate air supply to the lobby.

(3) **Mechanical Inlet and Natural Extract**—This method is suitable for certain types of factory, offices, boilerhouses, etc.

In the case of a factory the inlet may take the form of a series of fresh

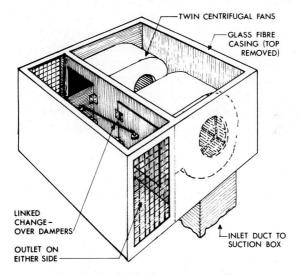

FIG. 16.18.—Duplex lavatory extract fan.

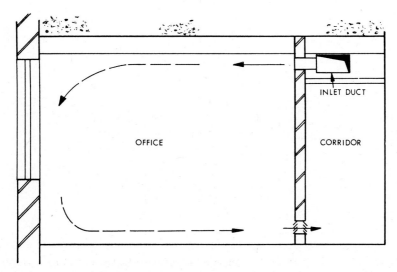

FIG. 16.19.—Mechanical inlet and natural extract.

air unit heaters with extract by natural roof ventilators, as in Fig. 10.12 (page 219).

In the case of offices and similar rooms a ducted system for the inlet may be used, delivering the air into the room at high level, and with louvred openings at low level, allowing the extract air to pass out into the corridor (Fig. 16.19). Provision must be made for the corridor to connect to a space of free escape, such as a staircase connecting to the street, or to a natural vent shaft open at the top, subject to the requirements of the Fire Prevention Officer. The Code of Practice, *Means of Escape in case of Fire*, published in 1974 by the Greater London Council requires that systems of mechanical ventilation should be so designed that the 'normal movement of air is directed away from routes of escape'. This means, effectively, that the type of system illustrated in Fig. 16.19 is no longer favoured.

Single offices, canteens and the like, may be served by a cabinet type fan convector, one type of which is shown in Fig. 16.20. This may have provision for filtering, as well as heating, and may be set to introduce fresh air or recirculate. If provided with cooling coils it becomes a unit air-conditioner, referred to on page 382.

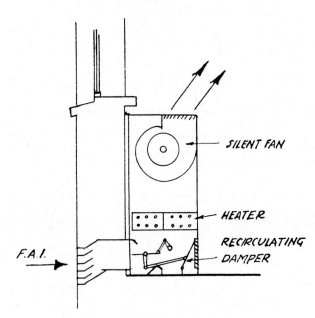

FIG. 16.20.—Cabinet type unit ventilator.

Ventilation of Boilerhouses—Boilerhouses are a special case owing to the air consumed in combustion. In any boilerhouse provision must be made for this air to enter. The volume may be estimated in the manner given in

Chapter 4 (page 78). Where a boilerhouse is above ground this air often provides adequate ventilation without augmentation. In the case of a confined basement boilerhouse, additional air supply may be necessary to keep the temperature down to a reasonable limit. The amount of heat liberated and volume of air necessary can be calculated.

It is then preferable to arrange for this air to be provided by mechanical inlet rather than by mechanical extract, since the latter may tend to create a negative pressure counteracting the effect of the flue. With mechanical inlet, provision must be made for the surplus extract air to escape by natural means such as by a shaft adjacent to the chimney.

It should be noted that openings provided for air supply to a boilerhouse act also as a route for the escape of noise. The provision of acoustic louvres to such openings or other means to avoid what can be a major nuisance should be considered.

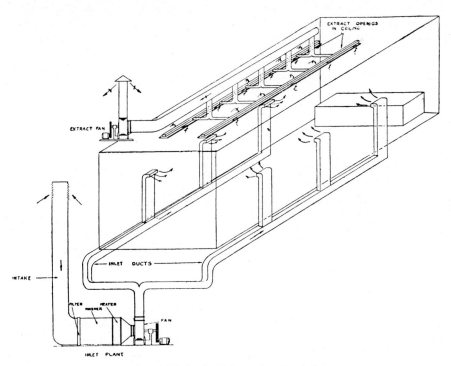

FIG. 16.21.—Mechanical inlet and mechanical extract.

(4) **Mechanical Inlet and Extract**—This method of ventilation is capable of the widest application, as distribution and pressure are both under control, as is also temperature. Air filters may be included for the cleaning of the air. It is particularly applicable to theatres, cinemas, restaurants, dance halls, banqueting halls, smoking rooms, libraries, offices, canteens, kitchens, etc. Fig. 16.21 shows a diagrammatic arrangement.

In all these cases both inlet and extract would most probably require to be served from fans at a distance by means of ducts.

In normal living rooms, offices, etc., where no fumes or smells are generated, the extract should be less than the inlet, so that air movement is outward not inward; it is then usually - -	75 to 90 per cent of inlet.
Where fumes arise, and a negative pressure in the room is necessary, as in laboratories, kitchens, shops, etc., the extract should exceed the inlet, usually taken at about - - - - -	110 to 120 per cent of inlet.

There are a great number of combinations of inlets and extracts with different kinds of fans, with or without ducts, etc., to suit various purposes, but enough has no doubt been said to indicate the general principles and possibilities.

From what has been said earlier it will be apparent that, where mechanical extract is employed for industrial buildings, the problem of introducing replacement air can as a rule only be met by mechanical inlet. All too often the fact is ignored that if air is removed it must be replaced, and this involves warming in winter. If no provision for inlet is made in the design, all sorts of subterfuges have to be made later, often with highly unsatisfactory results. One solution to the problem is to provide fresh-air unit heaters, as already mentioned. These may serve also as the means of warming the building, being changed over to recirculation for warming-up.

Where local relief on the working zone necessitates inlet air being directed to specific areas, ducted units may be employed, as shown in Fig. 10.12 (page 219).

Another method of warming inlet ventilation air is that of the direct oil-fired or gas-fired heater, one form of which is shown in Fig. 3.1 (page 50). Air distribution from large output units of this kind relies on a high-velocity discharge at high level and return at low level. If the unit is handling 100 per cent fresh air, there is no low-level return and temperature gradients will be high. Thus a substantial proportion of return air—say 50 per cent—is best allowed for, fresh air making up the other 50 per cent. The pattern of air flow will be similar to that in Fig. 2.7 (page 35).

Replacement air for certain heavy industries, such as foundries where contamination already exists, may be warmed by gas burners in the air stream. The volumes handled and the dilution are such that no hazard to health arises. This is a cheap method of dealing with vast air quantities as shown in Fig. 6.15 (p. 130).

DESIGN OF INLET SYSTEM

The design of the inlet ventilation system with ducts is dealt with in Chapter 18. Fans, heaters, filters, air-washers, controls, ducts, may all be similar to those described in the subsequent chapters.

The method of distributing the air will follow one of the systems mentioned in Chapter 18. With straight ventilation the upward system is more general than the downward.

Heat for Air—When inlet air is introduced at about room temperature, the heating system is relieved of the duty of warming infiltration air because leakage tends to be outward. Radiators, heating panels, etc., should therefore be sized for the fabric losses only. This leads to obvious economy and avoids the risk of over-heating.

In any system where air is the sole heating medium, there are two quantities of heat requirement involved: first, the heat necessary to raise the fresh-air supply from say $-1°$ C to a room temperature of say $20°$ C, and second, that necessary to off-set building heat losses through the structure. The mass flow of air will usually be determined by the ventilation requirement and thus the air-outlet temperature from the heater will need to be considerably higher than that required in the room. For instance, consider a space requiring a fresh-air supply of 1800 m³ per hour (500 litre/s) which has a building-fabric heat loss of 15 kW to maintain an internal temperature of 20° C when outside conditions are at $-1°$ C. The air supplied will require to be

$$\frac{15 \times 3600}{1800 \times 1 \cdot 22} = 24 \cdot 6 \text{ K above room temperature.}$$

That is to say the duty of the heater is to raise the air temperature

$$(21 + 24 \cdot 6) = 45 \cdot 6 \text{ K.}$$

Should the requirement lead to an unduly high leaving temperature from the heater, the addition of some recirculating units is called for. Thus, if unit heaters are being used, some would introduce fresh air and some would not: discharge temperatures not exceeding 50° C to 55° C should be aimed at.

PLENUM SYSTEM

Reference to this system is retained here more for historical interest than as indicating current practice. It was primarily a system of heating and only secondarily one of ventilation. Its application was mainly industrial but it lacked the flexibility generally demanded to-day. It was a fixed system which once installed could only be altered with difficulty. Furthermore, it was more costly and space-consuming than unit heaters and the like which have largely replaced it.

The system draws air from outside and, after heating, distributes it through a series of ducts to the space to be heated, as in Fig. 16.22. A slight

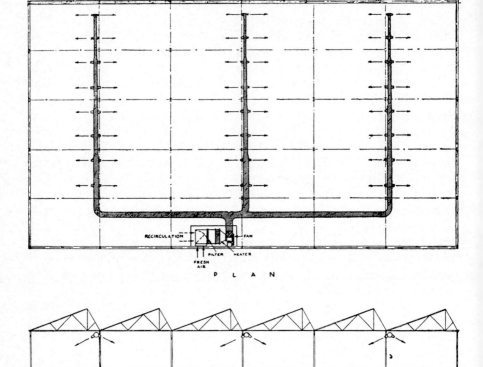

FIG. 16.22.—Typical plenum ventilation system.

pressure or plenum is supposed to be created within the building so that leakages should be outward rather than inward. The plant comprises an intake chamber with change-over damper—fresh to recirculating—an air filter, a heater battery (fed by steam or hot water for raising the air temperature to 60° C or so) and a fan and motor. The ducts in the Figure are shown overhead, but in some cases underground ducts have been used. Main trunks require insulating.

It will be clear that, when using fresh air, the heat put into the air to raise it to room temperature is lost; only the additional heat above this to meet fabric losses is useful. On the other hand, if recirculating, the supposed advantage of plenum no longer exists. The system was in fact susceptible to the vagaries of wind, making for patchy heating—under-heating on the windward side and overheating on the leeward side.

As mentioned, for a variety of reasons the plenum system has fallen out of use and it is not proposed to discuss it further.

PUNKAH AND DESK TYPE FANS

Fans of this type merely stir round the air of the room, and in so doing cause a rapid movement over the skin with resultant evaporation and cooling, which accounted for their widespread use in the tropics before the days of air-conditioning (Fig. 16.23). They should not be considered as providing ventilation but, if a ventilating system lacks cooling, they can give some relief.

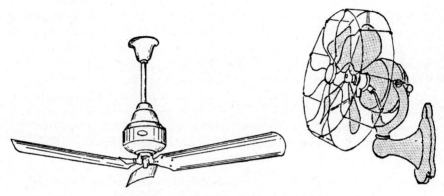

FIG. 16.23.—Punkah fan—Wall mounted fan.

CHAPTER 17

Air-Conditioning

THE SCIENCE OF AIR-CONDITIONING may be defined as that of providing and maintaining a desirable internal atmospheric environment irrespective of external conditions. As a rule 'ventilation' involves the delivery of air which may be warmed, while 'air-conditioning' involves delivery of air which can be warmed or cooled and have its humidity raised or lowered.

The desired atmospheric condition usually involves a temperature of 18° to 22° C in winter and 21° to 24° in summer; a relative humidity of about 40 per cent to 60 per cent; and a high degree of air purity. This requires different treatments according to climate, latitude, and season, but in temperate zones such as England it involves:

In Winter—A supply of air which has been cleaned and warmed. As the warming lowers the relative humidity, some form of humidifying plant, such as a spray or a steam injector, with preheater and main heater whereby the humidity is under control, is generally necessary.

In Summer—A supply of air which has been cleaned and cooled. As the cooling increases the relative humidity, some form of dehumidifying plant may be an essential. This dehumidifying is generally accomplished by exposing the air to cold surfaces or cold spray, whereby the excess moisture is condensed and the air is left saturated at a lower temperature. The temperature of the air has then to be increased, to give a more agreeable relative humidity, which can be done by warming or by mixing with air which has not been cooled.

Dehumidifying can also be brought about by passing the air over certain substances which absorb moisture. Thus, in laboratories, a vessel is kept dry by keeping a bowl of strong sulphuric acid in it or a dish of calcium chloride, both of which have a strong affinity for moisture. Silica-gel, a form of silica in a fine state of division exposing a great absorbing surface, is used also for drying air on this principle, but this process is complicated by the need for re-generation of the medium by heat and subsequent cooling, and is not generally used in comfort air-conditioning applications.

The application of air-conditioning may be considered necessary to meet a variety of circumstances:

1. Where crowds of people congregate such as in restaurants, cinemas, theatres and the like.

2. Where work has to be carried on in a confined space, the task being

of a high precision and intensive character, such as in operating theatres, instrument assembly shops and the like.

3. Where the exclusion of air-borne dust is essential.

4. Where the process to be carried out can only be done efficiently within strictly controlled limits of temperature and humidity.

5. Where the type of building and usage thereof involves considerable heat gains such as in multi-storey office blocks with large glass areas subject to solar gain, and including heat-producing office machinery, computers, intensive electric lighting, etc.

6. In a great variety of conference rooms, lecture theatres, laboratories and animal houses.

7. The core areas of modern buildings planned in depth, where the accommodation in the core is remote from natural ventilation and windows and is subject to internal heat gains from occupants, lighting, etc.

In tropical and sub-tropical countries, air-conditioning is primarily required to reduce the high ambient temperature to one in which working and living conditions can be tolerable. In the temperate maritime climate of the British Isles and in similar parts of the world, long spells of warm weather are the exception rather than the rule, but modern forms of building and modern modes of living and working have produced conditions in which, to produce some tolerable state of comfort, air-conditioning is the best answer. Thus we find buildings of the present day incorporating to a greater or lesser extent, almost as a common rule, some form of air-conditioning. This great variety of applications has produced an almost equally great variety of systems, although all are fundamentally the same in basic principle: that is, to achieve a controlled atmospheric condition both in summer and winter, as referred to earlier, using air as the medium of circulation and environmental control.

The installation of complete air-conditioning in a building as a rule eliminates the necessity for heating by direct radiation, and it naturally incorporates the function of ventilation, thus eliminating the need for opening windows or reliance on other means for the introduction of outside air.

All air-conditioning systems involve the handling of air as a means for cooling or warming, dehumidifying or humidifying. If the space to be air-conditioned has no occupancy, no supply of outside air is necessary, that inside the room being continually recirculated. In most practical cases, however, ventilation air for occupancy has to be included and in the design for maximum economy of heating and cooling, this quantity is usually kept to a minimum depending on the number of people to be served. Thus, in most instances it will be found that the total air in circulation in an air-conditioning system greatly exceeds the amount of outside air brought in and exhausted. Where, however, it is a matter of contamination of the air, such as in a hospital operating theatre, or where

N

some chemical process or dust-producing plant is involved, one hundred per cent outside air may be needed and no recirculation is then possible.

With a certain design of plant it is possible to arrange for 100 per cent outside air to be handled during periods of medium weather, such as in spring and autumn, when neither cooling nor heating is required, or at any other time when it can do useful cooling. Fig. 17.1 shows, for various months of the year, the percentage of day time hours when the outside temperature at Croydon is below 13° C and is thus available for a cooling duty. Meteorological data for other stations in Britain show similar availability.

The basic elements of air-conditioning systems of whatever form are:

Fans for moving air;
Filters for cleaning air, either fresh or recirculated, or both;
Refrigeration plant connected to heat exchange surface, such as finned coils or chilled water sprays;
Means for warming the air, such as hot water or steam heated coils or electrical elements;
Means for humidification; and/or dehumidification;
A control system to regulate automatically the amount of cooling or warming.

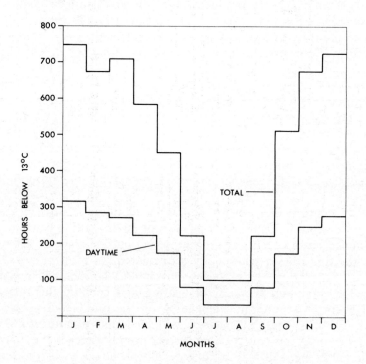

Fig. 17.1.—Annual hours when 'free cooling' is available (Croydon).

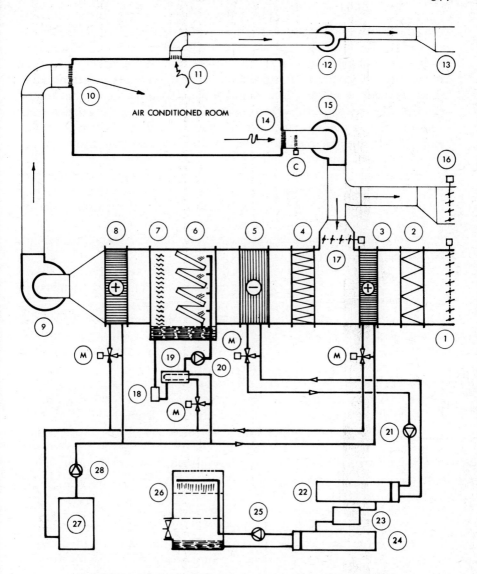

1, Outside air intake. 2, Self cleaning air filter. 3, Pre-heater for fog elimination. 4, Electrostatic or dry fabric type air filter. 5, Chilled water cooling coil. 6, High efficiency air washer. 7, Eliminators. 8, Re-heater. 9, Supply fan. 10, Conditioned air supply. 11, Smoke extract in ceiling. 12, Smoke extract fan. 13, Discharge to outside. 14, Extract and recirculation. 15, Main extract fan. 16, Discharge to outside. 17, Recirculation. 18, Spray-water strainer. 19, Spray-water heater. 20, Spray-water pump. 21, Chilled water pump. 22, Water chiller (evaporator). 23, Multi-stage refrigerant compressor. 24, Shell and tube condenser. 25, Condenser water pump. 26, Forced draught cooling tower. 27, Hot water boiler. 28, Hot water pump. *M* denotes motorised valves. *C* denotes temperature and humidity control points.

FIG. 17.2.—Central plant system with air washer.

TRADITIONAL SYSTEMS

Central Plant—The type of system shown in Fig. 17.2 is suitable for air-conditioning large single spaces, such as theatres, cinemas, restaurants, exhibition halls, or big factory spaces where no sub-division exists. The manner in which the various elements just referred to are incorporated in the plant will be obvious from the caption. It will be noted that in this example the cooling is performed by means of chilled water-cooling coils. The humidification is by means of a capillary cell air-washer. This unit would only run when humidification was required such as in winter.

In an alternative version of a central plant system, the cooling coil is connected directly to the refrigerating plant and contains the actual refrigerating gas. On expansion of the gas leaving the condenser, cooling takes place and hence this system is known as a 'direct expansion system'. It is suitable for small to medium size plants. Humidification could be by means of a direct steamjet or water spray into the air stream, but it might equally be in the form shown in Fig. 17.2.

In Fig. 17.2 it will be noted that there is a separate extract fan shown exhausting from the ceiling of the room. This would apply particularly in cases where smoking takes place, such as in a restaurant, to remove fumes which might otherwise collect in a pocket at high level. Sometimes this exhaust can be designed to remove the quantity equivalent to the outside-air intake, in which case the discharge shown to atmosphere from the return air fan would not be necessary.

Where there are a number of rooms or floors in a building to be served, it is necessary to consider means by which the varying heat gains in the different compartments can be dealt with. Some rooms may have gains, and others none; some may be crowded and others empty; again, some may contain heat-producing equipment. Variations in requirements of this kind are the most common case with which air-conditioning has to deal and for this a simple central system is unsuitable. The ideal, of course, would be a separate system for each room but this is rarely practicable unless the individual spaces are very large or important.

Zoned Systems—There are a number of ways in which a central plant, broadly on the lines of that previously described, may be modified in design to serve a number of groups of rooms or zones. To take the simplest case, if such a plant served one large space of major importance and a subsidiary room having a different air supply temperature requirement, it would be possible to fit an additional small heater or cooler, or both, on the branch air duct to the subsidiary room. Separate control of temperature would thus be available.

An extension of this principle, to serve two or more zones more or less equal in size from a single plant, may be achieved by deleting the re-heater of Fig. 17.2; dividing the supply fan outlet into the appropriate number of ducts and fitting a separate re-heater to each. Since, however, the output of the central plant cooling coil will have to be arranged to meet the demand of whichever zone requires the maximum, extravagant use of reheat will result and lead to uneconomic running costs.

The plant illustrated in Fig. 17.2 is arranged on the 'draw-through' principle, the various components being on the suction side of the supply fan. It is, of course, possible to adapt this sequence to produce a 'blow-through' arrangement with the fan moved to a position immediately following the secondary filter. With such a re-disposition, what is known as a *Multizone* arrangement may be produced, as illustrated in Fig. 17.3. Here, the fan discharge is directed through either a cooling coil or a heating coil into one or other of two plenum boxes, a 'cold deck' or a 'hot deck'. Each building zone is supplied with conditioned air via a separate duct which, at the plant, is connected to both plenum boxes. A system of interlinked dampers, per zone, is arranged such that a constant air supply is delivered to each zone duct which may be all hot, all cold, or any mixture of the two as is required to meet local demands.

An alternative approach to the problem of providing local control to a variety of building zones having differing demands involves the provision of a re-circulating fan-coil unit, having a booster re-heater or re-cooler, within each zone. Outside air, in a quantity suitable to provide for occupancy, is conditioned as to temperature dependant upon outside conditions and corrected for moisture content by means of what is, in effect, a

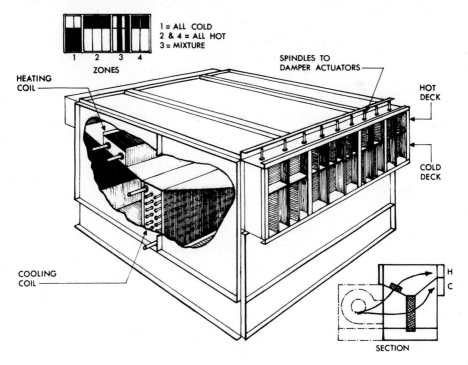

FIG. 17.3.—Multizone plant showing damper arrangement.

conventional central plant. This supply is delivered to the local zone plant where it is heated or cooled as may be required to suit the zone conditions. Fig. 17.4 shows two alternative local plant arrangements, suitable in this case to provide for conditions within the apparatus area of a major telephone switching centre. Note how integration of the air conditioning system with the structure has been arranged. (See Plate XXVI, facing p. 247.)

For a more conventional application to an office building, Fig. 17.5 shows plants arranged in floor service rooms. In this case, air recirculation from the individual rooms passes through louvres above the doors into the corridor and thence back to the zone plant. Whilst such an arrangement has the great merit of simplicity, coupled with relatively low cost, current thinking and Fire Regulations disapprove the use of a corridor 'means of escape' as an airway on the grounds that fire and/or smoke generated in any one room could lead to disruption and panic. It is now, in consequence, more usual to provide a quite independent return-air collection system, via duct branches from each room, back to the local zone plant.

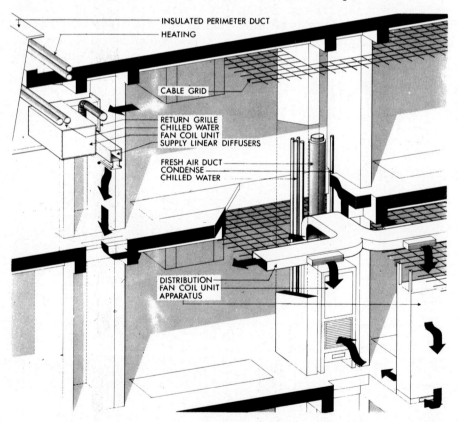

FIG. 17.4.—Fan-coil unit with ducted outside air.

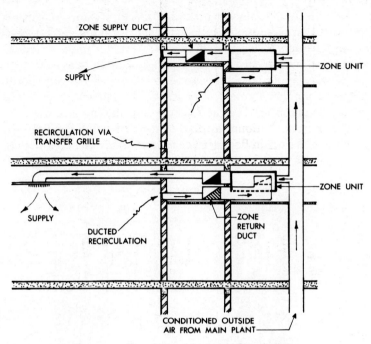

FIG. 17.5.—Zoned system in multi-storey building.

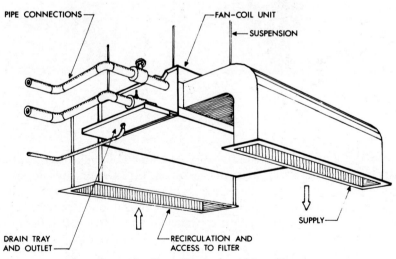

FIG. 17.6.—Fan-coil unit above false ceiling.

Fan Coil Units—If, for a zoned system, we consider each room as a separate zone, distributing ductwork from the recirculating unit may be eliminated, and we have what is known as the 'Fan Coil' system. Fan coil units may be arranged in a variety of ways; one is as shown in Fig. 17.6, where the unit is mounted overhead. Fig. 17.7 shows a cabinet type arranged with component parts stacked vertically and designed specifically for easy application to existing buildings. Fan coil units generally have been developed with silent-running fans requiring little maintenance, and they are of wide application. It is usual to control these units thermostatically by switching the fan motor on and off, but an alternative is to control the water circulation leaving the fan running. Zoning of the water circulations is important, bearing in mind that under some conditions certain aspects of the building may require warming whilst others require cooling.

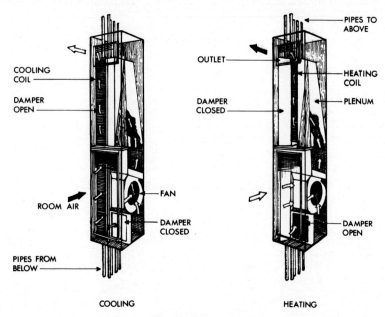

FIG. 17.7.—Vertical type fan-coil unit (Whalen).

An alternative and obviously cheaper method of application of fan coil units is where the fresh air supply is taken direct from outside and the central plant and ducting therefrom is eliminated. In this case, as shown in Fig. 16.20, each unit placed beneath the window has an aperture through the wall to outside, and the unit contains an air filter. Chilled or warmed water circulating piping is connected and the unit is under local thermostatic control. Humidity control is not specifically catered for.

HIGH VELOCITY SYSTEMS

Traditionally, until the mid 1950's, air conditioning systems were designed to operate with duct velocities of not much more than about 8 to 10 m/s and fan pressures of 0·5 to 1 kPa. With the advent of high rise buildings and, concurrent with their introduction, demands for improved working environments coupled with *less* space availability for services there was a requirement that tradition be overthrown. This situation led to a radical rethink and to the introduction of a number of new approaches to air conditioning design using duct velocities and fan pressures twice and more greater than those previously in use.

Whilst the principal characteristic of the new generation of systems relates to the methods adopted for distribution of conditioned air and exploits these to the full, the principles previously described in this Chapter remain unchanged. As before, the conditioning medium may be all-air or air-water dependent upon a variety of circumstances. Fig. 17.8 shows comparative space requirements for the alternative media.

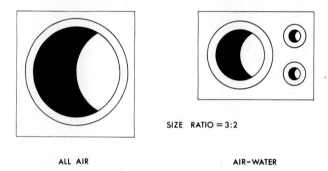

SIZE RATIO = 3:2

ALL AIR   AIR–WATER

FIG. 17.8.—Space requirements for all-air and air-water systems.

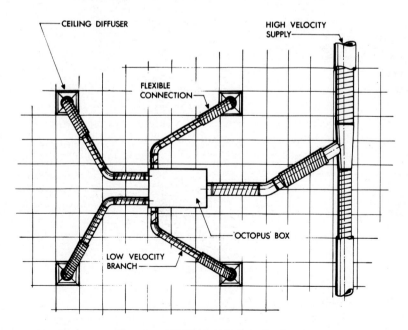

CEILING DIFFUSER

HIGH VELOCITY SUPPLY

FLEXIBLE CONNECTION

'OCTOPUS' BOX

LOW VELOCITY BRANCH

FIG. 17.9.—Single duct 'octopus box'.

Simple All-Air Systems—These, the most primitive of high velocity systems, differ only from their traditional counterparts in the terminals used to overcome problems arising from noise generated in the air distribution system. To this end, a variety of 'single-duct' sound attenuators has been developed to provide for the transition between a high velocity distribution system and air outlets local to the conditioned spaces. The happily named 'octopus box' illustrated in Fig. 17.9 is typical of equipment produced for this purpose. It consists, in principle, of an acoustically lined chamber provided with a sophisticated air volume damper or 'pressure reducing valve': such dampers are in some instances fitted with self-actuated devices arranged so that they may be set to provide constant output volume under conditions of varying input pressure. Such devices are a considerable aid during the commissioning process.

Simple all-air systems will provide adequate service in circumstances where the load imposed is either constant or will vary in a uniform manner for all the area served.

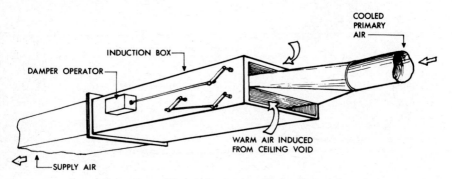

FIG. 17.10.—All-air induction box (Barber Coleman).

All-Air Systems: Induction—For the particular case where air is returned to the central plant via a ceiling void and, further, where lighting fittings (luminaires) are arranged such that the bulk of the heat output (which may be as much as 80 per cent) is transferred to this return air, an all-air induction system may be used.

With such an arrangement, conditioned air is ducted to induction boxes mounted in the ceiling void, as shown in Fig. 17.10. Each box incorporates damper assemblies or other devices which, under the control of a room thermostat, act to permit the conditioned air flow to induce a variable proportion of warm air from the ceiling void into the discharge stream. Re-heat and consequent local control is thus achieved such that, with one type of unit, the cooling capacity may be controlled down to about 50 per cent of maximum.

All-Air Systems: Variable Volume—The traditional approach to air conditioning design placed, as a first principle, insistence upon the concept

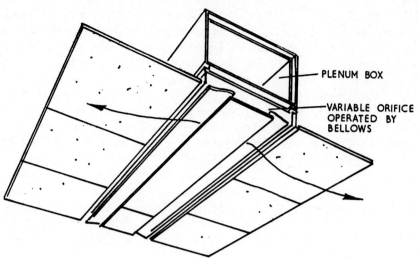

FIG. 17.11.—Variable volume strip diffuser (Carlyle).

of maintaining air discharge to the spaces served at constant volume. Load variations were catered for by adjustment to air temperature. This axiom arose, no doubt, from the known sensitivity of building occupants to air movement and, furthermore, from the relative crudity of the air diffusion equipment available.

With the advent of terminal equipment not only more sophisticated but also with performance characteristics backed by adequate test data, circumstances have changed. The activities of B.S.R.I.A. (late H.V.R.A.) and of a variety of manufacturers in this area must be applauded. Hence

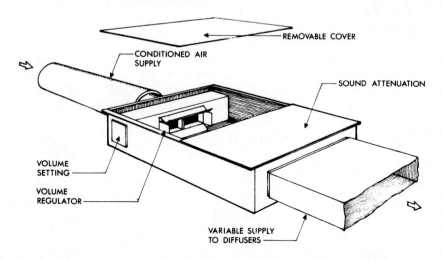

FIG. 17.12.—Variable volume terminal unit.

the availability of potential for abandonment of the traditional approach.

In principle, the variable volume system may be considered as a refinement to the simple all air system whereby changes in local load conditions are catered for not by adjustment of the temperature of the conditioned air delivered, at constant volume, but by adjustment of the volume, at constant temperature. This effect may be achieved by means of metering under thermostatic control of the air quantity delivered either to individual positions of actual discharge, as shown in Fig. 17.11, or to groups of such positions via a terminal unit of the type illustrated in Fig. 17.12. (See Plate XXVII, facing p. 416.)

In such cases, a true variable volume arrangement is possible since the effect of reduced output at the terminal units or at the discharge positions can be sensed by a central pressure controller and thus arranged to operate devices to reduce the volume output of the central plant correspondingly. Economies in overall operation in energy consumption and in cost will thus result.

An alternative form of terminal, as Fig. 17.13, operates upon a different principle. Here, variations in volume at the point of discharge are achieved by diversion of some part of the conditioned air supply either directly back to the main plant or indirectly thence via a ceiling void. This diversion, in one type of equipment, is in pulses under a self operating controller, the conditioned air supply from the central plant being directed all to the room or all to return for time periods varying according to the load. With this type of terminal, the volume of conditioned air supplied by the central plant is constant and although demand upon the central plant for cooling capacity will decrease as the proportion of air diverted rises, no economy in fan power is possible.

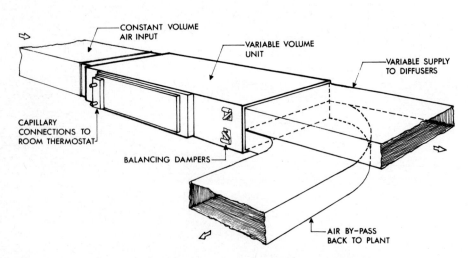

Fig. 17.13.—Variable volume pulsating flow unit (Biddle).

If an adequate supply of outside air is to be maintained and problems of distribution within the conditioned space avoided, volume cannot be reduced beyond a certain level. Good practice suggests that minimum delivery should not be arranged to fall below about 50 per cent of the designed quantity. For internal zones, such a limitation will allow adequate control to be available but in the case of perimeter zones, where conduction and solar gain form a high proportion of the design load, it may be necessary to introduce some level of re-heat to augment capacity control by volume reduction.

Whilst most variable volume terminal units can be adapted to incorporate re-heaters, the required effect may equally well be achieved via control of a constant volume perimeter system or even space heating units either of which may, in any event, be required at windows to deal with down draughts.

All-Air Systems: Dual Duct—The multi-zone system previously described (page 379) is arranged to mix, at the central plant, supplies of hot and cold air in such proportion as to meet load variations in building zones. An extension of this concept would be that an individual mixed-air duct was provided for each separate room in the building but this, on grounds of space alone, would not be practicable. The same degree of control may however be achieved by use of the dual-duct system where the mixing is transferred from the central plant to either individual rooms or small groups of rooms having similar characteristics with respect to load variation.

Use is made of two ducts, one conveying warm air and one conveying

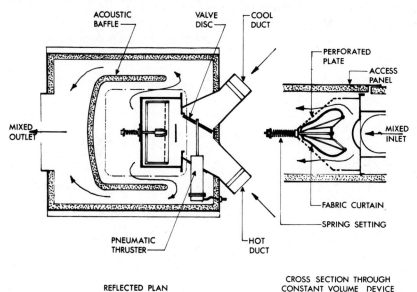

Fɪɢ. 17.14.—Dual-duct blender unit.

cool air, and each room contains a blender so arranged with air valves or dampers that all warm, all cool, or some mixture of both is delivered into the room. Fig. 17.14 illustrates such a unit, incorporated therein being a means for automatically regulating the total air-delivery such that, regardless of variations in pressure in the system, each unit delivers its correct air-quantity. Referred to as 'constant volume control', this facility is an essential part of such a system. (See Plate XXVIII, facing p. 416.)

Owing to the fact that air alone is employed, the air-quantity necessary to carry the cooling and heating load is greater than that used in an air-water system. Air-delivery rates with the dual-duct system are frequently of the order of 5 or 6 changes per hour, compared with the $1\frac{1}{2}$ to 2 required for introduction of outside air. Air movement within the room from a dual-duct system is generally augmented to some extent by the natural induction of air delivered from the outlet of a below-sill mounted blender unit, depending on the outlet velocity.

Owing to the considerable quantity of air in circulation throughout the building, dual-duct systems usually incorporate means for recirculation back to the main plant and this involves return air ducts and shafts in some form.

An advantage of the dual-duct system is that any room may be warmed or cooled according to need without zoning or any problem of change-over thermostats. Furthermore, core areas of a building, or rooms requiring high rates of ventilation, may equally be served from the same system, no separate plants being necessary.

Fig. 17.15 shows the plant arrangement of a system in one form, though there are variations of this using two fans, one for the cool duct, one for the warm duct. (See Plate XXIX, facing p. 417.)

To avoid undue pressure differences in the duct system, due for in-

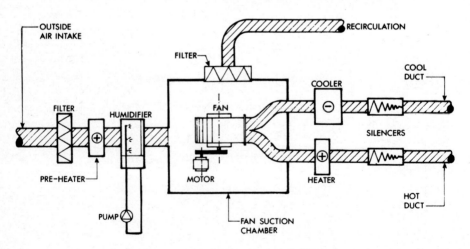

FIG. 17.15.—Dual-duct plant arrangement.

stance to a greater number of the units taking warm air than cool air, static pressure control is usually incorporated so as to relieve the constant pressure devices in the units of too great a difference of pressure, such as might otherwise occur under conditions where the greater proportion of units are taking air from one duct than from both.

In special circumstances, a dual-duct system may be provided with variable volume terminal equipment to combine the best features of each system. With such an arrangement, when volume has been reduced to the practical minimum, control is achieved by re-heat from the hot duct supply.

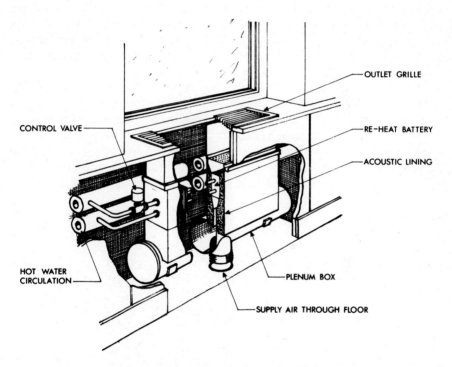

FIG. 17.16.—Terminal re-heat unit (Fläkt).

Air-Water Systems: Terminal Re-heat—As the name suggests,

terminal re-heat is a refinement added to the simple all-air system to provide potential for control where local load conditions within the spaces served vary in an irregular way. For instance, a group of rooms in a hospital or a laboratory might be subjected to changes in occupancy level, solar gain or use of heat producing equipment: a simple system could be controlled only to meet either the worst or the average load imposed.

Terminals of the type fitted to the simple all-air system previously described are therefore equipped with heat exchange coils such that thermostatic control, room-by-room, may be achieved by adjusting the common cooled air supply temperature to suit individual room loadings. A typical terminal unit is as shown in Fig. 17.16.

Needless to say, the use of energy to re-heat air cooled at some expense in a central plant is uneconomic. The terminal re-heat system therefore has only limited application.

Air-Water Systems: Induction—As the name implies, the principle of induction is employed in this system as a means of providing an adequate air circulation within a conditioned room. Primary air conditioned in a central plant is supplied under pressure to terminal units, generally placed below the window, each of which incorporates a series of jets or nozzles as shown in Fig. 17.17. The air induced from the room flows over the cooling or heating coils and the mixture of primary and induced air is delivered from the grille in the sill. The induction ratio is from three to one to six to one. The coils are fed with circulating water which, in the so-called *change-over* system, is cooled in summer and warmed in winter, an arrangement more suited to sharply defined seasons than the unpredictable long springs and autumns of the British climate. In Fig. 17.18 (*a*) and (*b*), control is achieved by an arrangement of dampers, such that the return air from the room is either drawn through the coils for heating or cooling, or the coils are by-passed to a greater or lesser extent. In the type shown in Fig. 17.18 (*c*), control is by variation of water flow: increase in water flow is required to lower temperature during the cooling cycle, and

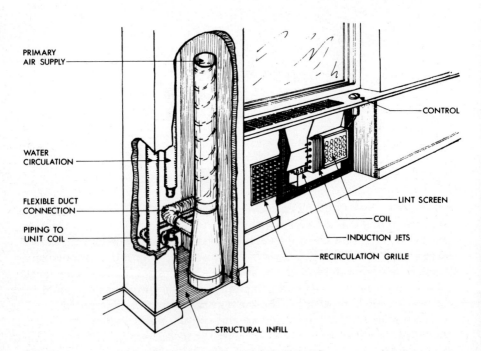

FIG. 17.17.—Air-water induction unit.

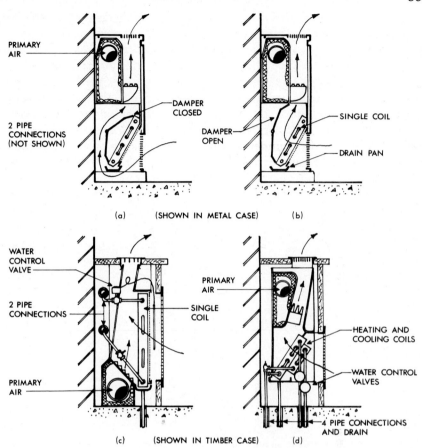

FIG. 17.18.—Alternative control arrangements for induction units.

increase in flow is required to raise the temperature in the heating cycle. There is thus required some means of change-over of thermostat operation according to whether the winter or summer cycle is required. (See also Plate XXX, facing p. 417.)

An alternative method, using the so-called *non-change-over* system, avoids this problem by always circulating cool water through the coils of the induction units and varying the primary air temperature according to weather only. Thus, throughout the year, the heating or cooling potential of the primary air is adjusted to suit that component of demand imposed upon the system, or zone of the system, by orientation, by outside temperature or by wind effect. Any other variant—solar radiation, heat from lighting or occupancy—will necessarily produce a local heat *gain* and the sensible cooling needed to offset this will be provided by the capacity of the unit coils under local control. In so far as such an arrangement acts as 'terminal re-cool' in winter, it is uneconomic in terms of energy wastage.

A variation of the two-pipe induction system is the three-pipe, in which both warm and cool water are available at each unit, with a common return, and the control arrangement is so devised as to select from one or the other. Likewise, in the main system the return is diverted either to the cooling plant or to the heating plant, according to the mean temperature condition. The ideal system would be with four pipes, but it would obviously be correspondingly expensive. A unit having two coils, one for heating and one for cooling, is shown in Fig. 17.18 (*d*).

Fig. 17.19 shows a type of induction unit continuous in length and low in height which had many applications but which is now believed to be out of production following a product-line 'rationalisation'.

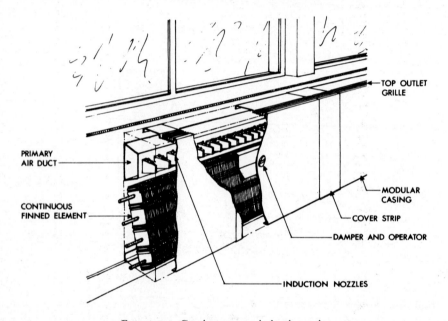

FIG. 17.19.—Continuous type induction unit.

Induction systems might be expected to be noisy, due to the high velocity air issuing from the jets, but the units have been developed with suitable acoustical treatment such that this disadvantage does not arise in practice. For reasons of economy, both in cost and space, ducts with this system are designed on high velocity principles. Air speeds of 15 to 25 m/s in the ducts are common, compared with the conventional speeds of the order of 5 m/s. Owing to the higher pressure at the fan, consequent on the use of high velocities, a silencer immediately after leaving the central plant is usually incorporated with further acoustic treatment in the ductwork distribution system as may be necessary.

Fig. 17.20 is a simplified diagram of an induction system showing the primary conditioning plant, the primary ducting and the water circulation.

The heat-exchanger shown is for warming the water circulated to the units, and this would be fed from a boiler or other heat source. It will be noted that the chilled water supply to the coils of the induction units is arranged to be in the form of a subsidiary circuit to that serving the main cooling coil of the central plant. Such a system has the advantage of providing a degree of 'free cooling' when the outside air supply to the central plant is at low temperature during winter. Furthermore this circuit arrangement provides an in-built protection against excessive condensation on the unit coils with the result that local drain piping therefrom can in most cases be dispensed with.

The induction system involves the distribution of minimum primary

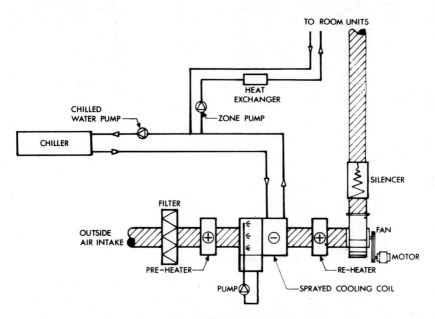

Fig. 17.20.—Induction system plant arrangement.

air, often as little as $1\frac{1}{2}$ to 2 air changes/hour, and has been widely applied to low-cost multi-storey office blocks or hotels where in either case there is a large number of separate rooms to be served on the perimeter of a building. Current practice suggests that, provided application of the system is confined to peripheral areas not deeper than 4 m, with relatively low occupancy, satisfactory service will result. Interior zones of such buildings or rooms where a higher rate of ventilation is required are usually dealt with independently by conventional systems.

Induction systems inherently cause any dust in the atmosphere of the room to be drawn in and over the finned coil surfaces, and, to prevent a build-up of deposit thereon, some form of coarse lint screen, easily removable for cleaning, is usually incorporated.

Air-Water Systems: Heat Recovery—An interesting development of the unit air-conditioner system was first made available under the proprietary name 'Versatemp', but other versions are now available. Fig. 17.21 illustrates diagrammatically the principle of this system. The various rooms throughout a building are each equipped with a unit containing a hermetic refrigerating compressor. The compressor, with its evaporator and condenser, either cools the room or warms it according to the requirements of the local control system.

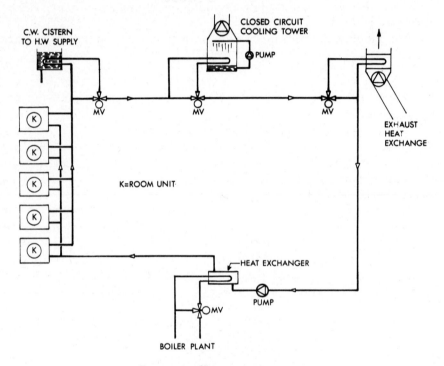

FIG. 17.21.—Heat recovery system.

When operating as a cooler, the circulating water system removes heat from the condenser, the water in turn passing to a closed circuit cooling tower on the roof as shown in the diagram. When serving as a heater (heat pump), the functions of the evaporator and the condenser are reversed, and the circulating water acts as a base heat source. The circulating flow is kept at a constant temperature, low grade heat being provided as necessary via the boiler and heat exchanger and cooling, on the other hand, by evaporation at the tower. When cooling is not required, the tower is by-passed in the manner indicated. Potential exists over much of the year for both cooling tower and boiler plant to be out of use, a balance being achieved between heat gains in some rooms and heat losses in others.

A typical system of this type (Plate **XXXI**) operates with a water supply presented to the units at a temperature of 27° C, year round, with rejection at 34° C when cooling and 18° C when heating.

Systems of this type are reported to have been applied to existing buildings, use being made of original heating system pipework, the radiators being removed and heat recovery units substituted. A cooling tower has, of course, been added at roof level.

OTHER METHODS

Self-contained Air-Conditioners—This heading covers a separate field in that the units concerned would more properly be described as Room Coolers. Such units are commonly complete in themselves, containing compressor, air filter, fan and cooling coil. Electric resistance heaters may be incorporated for winter use and, rarely, means for humidification. Fresh air can be introduced if required. Being of unit construction, alternatively described as 'Packaged', they are not tailor-made to a specific job and may thus be expected to be economical in first cost. In some cases, a so-called 'split type' of unit may be found where the condenser and compressor are remotely mounted from that part of the equipment which serves the room concerned. Bulk and noise at the point of use are thus much reduced.

Unit air-conditioners of small size generally have the condenser of the refrigerator air-cooled, but in larger sizes the condenser may be water-cooled in which case water piping connections are required. Apart from this, the only services needed are an electric supply and a connection to drain to conduct away any moisture condensed out of the atmosphere during dehumidification. Compressors in most units are now hermetic and

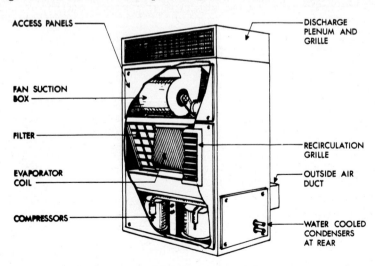

FIG. 17.22.—Self-contained air conditioner.

are therefore relatively quiet in running.

Sizes vary from small units suitable for a single room, sometimes mounted in a window or in a cabinet, as in Fig. 17.22, and the range goes up to units of considerable size suitable for industrial application, in which case ducting may be connected for distribution.

Panel Cooling—A fundamentally different approach to the whole matter of temperature control and ventilation, covered by the general term 'Air-Conditioning', is a system in which surfaces of the ceiling are cooled by chilled-water circulation for the removal of heat gains, leaving the air-distribution system the sole purpose of ventilation and humidity control. Such a system has been applied to the Shell building in London.

The essential feature of the system* is that the ventilation air shall be conditioned such that its dew-point is below the surface temperature of the cooling surfaces in the room, otherwise condensation might occur thereon. The cooling surfaces in the case in question comprise normal perforated aluminium trays with insulation above, clipped to a circulating water-grid fed with chilled water in summer and with warmed water for heating in winter. Air volumes are thereby reduced, as the sole purpose of the ventilation system is to provide air for occupancy, and the conditioning of same is to a set state. Fig. 17.23 gives an illustration taken from the above paper indicating the principles of the system. Bearing in mind that a ceiling treatment of this kind also incorporates acoustical absorption, an appraisal of overall costs needs to take this and all other factors into account.

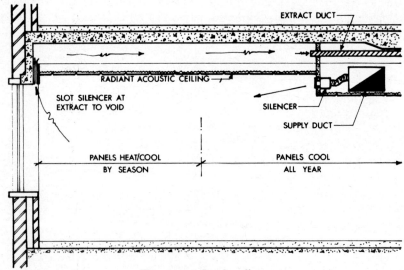

FIG. 17.23.—Panel cooling.

* Jamieson, H. C. and Calland, J. R., *The Mechanical Services at Shell Centre*. J.I.H.V.E., 1963. 31. 1

CHAPTER 18

Principles of Air-Conditioning Design

IT IS NOW PROPOSED TO CONSIDER the fundamental principles underlying the design of an air-conditioning system. These principles are the same no matter what particular form the system may take, but the degree of accuracy necessary to be achieved will depend on the application and the sophistication of controls. For instance, a special research laboratory may require minimum tolerance in conditions, whereas less strict limits would be acceptable for comfort-conditioning in the case of a department store.

First to be considered is the general case, as applied to a central air-conditioning system for a single large space, and this is followed by some notes on how these general principles are applied to certain of the specific types of apparatus already discussed. Design data has been built up around each of the particular forms of equipment mentioned, and it would be beyond the scope of this book to explore each one in detail.

HEAT GAINS

Heat gains are the opposite of heat losses. In designing a heating system for a building we are chiefly concerned with the losses. When designing an air-conditioning system we are concerned with the gains, although as the same system will probably serve for heating, the losses will also be involved.

Heat gains occur in warm weather due to heat transfer through the building fabric by conduction and by infiltration and ventilation: radiation from the sun is received throughout the year, particularly on clear days. Internal heat gains occur due to occupancy, lighting, and any other heat producing equipment within the space irrespective of weather.

The gains so outlined are *sensible heat gains* and the quantity of conditioned air which must be provided to combat them is directly proportional to the difference between the supply air temperature and that to be maintained in the room. The temperature rise which can be permitted may be limited to 6 to 8 K owing to the difficulty of mixing cool entering air with warmer room air without draught. The mass air flow required to maintain a desired temperature is simply arrived at by the sum:

$$\text{Mass flow air (kg/s)} = \frac{\text{heat gains (kW)}}{\text{design temperature rise (K)} \times \text{specific heat capacity of air (kJ/kgK)}}$$

397

Latent Heat Gains, on the other hand, do not cause temperature rise. These are due to moisture exhaled in the breath and by perspiration of occupants, by moist air entering the space due to infiltration and by any steam- or vapour-producing equipment within the space. The latent gains are to be treated separately from the sensible gains; the mass flow of air determined for the latter will usually be found to cause only a small humidity increment, but in an extreme case limitation of the increment to an acceptable figure may override that sufficing for sensible gains.

Solar Gains—The most important heat gain is that from the sun. Heat from the sun on a clear day at midsummer falling on a horizontal surface is given as 910 W/m² in latitude 52° N. In addition, diffuse radiation from all parts of the sky is given as 50 W/m². Of these amounts,

(a) some is reflected back into space;

(b) some warms the surface of the material of roofs and walls whereby external air in contact removes heat by convection;

(c) some is transmitted through the material by conduction constituting a heat gain in the room;

(d) of that falling on windows, part is reflected back, part warms the glass, but the greater part passes through in the form of radiant heat, so elevating the temperature of surfaces of structure and contents.

Referring to these effects in greater detail:

(a) and (b). The proportion of solar heat absorbed depends on the nature and colour of the surface. Thus, if a perfect 'black body' is taken as 1, the absorbance of white is 0·3 to 0·5 and of a polished metallic surface 0·1 to 0·4. Most building materials such as brick and concrete will absorb 0·5 to 0·8.

(c) The rate of heat flow through the material will depend on its conductivity and use may be made of the usual U factors. The temperature difference outside to inside will, however, be influenced by the heat absorbed from the sun. If we consider first the design-air temperatures outside t_o and inside t_i and the sun-heat increment t_s due to radiation absorbed by the surface, the overall difference causing flow of heat through the material is then $(t_o - t_i) + t_s$. An instantaneous rise of temperature would occur with a material of negligible heat capacity but, for a material of some mass, a time lag is involved. If a material is of sufficient thickness and mass, it is obvious that, as the sun moves round in its path, radiation falling on a wall in the early part of the day may not penetrate to the inside until long after the surface is in shadow, when most likely the curious phenomenon will occur of heat stored in the core of the wall flowing out in both directions. In some cases the space may have ceased to be occupied before the peak heat gain arrives.

Table 18.1 gives sun-heat temperature increments for latitude 52° N (London) for the month of August, for various orientations and for two surface colours. Table 18.2 gives the time lag applicable to various materials and thicknesses. The decrement factor f depends on the thickness of the material and applies to the overall heat gain through the material, i.e.:

$$U\{(t_o - t_i) + t_s\}f$$

TABLE 18.1
SUN HEAT TEMPERATURE INCREMENTS (K)
52° N—August

x = Dark Surface y = Light Surface

Position		British Summer Time (G.M.T. + 1).									
Orientation	Surface	0800	0900	1000	1100	Noon	1300	1400	1500	1600	1700
Horiz.	x	4·4	10·1	15·4	18·8	21·1	21·9	21·1	18·8	15·4	10·1
Horiz.	y	0·8	4·2	6·9	8·9	10·2	10·6	10·2	8·9	6·9	4·2
N	x	2·4	3·4	4·3	4·8	5·2	5·3	5·2	4·8	4·3	3·4
N	y	1·3	1·9	2·4	2·7	2·9	3·0	2·9	2·7	2·4	1·9
NW	x	2·2	3·4	4·3	4·8	5·2	5·3	5·2	5·0	10·8	16·2
NW	y	1·2	1·9	2·4	2·7	2·9	3·0	2·9	2·8	6·0	9·0
W	x	2·2	3·4	4·3	4·8	5·2	6·2	14·6	22·7	27·6	29·3
W	y	1·2	1·9	2·4	2·7	2·9	3·4	8·1	12·6	15·3	16·3
SW	x	2·2	3·4	4·3	6·7	14·4	22·3	27·7	30·9	30·8	27·1
SW	y	1·2	1·9	2·4	3·7	8·0	12·4	15·4	17·2	17·1	15·1
S	x	4·5	11·9	19·1	24·3	27·9	29·3	27·9	24·3	19·1	11·9
S	y	2·5	6·6	10·6	13·5	15·5	16·3	15·5	13·5	10·6	6·6
SE	x	20·3	27·1	30·8	30·9	27·7	22·3	14·4	6·7	4·3	3·4
SE	y	11·3	15·1	17·1	17·2	15·4	12·4	8·0	3·7	2·4	1·9
E	x	25·5	29·3	27·6	22·7	14·6	6·2	5·2	4·8	4·3	3·4
E	y	14·2	16·3	15·3	12·6	8·1	3·4	2·9	2·7	2·4	1·9
NE	x	17·4	16·2	10·8	5·0	5·2	5·3	5·2	4·8	4·3	3·4
NE	y	9·7	9·0	6·0	2·8	2·9	3·0	2·9	2·7	2·4	1·9

Flat roofs are subject to solar absorption throughout daylight hours, but, again, the greater the mass the longer is heat gain to the room deferred. Sun-heat temperature increments for flat roofs are given in Table 18.1. Insulation under a roof does not affect the time lag appreciably but it does influence the U value.

Heat gains through sloping roofs depend on direction of slope and pitch. The values of sun-heat increments for flat roofs may be adjusted according to the cosine of angle of incidence.

With regard to Tables 18.1 and 18.2, these are derived from *Guide* data and provide an approximate solution. For a more complete treatment applying to other latitudes and months of the year, the *Guide, Section A6*, should be consulted.

TABLE 18.2

ADJUSTMENT FACTORS TO SOLAR GAINS THROUGH
SOLID MATERIALS

Construction	Adjustments	
	Time lag (hours)	Decrement Factor f
Light frame, sheeted and lined	$\frac{1}{2}$	1·0
Brickwork		
105 mm bare	$3\frac{1}{2}$	0·8
105 mm internally lined	4	0·7
220 mm bare	8	0·4
220 mm internally lined	$8\frac{1}{2}$	0·3
Concrete		
150 mm bare	5	0·6
150 mm internally lined	$5\frac{1}{2}$	0·5
150 mm externally lined	6	0·4
200 mm bare	6	0·5
200 mm internally lined	$6\frac{1}{2}$	0·4
200 mm externally lined	7	0·3

Thermal resistance of lining assumed to be less than
0·4 m K/W

(*d*) Glass is by far the most important route by which solar heat
enters a building and, as pointed out, it is without time lag. The effect in
the room may not be instantaneous however, depending on the nature and

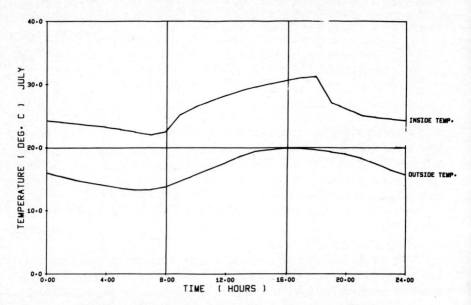

FIG. 18.1.—Computer plot of temperatures without air conditioning.

mass of internal structure, furniture or other contents. This subject has been explored by Loudon* in evolving a method of forecasting summertime temperatures in buildings without air-conditioning. By use of a suitable computer program, a plot may be produced for any given room, as shown in Fig. 18.1. For the case illustrated, conditions within the room would be quite unacceptable. The method involves the concept of environmental temperature (referred to on page 37) thus taking account of mean radiant temperature as well as air temperature. The method also involves the use of the term 'admittance'. Admittance of a building component is a measure of its ability to smooth out diurnal temperature swings in the building.

TABLE 18.3

COOLING LOAD DUE TO SOLAR GAIN

Through Vertical Glazing (12 hour plant operation)
for latitude 52° N. G.M.T. +1. W/m².
(Without sun protection or with heat-absorbing glass or
with external shades, using factors in Table 18.5(a))

Date	Orientation	Light Construction						Heavy Construction					
		0800 hrs	1000	1200	1400	1600	1800	0800	1000	1200	1400	1600	1800 hrs
	N	122	119	134	134	119	122	159	118	131	136	131	118
	NE	404	273	145	145	131	107	317	295	172	157	152	139
	E	510	478	285	155	140	116	373	431	348	190	185	172
June	SE	360	460	405	221	125	101	263	377	391	308	181	168
21	S	87	247	371	371	247	87	130	185	295	339	295	185
	SW	101	125	221	405	460	360	165	183	196	323	406	392
	W	116	140	155	285	478	510	186	204	217	222	380	463
	NW	107	131	145	145	273	404	151	169	182	187	203	326
	N	47	73	87	87	73	47	67	68	82	87	82	68
	NE	282	178	97	97	82	56	190	217	107	112	107	93
	E	416	435	244	111	96	70	261	385	315	153	148	135
Aug.	SE	340	487	440	251	103	77	227	388	418	335	182	157
and	S	102	306	440	440	306	102	137	233	353	400	353	233
Apr.	SW	77	103	252	440	487	340	146	166	191	343	426	396
	W	70	96	111	244	435	416	128	149	162	167	329	399
	NW	56	82	97	97	178	282	84	104	118	123	118	228
	N	25	54	69	69	54	25	21	44	58	63	58	44
	NE	150	104	73	73	58	29	33	135	70	75	70	56
	E	259	378	213	86	71	42	71	300	266	114	109	94
Sept.	SE	241	472	453	273	84	56	112	346	411	337	189	135
and	S	108	332	479	479	332	108	133	248	383	434	383	248
Mar.	SW	56	84	273	472	472	241	112	135	189	337	411	346
	W	42	71	86	213	378	259	72	94	109	114	266	300
	NW	29	58	73	73	104	150	33	56	70	75	70	135
Correction factor to peak load for 24 hour plant operation		0·96						0·80					

* *Summertime temperatures in buildings without air conditioning.* J.I.H.V.E. 1970. 37. 280.

Using the same conceptions, the *Guide* presents data for cooling load due to solar gains through windows on the basis of two forms of internal construction: *light* (i.e. demountable partitions and suspended ceilings) and *heavy* (i.e. brick or block partitions). Concrete floor slabs are assumed in both cases. Tables 18.3 and 18.4 are representative of part of these data for various orientations, seasons and times. Table 18.3 relates to un-shaded windows or windows with external shades. Table 18.4 relates to windows with internal shadings. Table 18.5 (*a*) gives shading factors apply-ing to the former and Table 18.5 (*b*) gives factors applying to the latter.

The data assume that design is on the basis of environmental tempera-ture, but, if used for air temperature, a small inherent margin will be included.

Scattered or diffuse radiation from earth and sky is included in the values given in these Tables; but, where a ground floor abuts on to a pavement subject to solar gain, some addition to the window gain seems jus-tified to allow for upward reflection. An addition of ten per cent is suggested.

It is perhaps worth noting that some recent all-glass buildings, even in Great Britain, are virtually 'un-airconditionable' (and in consequence virtually uninhabitable) unless measures are taken to exclude excessive solar gains.

External shades or sunbreaks are preferable to any form of internal shading, but they have in the past been considered as involving difficulties in maintenance; new materials and new methods of blind operation are beginning to solve this problem, however. Heat-reflecting glasses depend on finely divided metal incorporated in the glass and give a better per-formance than so-called 'heat-absorbing glass', due to the fact that they reflect radiation rather than absorb it. This feature of high reflection has, however, been known to cause serious complaint from occupants of neigh-bouring buildings.

In addition to solar gain by direct transmission of radiation, the air-to-air conductance must be allowed for, being the product of area, U value and air-temperature difference outside to inside. Double glazing in this case will show to advantage over single glazing. U factors may be taken as for heating, although some adjustment may be made for a higher external-surface resistance in summer.

Infiltration—This is the natural uncontrolled movement of air through an air-conditioned space due to opening of doors and minute crackage at fenestration. An allowance of one half air change per hour appears to cover this sort of leakage, but if double windows and impervious linings are used, this may be reduced to one quarter air change per hour. Special cases may occur in industrial buildings where doors must be open for transport of goods, etc. Air locks at all openings to unconditioned spaces are essential if conditions are not to go out of control.

Ventilation—The air-conditioning plant will of necessity be the means of introducing ventilation air, the rate depending on the number of occupants

TABLE 18.4

COOLING LOAD DUE TO SOLAR GAIN

Through Vertical Glazing (12 hour plant operation)
for latitude 52° N. G.M.T. +1. W/m².
(With internal sun protection using factors in Table 18.5(b))

Date	Orientation	Light Construction						Heavy Construction					
		0800 hrs	1000	1200	1400	1600	1800	0800	1000	1200	1400	1600	1800 hrs
June 21	N	73	71	81	81	71	73	73	72	81	81	72	73
	NE	261	170	82	82	72	56	252	168	85	85	76	60
	E	328	305	172	82	72	55	318	297	172	88	79	63
	SE	225	294	256	128	62	45	221	286	250	131	69	54
	S	40	151	236	236	151	40	47	150	230	230	150	47
	SW	45	62	128	256	294	225	54	69	131	250	286	221
	W	55	72	82	172	305	328	63	79	88	172	297	318
	NW	56	72	82	82	170	261	60	76	85	85	168	252
Aug. and Apr.	N	26	44	54	54	44	26	27	44	53	53	44	27
	NE	184	112	54	55	45	27	177	110	57	57	47	31
	E	268	281	149	56	46	28	259	272	148	62	52	36
	SE	210	312	280	149	46	28	208	303	273	151	55	38
	S	46	187	280	280	187	46	54	186	273	273	186	54
	SW	28	46	149	280	312	210	38	55	151	273	303	208
	W	28	46	56	149	281	268	36	52	62	148	272	259
	NW	27	45	55	55	112	184	31	47	57	57	110	177
Sept. and Mar.	N	13	33	43	43	33	13	14	33	42	42	33	14
	NE	97	66	44	44	33	14	94	64	44	44	34	16
	E	165	247	133	45	34	15	160	237	131	49	39	21
	SE	144	304	291	167	36	16	145	294	282	166	44	25
	S	48	203	305	305	203	48	57	202	297	297	202	57
	SW	16	36	167	291	304	144	25	44	166	282	294	145
	W	15	34	45	133	247	165	21	39	49	131	237	160
	NW	14	33	44	44	66	97	16	34	44	44	64	94
Correction factor to peak load for 24 hour plant operation		1·00						0·95					

TABLE 18.5

FACTORS FOR SHADING

(a) *Applying to Table 18.3*
Type of glass unshaded

	Glazing	
	Single	Double
Clear - - - - - - -	1·00	0·88
Lightly heat absorbing - - - -	0·75	0·56
Densely heat absorbing - - -	0·61	0·41
Heat reflecting gold (sealed unit when double) - - -	0·41	0·32
Lacquer coated, grey - - - -	0·79	—
External shades		
Dark green open weave plastic - - -	0·33	0·26
Canvas roller blind - - - -	0·22	0·16
White louvre sunbreaker, blades at 45° -	0·19	0·14
Dark green miniature louvre - - -	0·18	0·13

(b) *Applying to Table 18.4*

	Single	Double
Mid-pane Venetian blind white - -	—	0·60
Internal		
Dark green open weave plastic - -	1·32	1·25
White Venetian blind - - - -	1·00	1·02
White cotton curtain - - - -	0·77	0·82
Cream holland linen blind - - -	0·61	0·71

and whether smoking is permitted or not. Rates vary from a minimum of 0·005 m³/s (5 litre/s) per occupant, without smoking, to 0·018 m³/s (18 litre/s) assuming heavy smoking. Where contamination is concerned, as in a hospital operating theatre, 100 per cent outside air is required: i.e. no recirculation. The same applies in any industrial application where fumes or dust are produced.

Another special case occurs in department stores owing to the very uncertain and variable number of occupants. Investigations carried out by one of the revisers* show that, from actual counts, concentrations vary

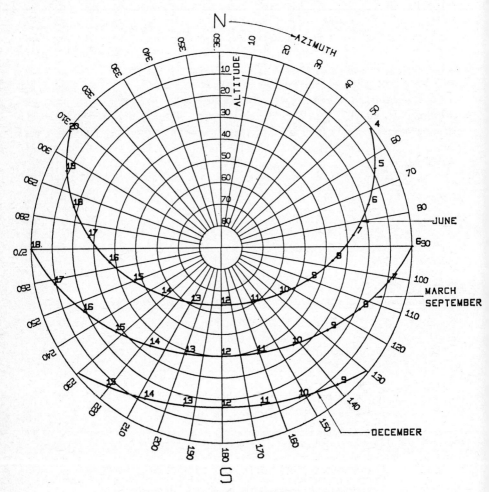

FIG. 18.2.—Diagram of sun positions at Latitude 50 N.

* Martin, P. L., *A study of occupancy in chain stores.* J.I.H.V.E. 1970. 38. 99.

from four persons per 10 m² to ten persons per 10 m² at peak periods. The outside-air requirement is given at 5 litre/s per occupant.

The cooling load and heating load for ventilation air may be estimated on the basis of enthalpies, as is discussed later in this Chapter in the section on psychrometry.

Sun-Path Diagrams—These diagrams are so drawn as to enable altitude and azimuths to be read graphically for various periods of year and times of day. A sample for 50° N is illustrated in Fig. 18.2, as plotted by computer. Similar diagrams for other latitudes are available in the form of overlays.*

A miniature plan of the building under consideration may be placed at the centre of the diagram, orientated correctly, and by inspection it is possible to note the likely peak incidence of solar radiation throughout the day on the various faces of the building. It will be clear that easterly faces will receive solar radiation first, the south-east and south next, meanwhile the east side will be going into shadow and so on.

For the purpose of assessing coincident cooling load, a schedule may be made of the various rooms or areas of the building and, taking each in turn, its own peak period may be established. From the diagram, co-

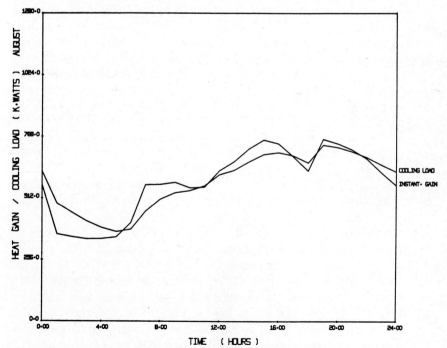

FIG. 18.3.—Computer plot of heat gain and cooling load.

* Petherbridge, P., *Sunpath diagrams and overlays for solar heat calculations*. B.R.E. 1969. *Research Series, No. 39.*

incident loads on various faces may be observed and, from this, the likely maximum throughout the day. This maximum may vary at different seasons of the year, as discussed in Chapter 23. Whilst this exercise is fairly straightforward in the case of a square or rectangular building, it becomes more complicated in a plan with re-entrant angles or internal courts where the building itself creates shadows on certain faces at certain periods. A model can be most helpful here. In the ultimate solution of a large scale project, calculation by computer is the obvious answer. Fig. 18.3 illustrates a computer plot output showing the cooling load for a complete building of about 750 kW loading, and its displacement from peak heat gain.*

A second use of the sun-path diagram is in the design of sunbreaks. The angles of incidence may be plotted on the plan and separate diagrams set up for the various aspects. Both horizontal and vertical sunbreaks may be employed, or a combination of both. Much architectural ingenuity is possible here. Adjustable sunbreaks are another possibility to overcome the disadvantage of completely obscuring the view. Such have indeed been used successfully in a building in Manchester.

The sun-path diagram may further be used to calculate solar gain from first principles, given the solar intensity, transmittance or absorbance of materials, and from the diagram determining the angles of incidence hour-by-hour and day-by-day. The task, however, is unnecessary if use is made of the data in the *Guide* or of the further boiled-down versions contained in the Tables in this chapter, which are admittedly limited to specific parameters. These are intended to indicate the process rather than the detail.

Design Temperatures, Summer—*External, U.K.:* A commonly accepted design condition is 28° C D.B., 20° C W.B., but 30° C D.B. is often used for conditions in London in order to take account of the 'Heat Island' effect arising from traffic etc. Infrequent occasions occur of higher temperatures, but are disregarded for design purposes. *Internal, U.K.:* It is usual to distinguish between conditions of continuous and transient occupancy, 21° C D.B. and 24° C D.B. being the commonly accepted levels respectively, with 50 per cent saturation in each case (50 per cent relative humidity, in common parlance).

Heat from Occupants—The heat emitted from the human body depends on activity (see p. 40), temperature, relative humidity and air-movement.

Fig. 18.4 shows, for a man seated at rest, how the sensible, latent and total heat varies with the dry bulb temperature for average humidities in still air. The sensible heat increases by about 20 per cent with air-movements up to 0·5 m/s. The latent heat is affected only slightly with air-movement under about 18° C, above which temperature the air-movement displaces the point at which the curve begins to rise, e.g. at 0·5 m/s the rise in evaporation rate begins at about 26° C.

---

* The authors are indebted to Oscar Faber and Partners for the use of their computer program 'COLO' to produce this illustration.

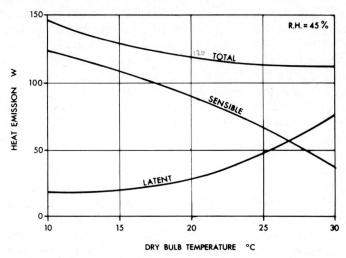

FIG. 18.4.—Latent, sensible and total heat for man seated at rest.

Heat from Lights, Motors, etc.—This is dealt with fully on page 41. The energy input to any machine or heat-producing appliance within the space (including lighting) constitutes a heat gain and can be evaluated in kilowatts. Exceptions are where the energy produced, as in an electric motor, is dispersed outside the space—as possibly in a conveyor system or air compressor. In such cases only the motor inefficiency comes out in heat in the space. Another exception is where heat from lighting is removed by arranging extraction from and around the lighting fitting—a practice which is becoming increasingly common and upon which subject much research remains yet to be evaluated (see page 528).*

Having established the total heat gain in the room itself, the mass flow of air to be delivered to the room is found, as has already been shown, by dividing this figure by the specific heat capacity of air × temperature rise.

PSYCHROMETRY

Psychrometry concerns the behaviour of mixtures of air and water vapour, and a knowledge of it is necessary in any air-conditioning calculations. The general principles were referred to on pp. 6 and 7. A complete study is outside the scope of this book and is dealt with in many text-books on thermodynamics.

The properties of a mixture of air and water vapour may be summarized as follows:

Dry-Bulb Temperature (D.B.) is the temperature of the air as indicated by an ordinary thermometer.

* Bedocs, L. and Hewitt, H., *Lighting and the thermal environment*. J.I.H.V.E. 1970. 37. 217

Wet-Bulb Temperature (W.B.) is a temperature related to the humidity of the air and to the dry-bulb temperature. If a thermometer has its bulb kept wet, as by a wick surrounding it, the evaporation of the water will take heat from the mercury which will consequently contract and indicate a lower temperature than in its dry-bulb condition. The rate of evaporation from the wetted bulb depends on the humidity of the air, i.e. very dry surrounding air will cause more rapid evaporation than moister air at the same dry-bulb temperature. The difference between dry- and wet-bulb temperatures can thus be used as a measure of the humidity. This difference is known as the *Wet-Bulb Depression.*

The rate of evaporation, and consequently the depression, depends also on whether the wetted bulb is exposed to still or moving air. In the latter case the film of moisture-laden air round the bulb is removed more rapidly than in the former, so that the depression is greater. Some systems of psychrometry adopt one condition as a standard, some the other. It is immaterial which is used provided the instruments are suitably calibrated, so that a given pair of dry- and wet-bulb readings can be accurately converted into a measure of the actual humidity.

Sling Wet Bulb is the temperature indicated by a sling or whirling psychrometer (Fig. 18.5), consisting of two thermometers mounted in a frame which can be whirled by hand. One thermometer has a wetted sleeve slipped over the bulb. Assmann has found that above an air-speed of 2·5 m/s and up to at least 45 m/s the depression is independent of the air-speed.

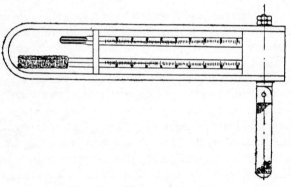

FIG. 18.5.—Sling psychrometer.

Another type of psychrometer is the Assmann type in which the air is drawn, by means of a small fan, over thermometers mounted in a tube. In this intrument the wet bulb is kept moist by means of a wick in a reservoir.

Screen Wet Bulb is the temperature indicated by a wet-bulb thermometer in stationary air, and is so called because the standard instrument is housed in a louvred box called a Stevenson Screen.

Another approach to measurement of moisture content is the conception known as *Temperature of Adiabatic Saturation.* This is the temperature which would be attained by moist air in intimate contact with a water surface, and in equilibrium conditions, assuming that no heat is gained or lost to an external source in the process of attaining equilibrium.

This process is a close approximation to that taking place in the wetted wick of a wet-bulb thermometer, and at ordinary atmospheric temperatures and humidities the Sling Wet Bulb and Adiabatic Saturation Temperature differ very little, which has led to the latter being adopted in many psychrometric systems in place of the true wet-bulb temperature.

This substitution has the great advantage that conditions of equal Adiabatic Saturation correspond to equal Total Heat, i.e. Adiabatic Saturation temperature becomes a criterion of Total Heat.

Dew-point Temperature is the temperature to which a mixture of air and water vapour must be reduced to produce condensation of the vapour. At the dew-point the air is said to be *saturated*, and in this condition the dry-bulb and wet-bulb temperatures both coincide and are equal to the dew-point temperature.

Moisture Content is the mass of water vapour contained in unit mass of the mixture. This is expressed in kg per kg of *dry air*, and not per unit of the mixture.

Vapour Pressure is the partial pressure exerted by the water vapour, and is expressed in kilopascals (kPa).

It follows from the above that the vapour pressure at any condition is dependent only upon the absolute humidity, and is independent of the pressure of the mixture.

Relative Humidity (R.H.) is the ratio of actual vapour pressure to the vapour pressure exerted when the air is saturated at the same temperature, expressed as a percentage. At the saturation temperature the relative humidity is 100 per cent, and dry bulb, wet bulb and dew-point are the same.

At ordinary humidities and at temperatures up to 40° C the vapour pressure is substantially proportional to the absolute humidity, from which it follows that the relative humidity, as defined above, is approximately equal to the ratio of the absolute humidity at the given condition to that of saturation at the same D.B. temperature. This ratio is referred to as *percentage saturation*.

Total Heat or Specific Enthalpy is the sum of the sensible and latent heats of the air and of the moisture contained in it at any given condition, reckoned from an arbitrary datum of 0° C. Specific enthalpies are referred to in kJ/kg of dry air.

Volume—The relationship between pressure, volume and temperature has been referred to on p. 12.

Specific volume is the volume of the mixture in m³ per kg of dry air. Air containing water vapour is less dense than dry air, hence the volume occupied is greater per unit mass.

Table 18.6 gives data regarding volume, mass of saturated vapour, and total heat of dry and saturated air.

Barometric Pressure—The standard atmospheric pressure now adopted in SI units and in the *Guide* is, at sea level, 101·325 kPa = 1013·25 mbar = 1·01325 bar. The bar may thus be taken as the near equivalent of atmospheric pressure. The psychrometric properties of mixtures of air and water vapour given in the *Guide* and referred to later in this chapter may be taken as accurate for all practical purposes from 95 kPa to 105 kPa. As a matter of interest the relationship between altitude and pressure is as follows:

Altitude (m)	500	1000	1500	2000	2500	3000
Barometric pressure, kPa	95	89·5	84·5	79·5	74·5	70

TABLE 18.6

PROPERTIES OF DRY AND SATURATED AIR AT
VARIOUS TEMPERATURES

Temperature °C	Specific Volume m³ per kg Dry Air	Saturation Moisture content kg/kg Dry Air	Specific Enthalpy (Total Heat) kJ/kg Dry Air	
			Dry Air	Saturated Air
0	0·77	0·003 8	0·00	9·47
2	0·78	0·004 6	2·01	12·98
4	0·79	0·005 0	4·02	16·70
6	0·79	0·005 8	6·04	20·65
8	0·80	0·006 7	8·05	24·86
10	0·80	0·007 6	10·06	29·35
12	0·80	0·008 7	12·07	34·18
14	0·81	0·010 0	14·08	39·37
16	0·82	0·011 4	16·10	44·96
18	0·82	0·012 9	18·11	51·01
20	0·83	0·014 7	20·11	57·55
22	0·83	0·016 7	22·13	64·65
24	0·84	0·018 9	24·14	72·37
26	0·84	0·021 4	26·16	80·78
28	0·85	0·024 2	28·17	89·96
30	0·86	0·027 3	30·18	99·98
32	0·86	0·030 7	32·20	111·0
34	0·87	0·034 6	34·21	123·0
36	0·87	0·038 9	36·22	136·2
38	0·88	0·043 7	38·24	150·7
40	0·89	0·049 1	40·25	166·6

From the *Guide, Section C1*: at atmospheric pressure of 101·325 kPa

Psychrometric Chart—The relationships between the various properties of a mixture of air and water vapour may be expressed in the form of Tables, such as those published in the *Guide, Section C1*, or the chart based thereon, as in Fig. 18.6.*

The method of using the chart is indicated by the diagram, Fig. 18.7.

* Copies of the chart in pad form are obtainable from C.I.B.S., price £2.50. Charts for non-standard pressures, computer drawn, are published by Troup Publications Ltd.

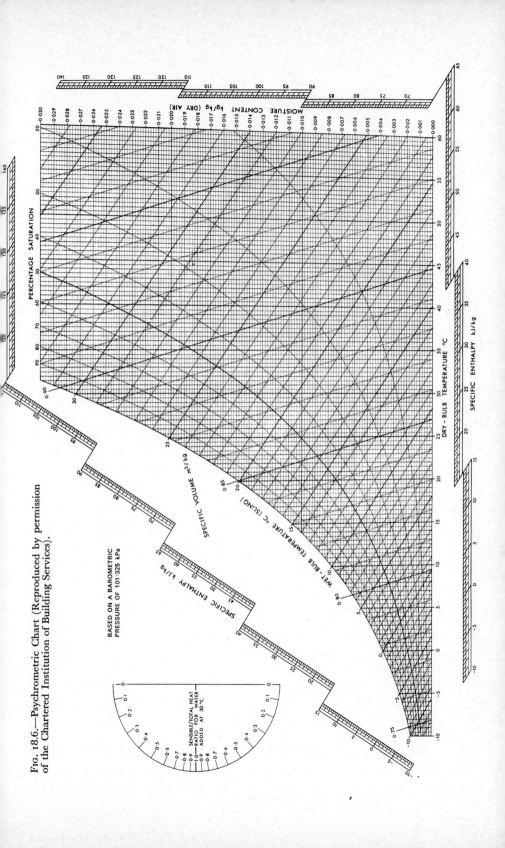

Fig. 18.6.—Psychrometric Chart (Reproduced by permission of the Chartered Institution of Building Services).

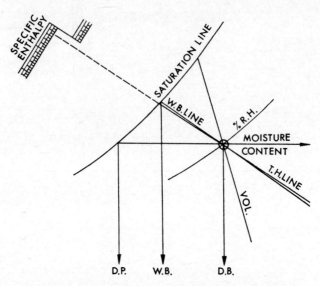

FIG. 18.7.—Example of the use of the Psychrometric Chart.

Point marked is 20° C D.B., 50% Sat. Other data
read from the chart are:

W.B.	= 13·9° C
Dew Point	= 9·4° C
Moisture content	= 0·0074 kg/kg dry air
Specific Enthalpy	= 38·8 kJ/kg
Specific volume	= 0·84 m³/kg dry air

The chart is constructed as follows:

Dry Bulb—Evenly divided horizontal scale with divisions of 0·5° C. Read vertically to the required point.

Moisture Content—Evenly divided vertical scale reading in kg per kg dry air. Read horizontally.

Percentage Saturation—Indicated by the curves for each 10 per cent. Intermediate percentages can be otained by interpolation.

Wet Bulb—Sloping lines for each 1° C.

Specific Enthalpy or Total Heat—Obtained by drawing a line through the required point parallel to the nearest T.H. line, and reading onto the inclined T.H. scale.

Dew-point—This depends only on moisture content. Hence, through the required point draw a horizontal line to cut the saturation line. From the intersection read vertically onto the D.B. scale.

Volume—Sloping lines of equal volume are given in m³/kg dry air.

The chart and Table 18.6 have been used in the *example* below.

Unit for Cooling Loads—The cooling load was expressed in the past in terms of the 'ton of refrigeration', which was the heat required to melt one

U.S. ton of ice (i.e. 2000 lb.) at 0° C per twenty-four hours (latent heat of fusion of ice = 144 Btu/lb.). Thus, (2000 × 144)/24 = 12 000 Btu/hour. In SI units, cooling loads are expressed simply in kW: (1 ton refrigeration = nearly 3·5 kW).

DESIGN OF PLANT FOR SUMMER COOLING AND DEHUMIDIFICATION

Having arrived at the maximum hourly heat gain for the spaces to be conditioned, it is necessary to calculate the conditions to be maintained in the plant, and the capacity of the various components, fans, cooling coils, heater batteries, water chillers etc.

Method—The following is a summary of the steps in the process which have to be calculated:

(a) Heat gains and solar gains through the structure.
(b) Infiltration gain.
(c) Internal heat gains.
(d) The mass flow of ventilation air and hence the cooling for same.
(e) The mass flow of air required for a given temperature rise, inlet to outlet; this is derived from the total sensible gains (a), (b) and (c).
(f) The recirculation mass flow, which equals (e) minus (d), and from whence the total heat of the mixture is found.
(g) The plant exit condition D.B., which is derived from room D.B. minus rise in room minus allowance for duct and plant gain. The moisture at plant exit derives from the moisture at room condition minus the increment due to occupants and infiltration.
(h) The apparatus dew-point, which follows from (g).
(j) The cooling load, which is equal to the product: mass air flow through plant and heat removed as between T.H. of entering mixture and T.H. at dew-point.
(k) The reheat load, refrigeration load and fan duties.

Example—

Basis of Design

Building	= Public Hall
Seating capacity	= 500
Lighting	= 20 kW
Cube	= 3000 m³
Internal conditions summer	= 23° C D.B.
50% Saturation	= 16·4° C W.B.
T.H.	= 45·79 kJ/kg
External conditions summer	= 30° C D.B.
	20° C W.B.
T.H.	= 56·71 kJ/kg
Outside air per occupant	= 0·005 m³/s

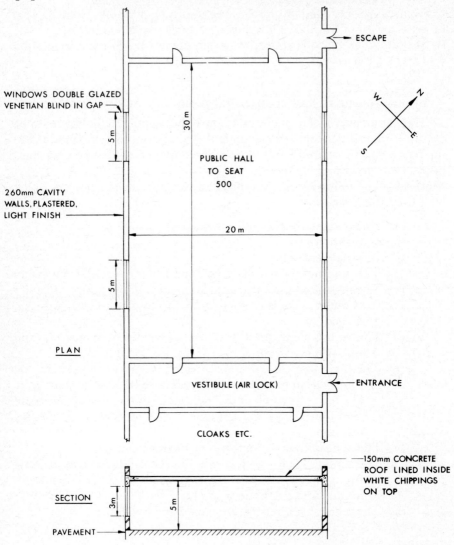

FIG. 18.8.—Plan and section of building used in Example.

Fig. 18.8 gives a plan and section of the building, its construction and orientation.

The orientation of the building presents two faces to solar gain.

From Table 18.4 it is clear that the SW glass will be subject to the greatest heat gain at 1600 hours in August and September.

Cooling loads taken for August may be arrived at as follows:

Glazing. (Double glazing; Venetian blind in gap).

Area NE = 30 m²
Area SW = 30 m²

From Table 18.4 (internal construction taken as 'light')

for NE glazing   45 W/m²
for SW glazing   312 W/m²

| Cooling load | 30×45 | = | 1350 W |
| | 30×312 | = | 9360 W |

Allow reflected radiation from pavement, SW
    side at 10%   =   936 W

                                          11 646 W

Shade factor (Table 18.5(b))   =   0·6   7·0 kW

Air to Air. Take U   =   2·8 W/m² K
    Temp. diff.   7 K
    Area   60 m²
               $60 \times 7 \times 2\cdot8$ =   1·2 kW

                                           8·2 kW

Walls

Construction: 260 mm cavity, lined. Surface light, U = 1·0 W/m² K.
From Table 18.2, time lag will be about 8½ hours and decrement factor f = 0·3.
Time considered for calculation is 1600 hrs.; so, allowing for time lag, we read sun heat increment 8½ hours earlier (say 0800 hrs.).
Thus

NE Wall (from Table 18.1), increment   =   9·7 K
Area   = (30×5) – windows   =   120 m²

$$\frac{120 \times 1\cdot0 \times 0\cdot3}{1000}\{(30-23)+9\cdot7\} \qquad = 0\cdot6\,kW$$

SW Wall (from Table 18.1), increment   =   1·2 K
Area, as before,   =   120 m²

$$\frac{120 \times 1\cdot0 \times 0\cdot3}{1000}\{(30-23)+1\cdot2\} \qquad = 0\cdot3\,kW$$

                                           0·9 kW

Roof

Construction: 150 mm concrete, lined internally.
Surface light, U = 1·7 W/m² K
From Table 18.2, time lag will be about 5½ hours and decrement factor f = 0·5.
Time considered for calculation is 1600 hrs.; so, allowing for time lag, we read sun heat increment 5½ hours earlier (say 1030 hrs.).
Thus, from Table 18.1, interpolating, increment   =   7·9 K
Area = (30×20) = 600 m²

$$\frac{600 \times 1\cdot7 \times 0\cdot5}{1000}\{(30-23)+7\cdot9\} \qquad = 7\cdot6\,kW$$

Infiltration

Allow half airchange per hour
Cube = 3000 m³
Specific volume dry air at room condition 0·85 m³/kg

$$\frac{0·5 \times 3000}{0·85} = 1765 \text{ kg/hr}$$

At 30°C   T.H. dry air external   30·18
At 23°C   T.H. dry air internal   23·14

difference   7·04 kJ/kg

$$\frac{1765 \times 7·04}{3600}$$ = 3·5 kW

Gain due to moisture difference of T.H.

TH, 30°C DB 20°WB.
TH 23°C DB 50% RH. $(56·71 - 45·79)$ = 10·92

$$\frac{1765 \times (10·92 - 7·04)}{3600}$$ = 1·9 kW

Internal Heat Gains

500 occupants @ 88 W sensible (from Fig. 18.4)   = 44 kW
20 kW lighting   = 20 kW   64 kW

Gain due to moisture
500 occupants @ 30 W latent (from Fig. 18.4)   =   15 kW

(*Note*: Heat gains due to occupants assume they are seated at rest. If for
instance they were dancing, the heat gain would increase, but the
number might be less.)

Total heat gains sensible
8·2 + 0·9 + 7·6 + 3·5 + 64   =   84·2 kW

Total gains due to moisture
1·90 + 15·00   =   16·9 kW

Ventilation Air

Outside air 500 occupants @ 0·005 m³/s   = 2·5 m³/s

$$\frac{2·5}{0·85}$$ = 2·94 kg/s

Sensible heat gain, by difference, as for infiltration
2·94 (30·18 − 23·14)   = 20·7 kW

Gain due to moisture, as for infiltration
2·94 (10·92 − 7·04)   = 11·4 kW

Plate XXVII. Variable volume system serving troffers at luminaires (see p. 386)

Plate XXVIII. Dual duct blender terminal units below windows (see p. 388)

Plate XXIX. Plant for dual duct system
(see p. 388)

Plate XXX. Induction unit below window
within timber casing (see p. 391)

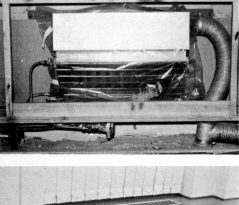

Plate XXXI. Heat recovery unit below
window in rehabilitated building (see
p. 395)

Total Load

It is of interest to consider the relation of the components of the total load for, whilst no building can be said to be 'typical', the significance of the internal gains and ventilation gains is worth noting as is the ratio between the sensible gains and latent gains.

Sensible gains	%	%	%
Solar			
Glazing	5·2		
Walls	0·3		
Roof	3·1	8·6	
Air-to-Air			
Glazing	0·9		
Walls	0·4		
Roof	2·7	4·0	
Internal			
Lighting	15·0		
Occupancy	33·0	48·0	
Ventilation			
Infiltration	2·6		
Outside Air Supply	15·5	18·1	78·7*
Latent Gains			
Internal			
Occupancy		11·3	
Ventilation			
Infiltration	1·4		
Outside Air Supply	8·6	10·0	21·3
			100·0

Air Quantity

Permissible temp. rise between room
inlet and outlet, assume   = 8 K
Temp. of inlet to room is thus   = $23 - 8$ = $15°$ C
Specific heat capacity of air take as 1·0 kJ/kg

Then air required $= \dfrac{84·2}{8 \times 1}$ = 10·53 kg/s

Specific volume,
dry air at 15° C   = 0·815 m³/kg
Fan duty   = 10·53 × 0·815   = 8·58 m³/s (8580 litre/s)

Air change rate $= \dfrac{8·58 \times 3600}{3000}$ = 10·3 per hour

* This total does not include any reheat, fan, pump or other plant gains which may be quite significant.

Of the mass of air provided, 10·53 kg/s, 3·0 is from outside (28%), 7·53 is recirculated (72%)

T.H. of mixture

$$\begin{aligned} \text{Room} &= 45\text{·}79 \times 7\text{·}53 = 345 \\ \text{Outside} &= 56\text{·}71 \times 3\text{·}0 = 170 \end{aligned}$$

$$515 \div 10\text{·}53 = 48\text{·}9 \text{ kJ/kg}$$

Plant Exit Condition

This will differ from the room inlet by heat gain through the duct walls (thermally insulated to reduce it as much as possible) and also by heat gain from the energy input by the fan. These may be assessed in detail but for present purposes will be taken as 12½ per cent of sensible gain, thus accounting for 1 K rise.

Hence:

Plant exit condition = 15 − 1   = 14° C D.B.
Moisture at room condition (from Fig. 18.6) = 0·0089 kg/kg
Latent load = 16·9 kW and taking air
quantity of 10·53 kg/s with latent
heat at 2450 kJ/kg, moisture
increment

$$= \frac{16\text{·}9}{10\text{·}53 \times 2450} = 0\text{·}0007 \text{ kg/kg}$$

Thus air must leave plant at   0·0082 kg/kg
which, at 14° C D.B. (from Fig. 18.6), will be
83% saturation, T.H. = 35·1 kJ/kg and   = 12·4° C W.B.

Apparatus Dew-point

Moisture content of 0·0082 kg/kg corresponds to a
dew-point of 11·2° C and the corresponding T.H.   = 32·2 kJ/kg

Cooling Load

Mass flow air through plant   = 10·53 kg/s
    T.H. mixture   48·9 kJ/kg
    T.H. dew-point   32·2

    difference   = 16·7

$$10\text{·}53 \times 16\text{·}7 \qquad\qquad = 176 \text{ kW}$$

Washer data

Outlet D.P.   11·2° C
Differential, water outlet to D.P., say   2·0

                                       9·2° C
Allow temp. rise in washer, say   3·0

Chilled water inlet temp.   6·2° C

Water mass flow through sprays
(specific mass, 4·2 kJ/kg K)

$$\frac{176}{3\cdot0\times4\cdot2} = 14\ \text{kg/s} \qquad = 14\ \text{litre/s}$$

Chilled water from refrigerating plant (allowing 10% heat gain in pump and piping) say 5·7° C.

Reheat Load

T.H. at plant outlet condition	= 35·1 kJ/kg
T.H. at apparatus dew-point	= 32·2
difference	= 2·9 kJ/kg

Air mass flow   = 10·53 kg/s
Heat input 10·53 × 2·9   = 30·5 kW

Refrigeration Load

Cooling load as above	176 kW
Heat gain in pump and piping, say 10%	18
Compressor duty	= 194 kW

Fan Duties

Inlet fan as before   = 8·58 m³/s (8580 litre/s)

Exhaust/Recirculation fan
  Exhaust (equal to outside air)
    3 × 0·85   = 2·55 m³/s (2550 litre/s)
  Recirculation
    7·53 × 0·85   = 6·4 m³/s (6400 litre/s)

Total Exhaust/Recirculation   = 8·95 m³/s (8950 litre/s)

(The increase over inlet is due to the rise in temperature. In practice, some excess of inlet over outlet would be allowed to reduce infiltration.)

Fig. 18.9 shows, graphically, the psychrometric changes taking place in the above example on a small area of the chart. Fig. 18.10 illustrates the conditions and other data on a plant diagram.

Cooling by Cold Coils—Cooling by this method has already been described (see Fig. 17.2).

To avoid cooling all the air down to a low dew-point for dehumidification and subsequently reheating, advantage may be taken of a characteristic of coil surfaces whereby moisture can be deposited on the chilled surface, without the bulk of the air being reduced to the same temperature. This is achieved by keeping the surfaces at the lowest possible temperature consistent with absence of freezing.

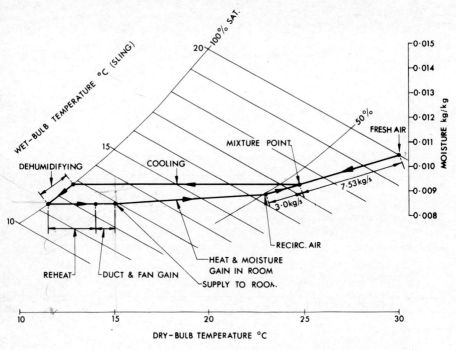

FIG. 18.9.—Diagram of psychrometric changes (see Example).

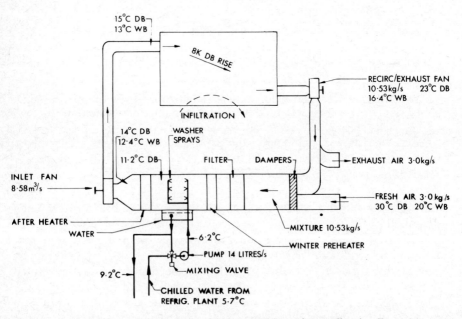

FIG. 18.10.—Air-conditioning plant showing conditions when cooling (see Example).

In the example above, if chilled water coil surface be used instead of an air washer, the calculation of refrigeration load would be as follows:

Mixed air inlet to coil
   3 kg/s at 30° C   = 90
   7·53 kg/s at 23° C   = 173·2

Similarly   263·2 ÷ 10·53 = 25° C D.B.
   T.H.   = 48·9 kJ/kg
   D.P.   = 12·8° C

Outlet from coil (as from washer)
   D.B.   = 14° C
   W.B.   = 12·4° C
   T.H.   = 35·1 kJ/kg
   Diff. in
      T.H. = (48·9 − 35·1)   = 13·8 kJ/kg

Mass flow   = 10·53 kg/s

Cooling load = 10·53 × 13·8   = 145 kW
   Heat gain in pump and piping 10% =   15 kW

Compressor duty   = 160 kW

This compares with 194 kW with the washer system.

Coils cooled by direct expansion would be calculated similarly, except that the surface temperature would be lower.

Cooling by Sprayed Coils—In this system unchilled and recirculated water is sprayed over the chilled coil surface, thus combining both the previous methods. The water serves to flush the coils of dirt accumulations but adds to resistance to air flow. The balance between the respective duties performed by the coil surface and the water spray are inherent in a particular design, depending on coil surface area, fin spacing and water quantity. It has the merit of compactness compared with a washer, and provides means for winter humidification which cooling coils alone do not.

DESIGN OF PLANT FOR WINTER WARMING AND HUMIDIFYING

The operation in winter may be with partial recirculation, as in summer, or with 100 per cent fresh air. The plant might be arranged so as to be suitable for the latter method of operation, in average spring and autumn weather, when neither cooling nor heating is required. In cold weather recirculation might be used for economy of running.

The following calculation assumes 100 per cent fresh air at 4° C, below which it is assumed that recirculation would be used. This is often done by an outside control operating motorized dampers to vary the proportion of fresh to recirculated air.

Example—

Basis of Design

Fresh air	4° C sat.	T.H.	16·7 kJ/kg
Room (fabric losses offset by direct heating:			
room empty)	20° C 60% sat.	T.H.	42·6 kJ/kg
D.P. of room condition = 12° C		T.H.	34·2 kJ/kg

Mass flow of air as for cooling $\qquad$ = $\qquad$ 10·53 kg/s

Plant duty

Pre-heater or spray-water heater

$(34·2 - 16·7) \times 10·53$ $\qquad$ = $\qquad$ 184 kW

Note: The washer performs adiabatically; thus T.H. at entry
is the same as at leaving, following a constant wet bulb
line of 12° C.

After-heater

$(42·6 - 34·2) \times 10·53$ $\qquad$ = $\qquad$ 88 kW

$\qquad\qquad\qquad\qquad\qquad\qquad\qquad\qquad\qquad\qquad$ 272 kW

Moisture added

Room condition	0·0088 kg/kg
Outside ,,	0·0050 ,,
	0·0038 ,,

$10·53 \times 0·0038 = 0·04$ kg/s $\qquad$ = 144 litre/hr

(The above assumes the air washer 100% efficient, i.e. saturating the air at the dew-point. Some allowance for inefficiency would in practice be made depending on the form of washer.)

Steam Humidification—Where cooling is by cooling coils, humidity may be increased by direct injection of vapour into the air stream.

Where some slight trace of odour is objectionable, the vapour should be generated indirectly by electric immersion heater, hot water or steam coil; steam direct from a steam boiler is liable to carry smell. Steam in contact with hot copper or steel acquires smell, but tinned copper seems to leave vapour odourless. The method of injection may be by one of the proprietary perforated pipes with the steam jacket designed to pass only dry steam. The point of injection should be after the cooling and heating coils. In the example above, the element generating vapour would have an input (taking latent heat at 2500) of

$0·04 \times 2500$ $\qquad\qquad\qquad\qquad\qquad$ = 100 kW

Air heater.   Temp. rise 16 K
Specific heat capacity of air   = 1 kJ/kg K
Mass flow 10·53 kg/s

$10·53 \times 16 \times 1$ $\qquad\qquad\qquad\qquad\qquad$ = 169 kW

Total heat requirement $\qquad\qquad\qquad\qquad$ 269 kW

This compares with 272 kW in the previous example, the difference being due to roundings up.

BASIS OF CALCULATIONS APPLYING TO OTHER SYSTEMS

Single Duct—Terminal Reheat Systems—The central plant is designed to provide full conditioned air quantities to all areas served, to meet peak demands. No allowance for diversity of load can be made. The terminal reheat equipment is designed to cater for local differences between maximum and minimum load conditions.

Whilst this system has the advantage of offering good temperature control, it is inherently wasteful of energy. Low humidity conditions may arise.

Fan Coil Systems, ducted outside air—These are designed on similar lines to those applied to induction systems in cases where the fresh air is delivered to the room space via the fan coil unit.

Where fresh air is ducted independently, it may be supplied year-round at constant temperature at, or a degree or two less than, room condition. In this case the fan coil unit which should be three of four pipe will require to be designed to deal with the whole of the heat gains or losses occurring in the space served.

Fan Coil Systems, local outside air—Such systems which are, in effect, no more than individual room heater/coolers must be arranged to deal with the whole conditioning load, outside air being provided either directly to the units or via opening windows etc. No direct control of room humidity can be achieved although de-humidification will occur.

Induction System—The primary air supply, conditioned at the central plant, is designed to provide an adequate volume for ventilation purposes and humidity control. Temperature is varied to suit external weather conditions and may be further adjusted to take account of solar gain.

The coil or coils in the room units are designed to deal with local sensible heat gains or losses. Chilled water temperature should be selected to avoid condensation on the coil.

Double Duct Systems—The central plant is designed to provide a supply of both cool and warm air which are distributed in parallel to individual terminal units. The cool air duct provides an air quantity adequate in volume and temperature to meet the maximum anticipated load of heat gain to the building fabric and from the outside air supply.

The warm air duct provides a supply adequate in temperature to meet the design heat loss, the volume usually being allowed as 75 per cent of that in the cool duct.

Variable Volume Systems—In the case of the true variable volume system, the air quantity provided by the central plant will be limited to that required to meet the maximum co-incident load. It will be reduced from that design volume for all part load conditions. Supply temperature will be constant and determined by the peak cooling load.

For the alternative approach, where air volume varies only at the point of supply to the conditioned space, the surplus being by-passed directly or indirectly to the return air system, the central plant will operate at constant volume.

Normal practice suggests that air supply to individual rooms should not be reduced by more than 50 per cent and, if minimum load is less than this, some form of temperature adjustment will be necessary, possibly at the central plant but otherwise by local or zonal re-heat.

AUTOMATIC CONTROLS

Automatic controls are an essential part of any air-conditioning system. The wide variety of purposes for which such plants are required, the number of different systems which it is possible to select, and the multitude of types of control equipment available require a complete book in themselves to cover adequately.

It is possible here only to indicate the main principles involved, some of the chief component items of apparatus commonly used being described elsewhere (Chapter 21). From this it will be possible to understand some of the problems concerned and how they may be tackled. Other permutations and combinations of system and equipment may be built up to suit any variety of circumstances.

Seven main types of plant will be considered and reference to the diagrams will be necessary along with the following descriptions.

Central Plant System with Air Washer—This is the type of plant envisaged in the basic cooling and heating calculations taken earlier, as illustrated in Fig. 18.11.

Dew-point
1. In summer thermostat T_1, which is a wet-bulb type, will operate damper motors M_1, M_2 and M_3 such that minimum outside air is admitted to the plant. Dew-point is controlled by thermostat T_2 operating mixing valve V_1 to provide an appropriate blend of chilled and tank return water to the washer sprays.
2. In mid-season, when thermostat T_1 senses an outside wet bulb temperature less than that required in the conditioned space, damper motors M_1, M_2 and M_3 will be operated such that maximum outside air is admitted, thus taking advantage of available 'free cooling'. Dew-point will continue to be controlled by thermostat T_2 operating mixing valve V_1 as for summer.
3. In winter, with outside wet bulb temperature well below that required in the conditioned space, mixing valve V_1 will be in the fully recirculating position and dew-point control, by thermostat

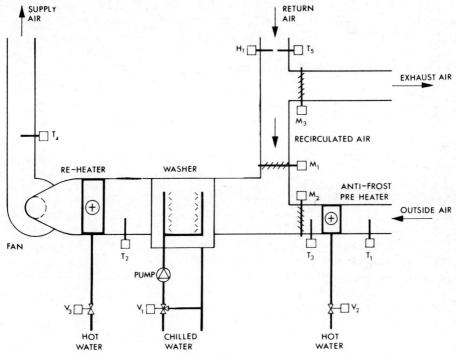

FIG. 18.11.—Central plant air conditioning with air washer.

T_2, will be effected by operation of damper motors M_1, M_2 and M_3 to blend outside and recirculated air. An anti-frost pre-heater in the outside air supply will have a thermostat T_3 operating valve V_2 to maintain a temperature of say 4° C.

4. In order to provide for some variation in latent load in the conditioned space and thus for adjustment of dew-point, humidistat H_1 placed either in the space or in the return air duct, may be arranged to re-set thermostat T_2.

Note. This control arrangement could equally well be applied where a cooling coil is used in lieu of an air washer, valve V_1 controlling the chilled water supply to the coil.

Dry-bulb
1. The final dry-bulb temperature leaving the plant will be controlled by thermostat T_4 operating valve V_3 to admit heat (hot water or steam) to the re-heater.
2. In order to provide for some variation in sensible load in the conditioned space and thus for adjustment of dry-bulb temperature leaving the plant, thermostat T_5 placed either in the space or in the return air duct may be arranged to re-set thermostat T_4.

Note. In order to avoid wasteful use of cooling and re-heating, in the
event of prolonged periods of less than peak load, a by-pass may be introduced around the washer as shown in Fig. 18.12. In this case thermostat
T_4 (re-set by thermostat T_5) would control damper motors M_1 and M_4 to
divert a suitable proportion of recirculated air.

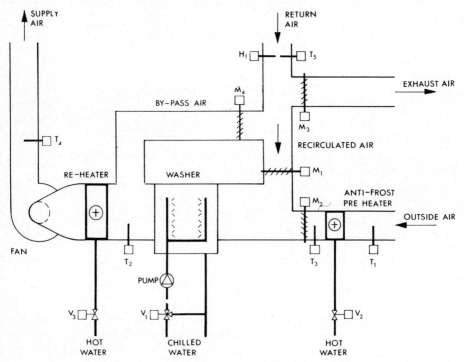

FIG. 18.12.—Central plant air conditioning, cooler by-pass.

Central Plant System with Cooling Coil

—This type of plant is con-
trolled in a manner similar to that used where an air washer is incorporated.
For example, however, consider the arrangement shown in Fig. 18.13
where no recirculation of air is permitted as in the case of a hospital.

Dew-point

1. In summer, dew-point is controlled by thermostat T_2 operating
   valve V_1 to admit chilled water to the cooling coil. The spray
   pump may be run in order to improve heat transfer or may be idle.
2. In mid-season, when humidification is required, the spray pump
   will be running and with decreasing outside wet-bulb temperature,
   thermostat T_2 will close valve V_1 such that the spray will provide
   adiabatic cooling with no requirement for a chilled water supply.
3. In winter, thermostat T_2 will act to control V_2 and admit heat (hot
   water or steam) to the pre-heater and V_1 also as may be required.

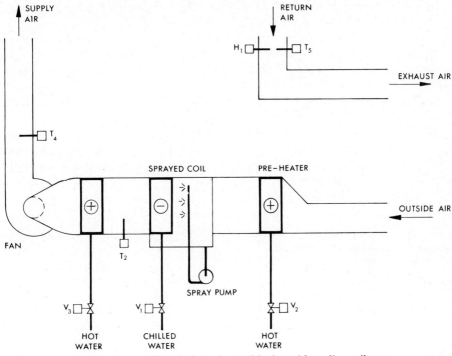

FIG. 18.13.—Central plant air conditioning with cooling coil.

4. As before, humidistat H_1 will be arranged to re-set thermostat T_2 in order to provide for variation in latent load in the conditioned space.

Dry bulb

1. The final dry bulb temperature leaving the plant will be controlled, as before, by thermostats T_4 and T_5 acting together to admit heat (hot water or steam) to the re-heater.
2. In cases where the relative humidity in the room is unimportant, economy in operation may be achieved by arranging for thermostat T_5 to control valves V_1 and V_2 in sequence.

Zoned System—In this case the central plant delivers air at a constant outlet condition and the zones are controlled individually.

Terminal Re-heat System—In this case, all or the bulk of the required re-heating to achieve control of dry-bulb temperature is transferred from the central plant to individual rooms within the conditioned space. Thus, the output of the central plant is controlled to meet the peak cooling load likely to occur at any one time in any individual room.

Within each room, a thermostat will be arranged to control admission of heat to the re-heater incorporated in the terminal unit. This re-heater would usually take the form of a hot water coil provided with a control valve but electric resistance heaters are sometimes used.

If room load variations from the peak are small, a conventional central plant re-heater may be used to deal with the common re-heat load. In such cases, the temperature of the conditioned air delivered may be such that it can be discharged into all rooms without creating problems and re-heat may thus be provided by a perimeter heating system fitted with suitably responsive room-by-room controls.

Fan Coil Systems—The principal difference between those systems previously described and a fan-coil arrangement is that where a ducted air supply is provided, this handles outside air only and not the full conditioned air quantity.

Where fan-coil units are large and are, in effect, themselves recirculating air-conditioning plants serving significant building zones, as shown in Fig. 18.14, the primary air supply from the central plant will be con-

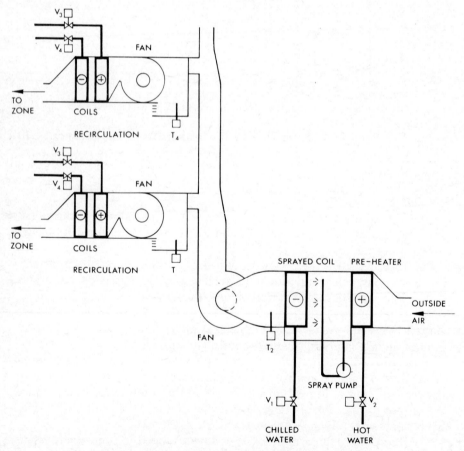

FIG. 18.14.—Zoned air conditioning with large fan coil units.

trolled at constant condition year-round. Controls will thus be as described for a central plant system, the re-heater being omitted. Each zone
plant will be provided with cooler and re-heater coils fitted with control
valves and these will be operated separately or, more usually, in sequence
by a thermostat in the fan-coil outlet. This may be re-set as before by a
further thermostat fitted in either the conditioned space or the return air
duct.

Fan-coil units of smaller size where fitted in individual rooms, will be
served by a central system controlled much as described for induction
units in a later paragraph. Individual units may be controlled by room
thermostats operating a valve or valves in the water connections which

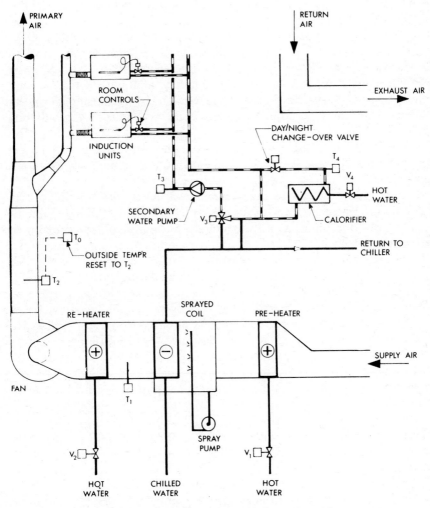

FIG. 18.15.—Induction system. Principles of control.

may be two-three-or four-pipe. Manual on/off control to the fan is often provided.

Induction Systems—There are two principal methods of control applied to induction systems, the 'changeover' and the 'non-changeover'. The former is more appropriate to climates having distinct seasons and need not concern us here. Two-, three- and four-pipe units are available to provide progressively better availability for control of space conditions: the two-pipe system is the more common for use in Great Britain.

The arrangement shown in Fig. 18.15 illustrates the controls of the non-changeover type having two-pipe water connections. Water is kept constant at say 10° C both in summer and winter by means of the mixing valve in the 'secondary' water circuit as shown. The water is cooled by a water-chiller in summer, but in winter by the entering fresh air. The whole of the chilled water is circulated through the cooling coils without control, thus ensuring that in summer the dew-point (about 7° C) of the primary air to the units is below the temperature of the coils in the room units, in order to prevent the latter from gathering condensation.

The primary air temperature is maintained at about 16° C in summer but in winter is varied according to the weather, by the outside thermostat shown, which may be adjusted to take some account of solar radiation, up to about 60° C.

To maintain warming during winter nights without the fan running, the calorifier is brought into service allowing the units to act as natural convectors.

Each induction unit in the system shown has a direct-acting control valve in the water circuit with the actuating element in the induction air-stream from the room. Thus, in summer, assuming no heat gains in the room, the valve will be closed, the air only maintaining room temperature. When sun or internal heat gains cause the room temperature to rise, the valve will open and the coil will lower the circulating air temperature. In winter the primary air will warm the room, but, should the temperature mount too high, the control valve will open to admit cool water, so correcting the temperature. To avoid variation in pressure, should a large number of control valves close, a constant pressure differential control valve is included as a by-pass, as shown.

An alternative arrangement provides local control by means of a damper arrangement within the unit which adjusts the direction of the induced air such that part or all is diverted past the cooled coil. This damper may be controlled manually or automatically by thermostat. Some such units make use of the pressure of the primary air to operate the damper but others rely upon an independent supply of compressed air to actuate the controller.

Moisture content is controlled in summer by the chilled coils and in winter by the humidifier, to a constant dew-point. Alternatively, a sprayed coil system is used, controlled in the same manner.

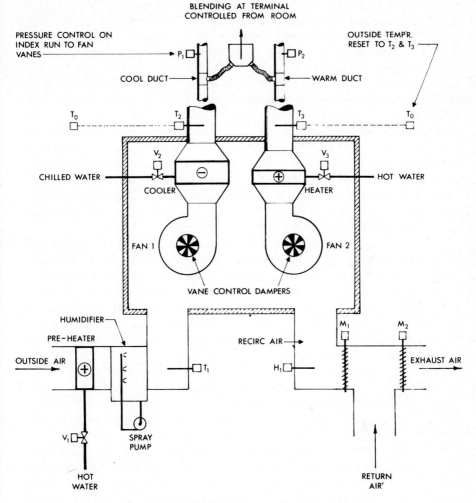

FIG. 18.16.—Dual-duct system. Principles of control.

Dual-Duct System—Fig. 18.16 illustrates the principle of the control system applying to a dual-duct system where two fans are included, one for the warm duct and one for the cool.

The temperature of the cool duct is varied according to the external temperature from say 7° C in summer to say 16° C in winter. In summer it is cooled by the chilled water coil: in winter the heater comes into use as necessary. The temperature of the warm duct is also varied according to the outside temperature, from say 21° C in summer to say 38° C in winter.

It will be noted that, in the arrangement shown, the return air is brought into the fan suction-chamber on the side nearest the warm duct, the return air being normally nearer to the warm duct condition.

Other controls indicated are concerned with maintaining air pressures within certain limits by damper operation. Dehumidification in summer is performed by the cooling coil, without specific control, and, in winter, humidity is increased as necessary by the sprays or steam jets controlled from return air condition.

The blender units are controlled from the rooms individually, or in groups thermostatically, by adjustment of an internal mixing damper or dampers.

Variable Volume System—A true variable volume system, in its simplest form, will have a central plant controlled in most respects in a manner similar to that first described here. Particular attention will be paid however, in the interests of energy saving, to control the fan operation such that volume variation at terminal units is reflected by volume reduction at the central plant. This aspect of control is dealt with in Chapter 19 (page 444).

In the simple form of the true system, control of space conditions and in particular those in internal zones of deep plan buildings is achieved by variation in the quantity of air supplied rather than by changes in temperature. This control, by room thermostat operation upon electrical or pneumatic devices, acts to reduce the volume of conditioned air delivered either to zone or individual room at less than design peak load. Such devices may be integral to the actual terminal diffusers or incorporated in regulators serving a number of such diffusers.

Good practice suggests that the volume of air supplied to meet the designed full load cannot be reduced under control by more than about 50 per cent if room distribution difficulties are to be avoided. In consequence, control by terminal re-heat may be necessary should local demand for cooling fall below that level: hence, room or zone terminals may be supplied from a dual-duct system or be themselves provided with re-heat batteries supplied by hot water or, rarely, with electrical resistance heaters. For perimeter zones, the necessary re-heat may be controlled via any form of convective heating system there installed if this can be arranged to provide adequate response.

Systems which are controlled to vary the volume of air delivered to rooms or zones whilst the central plant volume remains constant do not offer the same facilities for energy conservation. In principle, such systems are controlled such that some varying proportion of the designed peak-load air quantity is diverted from the terminal diffuser to be returned directly or via a ceiling void to the central plant. One such system uses what is called 'fluidic' control to avoid the possibility of air distribution problems in the conditioned space by delivering air at full volume in pulses, the duration of which varies according to the reduction from full load. The controllers are self operated by small pressure differences arising from the flow diversions.

Controls Generally—This brief precis of the control sequences which may be applied to air conditioning plant does no more than scratch the surface of the subject. There are ten times as many methods of control as there are systems and, as was emphasised in the preamble to this subject, a whole book would be needed to cover the combinations adequately. The reader is directed to the *Guide, Section B11* for information upon the fundamentals and to Chapter 19 (page 463) for a brief description of some of the components commonly used.

Fans and Air Treatment Equipment

THE FAN IS THE one item of equipment which every mechanical ventilation and air-conditioning system has in common. A fan is simply a device for impelling air through the ducts or channels and other resistances forming part of the distribution system. It takes the form of a series of blades attached to a shaft rotated by a motor or other source of power. The blades are either in the plane of a disc (propeller fan) or in the form of a drum (centrifugal fan).

There is as yet no other practicable commercial method of moving air for ventilation purposes, but fans in general suffer from various disadvantages such as low efficiency and noise. It is the latter which is probably the most troublesome to designers, and to which much careful attention must be given if silent running is to be achieved in the system as a whole: factors involved are air speeds, fan speeds, duct design, materials of construction, acoustical treatment of ducts and provision for absorption of vibration.

FAN TYPES AND PERFORMANCE

Static Pressure—The purpose of any fan is to move air. When air is moved in a duct or through a filtering, heating, cooling or washing plant, a resistance to flow is set up.

The air is slightly compressed by the fan on its outlet side, so setting up a *static pressure* in the duct or plant. This pressure is tending to 'burst the duct', and may be read by means of a U-tube partly filled with water, connected at right angles to the air stream at any point in the duct. It is called a 'side gauge' (Fig. 19.1 (a)).

On the suction side of the fan the static pressure is negative with respect to the surrounding atmosphere, tending to collapse the duct.

As the air proceeds along the duct from the fan its compression is gradually released until at the end of the duct open to atmosphere the air is at atmospheric pressure. This falling away of the static pressure proportionately with the length of travel is called the *resistance* of the duct. Similarly all obstructions, such as heaters, filters, etc., cause a loss of pressure when air is passing through them.

It should be noted that as the static pressure becomes reduced, the air in effect expands such that pressure × volume = a constant (or nearly so, as explained earlier). This expansion therefore signifies an increase in velocity of the air if the size of the duct is unchanged.

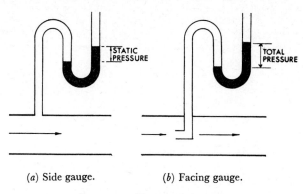

(a) Side gauge.   (b) Facing gauge.

FIG. 19.1.—Air pressure gauges.

Velocity Pressure—The fan, in addition to generating static pressure, supplies the force to accelerate the air and give it velocity. This force is termed the *velocity pressure*, and is proportional to the square of the velocity. It is measured by a U-tube connected to a pipe facing the direction of air flow in a duct, etc. It is called a 'facing gauge' (see Fig. 19.1 (b)). But, obviously, the pressure so measured will in addition include the static pressure which occurs throughout the duct, as mentioned earlier, and the reading so obtained will thus be the *total* pressure. Thus, the velocity pressure alone may be found by deducting the static pressure from the total pressure reading, or by connecting one side of the U-tube to the facing gauge and the other to the side gauge provided the two gauges are at the same point. A Pitot tube, as illustrated in Fig. 19.5, combines a side gauge with a facing gauge in one standard instrument.

If a fan discharges into an expanding duct (Fig. 19.2), the velocity will obviously decrease as the distance from the fan increases, and at the same time the velocity pressure will be converted into static pressure (not at 100 per cent efficiency, but about 75 per cent if the expansion is sufficiently

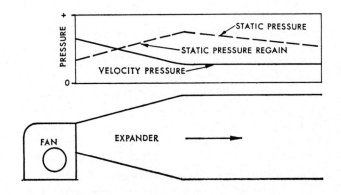

FIG. 19.2.—Velocity and pressure changes to expander.

gradual). If the fan discharges into a large box (Fig. 19.3), from which at some point a duct connects, the fan velocity pressure will be entirely lost in eddies, and at the duct entrance must be recreated by a corresponding reduction in static pressure.

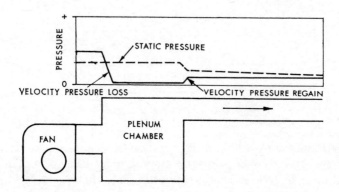

FIG. 19.3.—Velocity and pressure changes to plenum box.

Total Pressure—The sum of the static and velocity pressure is called the *total pressure.*

Fan Pressure—In all air flow considerations as affecting resistances of ducts, plant, etc., it is the static pressure alone which concerns us, as this is the pressure which changes with such restrictions. It is the static pressure set up by a fan which is, therefore, a criterion of its performance. The velocity pressure, if taken as supplementing the fan duty, may be more misleading than useful, owing to the uncertainty of friction losses which occur at points of varying velocities. The velocity pressure is more generally not recovered, though sufficient must remain at the duct termination to eject the air at the required speed.

Where, however, by careful design of the fan discharge expander, the velocity pressure is converted to static pressure (probably to the extent of about 75 per cent), this additional pressure may be reckoned as augmenting the static pressure of the fan.

The pressure generated by a fan may be better understood by study of Fig. 19.4.

    1. The total fan pressure is defined as the algebraic difference between the mean total pressure at the fan outlet and the mean total pressure at the fan inlet.

    2. The total pressure on the suction side, as will be seen from this Figure, is TP_S, i.e. the negative pressure AO minus the velocity pressure equivalent to AB.

    3. The total pressure TP_D on the discharge side is similarly the static pressure OC plus the velocity pressure CD.

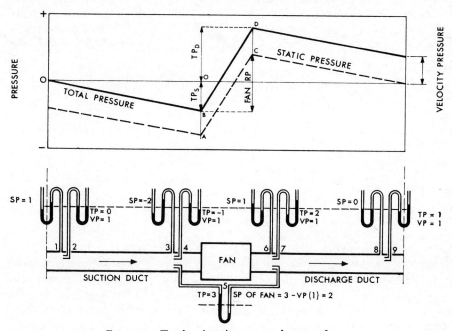

FIG. 19.4.—Total and static pressure change at fan.

It has previously been explained that we are concerned only with the resistance pressure set up by the fan and this, as may be seen, is the difference in pressure of points B and C.

We can arrive at the static pressure by measuring the total pressure of the fan as by the U-tube 5, and deducting therefrom the velocity pressure as given by difference between gauges 6 and 7. Such a method is valid only if the velocity in suction and discharge ducts is the same.

The other U-tubes indicate the pressures at the various points along suction and discharge ducts, and their meaning will be apparent.

If a fan has suction ducting only, the static pressure produced for overcoming friction of ducting will be represented by OB (a negative pressure), since the discharge will be at atmospheric pressure. Similarly, if the fan has only discharge ducting, the static pressure will be represented by OC, the suction being at atmospheric pressure.

Fan Characteristics—A comparison of the operation of fans of various types is best understood by studying their characteristic curves. For this purpose consider a fan connected to a duct with an adjustable orifice at the end, as in Fig. 19.5 (a). Pressures are measured by water gauges connected to a standard Pitot tube, the end of which is shown in (b). The perforated portion gives the static pressure, and the facing tube the total pressure.

If the fan is running with the orifice shut, no air will be delivered. Static

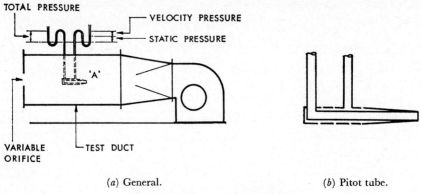

(a) General.   (b) Pitot tube.

FIG. 19.5.—Fan test arrangements.

pressure will be at a maximum, and velocity pressure nil. As the orifice is opened the static pressure will fall and the velocity pressure increase until, with the orifice fully open, the static pressure will be negligible and velocity pressure at a maximum. Over this range the power required to drive the fan will have increased from minimum to maximum, and perhaps will fall away as the total pressure falls off. The power required to drive the fan, if 100 per cent efficient, would be:

$$\text{Air power (W)} = \text{Volume (m}^3/\text{s)} \times \text{total pressure (Pa)}$$

The mechanical efficiency of a given fan will be the ratio between this Air power and the actual watts absorbed by the drive (Fan power): this will depend on design, type of fan, speed, and proportion of full discharge. In the days of Imperial measure it was easier to distinguish between these criteria, but we shall no doubt become accustomed to the new terminology as more manufacturers produce catalogues in coherent terms (as they, belatedly, come to understand S.I. units).

If the static pressure is used, the efficiency derived will be *static efficiency*; if the total pressure is used the efficiency will be *total efficiency*.

The standard air* for testing fans is taken at a temperature of 20° C, a density of 1·2 kg/m³ and a pressure of 101·325 kPa (1·013 bar). Any fan at constant speed will deliver a constant volume at any temperature; as the temperature varies the density will increase or decrease proportionately with the absolute temperature, hence the power input will vary in the same ratio. With increase of temperature the power will be reduced and vice versa. Similarly, decrease of atmospheric pressure (as in the case of a fan working at high altitude) will cause a reduction in power and conversely.

If a fan running at a certain speed is rearranged to run at some higher speed, the system to which it is connected remaining the same, the volume

---

* For details of standard fan testing see B.S. 848.

will increase directly as the speed; the total pressure will increase in the ratio of the speeds squared and the power input will increase in the ratio of the speeds cubed.

Characteristic Curves—Fans are of two main types, each with sub-divisions as follows:

 1. Centrifugal type:
  (*a*) Multivane, forward bladed.
  (*b*) „ radial „
  (*c*) „ backward „
  (*d*) Paddle wheel.
The three types of runner, (*a*), (*b*) and (*c*), are shown in Fig. 19.6.
 2. Propeller type:
  (*a*) Ordinary propeller or disc fan.
  (*b*) Axial flow.

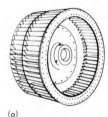

(a)   (b)   (c)

(*a*) Forward curved. (*b*) Radial. (*c*) Backward curved.

FIG. 19.6.—Runners in centrifugal fans.

We will consider their characteristic curves and discuss their applications. The curves for static pressure, power input, and static efficiency are drawn from tests at constant speed, as already described. The base of the curve is percentage of full opening of the orifice. The vertical scale is percentage of pressure, efficiency or power input.

Fig. 19.7 (*a*)–(*d*) gives typical curves for the four main types. Paddle-wheel fans are not given, as they are now little used in ventilating work on account of their noise, being confined chiefly to dust removal and industrial uses. Radial bladed fans have characteristics which are similar to those of the backward curved type but without the advantage of power limitation. They are not commonly used in ventilation applications.

Forward Curved—It will be noted that the forward-bladed centrifugal fan most commonly used in ventilation reaches a maximum efficiency at about 50 per cent opening, where at the same time the static pressure is fairly high. Fans are generally selected to work near this point. It will also be observed that the power curve rises continuously. Thus, if in a duct system the pressure loss is less than calculated, the air delivered will be more and the power absorbed more, which will lead to overloading of the motor.

Backward Curved—This type of fan runs at a higher speed to achieve the same output as a forward curved. The efficiency reaches a maximum at

P   K.H.A.C.

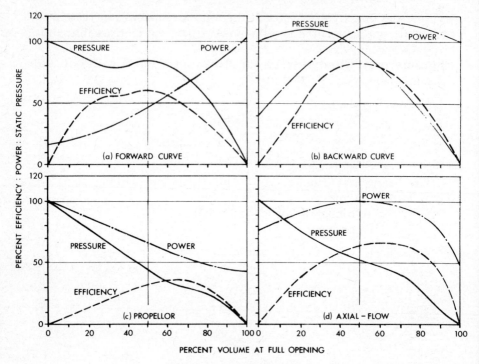

FIG. 19.7.—Fan characteristics.

about the same point, and the power, after reaching a peak, begins to fall. This is called a *self-limiting characteristic*, and means that if the motor is installed large enough to cover this peak it cannot be overloaded. This is often useful in cases where the pressure is variable or indeterminate. The pressure curve is smooth without the dip of the forward curved; for this reason this kind of fan is to be preferred where two fans are working in parallel. The forward curved type under such conditions is apt to hunt from one peak to the next, so that one fan may take more than its share of the load and the other much less. The backward-bladed fan is often made with aerofoil blades, so raising efficiency. It is much used in high velocity systems where high pressures are required. (See Plate XXXII, facing p. 470.)

Ordinary Propeller—From the curves it will be observed how the pressure falls away continuously, and the static efficiency reaches but a low figure. Thus, this type is unsuitable where any considerable run of ducting is used. To generate any appreciable pressure its speed becomes unduly high, and hence the fan is noisy. Its main purpose is for free air discharge where its velocity curve would rise towards a maximum at full opening. It should be noted that the fan power is a maximum at closed discharge, and, as the motors supplied with these fans are not usually rated to work at such a condition, the closing or baffling of the discharge or suction may cause overloading.

Axial Flow—This type shows a great improvement over the ordinary propeller fan, both as regards efficiency and pressure. The fan-power curve is self-limiting. Hence, these fans may safely be used in conjunction with a system of ductwork, being often more convenient than a centrifugal, particularly for exhausting. They run at higher speed than a centrifugal fan to produce a given pressure, and are liable to be more noisy: this may be overcome, to some extent, by the use of a silencer.

Fan Arrangements and Drives—Centrifugal fans may be *open* or *cased*. When open they can only be used for exhausting, and the discharge is tangential from the perimeter of the impeller, as might be suitable in a large roof turret.

The usual arrangement is the cased type, and the suction is then either on one side, as in Fig. 19.8 (*a*) with single inlet, or both sides as Fig. 19.8 (*b*) with double inlet. The single inlet is the more usual. The double inlet double-width fan is useful where large volumes are concerned, as it gives double the capacity of the single inlet with the same height of casing.

Fans are almost invariably driven by electric motor except for cases, such as underground car parks, where there may be a statutory requirement for independent drive from a petrol or diesel engine. Fig. 19.8 (*c*) shows a typical arrangement with the fan impeller mounted on a shaft extension of the motor. This is a compact arrangement, but generally used for small or medium-sized fans only.

A motor direct-coupled to a fan with a flexible coupling is illustrated in Fig. 19.8 (*d*). The fan shaft runs in its own bearings independently of the motor. This is obviously to be preferred for heavy duty and for large fans. The motor can be removed and replaced without affecting the fan.

The arrangement shown in Fig. 19.8 (*e*), where the motor drives the fan via pulleys and vee-belts has the great advantage that the *motor* speed may be a standard, such as 16 rps* or 24 rps, whilst the *fan* speed is that best suited to the duty. A further advantage is that if on testing it is found that the pressure loss of the system is less or more than allowed for, the fan duty can be corrected by merely changing the pulleys.

It will be noted that in the illustrations different positions of the discharge openings in relation to the suction eye of the fan in each case are given. It is usually possible to obtain a fan with its discharge at any angle, vertical, horizontal top, horizontal bottom, downwards, and intermediately at an angle of 45°.

Axial Flow Fan Arrangements—This type of fan, it is stated, can be designed to give static pressures up to 500 Pa within the limits of quiet running. The maximum peripheral speed should not normally exceed 37 m/s for this condition.

These fans can be built in one, two or three stages, to obtain increased pressure, the volume remaining the same. Alternatively they may have two blades made counter-rotating.

* rps = revolutions per second.

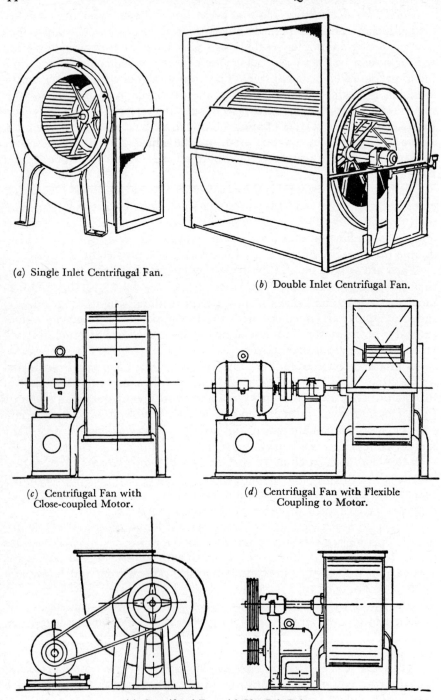

(*a*) Single Inlet Centrifugal Fan.

(*b*) Double Inlet Centrifugal Fan.

(*c*) Centrifugal Fan with
Close-coupled Motor.

(*d*) Centrifugal Fan with Flexible
Coupling to Motor.

(*e*) Centrifugal Fan with Vee Belt Drive.

Fig. 19.8.—Arrangement of centrifugal fans.

Such fans, illustrated in Fig. 19.9, may have blades of specially treated wood, similar to an aeroplane propeller, or be of diecast aluminium; these are shaped to aerofoil section and may be designed for any angle and any number of blades from two to twelve.

Many such fans are manufactured to permit adjustment of the pitch angle of the blade. Since this angle, for a given fan diameter and speed, determines the volume delivered it follows that adjustment facilities permit output to be matched to the duty required with some precision. Fig. 19.9 (c) and (d) show how pitch may be adjusted.

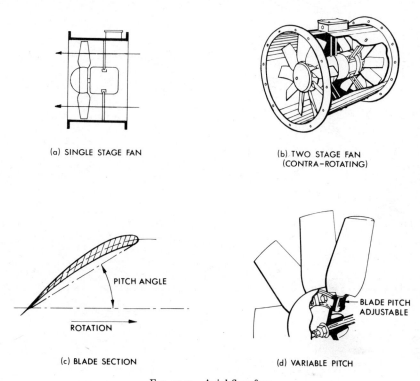

(a) SINGLE STAGE FAN

(b) TWO STAGE FAN
(CONTRA–ROTATING)

PITCH ANGLE

ROTATION

(c) BLADE SECTION

BLADE PITCH
ADJUSTABLE

(d) VARIABLE PITCH

Fig. 19.9.—Axial flow fans.

Fan Duties—The range of fan types, speeds, pressures, and volumes is too great for any indication to be given here of sizes, duties, power requirements, etc., or of the problem of motor types suitable for fan drives.

Enquiries to fan-makers should give the fullest information possible about any system, as there are many hidden points to be watched in the selection of fans which render mere catalogue reference insufficient.

Where it is necessary, as in the case of a variable volume air conditioning system, to exercise control over fan output this may be achieved in a number of ways. In the days when direct current was available, variation in drive motor speed could be arranged conveniently but, with a.c. supply, reliance must be placed upon gear or similar drives which are less convenient and relatively inefficient.

If throttling dampers are introduced into the system, volume reduction will be achieved at the expense of efficiency as illustrated in parts (a) and (c) of Fig. 19.10. In (a), the full volume operation of a system is represented by the intersection of the fan characteristic F_1 with the system characteristic C_1. The use of throttling dampers changes the *system* characteristic to C_2 and C_3 at 80 per cent and 60 per cent volume respectively. The resultant small savings in power absorbed are shown in (c).

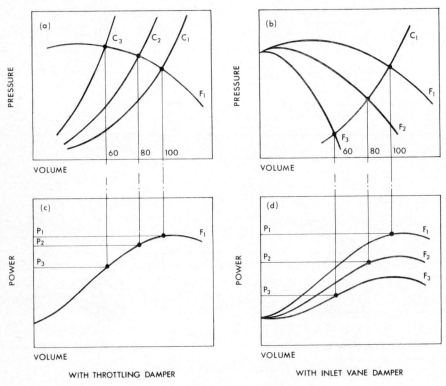

FIG. 19.10.—Volume control via dampers.

The use of radial inlet vane dampers fitted to the eye of a centrifugal fan provides a much more effective means of volume control since such devices act, by changing the air flow pattern at the fan inlet, to modify the performance of the fan impeller in much the same way as would result from change in the speed of drive. This effect is illustrated in parts (b) and (d) of Fig. 19.10. As before, F_1 represents full volume operation with the system characteristic C_1: operation of the radial inlet vanes changes the *fan* characteristic to curves F_2 and F_3, which represent 80 per cent and

60 per cent volume respectively, against a constant system characteristic curve. The resultant dramatic savings in power are shown in (d).

Where axial flow fans are used with systems demanding volume reduction, the facility of change in pitch angle of the blades may be automated. This permits both volume and power to be reduced to demand.

POSITION OF OUTSIDE AIR INTAKE

It might be thought that the purest supply of air would be obtainable at the roof of a building, but experience shows that in many instances this is not the case. At roof-level air is certainly free from street dust, but chimneys of the same or neighbouring buildings may, with certain states of the wind, deliver fumes into the intake.

An intake at low level, such as near a busy street, would be liable to draw in much road dust and exhaust fumes from motor vehicles.

If a point half-way up the elevation of a building can be found, this is probably the best. It must, however, be clear of windows where fire or smoke might occur, and particularly of lavatory windows.

There is no general solution to the problem of the outside-air intake position, as obviously every case requires examination of orientation, possible sources of contamination, position of air-conditioning plant, and so on.

AIR FILTRATION

Air Contaminants—Atmospheric air is contaminated by a variety of particles, such as soot, ash, pollens, mould spores, fibrous materials, dust, grit and disintegrated rubber from roads, metallic dusts and bacteria. The heavier particles may be such that under calm conditions they will settle out of their own volition. These are termed 'temporary'. The smokes, gases and lighter particulate matter remain in suspension and are termed 'permanent'.

The unit of measurement for dust particles is the micron; one millimeter = 1,000 microns. The human hair has a diameter of about 100 microns, and the smallest particle visible to the naked eye is about 15 microns. The smallest range of particles we need consider here is of the order of 0·01 to 0·1 micron, which is represented by smokes of various kinds, such as tobacco smoke. The upper range of particle size we need consider is about 15 microns.

Pollution in all its forms, and especially atmospheric pollution, is the subject of increasing public concern, though apart from smoke, fumes and soot it appears doubtful whether much of the other air-borne dusts and dirt are susceptible of reduction. Tests are regularly carried out and records kept of suspended black pollution material from which it is possible to deduce the following approximate table of contamination according to locality, expressed in milligrams per 100 cubic metres, as typical of winter conditions, when the greatest index occurs.

Typical average smoke index

Rural area - - - - 2
A very clean town - - - 4
Clean town - - - - 10
Average town - - - 20
Dirty town - - - - 30
Very dirty area - - - 40
Extreme pollution area - - 50

Necessity for air cleaning—If in a mechanically-ventilated or air-conditioned building air is blown in without some means for filtration, deposits of dust will be found to occur throughout the rooms, and the system of ducts will in itself become coated with solid matter. Heater batteries and fans will also become coated so that in time the efficiency of the system as a whole will fall off at an increasing rate.

Except in certain industrial applications, ventilation and air-conditioning systems therefore invariably include some means for filtration of the air.

The removal of the larger particles is, of course, a simple matter, since any mesh of fine enough aperture will arrest such particles. A plain mesh is however liable to become clogged very quickly, and hence is of little use for our purpose.

The finer material and the smokes are, however, much more difficult to arrest, and yet it is these which are largely responsible for the staining of decorations, the soiling of shirts and garments and, to some extent also, no doubt the bearing of harmful bacteria. Apart from fresh air, re-circulated air carries fluff from carpets, blankets and clothes, dust brought in on shoes and, in an industrial application, any dust resulting from the process.

The greater the degree of filtration, as a rule, the higher the cost of the equipment and the greater the space occupied. The selection of the best filter for a particular application therefore depends on whether great value is placed on a high degree of cleanliness or not.

Tests for Filters—*Weight test:* Under this method a carefully metered quantity of air is drawn through a filter paper from the unfiltered intake, and a similar quantity of air is drawn through another filter paper downstream from the filter. These are weighed on an accurate balance and a comparison of the two weights gives the gravimetric efficiency. The heavier particles, as explained, are the most easily collected and these constitute the greater part of the weight, hence even a poor filter will give a high efficiency of perhaps over 90 per cent by the gravimetric method.

Blackness test: Under this method equal quantities of air are drawn through filter papers in a similar manner to the above, and the resultant stains are viewed optically, and the reflected light measured by a light meter. A comparison of the relative reflected light values then gives the blackness test efficiency. This is a much more stringent test than the previous one, and many filters which may give over 90 per cent by the gravimetric method may well be less than 50 per cent on a blackness test.

It will thus be seen that the determination of the efficiency of a filter is by no means simple, as it depends on particle size and method of test. A test procedure is laid down in B.S. 2831 (1957) and certain standard dust sizes have been defined, being *British Standard Dusts*, numbers 1, 2 and 3; each of these has a range of particle sizes approximately as indicated at the base of Fig. 19.11. *Methylene Blue* in particle size closely resembles the distribution in atmospheric pollution, but this test can only be carried out in a laboratory.

In selecting a filter it is necessary to know by what method the maker's guarantee of test efficiency has been determined, and what type of dust was used, since it is unlikely that a determination was made under the particular conditions of atmospheric pollution obtaining in the case in question.

Practical Filters—Air filters fall into four main categories as follows:

(1) *Viscous Impingement.* Usually of some form of corrugated metal plates or metal coils or turnings or the like, in each case covered with a viscous oily liquid to arrest the particles on impingement. Other materials used instead of metal are glass fibres, similarly coated with a sticky fluid.

(2) *Fabric.* The material used in this type of filter is some form of textile, generally with a heavy nap, such as swansdown. The dirt particles are in this case arrested partly by being trapped in the interstices of the material, and partly by being caught on the fibres.

(3) *Electrostatic.* In this system of filtration dust particles entering the filter are subjected to an electrostatic ionising charge and, on subsequently passing through the parallel plates which are alternately charged and earthed, the particles are repelled by the charged plates and adhere to the earthed plates. The electrical charge is supplied from a power-pack containing the necessary transformers and rectifiers to produce the high voltage D.C. required.

(4) *Paper or Absolute.* This filter uses a special form of paper made usually from woven glass-fibre.

A further class of filter makes use of *activated charcoal* produced from coconut shell, and, by virtue of the high degree of porosity, is very active in absorbing odours, gases and the like.

Relative Efficiencies—Fig. 19.11* indicates trends of efficiencies which may be expected from the four types of filter above referred to, numbers 1 to 4. The tests are based for the very small particles on a count test, for

---

* Penrose, M. O., *Electrostatic Air Filtration, J.I.H.V.E.*, 1961, 29.113.

the middle range on a blackness test, and for the larger particles on a weight test.

It will be seen that filter 1, depending on impact, has no arrestance on a blackness test or count basis. Filter 2 has a blackness test efficiency of about 50 per cent for particles of about 1 micron, diminishing rapidly for the smaller particles. The Electrostatic filter, curve 3, has a blackness test efficiency of about 90 per cent whilst the paper filter 4 achieves virtually 100 per cent on a count basis for the smallest size of particles.

All filters vary in efficiency according to velocity of air through them.

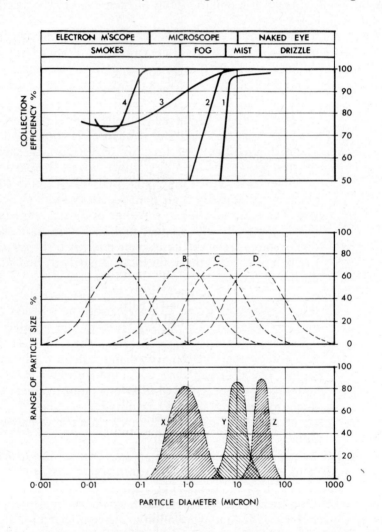

FIG. 19.11.—Air filter efficiencies. 1. Viscous coated metal, 2. Fabric, 3. Electrostatic, 4. Paper absolute. A = Count test, B = Blackness test, C = Weight test, D = Probable annual range of impurities, X = BS Test dust No. 1, Y = BS Test dust No. 2, Z = BS Test dust No. 3.

TABLE 19.1

CHARACTERISTICS OF FILTER MEDIA

Filter Type	Face Velocity m/s	Resistance, Pa	
		Clean	Dirty
Viscous Impingement			
Static oil coated panel	1·5–2·5	20–60	100–150
Automatic curtain	2 –2·5	30–60	100–125
Dry fabric or fibrous	1·3–2·5	25–180	125–250
Static panel or bag	2·5	30–60	100–175
Automatic roll			
Electrostatic	1·5–2·5	40–60	—
Absolute	up to 2·2	up to 250	up to 600

Velocities and resistances of various filters are given in Table 19.1.

In the case of fabric and paper filters, owing to the large surface areas involved, designs are usually based on forming the material into a zig-zag formation.

Filter Cleaning—When heavily charged with dirt, the resistance of most filters rises sharply, thus reducing air flow, and hence ventilation rate. The cleaning of filters is achieved in a variety of ways as follows:

Absolute Filters: These are discarded when dirty and replaced.

Washable Filters: These, consisting of a foam plastic element, are contained in metal frames with wire retaining strips as shown in Fig. 19.12. The material is flame retardant and, when fouled, is washed in detergent and re-used. The media have a long life.

Brush Filters: Eliminating the necessity for any special cleaning, these consist of 'flue-brush' elements contained in a segmented frame as shown in Fig. 19.13. The hair element is removed for either vacuum or other

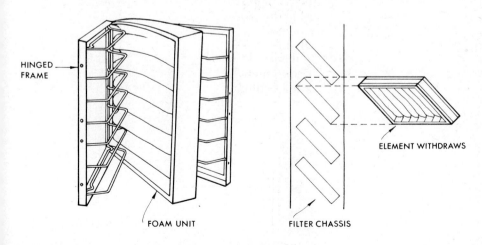

HINGED FRAME

FOAM UNIT

FILTER CHASSIS

ELEMENT WITHDRAWS

FIG. 19.12.—Washable filters.

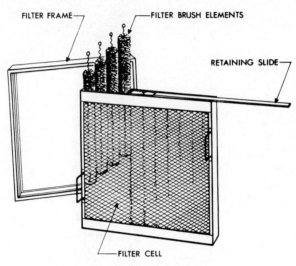

FIG. 19.13.—Hair filters.

simple cleaning and washed before replacement. The media have an indefinite life.

Fabric Filters: In one form, as in Fig. 19.14, these are mounted in frames and when dirty are thrown away. In another type, the filter material is of a glass fibre or other suitable base, and is supplied in rolls. The roll is horizontal and is gradually wound from one spool on to another on the principle of a camera film, see Fig. 19.15; this movement is achieved by motor drive controlled from a pressure differential switch across the filter, or from a timing device. Fresh filtering medium is unrolled only as it is required. Such equipment may, alternatively, operate with the rolls vertical but the edge sealing arrangements are then less effective

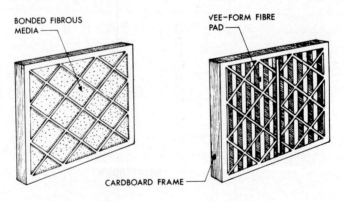

FIG. 19.14.—Fabric filters, single cell.

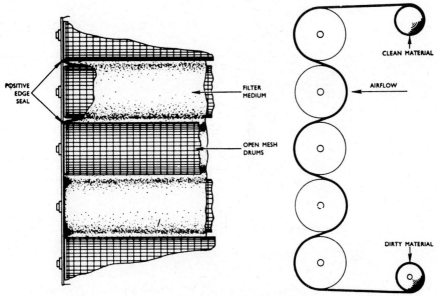

FIG. 19.15.—Automatic roller fabric filter.

and the drive may be less positive due to mechanical problems.

This type of filter will, in theory, require no labour for maintenance, excepting at the long intervals of possibly three to six months for the changing of a complete roll.

Bag Filters: The traditional bag filter has been superseded by a type using replaceable media and this is currently taking the place of the 'camera roll' fabric filter since it has better performance characteristics and requires less skill in maintenance. A common arrangement is shown in Fig. 19.16.

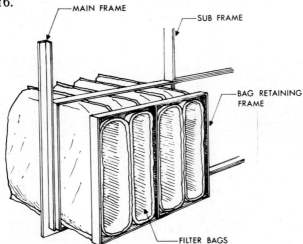

FIG. 19.16.—Bag type fabric filter.

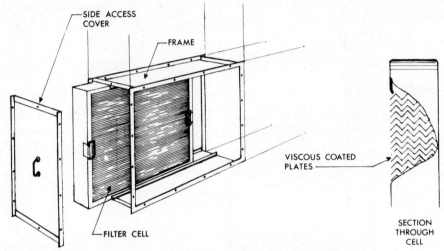

FIG. 19.17.—Viscous filter, single cell.

Viscous Impingement Filters: These may be in the form of cells which are removed for cleaning by hand, which are washed, re-oiled and replaced. They have also been developed on a crude self-cleaning principle and, in one form, the zig-zag plates are vertical, oil flowing over them from a pump drawing from the base tank at intervals. In another form the cells are on an endless chain dipping into the oil in the base tank and returning for re-use, see Fig. 19.18.

Electrostatic Filter: In this type of filter (Fig. 19.19), the dirt collects on the earthed plates and its removal is accomplished by washing with hot-

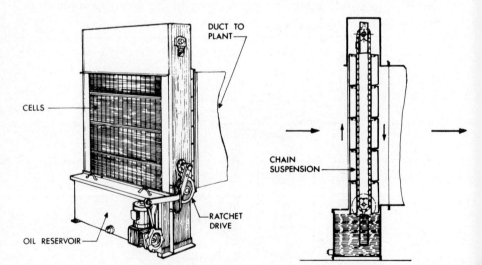

FIG. 19.18.—Automatic viscous roller filter.

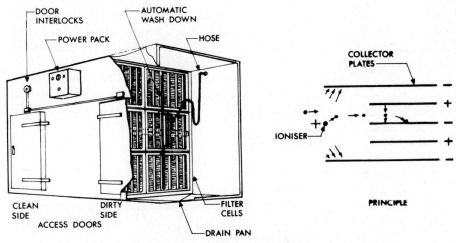

FIG. 19.19.—Electrostatic filter.

water jets. This may be done by hand or automatically. After cleaning, a period of drying is necessary before the filter can be put back into use.

Combination of Filters—Advantage may be taken of the properties of different types of filter in combination to achieve a high degree of filtration efficiency, with a minimum of labour in cleaning.

A logical combination is to use a coarse filter of viscous impingement

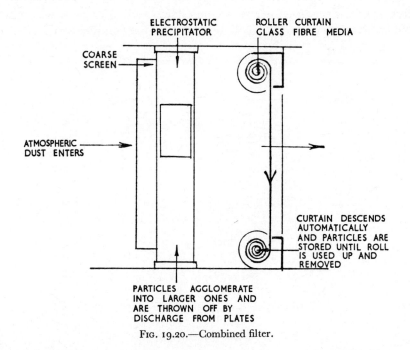

FIG. 19.20.—Combined filter.

type which will remove the larger particles constituting the great bulk of material to be collected, and to follow it with a more sophisticated unit. The latter will take out the fine particles which have passed through the coarse filter, hence renewal would be at long intervals.

There is a good case for placing a reasonably good filter downstream of an electrostatic filter. This is to guard against short periods when the electrostatic filter may be out of action, and to trap any quantity of dust which may blow off the main filter in the event of power pack trip-out or in shutting down the plant. It has been found that if a roller curtain or bag filter is used, which has a blackness test efficiency of some 30 per cent, the agglomerated particles which are collected on the electrostatic plates may be allowed to blow off and be caught as a continuous process, see Fig. 19.20. This combination may be found advantageous where a plant operates twenty-four hours a day and washing and drying time is not permitted, also where washing and draining facilities are difficult.

For clearance of fog only, the electrostatic and absolute filters are effective, or a combination of electrostatic plus fabric. Excessive collection of moisture from fog may cause an electrostatic filter to cut-out due to short circuit. A special heater may be used ahead of the filter to ensure dry air entering the filter and avoiding this trouble.

Resin Wool—A further type of filter of electrostatic type makes use of the property of resin-coated woollen fabric to generate a static charge when air is passed through the material. This naturally operates without external electric supply. On a blackness test basis this material is believed to have an efficiency of 90 per cent when new, tending to drop with age. A velocity through the material of about 0·1 m/s is recommended.

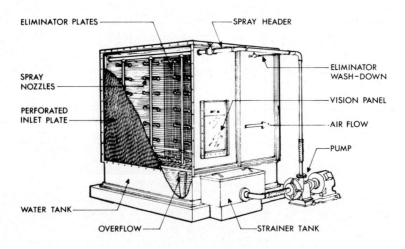

FIG. 19.21.—Air washer, single bank spray.

AIR WASHERS AND HUMIDIFIERS

The treatment of air by water sprays might be thought to be an effective way of air-cleaning. Unfortunately this is not the case, owing presumably to the greasy nature of soot and other contaminants. The washer will remove heavier and gritty material but, on a blackness test, its efficiency is low even when wetting agents are tried. The air washer's chief function is therefore in humidification and de-humidification.

An air washer, see Fig. 19.21, consists of a casing with tank formed in the base to contain water. Spray nozzles mounted on vertical pipes connected to a header deliver water in the form of a fine mist. The spray is projected either with or against the air current, or, where two banks of sprays are provided, in both directions, one with and one against the air stream. The water for the sprays is delivered by a pump, under a gauge pressure of between 0·2 and 0·3 kPa, the water being drawn from the tank through a filter. The casing has an access door and internal illumination, and may be of galvanised steel, or constructed of brick or concrete asphalted inside.

As the mist-laden air is drawn through the chamber it requires to have the free moisture removed, and for this purpose eliminator plates of zig-zag formation are provided to arrest the water droplets. Some makes precede these with scrubber plates, down which water is caused to run by a spray pipe at the top in order to flush down dirt which has collected from the air.

Galvanized eliminator plates are liable to rapid deterioration; a better material is copper. Glass plates with serrated ribs have been used successfully, and they are permanent.

A single bank air washer will not saturate the air more than that corresponding to about 70 per cent of the wet-bulb depression. A double bank washer may reach 90 per cent. For full air-conditioning and de-humidification, where it is desired to saturate at the dew-point, a two-bank washer is generally necessary with large spray nozzles to pass the necessary quantity of water for the temperature rise allowed. Spray nozzles vary in capacity from 0·05 to 0·25 litre/s, and should be easily cleanable and of non-corrodible metal.

The speed of air through an air washer is usually 2·5 to 3 m/s. The length is normally about 2·5 m, but is increased with the two or more banks of sprays needed for cooling—sometimes to as much as 4·0 m. On the inlet side straightening vanes, or a perforated grille, are to be advised in order to distribute the air evenly over the whole area.

The tank is kept filled by a ball-cock, with a hand valve for quick filling, and there is in addition a drain and overflow pipe. The pump is of normal type, preferably with insulated connections where noise may cause trouble, and arranged as to level so as to be flooded by the water in the tank, otherwise it will need priming.

Capillary Cell Type Washer—This type is shown in Fig. 19.22 and con-

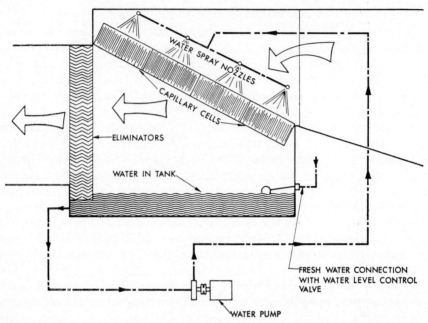

FIG. 19.22.—Air washer, capillary type.

sists of a series of cells inclined at an angle and containing glass fibres laid in straight formation to a depth of some 200 mm. One cell, size 500 × 500 mm, passes an air quantity of about 550 litre/s. Water at low pressure is caused to flow over the cells by flooding nozzles from a pump of low power consumption. The water and air have to negotiate the striations of the glass fibres together, and are thus intimately mixed so that nearly complete saturation is achieved. At the same time it is stated that the filtering efficiency is of a high order. It may equally be used with refrigerated water for a full air-conditioning system, since as much as 0·75 litre/s per cell may be passed through; the normal rate is about 0·15 litre/s.

Humidifiers—For plants which do not incorporate air washers, other means for winter humidification may be required. Controlled steam injection may be used either from a central boiler plant via a sparge pipe, Fig. 19.23 (*a*), or from a local packaged unit as Fig. 19.23 (*b*). The latter type of equipment is mounted on the air-conditioning plant and consists, in essence, of a small electrode steam generator sized to suit the moisture requirement of the plant.

In some instances a simple pan humidifier, consisting of a shallow water tray replenished by ball valve and heated by electric immersion elements, may be adequate. Alternatively some form of atomising device can be incorporated in the plant. These latter commonly operate via use

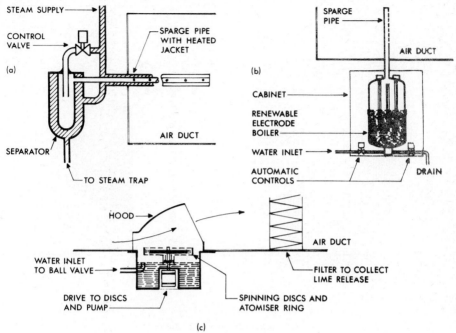

FIG. 19.23.—Humidifier arrangements.

of spinning discs as illustrated in Fig. 19.23 (*c*). It must be remembered that, in the process of atomisation and evaporation, any mineral salts dissolved or suspended in the water supply will be released and deposited: some water treatment may be required.

HEATERS

Air heaters when originally introduced were commonly of plain tubing, as shown in Fig. 19.24 (*a*). In order to economise in space and achieve greater output from a given amount of metal, finned heaters are now generally used as in Fig. 19.24 (*b*). Plain tubes are less likely to become choked with dirt than finned—a fate all too common. Plain tube heaters are suitable for use in fresh air intakes to prevent wet fog collection on fabric or electrostatic filters. They should in this case be of galvanised steel or other non-corrodible metal, but not copper.

Heaters are arranged in stacks or batteries with automatic controls preferably of modulating type to give a steady temperature of output.

If the heater is warmed by hot water, a constant temperature supply is required from the boiler or calorifier with pump circulation. The flow required being relatively large, a control valve of diverter type will preserve the circulation irrespective of the demands of the heater. If the heater is fed by steam, it will require the usual stop valve and steam trap and a means of balancing the pressures so that the condense may not be

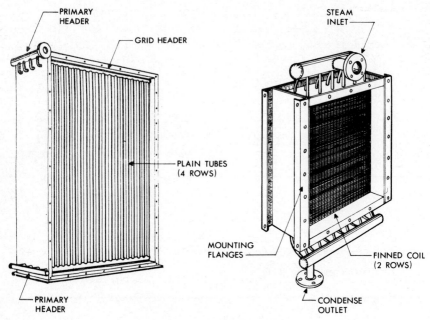

FIG. 19.24.—Air heater batteries.

held up by a vacuum caused by control valve shut-off.

Heaters may be enclosed in sheet-steel casings, 'packaged' with other elements into a prefabricated unit, or built into builders' work enclosures.

The number of rows of tubes depends on temperature rise, temperature and nature of heating medium, i.e. whether steam or hot water, and air speed. The face area depends on air volume and free area between tubes. This is again determined by the velocity, and it is usual to fix this arbitrarily beforehand, generally between 4 and 6 m/s through free area, or 2·5 to 3 m/s face velocity. For sizing of heaters reference is necessary to makers' data.

Air Cooling Coils—Cooling coil surfaces as used in air-conditioning generally perform two functions—to remove sensible heat, and to remove moisture or latent heat.

The sizing and temperature of operation depend on the sensible : latent ratio. If low relative humidities are required, the dew-point will be low, and hence the water temperature will be low. If sensible cooling only is required, the coil surface temperature will be kept above dew-point.

In any coil a certain proportion of air fails to come in contact with the cold surfaces, and thus a part may be chilled and dehumidified and a part remain unchanged. By correct selection of coil form the desired ratio may be obtained.

In direct-expansion systems, the actual refrigerant gas from the com-

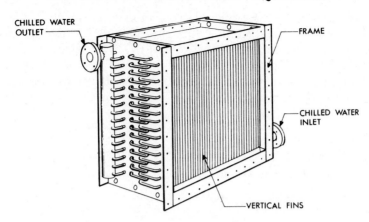

FIG. 19.25.—Cooling coils.

pressor is passed direct into the coils, which, in effect, form the evaporator of the refrigerating plant.

In a chilled-water system, chilled water is circulated through the coil, and this may be used down to about 3·3° C. Below this an anti-freezing mixture such as calcium chloride brine becomes necessary.

In cooling by refrigeration, the higher the evaporator temperature the less power is consumed for a given duty, hence it is desirable to keep the cooling coil temperature as high as possible consistent with the final air temperature required to meet design conditions. This applies whether the coil is used for direct expansion or with chilled water or brine.

Cooling coil surface is therefore usually designed for small temperature differences between water and air; for instance, for air cooled from 24° to 13° C, water may be at 10° inlet and 14·5° C outlet, (i.e. leaving above the air-outlet temperature). An air washer cannot give a comparable performance to this.

Coolers are generally of finned or block type (as in Fig. 19.25), comprising banks of small bore tubes threaded through plates between which the air passes, the whole being of copper tinned after fabrication or, more normally in present day practice, made up with copper tubes and aluminium fins. Coils are frequently arranged horizontally, with fins vertical, so as to facilitate drainage of condensation when dehumidifying.

PACKAGED AIR HANDLING PLANT

The various components previously described may be built up to provide a complete plant in a number of ways. They may be connected using sheet metal ducts or by incorporation within a masonry chamber: the former method tends to be untidy and space consuming and the latter is somewhat cumbersome and expensive if air tightness is ensured.

Current practice favours the use of factory packaged plant where individual components are housed in modular casings for site assembly. The better designs for such modules are so arranged, as shown in Fig. 19.26, that they may be assembled in a number of different ways and thus fitted to the building space available. (See Plate XXXIV, facing p. 470.)

A recent and logical extension of this principle has been the production of such modules in weather protected form such that they, themselves, are plant rooms for, say, roof mounting without the need for architectural enclosure. (See Plate XXXIII, facing p. 470.)

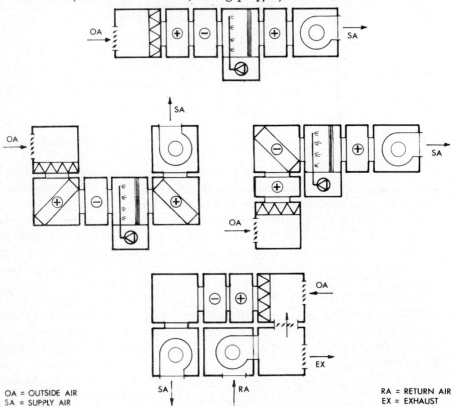

OA = OUTSIDE AIR
SA = SUPPLY AIR

RA = RETURN AIR
EX = EXHAUST

FIG. 19.26.—Packaged air handling plant arrangements.

DAMPERS

Control of air volume is effected by means of dampers, which are either:

    (*a*) Permanently set when the installation is tested, to give the designed volumes in each branch. These are commonly of butterfly or sliding type with some form of locking device.

    (*b*) 'Controllable' dampers for use in air-conditioning and ventilation installations, and adjusted, either manually or automatically, to suit varying conditions. Such dampers are generally of louvred type as in Fig. 19.27.

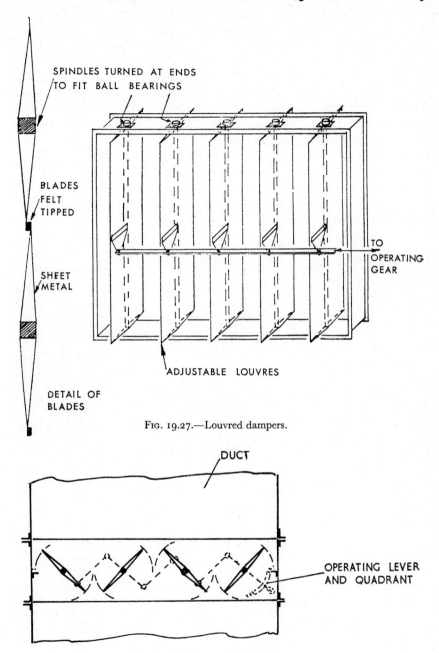

SPINDLES TURNED AT ENDS
TO FIT BALL BEARINGS

BLADES
FELT
TIPPED

SHEET
METAL

TO
OPERATING
GEAR

ADJUSTABLE LOUVRES

DETAIL OF
BLADES

FIG. 19.27.—Louvred dampers.

DUCT

OPERATING LEVER
AND QUADRANT

FIG. 19.28.—Opposed blade dampers.

Multi-louvre dampers having adjacent louvres contra-rotating, are much superior, from the regulating point of view, to the commonly used type where all vanes rotate in the same direction, as Fig. 19.28.

Where tight shut-off is essential, it is necessary for the dampers to be felt-tipped and to close on to a felted frame.

Fire Dampers—Where ducts pass through walls and floor slabs, fire authorities commonly require the insertion of fire dampers. One type of fire damper is shown in Fig. 19.29 (*a*) and consists of a heavy steel casing and damper normally kept open by the fusible link. In the event of fire the link melts and the damper closes. Another pattern, as Fig. 19.29 (*b*), takes the form of a number of substantial shutter blades retained out of the air stream by a fusible link as before.

A relatively new approach to the provision of fire dampers relies upon the insertion of a steel honeycomb into the air duct, as Fig. 19.29 (*c*), all surfaces of which are coated with an intumescent paint. This material has the property of swelling to many times its original volume when heated and thus forming a barrier to air flow.

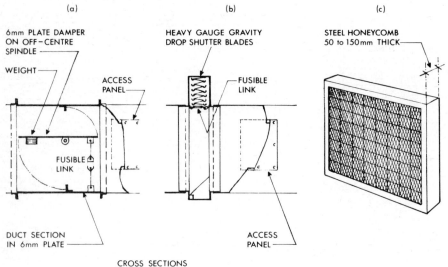

FIG. 19.29.—Fire dampers.

AUTOMATIC CONTROL APPARATUS

There are three main classes of controls appropriate to ventilation and, in particular, air conditioning systems: electrically operated, pneumatically operated, and hydraulic or oil operated. The choice will be determined by preference based on experience, size of project, cost, personnel in charge of plant, etc., but it should be a first principle that control arrangements are kept as simple as possible. Complications in design, however attractive, lead to complications in the maintenance which is all too often absent.

Electrical controls are best where inexperienced operators supervise the plant, but are equally suitable on any plant.

Pneumatic controls need skilled attention from time to time, and when properly adjusted give perhaps the most simple and gradual operation of all. They are often to be found on industrial plants for this reason, especially as such plants generally already include a compressed air system.

Hydraulic and oil controls have their special uses but their limitations.

The positions in which controlling elements, such as thermostatic bulbs, are fixed, require careful selection. Air passing through a plant may 'layer' seriously. Near the edges of a duct or plant it may similarly be different in temperature from that near the core. Again, there is the matter of heat or cold radiation from heater batteries and cooling coils or even from cooled spray water.

To obtain as nearly average results as possible, temperature sensitive or hygroscopic elements should extend well into the middle of the stream, and not merely project a few inches through the casing. The elements should be shielded against radiation by polished shields, left open or perforated away from the radiation, or they may be in a bleed-off duct.

ELECTRICAL CONTROLS

Electrical Thermostats—For on-off control these comprise a temperature sensitive element, either of volatile liquid tilting a mercury switch, or bi-metallic type closing point contacts. Fig. 19.30, 2-wire, simply opens or closes a circuit to stop or start a motor. The 3-wire type opens one circuit and closes another to start a motor or operate a valve in reverse.

These thermostats may be arranged to give a step-by-step action, opening or closing a series of contacts in sequence.

Modulating type—This works in conjunction with a modulating valve or damper motor as described later. The thermostat operates the moving arm of a potentiometer in a 'bridge' circuit.

Electrical Humidistat—The hygroscopic material here, Fig. 19.31, is either a bundle of hairs maintained in a state of tension, or a length of specially sensitive wood or pine-cone fibre. The expansion on increase of percentage saturation is arranged to open or close contacts or operate a potentiometer, as in the case of a thermostat.

Another form of humidity control is by wet and dry bulb thermostats

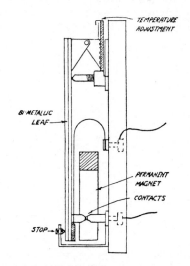

Fig. 19.30.—Electrical bi-metallic thermostat.

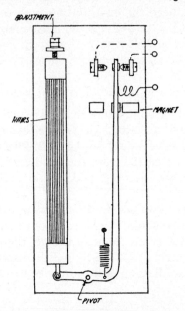

FIG. 19.31.—Electrical hair type humidistat.

working on a differential principle. These control for a constant difference which corresponds over small ranges with the relative humidity. The wet bulb is kept wet by a wick dipping in a reservoir of distilled water.

More sophisticated humidity controls convert change of moisture content directly into change of electrical resistance and are used in combination with electronic circuits.

Contactors—When a large machine such as a refrigeration compressor motor has to be started, it is necessary to interpose between the thermostat

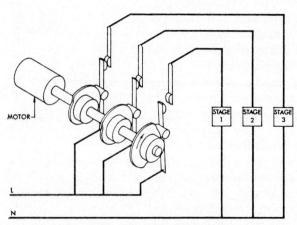

FIG. 19.32.—Electrical step controller.

and the heavy current switchgear a device which will operate with the small current handled by the 'stat. This is accomplished by a contactor, which is a common piece of electrical gear. The small current serves to close the solenoid circuit, which in turn pulls in the main switch.

Step Controller (Fig. 19.32)—It is sometimes necessary to operate from one thermostat a series of switches in sequence, as when two or three refrigeration compressors have to be started in turn with increases of load, or when the air heating is accomplished by electrical heaters requiring to be switched on in steps. For this purpose a step controller may be used as shown in the Figure. The switches are operated by cams fixed on a shaft driven through gearing from a small motor controlled by the thermostat or humidistat of modulating or step-by-step type.

Electrically-operated Valves—These control steam or water admitted to heaters and coolers, or refrigerant admitted to coils. Control in the *on-off* mode requires connection to a three-wire thermostat. The motor operating the valve contains limit switches which, at the end of half a revolution, cut off and re-set for reversal.

Another type is the *stalling motor*, which opens on making of circuit and closes on opening of circuit. It is self-closing on current failure, a desirable feature in some cases. This requires a two-wire thermostat only. Alternatively it may be arranged to operate in the opposite direction, i.e. opening when the switch closes and vice versa.

Electrically-operated Dampers—For operating louvres or dampers a *damper motor* is required. The power required to move dampers may be considerable and the motor must be chosen to suit. Sometimes the louvres will require to be sectionalized, each section being worked by one motor with linkage. Again, such damper motors may be on-off, i.e. open-shut, or they may be modulated to give settings in any intermediate position.

Electronic Controls—A development of the electrical modulating system is the electronic system, which dispenses with moving parts in the thermostats, such as sliding potentiometers. One such system is shown diagrammatically in Fig. 19.33. The temperature detecting element is, in effect, a resistance thermometer. Variations in resistance due to change of temperature cause a change in grid current in an electronic valve circuit, causing a much greater change in current passed by the valve. This may be magnified again by further valves and applied to operation of motorized valves, damper motors, etc. More recent developments replace valves with transistors.

A further development is the use of the *Thermistor* element for temperature control in an electronic circuit. A Thermistor is a resistance material having a negative characteristic; whereas most materials increase in resistance with increase in temperature, a Thermistor decreases in resistance. It furthermore gives a greater change in resistance for a given temperature change than other forms of element, and less magnification is necessary.

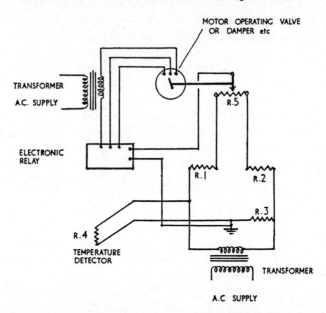

FIG. 19.33.—Electronic control circuit.

R1, R2, R3, are fixed resistances in bridge circuit. R4 is variable resistance depending on temperature. R5 is adjusted by motor to preserve balance.

Control Panels—The wiring to electrical controls on a big scheme becomes very involved, and it is usual to centralize all switches, pilot lights, etc., on a panel with labelling, so that the operation can be seen at a glance. This panel may also accommodate remote temperature and humidity indicators and recorders if required. A compact example is shown in Plate XXXV, facing p. 471.

PNEUMATIC CONTROLS

The compressed air for these controls, if derived from a central supply, is usually taken through a reducing valve at about 1 bar gauge. If the supply is provided independently a small air compressor and storage cylinder are required, automatically maintaining a constant pressure of perhaps 4 bar gauge and supplying again through a reducing valve so as to give a good storage of air for sudden demands.

The piping has to be carefully run to provide adequate drainage of the water which condenses out of the air. Traps are generally fitted at low points, and other steps taken to ensure dryness. Water may completely upset operation as also will oil—hence the use of oil-free compressors.

Where several plants exist, a central compressor may be used piped to each unit at high pressure, each having its own storage cylinder and reducing valve.

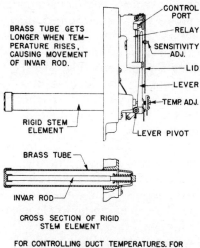

BRASS TUBE GETS LONGER WHEN TEM-PERATURE RISES, CAUSING MOVEMENT OF INVAR ROD.

CONTROL PORT

RELAY

SENSITIVITY ADJ.

LID

LEVER

TEMP. ADJ.

RIGID STEM ELEMENT

LEVER PIVOT

BRASS TUBE

INVAR ROD

CROSS SECTION OF RIGID STEM ELEMENT

FOR CONTROLLING DUCT TEMPERATURES. FOR CONTROLLING LIQUID, ELEMENT IS PROVIDED WITH PIPE THREADS.

FIG. 19.34.—Pneumatic immersion thermostat.

Pneumatic Thermostat (Fig. 19.34)—The type shown operates on the different rate of expansion of two metals, one a rod and the other a tube surrounding it. This movement opens and closes a pilot valve to control the main unit.

Pneumatic Humidistat—One type operates on the wet and dry bulb principle, comprising two normal thermostats connected so that when their differential changes, the pilot valve is caused to open or shut. Alternatively a hygroscopic material (Fig. 19.35) may be used.

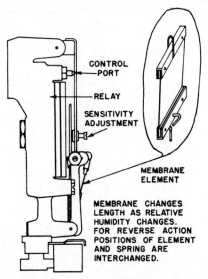

CONTROL PORT

RELAY

SENSITIVITY ADJUSTMENT

MEMBRANE ELEMENT

MEMBRANE CHANGES LENGTH AS RELATIVE HUMIDITY CHANGES. FOR REVERSE ACTION POSITIONS OF ELEMENT AND SPRING ARE INTERCHANGED.

FIG. 19.35.—Pneumatic membrane type thermostat.

Pneumatic On-Off Controls—The method of operation is the same as the above, but the pilot valve is so arranged as to remain open or shut against a spring, or with a lost-motion device until beyond a certain point of movement it closes or opens suddenly.

Pneumatic Actuating Equipment (Fig. 19.36) —*Pneumatic valves* operate by means of a diaphragm or copper bellows connected to the valve spindle. The air supply connects to the top of this and has a small orifice plate, or needle valve, allowing only a small quantity of air to pass. The diaphragm top also connects to the thermostat. When the pilot valve in the latter shuts, the air pressure builds up on top of the diaphragm and depresses the valve spindle to close or (in the reverse-acting type) to open the valve. When the pilot valve opens, the pressure on the diaphragm is released and the spring around the valve spindle forces the valve up again.

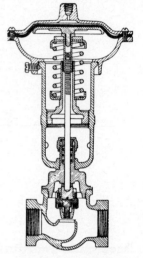

The air discharged from the pilot valve discharges to atmosphere. Any movement of the pilot valve causes the main valve to find a similar intermediate position, so giving a 'floating' control over the complete range. A pneumatic *mixing valve* operates in the

FIG. 19.36.—Pneumatic diaphragm valve.

same way, as will be appreciated.

Fig. 19.37—Pneumatic *damper motors* work on the same principle as for

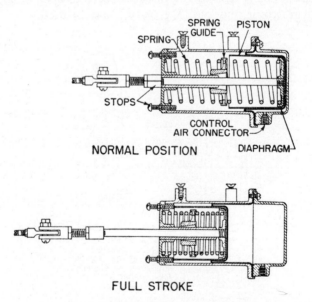

FIG. 19.37.—Pneumatic damper operator.

valves, except that the diaphragm is larger to give the necessary force to operate the damper.

Pneumatic Fittings—Each valve should have a control cock and pressure gauge. These are usually centralized on a board along with the thermometers and other instruments as with the electrical system, so that the operator can see the condition at a glance. The pressure gauges are connected to the diaphragm top in each case, so giving an indication as to whether the valve is open, shut, or in a mid-position. A main pressure gauge on the air supply will show whether or not the compressor is functioning.

Electro-pneumatic—A combination of electrical and pneumatic controls may be useful where room control is required. Pneumatic room 'stats are sometimes cumbersome, and may be replaced by electrical instruments which in turn control the air supply to pneumatic valves, etc., as before. This special application will be apparent, and need not be discussed further.

MISCELLANEOUS CONTROLS AND INSTRUMENTS

Static Pressure Control—This is a device to adjust main air flow to suit variable flow of a number of branches. If, for instance, volume control is applied to a zoned system to meet variable occupancy in the different zones, the main fan volume must be controlled; otherwise any branches left full open will receive a great excess of air if others are shut down. The static pressure controller is a pressure sensitive device connected to the main trunking system, and arranged, either by electric or pneumatic means, to regulate a set of louvre dampers in the fan discharge. Other methods have been devised for achieving the same object, notably *constant velocity control* which is preferable in some circumstances.

Instruments for Air-Conditioning—The results which are required to be achieved by an air-conditioning plant necessitate some supervisory equipment.

For a small direct expansion cooling plant for air-conditioning, say a small restaurant, perhaps nothing more than a thermometer on inlet and outlet is required. For a large plant, or a combination of several plants in a building, more extensive instruments are advisable.

For a multi-plant installation, the essential outlet conditions may be transmitted electrically to dials or recorders on a central control panel. A view of a comprehensive control and instrument panel is shown in Plates XVII and XXXV, facing pp. 214 and 471.

CHAPTER 20

Refrigeration for Air-Conditioning

THE COOLING NECESSARY FOR full air-conditioning is in nearly all cases effected by means of a refrigerating machine. Such a machine may be similar to the usual types of plant used for cold storage work, ice-making, etc., except that the temperature to be produced is not so low as in such applications.

Mechanical Refrigeration depends on the principle that a liquid can be made to boil at a low temperature if its pressure is reduced enough. To boil, it must be supplied with heat from an outside body or fluid. This outside fluid thus loses its heat and becomes cool. By choosing the right liquid, the heat for boiling can be extracted from the surroundings at low temperatures, without unduly low pressures: such liquids are known as refrigerants. The vapour given off on boiling is compressed (which makes it hot), is liquefied by removing this heat whilst still under pressure, and its pressure is then suddenly reduced, which brings it back to the state in which it started, whereby it can be made to boil at a low temperature.

A refrigeration plant (see Fig. 20.1), therefore, consists essentially of:

(a) A *compressor* to compress the refrigerating medium.

(b) A *condenser* to receive the compressed gas and liquefy it; the latent heat is taken out of the circuit by some external means. One method is to cool the condenser with circulating water, which may then be either run to waste or passed to a cooling tower for re-use. Alternatively, the cooling may be by air.

(c) An *expansion valve* in which the pressure on the liquid medium is reduced.

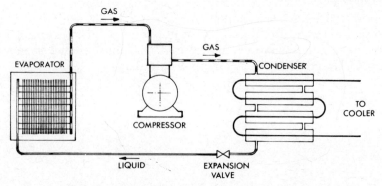

FIG. 20.1.—Principle of mechanical refrigeration cycle.

Plate XXXII. Backward bladed fan with inlet vane control (see p. 440)

Plate XXXIII. Packaged air handling plant and plant room (see p. 460)

Plate XXXIV. Packaged air handling plants serving hospital (see p. 460)

Plate XXXV. Composite desk type control panel with multi-point indicators (see p. 466)

Plate XXXVI. Shell for centrifugal machine being delivered to roof plant room (see p. 477

(d) An *evaporator* in which the medium re-evaporates, extracting heat from the surrounding material, i.e. from the cooling water or air in an air-conditioning plant, or from the brine where temperatures below freezing point of water are needed.

Application of Refrigeration—For use with air-conditioning using a washer and dehumidifier, the water from the latter is returned to the evaporator of the refrigerating plant generally at between 7 and 10° C depending on the dew-point to be maintained: in passing through the plant this is lowered by about 4 to 5 K. For the necessary heat transfer to take place, the refrigerant must be at some temperature below that of the water, but at the same time it must generally be slightly above freezing point. Thus, in a typical example, the following conditions might obtain:

Apparatus dew-point - - -	- 12° C
Washer outlet - - - -	- 9°
„ inlet - - - - -	- 6°
Water at evaporator outlet - -	- 5·5°

The refrigerant in the evaporator would in this case be maintained at about 1° C, giving 4·5 K differential for heat transfer. This small temperature potential means a very large cooling surface in the evaporator, and various devices have been developed to augment the transfer rate.

Where chilled water coils are used, water is circulated from the cooling plant in a closed system by a pump, and water temperatures down to about 3·3° C flow and 7° C return may be used.

Where brine is used in cooling coils to enable lower temperatures to be obtained (for example to achieve a low dew-point condition), temperatures may be taken down to – 7° C flow and – 3° C return, or lower as desired. The temperature to which the brine may be cooled is dependent on the strength of solution, data for which are obtainable from standard tables.

When a refrigerating machine is used in a direct-expansion air-conditioning system, the refrigerant is conveyed directly to the cooling coils in the airstream, and the surface temperature of these is dependent on the air-conditions required, on the form of coil surface and on the air speed over the coils. Refrigerant temperatures much below freezing point are inadmissible owing to the risk of freezing on the surfaces when dehumidification is being performed, as such freezing would block the air flow. An apparatus dew-point of 3° C is considered the minimum for direct expansion coils, to avoid frosting.

Refrigerating Cycle—The refrigeration cycle may be considered on a pressure-total heat (or enthalpy) diagram, as Fig. 20.2, which is drawn for Refrigerant 12.

Inside the curved envelope the medium exists as a mixture of vapour and liquid, and the increase of enthalpy from left to right on any pressure line within the envelope represents the increase in latent heat. Within this envelope lines of equal temperatures (isothermals) are horizontal.

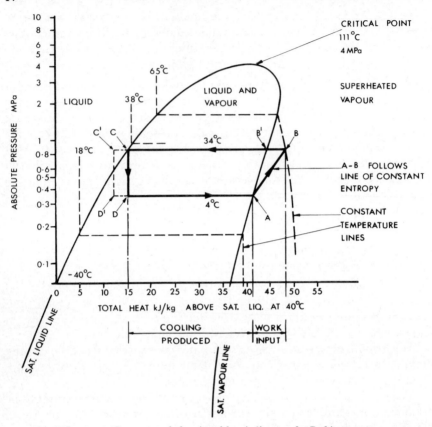

FIG. 20.2.—Pressure-enthalpy (total heat) diagram for Refrigerant 12.

Outside the envelope to the left the liquid exists at a temperature below its saturation temperature, and isothermals are almost vertical. To the right of the saturated vapour curve, the vapour exists in a superheated form, and isothermals curve downwards.

The *critical point* is that at which latent heat ceases to exist. It is not possible to liquefy a gas by pressure alone if it is above its critical temperature.

The refrigerating cycle may be represented by the figure ABCD;

A—B the gas is compressed causing rise in pressure and total heat which equals the energy put into the gas by the compressor—all in the superheat region. This takes place at constant entropy.

B—B′ is cooling the superheated gas in the condenser down to saturated vapour temperature.

B′—C is removing latent heat and condensing the gas to liquid in a the condenser.

C—D is pressure drop in the expansion valve, without change in total heat (adiabatic).

D—A is vaporization in the evaporator, latent heat—represented by increasing enthalpy—being drawn from the water or other medium being cooled. This is the cooling effect.

If the condenser is arranged to 'sub-cool' the liquid, say to point C', each unit mass of refrigerant in circulation will do more cooling work (D'A) and the cycle will be more efficient.

The ratio of cooling effect (as DA) to energy input (as AB), in terms of total heats, is termed *the coefficient of performance*. The smaller the range of pressure over which the cycle operates, obviously the less work will be required for a given cooling effect: hence, for economy of running, it is desirable to design for

(a) evaporator temperatures as high as consistent with other considerations (such as dew-point temperature in an air-conditioning application), and

(b) condenser temperature as low as possible. If cooling is atmospheric, the local weather records will decide the safe minimum to assume. Maximum cooling is generally wanted in hottest summer weather when the condensing arrangements are least efficient, so that caution is necessary in deciding on these operating temperatures.

Complete charts of properties of refrigerants are available in refrigeration literature for those who wish to pursue this matter further.

REFRIGERATING MEDIA

The factors affecting the choice of refrigerant will now be clear. A substance is required which can be liquefied at moderate pressures and which has a high latent heat of evaporation. The size of the compressor will then be kept down, and a relatively small amount of refrigerant need be circulated for a given amount of cooling. Such gases include ammonia, carbon dioxide, sulphur dioxide, and various organic gases.

Ammonia, whilst high in efficiency, and cheap, is ruled out in most cases for air-conditioning by the serious results which may follow a burst or leak. CO_2 calls for higher power input for a given capacity, much higher pressures and requires a skilled engineer for its operation. It is no longer used in air-conditioning work. SO_2 and methyl chloride are suitable only for small plants, but are not now used in air-conditioning applications.

Freon is a group name for a range of organic gases in which fluorine and chlorine are in combination with carbon in a number of different proportions. They are 'safe', non-toxic, colourless, non-inflammable and non-corrodible to most metals. Three of the forms in common use are:

R. 11 (trichloromonofluoromethane) CCl_3F, mainly used in centrifugal compressors for air-conditioning.

R. 12 (dichlorodifluoromethane) CCl_2F_2, used in piston compressors for air-conditioning and a wide range of other applications.

R. 22 (monochlorodifluoromethane) CHClF$_2$, used for low temperature work, as well as normal air-conditioning.

Air also may be used as a refrigerating medium. One method is to compress it to an absolute pressure of about 1·4 MPa (14 bar) and then, after removing the heat of compression, allow it to expand through a valve. At this pressure the air is not, of course, liquefied, and the latent heat is not therefore available. The specific heat being small, very large volumes of air must be handled to achieve the desired amount of cooling. In aircraft, the air cooling cycle is used in quite small turbo equipment running at very high speeds taking advantage of the extremely low temperature of the surrounding ambient for use in the condensing side. Another method is to use pressures high enough to liquefy the air; at least 20 MPa (200 bar)

TABLE 20.1

PROPERTIES OF REFRIGERANTS

	Refrigerant				
	Ammonia	R. 11	R. 12	R. 22	Water
Symbol	NH^3	CCl$_3$F	CCl$_2$F$_2$	CHClF$_2$	H$_2$O
Gauge Pressure (kPa)					
Condenser (30° C)	+1060	+24·8	+642	+1100	− 97
Evaporator (− 15° C)	+ 155	−80·6	+ 81·2	+ 195	—
Evaporator (5° C)	+ 404	−52·5	+254	+ 427	−100
Boiling Point (°C) (Standard Pressure)	− 33·3	+23·8	− 29·8	− 40.8	+100
Critical Temperature (°C)	1333	198	111	96	373
Volume of Vapour at − 15° C (m³/kg)	0.509	0·766	0·093	0·078	616
Latent Heat of Evaporation at 15° C (kJ/kg)	1320	198	162	218	2530
Theoretical unit Energy input per unit ÷ Energy output (kW/kW)	0.211	0·200	0·213	0·216	0·238
Coefficient of Performance (27° C to − 15° C)	4·75	5·00	4·69	4·65	4·20
Characteristics	Strong irritant. Forms explosive mixture under certain conditions	Odourless Non-toxic Non-irritant Non-inflammable Innocuous			Wet
Used in	Large Plants. Piston Machines.	Centrifugal Plants. All sizes of Piston Machines.			Centrifugal Plants. Steam Jet Plants.

is required. With either method expensive plant is necessary, but the weight penalty in aircraft applications is of course the principal criterion. For normal land use, bearing in mind the relative cost of plant required, air is not a practical choice as a refrigerant for air-conditioning.

Water is also used as a refrigerant in a manner described later.

The refrigerant most suitable for direct expansion into coils in the air-way is R.11, R.12, or R.22, the others all being objectionable owing to their smell, inflammability, or inefficiency. Table 20.1 gives the properties of R.11, R.12 and R.22, with ammonia and water for comparison.

TYPES OF REFRIGERATION PLANTS

Types of refrigerating plants are:
Piston or reciprocating.
Rotary (which is another form of piston but without valves and mainly used in small sizes).
Helical Screw.
Centrifugal.
Absorption.
Steam jet.

Piston Compressors—The reciprocating type of machine may consist of one, two, three or four cylinders, according to the load, suction and discharge valves being operated by suction and pressure only and not mechanically.

Fig. 20.3 shows a diagram of a piston-compression unit of the older slow speed type occupying considerable space for a given duty.

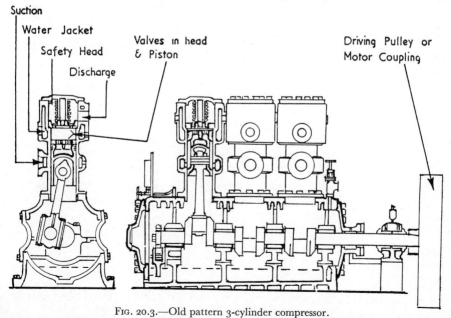

FIG. 20.3.—Old pattern 3-cylinder compressor.

The gland where the crank shaft passes through the casing is of rotary type and various systems are used for keeping this gas-tight. The cylinders are water-cooled. An oil separator is a necessity with ammonia machines to prevent oil from the crank case being carried over into the coils.

A more modern type of reciprocating compressor running at a higher speed is one in which the cylinders are arranged in V or W formation, as shown in Fig. 20.4. Four, six or eight cylinders may be arranged in one block, thus enabling a large duty to be obtained in a small space with marked freedom from vibration. Two such machines may be coupled to one motor on a common centre line, and in this way capacities up to 700 kW may be obtained in one set.

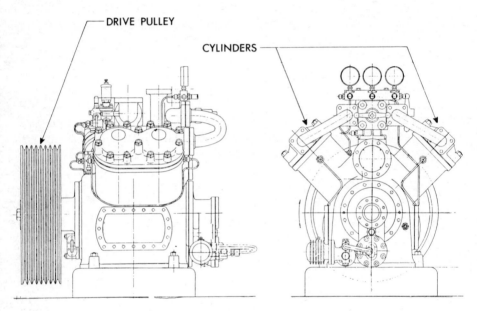

FIG. 20.4.—4-cylinder V type machine, 140 kW.

Totally enclosed Freon compressors are available in which the motor is contained within the compressor casing, so dispensing with the need for a gland. These are termed 'Hermetic' (see Fig. 20.5). Such sets are usually quiet-running and particularly suitable for self-contained air-conditioning units.

Multi-cylinder machines are sometimes provided with a bye-pass, or unloading valve, on one or more cylinders, enabling the load to be varied either by hand or automatically. Variation of output may also be achieved by speed regulation.

Centrifugal Compressors—The centrifugal compressor is used chiefly where large duties are required. The advantages are:

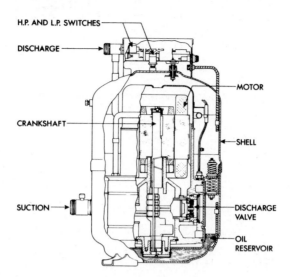

FIG. 20.5.—Hermetic compressor.

(1) Saving of space as compared with reciprocating machines.
(2) Absence of vibration (thus suitable for a roof-top plant chamber, see Plate XXXVI, facing p. 471).
(3) Reduced maintenance due to there being no wearing or reciprocating parts.
(4) Throttling to suit load gives corresponding power reduction.

Refrigerants used in centrifugal compressors are usually one of the Freon group, such as R. 11, which for air-conditioning temperatures operate at low pressures and over a small pressure range, thus reducing slip losses between blades and casing.

An important consideration is the turn-down range, i.e. the ability to follow load variations. The two-stage machine shown in Fig. 20.6 has a turn-down to 10 per cent of full load, thus making it highly flexible in operation.

In the diagram (Fig. 20.6) of this centrifugal plant, it will be noted that the condenser and evaporator are close-coupled to the compressor and usually form one complete unit. Plate XXXVII, facing p. 502, shows three such units. The capacity of each unit is 2·5 MW.

Helical Screw Compressor—Compression is produced by two rotating helical screws, the seal being achieved by oil. A retractable vane enables load variation in a simple manner over a wide range. This form of compressor, Fig. 20.7, may be used with a wide range of refrigerants up to a capacity of about 2 MW and offers the advantages of minimum vibration and low noise level.

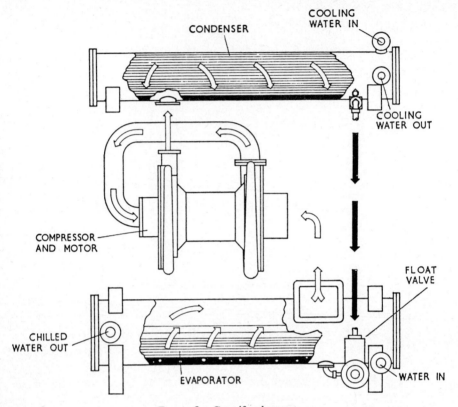

FIG. 20.6.—Centrifugal system.

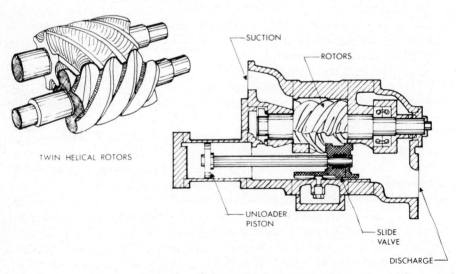

FIG. 20.7.—Helical screw compressor.

Absorption Plant—One form of absorption plant is shown in Fig. 20.8. It uses lithium bromide in solution with water and has no moving parts except pumps, the source of heat energy being either steam, hot water or gas. The heat to be removed by the condenser water is greater with this system than with mechanical plant. As a unit of nominal 350 kW capacity uses a nominal steam quantity of 220 g/s, the heat rejected to the cooling tower is some 900 kW, as compared with 540 kW for mechanical compression refrigeration. The plant works under a high vacuum.

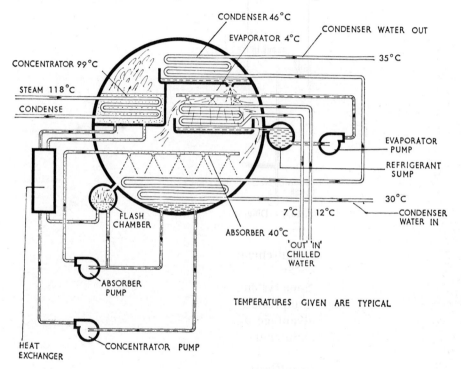

Fig. 20.8.—Diagram of absorption system.

Steam-Jet Plant—An interesting type of refrigerator using water as the medium is that shown in Fig. 20.9. Its operation depends on the possibility of causing water to boil at low temperatures under high vacua. Thus, at 7° C, water boils at an absolute pressure of 1 kPa: i.e., about one hundredth of atmospheric pressure.

The water used is the same as that circulated to the washer or cooling coil, so that temperature difference due to heat exchange as in a refrigerant evaporator is avoided. The absence of any special refrigerant is an advantage and an economy.

The heat input with this equipment is much greater than with the positive compression types owing to the inefficiency of jet compression. All the steam used has to be condensed and the heat rejected at the cooling

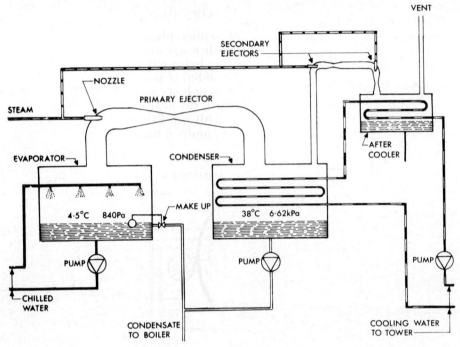

FIG. 20.9.—Diagram of steam-jet system.

tower. The condensing requirement is about five times that of a mechanical compression plant.

A variation of the same system, but using a centrifugal compressor in place of the steam-jet compressor, has also been developed. It avoids the above mentioned disadvantage of high heat input. The great difficulty with both types is the maintenance of the extraordinarily high vacuum for long periods.

Choice of Refrigeration Plant—The selection of refrigeration plant will depend on a number of factors, among which are: available space and location, whether water is available, whether noise is important and whether condenser heat must be rejected at a distance from the compressor.

Hermetic piston-type compressors are widely available up to 30 kW and are used in combination to make larger capacities. Such machines are obtainable built as a weatherproof unit complete with an air-cooled condenser. A low silhouette arrangement is available for installation on flat roofs. The refrigerant lines must be run to the evaporator located inside the building at a not too distant point.

Refrigeration units of all sizes may be water-cooled and they must be so arranged if there is no free access to open air. Up to about 15 kW they are sometimes cooled by fresh water run to drain, but larger sizes are connected to cooling towers. The cooling tower may be located a considerable distance from the refrigeration plant, for instance a plant in a basement may be connected to a cooling tower on the roof.

Evaporative condensers are available in sizes up to 1 MW. They give an increase in efficiency because one exchange of heat is eliminated as compared with a water-cooled condenser connected to a cooling tower. An evaporative condenser must be installed fairly close to the compressor, either inside the plant room or outside in the open. Its design must be carefully considered where capacity control is required and where winter operation is envisaged. It is used only with reciprocating compressors.

Open type or non-hermetic reciprocating compressors are available in sizes up to 750 kW and beyond. They may be connected to direct expansion air coolers or water chillers on the suction side and to air-cooled, water-cooled, or evaporative condensers on the discharge side. The large air-cooled condenser would only be used in dry or desert climates.

Centrifugal or turbo compression refrigeration plant is used in sizes from 500 kW to 5 MW and has been made up to 14 MW. This plant invariably includes a water or brine chiller and water-cooled condensers, using water from a cooling tower, or other economical source are normally incorporated.

The absorption refrigeration unit has been developed for use in air-conditioning and is attractive where a supply of cheap heat or exhaust steam may be freely obtained. Sizes of 300 kW to 2 MW are usual. It is finding a new field of development making use of natural gas as the heat source.

Steam jet refrigeration is not in wide use, but has been found economical in large industrial enterprises where exhaust steam is available.

REFRIGERATING PLANT COMPONENTS

Evaporators—The evaporator is tubular in form. The tubes may contain the refrigerant and the whole is immersed in the liquid to be cooled. In the *dry system* the evaporator coils are filled with vapour having little liquid present. When operated on the *flooded system* the liquid refrigerant discharges into a cylinder feeding the coils by gravity. As evaporation takes place the gas returns to the top of the cylinder and from there returns through the suction pipe to the compressor.

A more common form of evaporator for application to air-conditioning practice is the shell and tube type, used with a closed circuit system such as with cooling coil surface. It is necessary to include safety cut-outs to prevent icing up. The water is contained in the tubes and the refrigerant in the shell.

A development of the shell and tube type is the direct-expansion shell type evaporator in which the refrigerant is in the tubes and the water in the shell. The refrigerant tubes may be arranged in two or more circuits, each with its own expansion valve and magnetic valve on the liquid inlet to allow step control. Different arrangements of baffles in the shell control the water velocity over the tubes to improve heat transfer.

In the centrifugal machine the evaporator forms part of the unit, the water passing through the tubes and the refrigerant being dripped over the outside. Plate XXXVIII shows pipework associated with a plant of this type (two of four 4 MW units).

It is generally not necessary to resort to brine for air-conditioning purposes, as chilled water at about 4° C satisfies all normal requirements. Precautions against accidental freezing of the water in the evaporators include low suction pressure cut-outs and low water temperature cut-outs as well as water-flow switches.

Modern water-chilling plant does not generally require the addition of chilled-water storage to act as a 'flywheel', because it can be arranged in suitable steps of capacity control to suit the variations in load. Little complication is experienced with piston compressors having four to eight steps of control. Centrifugal units which turn down to 10 per cent can virtually run on pipeline 'losses' as can screw-type machines.

In cases where machines are limited to a small number of starts per hour, it is possible to fit a 'hot gas by-pass' valve which automatically passes hot gas from the condenser into the evaporator to provide some load. This can come in automatically at a predetermined point on the capacity controller travel and enables the machine to run continuously when there is no cooling load at all.

Chilled-water storage-tanks should be avoided in general because they require the chilled water to be produced at a lower temperature than is required to be used. This is because water at about 4° C does not stratify as in a hot-water calorifier but tends to mix. Some ingenious arrangements of weirs and baffles have been used in chilled-water tanks to overcome this difficulty, but they add complications to the system. At the present day a chilled-water tank would be justified in connection with perhaps two to four small single-step cooling units, so providing water chilling in steps.

Condensers—The evaporative condenser, consisting of coils of piping over a tank from which water is circulated and dripped over the pipes, is the simplest form of condenser. Its use is restricted to cases where the compressor can be near to the condenser, otherwise long lines of piping containing refrigerant under pressure are necessary.

More often in air-conditioning systems, the condenser takes the form of a heat exchanger of the shell and tube multi-pass type. Circulating piping from the condenser to the water cooler with these types alone traverses the building, and all equipment containing refrigerant is then confined to the plant room.

Air-cooled condensers are used in most small self-contained or split package systems and in duties of up to 350 kW in capacity, Fig. 20.10. They are often found convenient to use and economical in first cost in this country in sizes up to say 100 kW. A recent interesting development takes

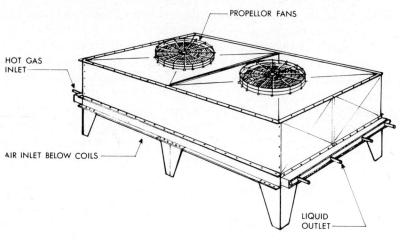

PROPELLOR FANS

HOT GAS INLET

AIR INLET BELOW COILS

LIQUID OUTLET

FIG. 20.10.—Air cooled condensers.

the form of a totally packaged unit comprising evaporator, centrifugal chiller and multiple-fan air cooled condenser as shown in Fig. 20.11. Sizes are limited to about 1 MW simply because this is the largest package which can be transported on a conventional low-load lorry.

Air-cooled condensers are often used in the tropics, despite the fact that they are bulky, in circumstances where the outside air temperature is

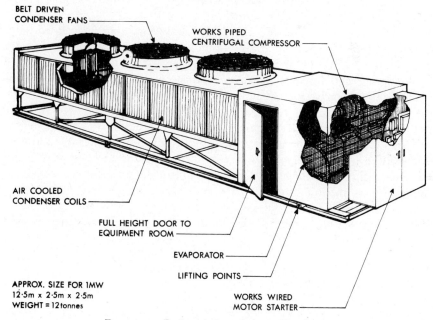

BELT DRIVEN CONDENSER FANS

WORKS PIPED CENTRIFUGAL COMPRESSOR

AIR COOLED CONDENSER COILS

FULL HEIGHT DOOR TO EQUIPMENT ROOM

EVAPORATOR

LIFTING POINTS

APPROX. SIZE FOR 1MW
12·5m x 2·5m x 2·5m
WEIGHT = 12 tonnes

WORKS WIRED MOTOR STARTER

FIG. 20.11.—Packaged air cooled centrifugal unit.

high. This apparent paradox arises from the simple fact that evaporative condensers and cooling towers require some measure of skilled maintenance, whereas the air-cooled condenser requires little attention other than simple basic cleaning and, of course, some lubrication. A further advantage is that it requires no water supply in those places where such may happen to be scarce.

Evaporative Coolers—The heat extracted by the refrigerating machine, together with the heat equivalent of the power input to the compressor, raises the temperature of the condenser water by an amount which is dependent upon the quantity of the water which is circulated through the condenser.

The lower the temperature of the condenser water the less power will be required to produce a given refrigerating effect; and it also follows, conversely, that with a given size of plant the greater will be the amount of cooling possible.

Water from a well or from the main supply will always be the coldest, the former at $12°$ C and the latter at about $18°$ C in summer. The quantity to be wasted, however, generally rules this method out. For instance, a **700 kW plant with power input of about 120 kW with a 10 K rise** through the condenser would require

$$\frac{700 + 120}{10 \times 4 \cdot 2} = 20 \text{ litre/s}$$

This at 2p per 1000 litres would exceed the cost of current for running the compressor by about three times.

Applications exist outside the U.K., however, where well water is more freely and cheaply available, which produce economical solutions. In one known instance in Europe, such well water is first passed through a pre-cooling coil integral to the air-handling plant prior to use in the condenser. Similarly, in some parts of the Caribbean, clear sea water is available via fissures in the coral which may be pumped in quantities of up to 1000 litre/s through specially designed condensers.

However, cooling the water by evaporation is more generally adopted. Evaporative coolers depend on the ability of water to evaporate freely when in a finely divided state, extracting the latent heat necessary for the process from the main body of water, which is then returned, cooled, to the condenser. In the case stated above, the consumption of water with an evaporative cooler (assuming no loss of spray by windage) would be only

$$\frac{820}{2258 \text{ (latent heat)}} = 0 \cdot 36 \text{ litre/s}$$

Evaporative coolers divide themselves into two categories, i.e., *Natural Draught* and *Fan Draught*. The former is represented by:

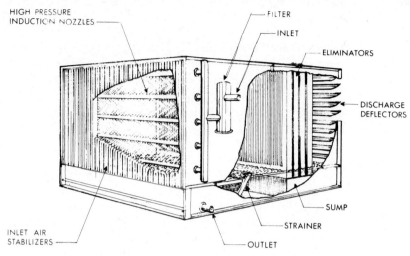

HIGH PRESSURE
INDUCTION NOZZLES

FILTER

INLET

ELIMINATORS

DISCHARGE
DEFLECTORS

SUMP

STRAINER

OUTLET

INLET AIR
STABILIZERS

FIG. 20.12.—Horizontal counterflow water cooler.

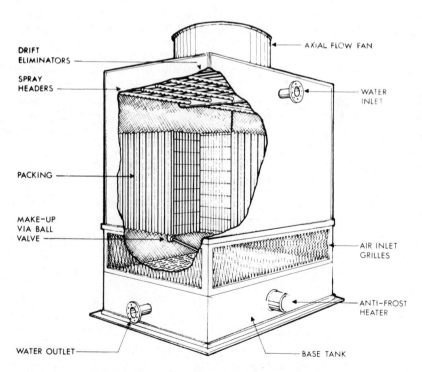

DRIFT
ELIMINATORS

SPRAY
HEADERS

PACKING

MAKE-UP
VIA BALL
VALVE

WATER OUTLET

AXIAL FLOW FAN

WATER
INLET

AIR INLET
GRILLES

ANTI-FROST
HEATER

BASE TANK

FIG. 20.13.—Vertical induced draught cooling tower.

(a) *The Spray Pond.* In this type the water to be cooled is discharged through sprays over a shallow pond in which the water is collected and returned to the plant. To prevent undue loss by windage the pond is usually surrounded by a louvred screen. Owing to the large area needed for the spray pond system and, moreover, the consequent probability of pollution, this type of equipment is seldom possible for air-conditioning applications.

(b) *Cooling Tower.* In this, the water is pumped to the top of a tower which contains a series of timber or specially designed plastic slats arranged so as to split up the water stream and present as large an area as possible to the air, which is drawn upwards due to the temperature difference, and by wind. The base of the tower is formed into a shallow tank to collect the water for return to the plant. Again, owing to its size and height, this type of cooler is not frequently used for air-conditioning.

(c) *Condenser Coil Type.* Where the refrigerating machine is near the point where an outdoor cooler may be used, the condenser heat exchanger may be dispensed with and the refrigerant is then delivered to coils outside, over which water is dripped by a pump. The water collects in a tank at the base and is recirculated. A louvred screen is usually necessary surrounding the coils.

Fan draught systems are more commonly used owing to the compact space into which they may be fitted. Where possible they are placed on the roof, but if this is impracticable they may be used indoors, or in a basement, with ducting connections for suction and discharge to outside. One system in Paris uses the air extracted from an underground car park for this purpose, thus economising in the overall power requirement of the engineering systems.

The following four methods are typical:

(a) *Air Washer Type*, Fig. 20.12. This is similar in some respects to an air washer containing several banks of sprays, base tank, and eliminators in the normal way. 'Packing' may be incorporated to increase the wetted surface exposed to the air stream.

(b) *Induced Draught Cooling Tower*, Fig. 20.13. In this the water is delivered to the top by the condenser pump, and delivers over a 'packing' of timber or plastic slats as with the natural draught type. An axial fan is arranged to draw air upwards over the slats at high velocity so that a much reduced area of contact is required.

(c) *Film Cooling Tower*, Fig. 20.14. The water is not sprayed or

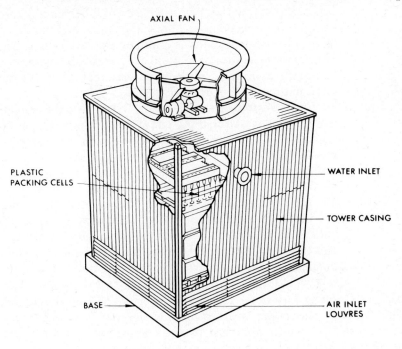

FIG. 20.14.—Film type cooling tower.

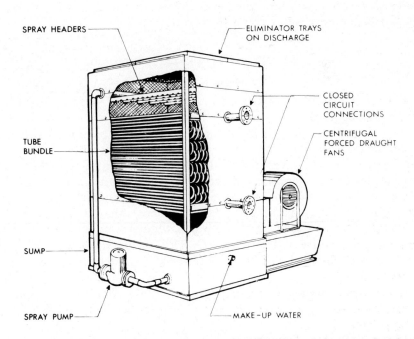

FIG. 20.15.—Closed circuit water cooler.

broken up in any way, but is allowed to fall from top header troughs down wooden or plastic slats arranged in egg-crate form. The water remains as a film on the surface of the slats. Higher air velocities than usual are permissible without risk of carry over of water, and yet air resistance is low due to the open nature of the surfaces.

(d) *Closed Circuit Water Coolers*, Fig. 20.15. In this case the condenser water is not evaporated but circulated through tube bundles which act, as it were, as the packing within the tower. In most other respects such coolers resemble induced draught units, and an independent water supply is required for the local evaporation circuit which is pumped to the tower top for spray or gravity fall over the tube bundles. Such coolers are useful for application to heat recovery systems where fouling of condenser water must be avoided. Due to the additional heat-exchange inherent to the circuit and consequent loss of efficiency, this type of tower tends to be relatively large in size.

Rating of Cooling Towers—The heat to be removed from the condenser cooling water by the cooling tower is equal to the sum of cooling load plus heat equivalent of power absorbed by the compressor. An approximate figure of 1·2 kW per kW of refrigeration may be used.

The quantity of water to be passed through the tower is dependent on the temperature drop allowable between inlet and outlet. A usual figure is 5 K and the inlet water flow will then be

$$\frac{1 \cdot 2}{5 \times 4 \cdot 2} = 0 \cdot 057 \text{ litre/s per kW of refrigeration.}$$

Part of this will be evaporated as above and made up by fresh water through the ballcock.

The temperature to which the cooling tower may be expected to cool the condenser water depends on the maximum wet bulb of the atmosphere and the design of the tower: the higher the efficiency the closer will be the water outlet to the wet-bulb temperature. A good efficiency will give a difference of about 3 K, so that, if the maximum wet bulb is taken at 21° C (the highest in England except on rare occasions), the cooling water will be brought down to 24° C, and, with 5 K temperature rise through the condenser, the outlet will be at 29° C. This temperature then, in turn, forms the basis for design of the compressor and condenser (e.g. condensing at about 38° C).

Certain new designs of cooling towers achieve considerable capacity in minimum space, an example of which is shown in Plate XXXIX, facing p. 503.

Condenser Water Treatment—As evaporation proceeds, there is a continuous concentration of scale-forming solids which may build up to such a degree as to foul the condenser tubes (as in the example shown in Plate XL, facing p. 503). Similarly, the evaporative surfaces of the cooler suffers a build-up of deposit.

In order to overcome this problem:

(a) a constant bleed-off of the water is required from the base tank of the cooling tower, the rate of which bleed-off may be calculated from the known analysis of the water, the evaporation rate and the maximum concentration admissible;

(b) as is common practice, the water can be treated by a regular chemical dosage which also generally contains an additive to prevent algae growth.

In the case of the closed circuit water cooler, the condenser tubes and the tower tube bundles do not suffer internal fouling but the external surfaces of the latter will, in time, become coated with solids. The need for water treatment of the spray water thus remains.

Packaged Cooling Towers—Whereas past practice was to use structural containment to form the tower proper and use purpose built internals, water distribution arrangements etc., cooling towers are now commonly factory fabricated for delivery to site in a minimum number of parts. Whilst this development reduces site works, there is a tendency to design to minimum size also with the result that the depth and capacity of the base tank is inadequate. This condition may lead to problems of overflow when the condenser pumps are stopped or, more seriously, to the formation of a vortex at tank outlet due to an insufficient water depth.

CHAPTER 21

Ductwork Design and Sound Control

HAVING DECIDED ON THE type of air system to be employed, made the calculations of air quantity and temperature, and considered the form of the central plant, it is now necessary to consider in more detail the characteristics of the ductwork system which will convey the conditioned air about the building. Most of the discussion which follows applies equally to ventilating or air-conditioning systems.

MEASUREMENT OF AIR FLOW

Pitot Tube—Reference has been made to the Pitot tube as a means of measuring static and velocity pressure in assessing fan performance. It may be used in its latter capacity to determine the velocity in a duct.

The relationship between speed and pressure is:

$$p_v = 0.5 \rho v^2$$

where p_v = Velocity pressure, Pa

ρ = density of fluid, kg/m³

v = Velocity, m/s

for standard air $\rho = 1.2$ kg/m³

thus $p_v = 0.6\,v^2$

Fig. 21.1 gives the relationship graphically for air velocities encountered in ventilation work. For other temperatures and pressures

$$\rho = 1.2 \left(\frac{\text{Working pressure kPa}}{101.325}\right)\left(\frac{293}{273 + t}\right)$$

where t = temperature of air °C and working pressure = atmospheric pressure (approximately 101.325 kPa) ± static pressure.

Micro-manometer—The pressures at low velocities are slight, as will be noted, and for the purpose of reading them an ordinary U-tube or inclined gauge is too insensitive. The micro-manometer, of which one type is shown in Fig. 21.2, is therefore necessary. In this an extended U-tube is used, being tilted by a micro-adjustment. The level is viewed through the magnifying eyepiece against a crosswire. The coarse reading is taken on the side scale, and the fine reading on the rotating dial, one revolution of which corresponds to one division of the scale. The liquid may be alcohol and the scale is calibrated in Pa or dPa. Other still more sensitive instruments are made.

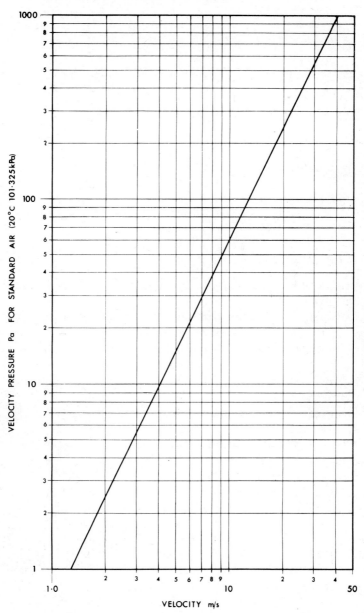

FIG. 21.1.—Velocity pressure related to air speed for standard air.

When measuring air speeds in a duct, however, the velocity varies across the duct, being greatest at the centre and least at the periphery. Thus, in a round duct it is necessary to take readings at a number of points in concentric rings of approximately equal area. The average speed multiplied by the area of the duct will give the volume of air passing. In

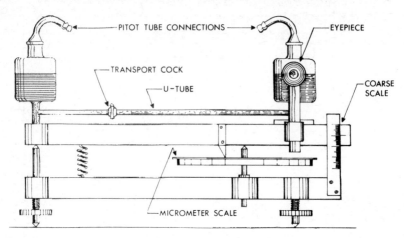

FIG. 21.2.—Micro-manometer.

the case of a rectangular duct the method is similar, except that the duct is divided into equal rectangles.

The Rotating Vane Anemometer—This instrument (Fig. 21.3) measures air speed by vanes which revolve as the air impinges on them. The dials, which are calibrated in metres, serve only to count the revolutions over a given time, such as one minute, taken by a stop-watch.

The lever at the top enables the gearing to be engaged and disengaged from the vane at the start and finish of the time. The knob shown on top is depressed to cancel the reading and return the dials to zero.

The instrument requires to be calibrated periodically, and a correction factor applied. The standard instrument is too insensitive for use below about 0·5 m/s. Above about 15 m/s a high speed type is used.

FIG. 21.3.—Rotating vane anemometer.

For general work it is a useful instrument, chiefly for measuring the air speed from or to openings, etc. In such case the instrument is placed about 30 mm from the grille and the speed is multiplied by the whole face area (regardless of free area) to obtain the volume in m³/s. Readings are taken at various points and averaged. It is not so useful for measurements in ducts, as the anemometer has to be introduced through a hole in the duct wall and is then difficult to manipulate.

It must be admitted that the anemometer requires very careful handling, and in the hands of an unskilled operator entirely erroneous results can be obtained.

The Hot Wire Anemometer—Originally developed for laboratory use, this type of instrument is particularly useful for measurement of low velocities such as occur in occupied spaces. The measuring head consists of a fine platinum wire which is heated electrically and inserted in the air stream. Air movement cools the wire and, by calibration, current flow may be metered to indicate the velocity sensed.

The Swinging Vane Anemometer—An instrument developed for site use, (the 'Velometer'), is shown in Fig. 21.4. It relies on the speed of air to deflect a shutter against a spring. The shutter causes the needle to move over the sector-shaped dial, reading direct in metres per second. It requires no timing.

Fig. 21.4.—Swinging vane anemometer.

When used direct, as in the illustration, for measuring speeds from or to gratings at low velocities, it reads from zero to 1.5 m/s. Higher ranges of speeds, 0–15 and 5–30 m/s, are covered by means of adaptors screwed into the aperture at one end. These may have flexible tube connection for reading at a distance, which, in turn, may be used with various special mouthpieces fitted to the end of the tube to read velocities in ducts etc. Altogether this is a suitable instrument for giving a quick approximate check on adjustments or for exploring velocities over an area.

AIR DISTRIBUTION IN DUCTS

Ducts for ventilation and air-conditioning are commonly constructed either of galvanized sheet steel or in 'Builders' Work'. The principles to be followed in the design of such ducts are:

(*a*) Avoidance of sudden restrictions or enlargements, or any arrangement producing abrupt changes of velocity.

(*b*) Bends to be kept to a minimum, but where required should have a centre-line radius not less than one and a half times the diameter or width of duct at the bend. Alternatively, deflectors should be fitted into the duct, of the type shown in Fig. 21.5, which permit of right-angled turns being used with much reduced pressure loss, and reduction of noise.

(*c*) Where branches occur, they should be taken off at a gradual angle before turning.

(*d*) Sharp edges should be avoided, as these will be the cause of noise which may travel a considerable distance through the duct system.

(*e*) Rectangular ducts should be as nearly as possible square, the more they depart therefrom the more uneconomic they become.

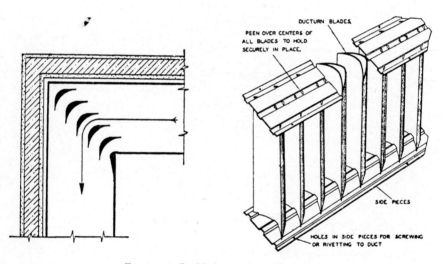

FIG. 21.5.—Prefabricated duct deflectors.

Galvanised Steel Ducts—The gauges of metal usually used are set out in a series of well presented documents* issued by the H.V.C.A. together with much information regarding constructional techniques for both low-pressure and high-pressure systems. (Plate XLI, facing p. 503.)

Ducts of galvanised sheet steel are often priced by weight. Sheet metal is made in metric thickness and the mass of metal will thus merely be thickness times area times 7800 kg/m³ (for steel) or whatever other specific mass is appropriate.

---

* Metric Specification DW. 141 combines those data previously issued in DW. 112, 121, 132 and 161. Heating and Ventilating Contractors Association.

Rectangular ducts are made up out of flat sheets by bending, folding and riveting and are erected in sections with slip joints or angle ring joints bolted. Larger sizes tend to drum and adequate stiffening is necessary. Sometimes a 'diamond break' is used to assist in stiffening, as Fig. 21.6, the sheets being pressed to provide, in effect, very shallow pyramid form. Where rectangular ducts are used in high velocity systems, stiffening is particularly important using bracing angles, tie angles, and internal tie rods for larger sizes.

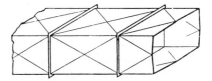

Fig. 21.6.—Diamond-break duct stiffening.

Circular ducts may equally be formed from flat sheet with folded or riveted seams and are inherently free from risk of drumming. Traditionally, the use of good quality circular ductwork was largely confined to industrial applications. Mass produced 'snap-lock' duct lengths were used for low cost domestic warm air heating, reliance being placed upon jointing tape for air-tightness.

The advent of circular ducts made from galvanized strip steel with a special locking seam, as in Fig. 21.7, has brought adoption of their use more generally. Sizes are available from 75 mm up to 800 mm diameter. Ducts of this type are particularly suitable for high velocity systems due to their rigidity, 'deadness' and air-tightness. Ranges of standard tees, bends, reducers and other fittings are made and these, together with lengths of straight ducting cut off on site as required, facilitate the task of

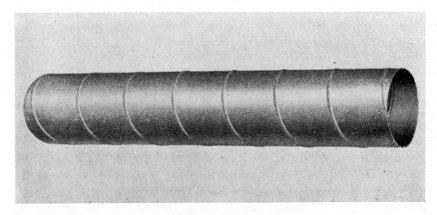

Fig. 21.7.—Spiral-wound circular duct.

erection compared with the tailor-made methods previously adopted.

Jointing is by means of a special mastic, with some riveting, and in best practice a 'heat shrink' plastic sleeve is used externally to seal the joint. Such sleeves are manufactured from a flexible cross-linked poly-ethylene tube coated internally with an adhesive: both tube and adhesive contract when heated by a propane torch.

New methods of jointing ductwork appear from time to time and a recent development from Scandinavia offers promise. Here, the duct fit-tings have a groove formed at each end which houses a sealing gasket and locking collar. Connection is made by pushing the duct and fitting to-gether which forces the locking collar into position to provide an air-tight joint.

Flat Oval or spirally wound rectangular ducts are currently coming into increasing use in air-conditioning. These are formed from spirally wound circular ducts to the profile illustrated in Fig. 21.8 and are available in sizes from 150 mm × 550 mm to 500 mm × 980 mm. A limited range of standard tees, bends and reducers is made. Such ducts have the advantage of being mass produced and can be used in conjunction with circular sections where space is limited. Site jointing, as before, is completed by a 'heat-shrink' sleeve.

Builders' Work Ducts—It is often possible, in the design of a building, to construct main ducts and large rising shafts as part of the fabric. Where this can be done it has the following advantages:

(*a*) Rigidity, and hence reduction of noise transmission.
(*b*) Permanence.
(*c*) Cheapness, in that the space very often exists in any case, and only requires to be made smooth and air-tight.
(*d*) Reduction of heat gains and losses, due to the heavy construc-tion as compared with metal.
(*e*) Accessibility for cleaning.

A disadvantage is the 'flywheel' effect of heavy constructions, and hence for applications where careful control of the air temperature is required, such ducts must be insulated *on the inside* and great care is needed to ensure air-tightness.

Fig. 21.9 shows an example of a duct constructed over a basement corridor and serving as a main distribution trunk from the plant to rising shafts about the building.

Ducts of Other Materials—Other materials used for ducts are:

(*a*) *Welded Sheet Steel* is mainly used in industrial work, or where the air-tightness of the joints is of great importance; such ducts may be 'galvanised after made'.
(*b*) *Copper, Aluminium, Stainless Steel.* Ducts constructed of these materials are used in special cases where permanence or a

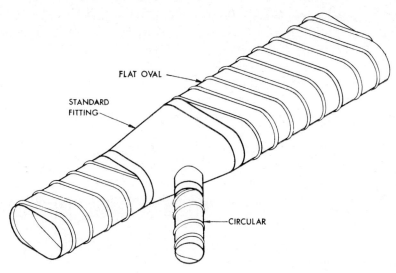

FLAT OVAL

STANDARD FITTING

CIRCULAR

FIG. 21.8.—Flat-oval duct and branches.

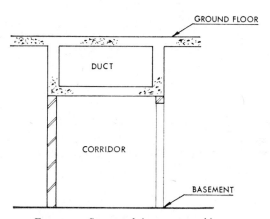

GROUND FLOOR

DUCT

CORRIDOR

BASEMENT

FIG. 21.9.—Structural duct over corridor.

high degree of finish is required, or where special corrosion problems arise.

(c) *P.V.C.* which is chiefly used for chemical fume extraction and tends to be expensive due to the cost of moulds etc. for fittings and special sections.

(d) *Glass fibre resin bonded* slabs can be cut and adapted at site, and jointed to form a smooth continuous duct. This technique is much favoured in the USA but has not been much used in this country although a H.V.C.A. specification exists (DW. 191) to codify construction methods. The material is, of course, suitable for low velocities only.

Duct Sizing for Normal Velocities—The most convenient method of sizing ducts is on the *Equal Pressure-loss* basis. For this purpose use may be made of Figs. 21.10 and 21.11 and Tables 21.1, 21.2 and 21.3.

The following procedure is adopted:

(1) On the plan of the duct system it is necessary to mark the volume of air to be delivered or exhausted at each outlet. These must be totalled back to the fan, including the sums brought in at each branch. These volumes are conveniently expressed in m³/s or litre/s.

(2) Establish the maximum velocity in the main duct leaving or entering the fan according to the type of building, etc., from Table 21.1. The velocities given in this table are arbitrary. The higher the velocity the greater the noise, hence low velocities are desirable for buildings where silence is required. Some experience is necessary in interpreting this aspect of the matter.

TABLE 21.1

MAXIMUM DUCT VELOCITIES FOR CONVENTIONAL SYSTEMS FOR VARIOUS TYPES OF BUILDINGS

Application	Main Ducts and Shafts	Branch Ducts
	m/s	m/s
Hospitals, Concert Halls, Theatres, Libraries, Film Studios, etc. - -	5	4
Cinemas, Restaurants, Assembly Halls, etc. - - - - -	7.5	5
General Offices, Dance Halls, Shops, Exhibition Halls, etc. - -	9	6
Factories, Workshops, etc. - - - - - - - -	10 to 12	7.5

(3) From the chart in Fig. 21.10, select the velocity line for the main duct, and find the point of intersection with the air volume carried in the main. The vertical line passing through this point will give the pressure loss per metre run of duct by reference to the bottom scale, and at the same time the diameter of main duct, assumed to be circular.

(4) The sizes of the subsequent sections of the main duct on to the end are then read off at the point of intersection of the appropriate horizontal volume lines with the same vertical resistance line as for (3). It will be observed that as the volume reduces, the velocity also becomes less.

(5) The total pressure loss of the main duct is then calculated by tabulating thus:

(a) Section (Ref. Letter or No.)	(b) Volume litre/s	(c) Duct Dia- meter mm	(d) Length m	(e) Resist- ance per m from Chart Pa	(f) Pressure loss for Length (d) Pa	(g) Single Resist- ances Pa	(h) Total for Section Pa	(i) Progres- sive Total Pa

The values of single resistances for bends, etc., column (g), are obtained on the velocity pressure (v.p.) method. The pressure corresponding to the appropriate duct velocity is taken from Fig. 21.1. The loss in terms of velocity pressure for each particular bend, reducer, etc., is taken from Fig. 21.11.

(6) The branches are then dealt with thus:

Pressure available for branch = (total pressure loss of main duct) – (pressure loss from fan to branch, column (i)).

The length of the branch to its end is then measured and the loss per m length established.

With this new loss, duct sizes for each volume are read from Fig. 21.10 as before.

If this gives too high a velocity, a lower pressure loss must be selected, and the surplus pressure can then only be absorbed by dampening.

Sizes for the whole system in terms of circular ducts will then be marked on the plan. If rectangular ducts are used instead of round, the equivalent sizes to give equal pressure loss per m run may be taken from Table 21.2. The rectangular sizes may be selected to suit the positions which the ducts must occupy.

TABLE 21.2

EQUIVALENT DIAMETERS (d) OF RECTANGULAR DUCTS $(a \times b)$ FOR EQUAL VOLUME AND PRESSURE DROP.

$\dfrac{a}{b}$	x	$\dfrac{a}{b}$	x	$\dfrac{a}{b}$	x	$\dfrac{a}{b}$	x	$\dfrac{a}{b}$	x	$\dfrac{a}{b}$	x
1·0	0·908	1·6	0·710	2·2	0·622	2·8	0·557	3·4	0·510	4·0	0·475.
1·1	0·865	1·7	0·701	2·3	0·608	2·9	0·548	3·5	0·504	4·2	0·465
1·2	0·829	1·8	0·683	2·4	0·597	3·0	0·540	3·6	0·498	4·4	0·455
1·3	0·798	1·9	0·665	2·5	0·586	3·1	0·532	3·7	0·491	4·6	0·447
1·4	0·770	2·0	0·650	2·6	0·576	3·2	0·524	3·8	0·486	4·8	0·438
1·5	0·745	2·1	0·635	2·7	0·566	3·3	0·517	3·9	0·480	5·0	0·431

$$d = 1·265 \left[\frac{(ab)^3}{(a+b)} \right]^{0.2}$$ and thence $b = xd$ where b is the shorter side

Table 21.2 is based on the formula stated where

d = diameter of duct.

a and b = dimensions of rectangular duct.

It is advantageous when sizing a long run of ducting with air supply grilles on its full length, to take static regain into account. This is done by so reducing the velocity in the duct in stages that the regain of static pressure compensates for the frictional loss in the duct to that point. This gives approximately equal pressures at each grille, and assists regulation. The regain is usually taken as 50 per cent of the change in velocity head.

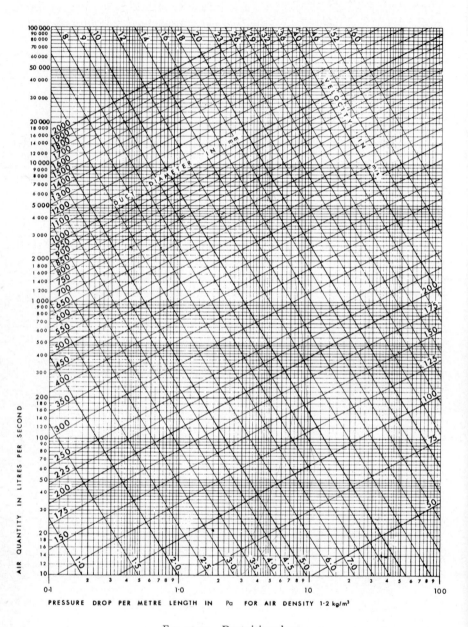

FIG. 21.10.—Duct sizing chart.

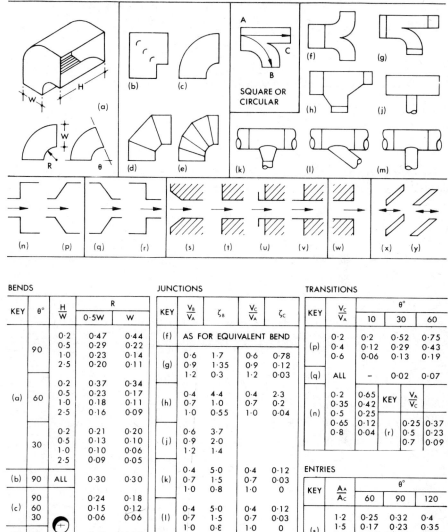

BENDS

KEY	θ°	H/W	R 0·5W	R W
(a)	90	0·2	0·47	0·44
		0·5	0·29	0·22
		1·0	0·23	0·14
		2·5	0·20	0·11
	60	0·2	0·37	0·34
		0·5	0·23	0·17
		1·0	0·18	0·11
		2·5	0·16	0·09
	30	0·2	0·21	0·20
		0·5	0·13	0·10
		1·0	0·10	0·06
		2·5	0·09	0·05
(b)	90	ALL	0·30	0·30
(c)	90		0·24	0·18
	60		0·15	0·12
	30		0·06	0·06
(d)	90		0·45	0·32
(e)	90		0·45	0·34

JUNCTIONS

KEY	$\frac{V_B}{V_A}$	ζ_B	$\frac{V_C}{V_A}$	ζ_C
(f)	AS FOR EQUIVALENT BEND			
(g)	0·6	1·7	0·6	0·78
	0·9	1·35	0·9	0·12
	1·2	0·3	1·2	0·03
(h)	0·4	4·4	0·4	2·3
	0·7	1·0	0·7	0·2
	1·0	0·55	1·0	0·04
(j)	0·6	3·7		
	0·9	2·0		
	1·2	1·4		
(k)	0·4	5·0	0·4	0·12
	0·7	1·5	0·7	0·03
	1·0	0·8	1·0	0
(l)	0·4	5·0	0·4	0·12
	0·7	1·5	0·7	0·03
	1·0	0·8	1·0	0
(m)	0·4	1·1	0·4	0·12
	0·7	1·3	0·7	0·03
	1·0	1·4	1·0	0

TRANSITIONS

KEY	$\frac{V_C}{V_A}$	θ° 10	30	60
(p)	0·2	0·2	0·52	0·75
	0·4	0·12	0·29	0·43
	0·6	0·06	0·13	0·19
(q)	ALL	–	0·02	0·07
(n)	0·2	0·65		
	0·35	0·42		
	0·5	0·25		
	0·65	0·12		
	0·8	0·04		

(r) KEY	$\frac{V_A}{V_C}$		
	0·5	0·25	0·37
	0·7	0·23	0·09

ENTRIES

KEY	$\frac{A_A}{A_C}$	θ° 60	90	120
(s)	1·2	0·25	0·32	0·4
	1·5	0·17	0·23	0·35
	2·0	0·12	0·21	0·31
	2·5	0·12	0·2	0·3
(t)	0·5	(u) 0·5	(v) 0·95	

EXITS & LOUVRES (50% FREE AREA)

(w)	1·0	(x)	6·0	(y)	3·6

NOTE

FOR BENDS IN SERIES LESS THAN 10W APART
S FORM—ζ = 0·9 x SUM
U FORM—ζ = 0·6 x SUM

NOTE

ζ_B APPLIES TO VELOCITY AT B
ζ_C APPLIES TO VELOCITY AT C

NOTE

ζ FACTORS APPLY TO VELOCITY IN THE SMALLER AREA

FIG. 21.11.—Velocity pressure loss factors for duct fittings.

Where ducts are made of other materials, the resistance will be less or more than that of galvanized steel ducts as given in Table 21.3.

TABLE 21.3

FACTORS FOR DUCTS OF OTHER MATERIALS

(Basis: Taking resistance from Fig. 21.10 as 1·0 for galvanized steel, the pressure loss should be multiplied by the appropriate factor given below.)

Smooth Copper, Aluminium, or PVC, etc.	0·8
Smooth Cement	1·2
Rough Concrete or Good Brick	1·5
Rough Brick or Pre-cast Concrete	2·0

Fan Pressure—The static pressure to be produced by the fan is the sum of all resistances to air flow throughout the system as follows:

Loss
Pa

(1) Fresh air intake louvres or grille (estimate on v.p. method, Fig. 21.1) - ....

(2) Fresh air intake duct (size and pressure loss established as for other ducts). This must include loss in single resistances as before - - - - - ....

(3) Pressure loss of:
    (a) air filter if any - - - - - - - - - ....
    (b) heater or heaters - - - - - - - - - ....
    (c) washer if any - - - - - - - - - ....
    (d) cooling coils if any - - - - - - - - ....
        (The above will be determined from makers' data.)

(4) Convergence of fan suction - - - - - - - - ....
    (Use v.p. method.)

(5) Divergence of fan delivery - - - - - - - - ....
    (Use v.p. method.)

(6) Changes of section, if any, through plant - - - - - - ....
    (Use v.p. method.)

(7) Pressure loss of duct system estimated as already tabulated - - ....

(8) Pressure loss of final outlet grille - - - - - - - ....

        Static pressure of Fan - - - - - - ....

Extract System—Exactly the same summation of pressure loss is arrived at for an extract system as for an inlet. Items (3) and (6) in the previous calculation will disappear, and in place of item (1) must be included the pressure loss of the discharge duct and cowl or other outlet.

Ducts for High Velocity Systems—Velocities over about 10 m/s may be described as high velocity. Such systems work up to 20 m/s and are applicable to induction, dual-duct, and variable volume systems of air-conditioning where adequate provision for silencing is included.

Duct sizing may follow the method of equal pressure drop previously discussed for conventional velocity systems.

Circular ducts should be used in preference to rectangular, to avoid drumming and to facilitate airtightness. Adequate sealing of joints is

Plate XXXVII. Three 2·5 MW centrifugal machines with associated water pumps
(see p. 477)

Plate XXXVIII. Pipework erection to two of four MW centrifugal machines
(see p. 482)

Plate XXXIX. Packaged cooling towers (see p. 488)

Plate XL. Condenser tubes fouled as a result of poor maintenance (see p. 489)

Plate XLI. Ductwork to dual-duct system prior to insulation (see p. 494)

necessary. Branches are usually taken off at 45° and sharp edges are to be avoided. Spiral formed ducts referred to on p. 495 are commonly used for high velocity systems.

Thermal Insulation of Ducts—Ducts conveying cooled air in an air-conditioning installation are usually insulated on the outside so as to reduce temperature rise to a minimum, also to prevent surface condensation from the surrounding air. Such insulation must be 'vapour sealed', i.e. sealed against ingress of atmospheric air, otherwise the insulation will become quickly waterlogged and useless. Ducts conveying warm air in most cases equally require insulation to reduce loss of heat. Where ducts are run in the conditioned space, insulation may be unnecessary.

Transmission of heat through bare metal duct walls in W/m² per degree difference varies according to air velocity as follows:

Air velocity m/s	Transmission factor U (still air outside) W/m²K
2	5·6
4	6·2
6	6·5
8	6·8
10	7·0
12·5	7·1
15 and over	7·2

The effect of insulation is to reduce the loss such that air velocity in the duct affects the U value only to a minor extent, and the following coefficients may be adopted as covering the loss at any practical air speed:

Metal duct with glass fibre insulation of the following thickness	U W/m²K
25 mm	1·4
38 mm	1·0
50 mm	0·7

The economic thickness can be estimated for any particular case.

Materials for duct insulation must be such as not to support fire, and cork (at one time in common use) is no longer acceptable. Insulating materials for use in ventilating systems are now subject to Fire Test*: some suitable ones are: glass fibre neoprene-faced, polyurethane and mineral wool.

The gain or loss of temperature in air conveyed in a length of duct may be calculated from the following:

let mean temperature difference
    between inside and outside of duct $=\theta$
transmission coefficient of bare
    or insulated duct W/m² K $=U$
length of duct in m $=L$
perimeter of duct in m $=P$

---

* See publications by Fire Research Station and G.L.C. Report No. 4250 *Working Party on Fire Prevention.*

then heat loss in watts　　　　　　　　　　　　　　$= PLU\theta$
let mass of air carried by duct in kg/s　　　　　　　$= M$
and specific heat capacity of air in kJ/kg K　　　　$= 1 \cdot 0$
thus temperature difference between
ends of duct in question will be (K)　　　　　　　$= \dfrac{PLU\theta}{M}$

This is sufficiently true for practical purposes, though not strictly so when the difference of temperature along the duct is of such a magnitude as to bring the internal temperature of the duct closer and closer to that outside.

Where insulating material is used as a duct lining it may serve a double purpose, thermally and acoustically.

SOUND CONTROL

The increasing amount of mechanical plant in buildings brings with it the problem of noise. We here confine our attention to equipment such as fans, compressors, pumps and boilers, and the channels by which sound may be conveyed therefrom to other parts of the building.

Sound—It is necessary first to consider the mechanism by which sound is transmitted.

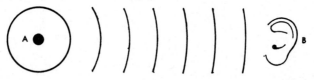

Fig. 21.12.—Sound transmission.

Vibrations at a point A set in motion the molecules of adjacent air such that a series of waves of compression and rarefaction radiate from the source spherically in all directions, a part reaching the ear at point B where the eardrum is set in like motion and the brain interprets the sensation as a sound, pleasant or otherwise. A *noise* may be defined as an unpleasant sound, or one that is overloud, or is monotonous or erratic.

The speed of sound depends on the properties of the medium, but may be taken at 331·46 m/s at 0° C and 101·325 kPa (N.T.P.).

Frequency—Sound from point A will most likely be of more than one frequency. The deepest note which the human ear can detect is about 15 Hz, and the highest about 20,000 Hz. Frequencies are divided into octaves, each octave being double that of the lower one. The frequency of Middle C is 262 Hz; thus C of the octave above is 524 Hz and so on. Fig. 21.13 shows standard frequency bands. Most sound sources set up harmonics, i.e. higher frequencies which are multiples of the fundamental.

Amplitude—The amount of energy imparted to the air at point A is termed the *Sound Power*, and is referred to in terms of *watts*. Sound power may be visualised as the amplitude at the source.

The amplitude, see Fig. 21.14, of sound X is greater than Y: X is a

APPROX SCALE OF WAVE LENGTH (mm)

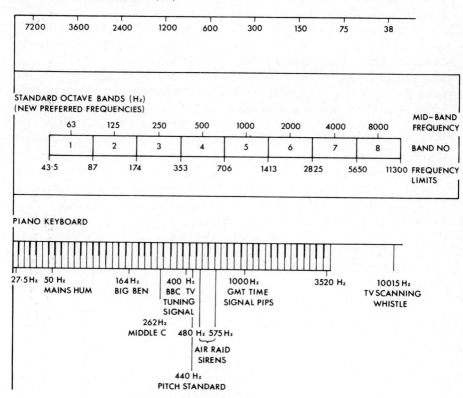

FIG. 21.13.—Standard octave bands.

loud sound and transmits more energy than soft sound Y. The energy involved is represented by power multiplied by time (Ws).

Sound power cannot be measured as such; it can be inferred mathematically from the sound received at a known distance in a sound-proof room.

Sound Received—The power of sound received is termed *Sound Intensity*

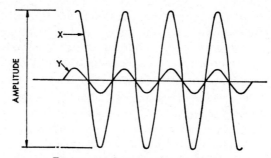

FIG. 21.14.—Sound power amplitudes.

and is measured in W/m². Although the sound produced at point A cannot be measured directly, the sound received by the ear—or by a sound meter—at point B can be expressed in terms of pressure variation.

The amount of variation in pressure of the sound waves reaching the receiver—termed the *Sound Pressure*—is measured in Pa.

The Decibel (dB)—The units referred to above are absolute values and are not convenient for general use. Units of reference have therefore been invented and are termed *bel* and *decibel* (10 dB = 1 bel). They are logarithmic, so that if two sounds differ in magnitude by 1 bel, the magnitude of one is ten times that of the other. Similarly, if the difference is 1 decibel, the magnitude of one will be about 26 per cent greater.

The decibel may be defined as $10 \times \log_{10}$ of the ratio between two absolute values such as levels of sound power, sound intensity and pressure.

Sound Power level (with reference to a datum of 10^{-12} watt)

$$= 10 \log_{10} \frac{W}{10^{-12}} \, dB$$

where W = actual sound power.

Sound Intensity level (with reference to a datum of 10^{-12} watt/m², taken to be threshold of hearing)

$$= 10 \log_{10} \frac{I}{10^{-12}} \, dB$$

where I = actual sound intensity.

Sound Pressure level (with reference to a datum of 2×10^{-5} Pa which corresponds to the reference intensity 10^{-12} watt)

$$= 20 \log_{10} \frac{P}{2 \times 10^{-5}}$$

where P = actual sound pressure level.

Decibels cannot be added arithmetically, but if two sounds of known decibel rating are taken separately, the decibel rating of the two taken together will be:

$$\log_{10} \text{(antilog of sound 1 in dB} + \text{antilog of sound 2 in dB)}$$

Use of a graph, Fig. 21.15, simplifies the task of addition of two dB ratings.

The Phon—This unit is based on the subjective reaction of the ear to loudness and it may be related to sound pressure level in dB by means of data which have been derived by experiments on individuals. Some values on this scale are:

Threshold of hearing	-	0	phons
Whispering	- -	20	,,
Average room	- -	40	,,
Busy street -	- -	50 to 70	,,
Pneumatic drill -	-	100	,,

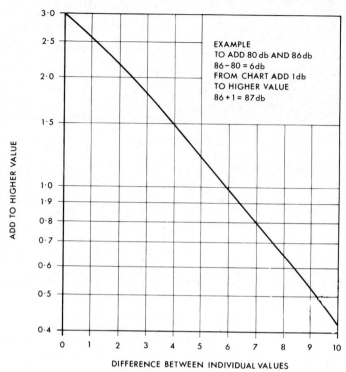

FIG. 21.15.—Addition of decibel ratings.

Noise Criteria—It is common experience that the human ear can tolerate noise of low frequency much more than of high frequency, even though of equal intensity. Thus it is necessary to consider frequency in relation to sound measurement and for this purpose the spectrum of frequencies has been split up into 8 octave bands as previously shown in Fig. 21.13.

Noise Criteria and Noise Rating curves (NC and NR) have been established on a subjective basis and each is given a number corresponding to the sound pressure level in decibels in the sixth octave band (1200–2400 Hz). NC curves were much used in the past but have now been superseded by NR curves for general use. For most practical purposes the two sets of curves may be regarded as interchangeable. Fig. 21.16 shows NR plots of equal tolerance for the eight frequency bands. Acceptable NR values may be classified as follows:

NR 25   Very quiet
NR 30   Normal living space
NR 35   Spaces with some activity
NR 40   Busy spaces
NR 45   Light industry
NR 50   Heavy industry

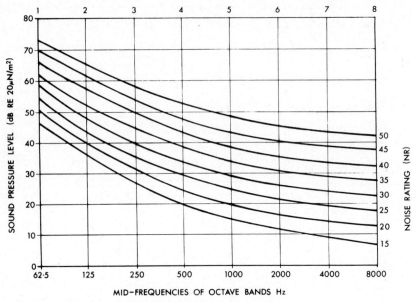

MID-FREQUENCIES OF OCTAVE BANDS Hz

Fig. 21.16.—Noise rating curves.

SOUND MEASUREMENT

Sound Meter—In one form this consists of a microphone coupled to a sensitive milli-ammeter and battery such that variations in sound pressure cause corresponding deflections of the needle of the instrument. By calibrating the dial in decibels a direct reading in dB can be obtained.

As pointed out, however, the ear does not evaluate sounds of different pitch in a linear manner and hence scales have been weighted in an attempt to correct for this by adding filters to the circuit.

Scale C is as Linear, but very high and very low frequencies
are suppressed.
Scale B is as C but with more low frequencies suppressed.
Scale A is further weighted to exclude all low frequency
sound intensities under about 55 dB.

Sound Analyser—In effect this is a sound meter, but so equipped with filters that measurement of sound pressure at any range of frequency is possible, thus enabling a sound 'spectrum' to be built up.

In using the three scales A, B and C for subjective assessment, scale A would be appropriate for low noise levels, bearing in mind the greater tolerance of the ear to low pitch sounds, scale B is for medium levels and C for loud noises. The greater the difference between A and C scale readings, the greater the importance of the low frequency component.

The Linear scale would be used to obtain a basis for calculation of absolute values.

Noise of Fans—The noise produced by fans is often quoted in makers' lists in decibels. It will be apparent from what has been stated that such a rating can only be a measure of the sound intensity level at a given point and the level will vary according to the distance at which the measurement is taken and other conditions. The sound spectrum is also highly important. It is to be assumed that the test takes place in an anechoic room, i.e. without reflection or extraneous noise. Given the sound power level of the fan at each frequency range, and ignoring any difference between suction and discharge, it is possible to determine the attenuation resulting from ducts, changes of direction, grilles and the like. If the result shows that too high a sound level from the fan remains, it is necessary to consider means for greater absorption as referred to later.

Generalised formulae have been published which enable the likely sound power level of a fan to be predicted in the absence of manufacturers' data. One such is as follows:

$$SWL = 10 + 10 \log Q + 20 \log P$$

where Q is the volume flow in litre/s, P is the static (resistance) pressure in Pa and SWL is the *overall* sound power level. To establish the individual sound power levels in the various frequency bands, correction factors must be subtracted as read from Fig. 21.17.

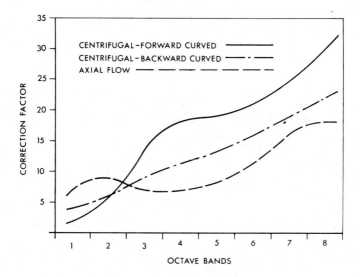

FIG. 21.17.—Correction factors for sound power of fans.

Noise in Rooms—The measurement of noise in rooms should be in conformity with some standard. Readings are commonly taken:

    not more than 1·2 m from floor;
    not less than 1 m distance from any grille or other ventilation device;
    and not more than 1 m/s air speed.

There is the question of background noise which is never absent except in a sound-proof room. It is necessary to measure the dB of the background, and then of the background plus the noise in question. By taking antilogs and the log of the difference, the dB level of the noise can be found (see **Fig. 21.18). If there is a difference of more than 10 dB the background can be ignored.**

The ear must, of course, always remain the final arbiter, as instrumental measurement, however careful, may be misleading. There may for instance be a single monotonous note which has only a slight effect in terms of dB. Or there may be an intermittent noise which goes quite unrecorded.

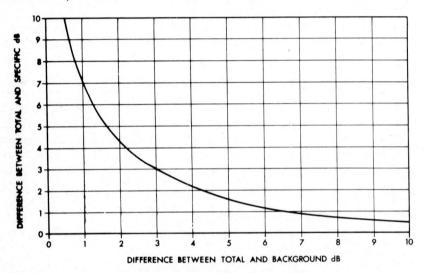

FIG. 21.18.—Subtraction of decibel values.

NOISE DISPERSAL

The less the noise produced, the easier is the problem of disposing of it. Nevertheless, all plant is liable to cause noise in greater or less degree.

Plant noises are of two kinds:

(*a*) Mechanical vibration which is transmitted to the building structure and thence to the occupied rooms.

(*b*) Airborne noise which likewise can enter the structure but which **will be more troublesome if conveyed via ventilation ducts to** the rooms.

(*a*) can be dealt with by mounting such plant on anti-vibration supports comprising springs, rubber blocks and the like. Manufacturers' data are available for the various types of systems. (*b*) is considered further here.

Absorption of sound is achieved by the application to surfaces of so-called acoustic materials such as acoustic tiles, and various soft materials such as expanded polyurethane.

Insulation of sound is achieved by interposing a barrier between the space where the noise is produced and some other space to be kept silent. In the case of the plant room mentioned in (*b*) above, the sound may be absorbed by covering the wall, ceiling and other surfaces with suitable material; but this is generally friable and liable to damage, hence the use of insulation might be preferable between say two faces of a hollow partition, or on top of a false ceiling. Usually this problem is not intractable and is covered under the general heading of 'Building Insulation' dealt with elsewhere.

Large scale boiler and refrigeration plant rooms give rise to special problems and must be considered in great detail.

Ducts—Considering now the transmission of noise by ducts, we may assume that the noise produced by a fan is conveyed equally via the suction and delivery. The air in passing through the system of ducts will lose some of the noise by attenuation, that is by the pressure waves being damped down due to friction on the duct walls, and at changes of direction.

Secondary noise may be generated in the ductwork by sharp edges, and particularly by dampers.

At the terminal end the grille or diffuser may also generate noise, and this at a point where no treatment is possible; hence correct selection of type and velocity of such items is inherent in good acoustical design.

When the air enters the room from the grille, it expands into free space which again accounts for some attenuation.

The attenuation in ducts depends on length and size: the longer the duct the greater the attenuation; the larger the duct the less the attenuation. The amount of attenuation may be calculated from the data given in Table 21.4 and the accompanying notes, but for fuller treatment the *Guide, Section B12*, should be consulted.

TABLE 21.4
ATTENUATION OF PLAIN DUCTS

Duct Item		Attenuation
		dB/m run
Straight ducts	150 × 150 mm	0·3
,, ,,	600 × 600 mm	0·15
,, ,,	1800 × 1800 mm	0·03
		dB/Bend
Round bends	75–375 mm depth	2
,, ,,	375–900 mm	1·5
,, ,,	over 900 mm	1
Branches	Acoustic energy is divided in ratio of duct areas.	

Where it is clear that residual sound will be too great, ducts may be lined with absorption material either throughout or for certain lengths, as Fig. 21.19.

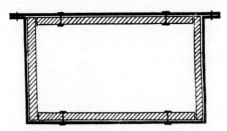

FIG. 21.19.—Acoustic insulation to ductwork.

An approximate prediction for attenuation by use of an absorption lining to ducts is given by the formula

$$\mathrm{dB/m\ run} = \left(\frac{D}{A}\right)\alpha^{1.4}$$

where D = perimeter of lined duct in m.
A = free area of duct in m².
α = absorption coefficient of lining material (Table 21.5).

This expression holds for lined lengths up to about 2 m only and has been found to over-estimate performance at high frequencies.

TABLE 21.5
ABSORPTION COEFFICIENTS OF SOME LINING MATERIALS

	Absorption coefficient at			
	125 Hz	500 Hz	2000 Hz	4000 Hz
Fibreglass resin bonded mat				
25 mm thick	0·1	0·55	0·75	0·8
50 mm thick	0·2	0·70	0·75	0·8
Stillite slabs - 25 mm thick	0·05	0·45	0·80	0·75
Expanded Polyurethane				
25 mm thick	0·25	0·85	0·90	0·90
Fibreboard perforated tiles - -	0·10	0·4	0·45	0·5
Burgess tiles backed fibreglass -	0·10	0·60	0·80	0·90

Alternatively, or in addition, specially designed silencers may be used inserted in the run of ducting. These are rated according to dB loss, but the specification of performance should be based on 'insertion loss' not on a test in free space. Silencers may be built up by splitters or egg crates, as in Fig. 21.20, and their design may be considered as if each section were a small duct lined with absorption material. As attenuation is inversely proportional to the cross-sectional area of the duct, it follows that the smaller the sub-divisions of the absorber the shorter the length required. On the other hand, any absorption device increases resistance to air flow and hence a reasonable balance must be kept.

Where silencers are fitted in plant rooms, it is particularly important to ensure that they are so sited that 'break back' of noise is not possible.

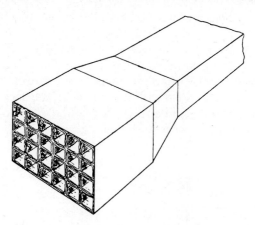

FIG. 21.20.—Honeycomb silencer for duct insertion.

It is of little use to fit an expensive attenuator to a piece of equipment and then allow a bare sheet metal duct connection to wander across the plant room where noise can re-enter and be conveyed about the building: such silencing equipment should either be followed by an insulated duct or be fitted immediately adjacent to the plant room wall. Fig. 21.21 illustrates this point.

Another application for silencers occurs where a number of rooms are connected to a common duct system either inlet or extract, and where it is essential that speech or sounds in one room are not heard in other rooms. In Fig. 21.22 (*a*), a sound in Room A has but a short path to rooms B and C. In Fig. 21.22 (*b*) the branch serving each room is treated acoustically, thereby avoiding any cross-talk. It will be noted that the 'shunt-ducts' mentioned in Chapter 16 (Fig. 16.17) may act to some extent as cross-talk attenuators.

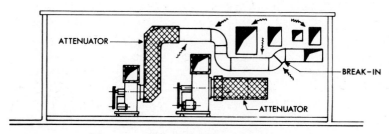

FIG. 21.21.—Noise 'break-in' in a plant room.

High-Velocity Systems—The high pressures much involved in high velocity systems necessitate special attention to acoustical design. The usual procedure is to provide a long silencer in the fan-discharge for absorption of fan noise, especially at the lower frequencies. Noise generated in ducts is dealt with by lining of selected lengths. With the induction system the jets have an important attenuation effect, but some acoustical treatment of the cabinet is often provided. In the case of the dual-duct and variable volume systems the blending cabinet is always acoustically treated. Where volume control or similar boxes are used to serve low velocity ducts, a silencer in the outlet duct is desirable.

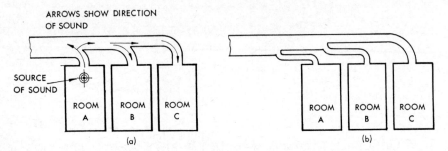

Fig. 21.22.—Cross-talk attenuation.

CHAPTER 22

Air Distribution

EFFECTIVE DISTRIBUTION OF AIR within the occupied space is the key to successful operation of a ventilation or air conditioning system. It is of little use to provide plant and ductwork distribution arrangements ideally suited to meet load demands if the methods used to introduce the air supply do not provide for human comfort or process needs.

Successful air distribution requires that an even supply of air over the whole area be given without direct impingement on the occupants and without stagnant pockets, at the same time creating sufficient air movement to cause a feeling of freshness.

This indicates what is probably the key to the problem of successful distribution: that unduly low velocities of inlet are to be avoided, just as much as excessively high ones and that distribution above head level not movement to ensure proper distribution over the whole area without draughts.

GENERAL PRINCIPLES

There are four general methods of air distribution.

(1) Upward.
(2) Downward.
(3) Mixed upward and downward.
(4) Lateral.

The choice of system will depend on:

(a) Whether simple ventilation or complete air-conditioning is employed.
(b) The size, height and type of building or room.
(c) The position of occupants and/or heat sources.
(d) The location of the ventilating plant, and economy of duct design.

Upward System—The air is introduced at low level and exhausted at high level as in Fig. 22.1 which shows a section through an auditorium, with mushroom inlets under the seats and riser gratings (see p. 530) in the gallery risers. The air is exhausted around the central laylight in the roof. This system was designed so as to be reversible, i.e. it may be worked as a downward system.

When working upwards the air appears to be somewhat 'dead', due to the very low velocity of inlet (about 0·5 m/s) necessary with floor outlets

515

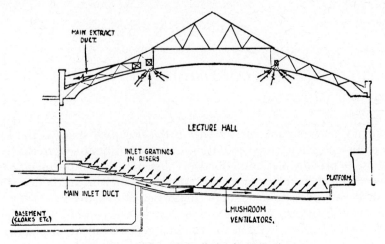

FIG. 22.1.—Upward ventilation in auditorium.

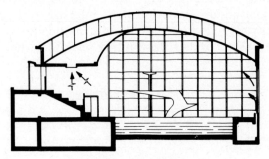

CROSS SECTION THROUGH POOL

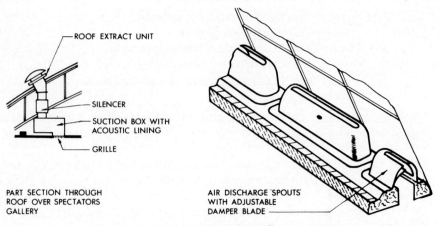

FIG. 22.2.—Upward ventilation in swimming pool hall.

to prevent draughts. When working downwards more turbulence is set up in the air stream, with a greater feeling of freshness.

The upward system is not, however, confined to one with floor inlets. The inlets may equally well be in the side walls, with extract in the ceiling as before. The limitation of an upward system is that in a large hall it may be difficult to get the air to carry right across without its picking up heat *en route* and rising before it reaches the centre.

The upward system is generally used with simple ventilation systems. When the air is cooled, as in a complete air-conditioning system, it will tend to fall too early, before diffusion, and thus cause cold draughts. The upward system however lends itself to simple extract by propeller fans in the roof in the case of a hall, factory, etc., and is thus generally the cheapest to install.

Another application of the upward system is the swimming pool hall, as shown in Fig. 22.2. Here the supply air is introduced through special plastic discharge spouts situated below a large area of glazing and is exhausted by specially treated roof-exhaust units.

On a smaller scale appropriate to single rooms, experimental work has been reported* which suggests that for winter use it is possible to introduce warm air via a long slot near floor level, as shown in Fig. 22.3, and by this means much reduce the temperature gradient within the room. Discharge velocities of up to nearly 12 m/s were used without reports of discomfort from occupants.

Downward System—In this type of system the air is introduced at high level and exhausted at low level, as in Fig. 22.4. It is commonly used with full air-conditioning where, due to the air admitted being cooled, it has a tendency to fall. The object of distribution in this case is so to diffuse the inlet that the incoming air mixes with room air before falling. Thus, the inlets shown in the figure as discharging downwards, in practice deliver

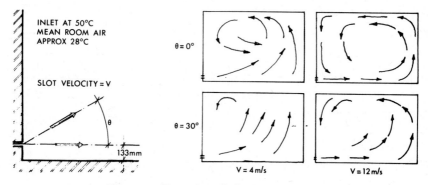

FIG. 22.3.—Upward ventilation for small rooms.

* Howarth, A. T., Sherratt, A. F. C., Morton, A. S., *Air movement in an enclosure with a single heated wall*. *J.I.H.V.E.* 1972.40.211.

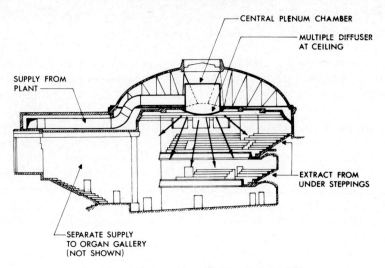

FIG. 22.4.—Downward ventilation in concert hall.

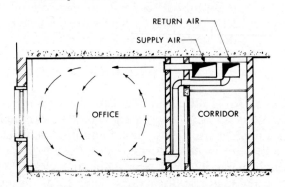

FIG. 22.5.—Downward ventilation in offices.

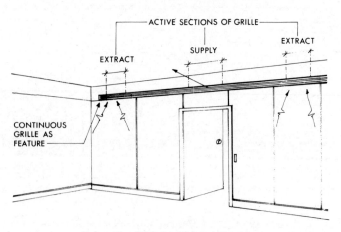

FIG. 22.6.—High level inlet and extract.

horizontally at sufficient speed to ensure that the air completely traverses the auditorium. Turbulence is thus caused with the desirable effect already mentioned (see Plate XLII, opposite p. 576). On a smaller scale, as applied to an office building, this system appears as in Fig. 22.5.

Provided the height of room is normal, the extract opening may be at high level as in Fig. 22.6. Short circuiting is avoided by the velocity of the inlet air carrying over to the far side of the room. Another possible arrangement is a variation of this, namely, 'downward-upward', as in Fig. 22.7(a).

An application of downward inlet with both downward and upward

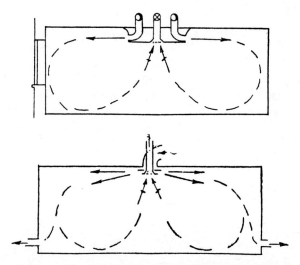

FIG. 22.7.—Downward-upward and combined arrangements.

extract, suitable for rooms of greater height, is shown in Fig. 22.7(b). This is usually adopted where smoking occurs and it is necessary to provide some top extract to remove the smoke. In this case the top exhaust is discharged to atmosphere by a separate fan and the low-level extract constitutes the recirculated air. Care must be taken to avoid placing low-level extract grilles near where people sit. If there is no alternative position, these grilles must be selected to operate at a very low velocity (about 0·75 m/s) through the free area, and well spaced out, so that big volumes do not occur at any one point.

Mixed Upward and Downward—Such a system is shown in Fig. 22.8. The principle will be clear from what has been described above. It is in effect an upward system giving good turbulence above head-level, with about 25 per cent extract at low level. The remainder of the extract is exhausted normally at the ceiling.

Lateral Ventilation—This arrangement may sometimes be necessary where dictates of planning preclude more orthodox solutions. Air is intro-

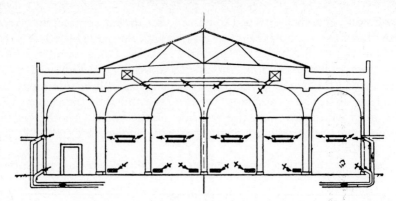

FIG. 22.8.—Mixed system of ventilation in restaurant.

duced near the ceiling on one side of a long low room with a smooth flush
ceiling, and is exhausted at the opposite end at the same level taking
advantage of the 'Coanda' effect which will be referred to in more detail
later (see Fig. 22.9). The inlet is at a high velocity and strong secondary
currents are set up in the reverse direction at the lower levels, as shown. It
is these secondary currents which are important in any distribution system,
causing turbulence as already mentioned.

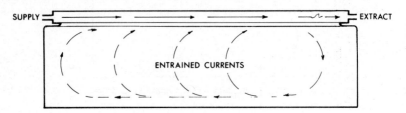

FIG. 22.9.—Lateral distribution.

AIR DISTRIBUTION FOR AIR-CONDITIONING

Air-conditioning usually involves the handling of large air quantities,
and one of the chief problems of successful design is how to introduce and
extract these quantities without giving rise to complaints of draught or of
causing noise. A great variety of methods has been adopted and there are
innumerable devices available to suit different conditions or architectural
tastes. Some of these are described in following paragraphs.

Perforated Ceilings—Probably the most completely diffused form of inlet
is the system which employs a perforated ceiling as shown in Fig. 22.10.
In this case air is discharged into the space above the ceiling, which is
usually divided up so as to ensure uniformity of distribution, and the air
enters the room through the perforations. Where extremely large air
quantities are involved the whole ceiling may be used, but in the normal
case only selected areas of panels serve as inlets, acoustic or other baffles

being positioned behind the 'non-active' panels. The ceiling inlet system destroys all turbulence, which may be a good thing in some cases but not in others. This system has been applied successfully in law-courts, department stores and in confined spaces such as radio commentators' boxes where no other distribution arrangements would be possible.

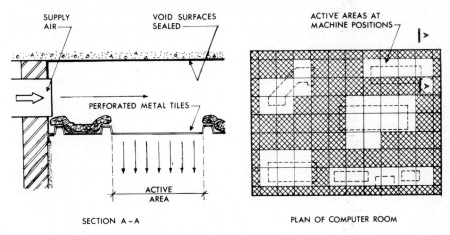

SECTION A-A   PLAN OF COMPUTER ROOM

FIG. 22.10.—Inlet via perforated ceiling.

Cone type Ceiling Diffusers—The type of ceiling diffuser shown in Fig. 22.11(*a*), has certain interesting characteristics. Fig. 22.11(*b*) illustrates the kind of flow pattern to be expected from such a unit and it will be noted that entrainment air is drawn up in the centre immediately under the diffuser, being caught up in the nearly horizontal delivery from the unit itself. Due to the fact that air is discharged radially, the velocity of inlet falls off rapidly as the distance from the centre increases, and hence this kind of unit may be used with temperature differentials up to 20 K.

There is a great variety of makes of this type of ceiling diffuser, all under different trade names and some with special features, such as:

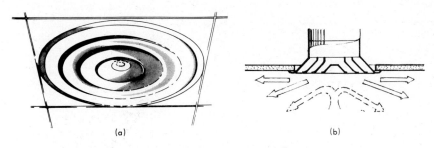

(a)   (b)

FIG. 22.11.—Cone type ceiling diffuser.

An adjustable arrangement whereby the inner cones can be raised or lowered in relation to the periphery, so causing a variation in the flow pattern to be achieved. Instead of the bulk of the air travelling horizontally, it is possible by this kind of adjustment to make it discharge vertically downwards or at any intermediate flow pattern desired. This might be of advantage in cases where it is desirable to cause the air to descend quickly, such as in a hot kitchen, rather than it should be dispersed and lost at high level.

In another type, the unit is square instead of circular, this sometimes being necessary to match ceiling tiles etc.

In other examples, the diffuser is flush with the ceiling instead of projecting.

Perforated-Face Ceiling Diffusers—This type, shown in Fig. 22.12, has been developed from the early 'pan' devices and this has characteristics similar to the cone pattern: it is often used to blend in with perforated acoustic-tile ceilings. The deflecting pan may be removed if a predominantly vertical air distribution is required.

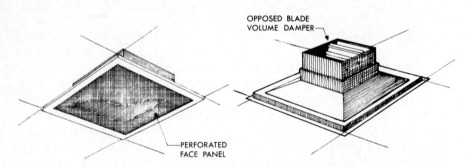

Fig. 22.12.—Perforated plate type ceiling diffuser.

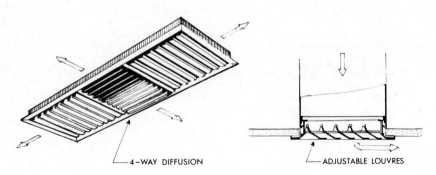

Fig. 22.13.—Multi-directional ceiling diffusers.

Multi-Directional Ceiling Diffusers—Such units provide facilities for positive directional control of air discharge, the blades being individually adjustable as illustrated in Fig. 22.13. They are useful for applications where partition change may occur or where movement of office machinery may create change in demand for air movement.

A typical layout of ceiling diffusers is shown in Fig. 22.14. It will be

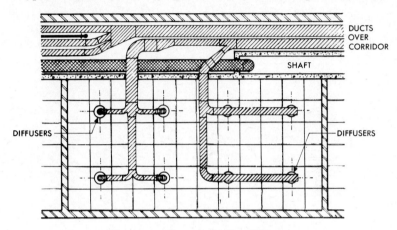

FIG. 22.14.—Layout of ceiling diffusers.

noted that use is made of the false ceiling space for the concealment of the connecting ducts, the final connections to the diffusers being of flexible ducting. To ensure that air distribution is effective, the room is divided into approximate squares with one diffuser to each. Alternatively, if it is necessary to use a less symmetrical spacing, segments of diffusers may be blanked off to avoid impingement of the air pattern on walls etc.

Line Diffusers—Ceiling—For use in open plan offices and in avoidance

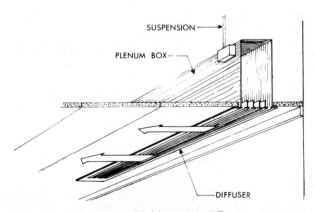

FIG. 22.15.—Multi-vane line diffuser.

of interference with ceiling pattern, the line diffuser in many different forms is commonly used. In its original form, Fig. 22.15, it was an adaptation of the continuous side-wall grille but took advantage of the 'Coanda' effect in that air flow directed across a plane surface has the tendency to retain contact until the velocity of discharge has decayed by entrainment of room air.

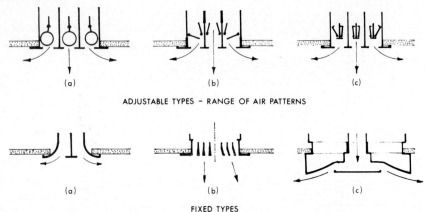

ADJUSTABLE TYPES – RANGE OF AIR PATTERNS

FIXED TYPES

FIG. 22.16.—Various forms of ceiling-integrated line diffuser.

Development of this type of diffuser, as shown in Fig. 22.16, takes many forms. As may be seen, facilities are available for adjustment of the air flow pattern. Individual sections of the continuous length are supplied by means of air plenum boxes which may, in turn, be integrated with Luminaires as Fig. 22.17.

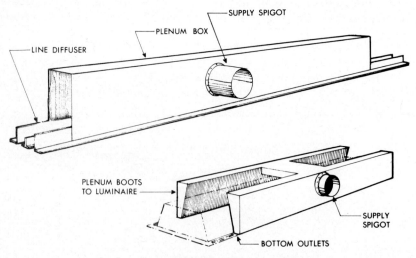

FIG. 22.17.—Plenum boxes for line diffusers.

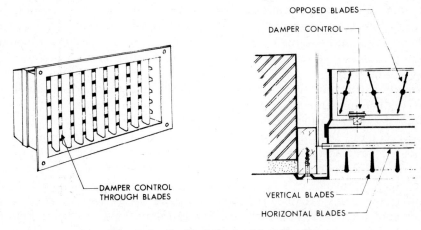

FIG. 22.18.—Double deflection side-wall grilles.

Side Wall Inlets—Where there is no false ceiling or other means of introducing the air through ceiling diffusers, it is necessary to adopt side wall inlets, this usually taking the form of a series of grilles distributed at intervals along the inner partition wall with ducts in the corridor false-ceiling behind. Each inlet is equipped with a grille, so designed as to enable the requisite quantity of air to be introduced without draught.

A common and effective form of grille is that known as the 'double-deflection' type, as shown in Fig. 22.18. In this form of grille there are two sets of adjustable louvres, one controlling the air delivery in the vertical plane and one in the horizontal plane. The vanes are usually independently adjustable by means of a special tool, and when once set are not altered. A variety of flow patterns can be achieved according to the width of room, aspect ratio and velocity.

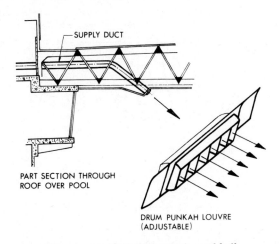

FIG. 22.19.—Punkah louvres in pool hall.

For side-wall applications however, much use is now made of the linear diffuser as illustrated, for ceiling application, in Fig. 22.15. By careful selection and making use of the 'Coanda' effect by direction of the air pattern towards the ceiling, good overall air distribution may be achieved from such diffusers.

Nozzles—Introduction of air by nozzles was, in the past, usually confined to small compartments such as ship's cabins and to aircraft, where it was desired to obtain the maximum effect under the direct and easy control of the occupant, both in the matter of quantity and direction: terminals of this type are known as 'punkah louvres'. Nozzles may however be a useful manner of introducing large quantities of air in some vast arena or high bay factory, where it would be impossible to conceive of too much air movement being created. In effect, the more the better. An example of this kind was the arrangement used to ventilate the Earls Court Exhibition building. Nozzles were specially designed to achieve a throw of about 40 m to the breathing zone.

Fig. 22.19 shows another but similar approach where 'drum' type punkah louvres have been used to introduce large air quantities into a swimming pool hall. For the auditorium illustrated in Fig. 22.4, the Usher Hall in Edinburgh, conditioned air was introduced solely from the central ceiling feature making use of 65 jet diffusers of the type shown in Fig. 22.20. Each handled approximately 280 litre/s at an outlet velocity of approximately 3 m/s.* (Plate **XLII**, facing p. 576).

Volume Control—Means for controlling as accurately as possible the volume of air issuing from each grille, diffuser or other inlet, are essential. Probably the commonest form of control is the opposed blade damper. For ceiling diffusers and the like, a variety of multi-louvre dampers are

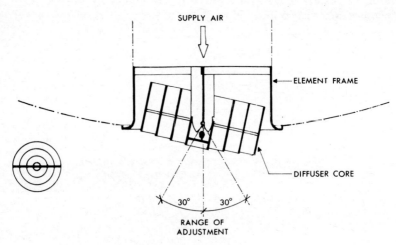

Fig. 22.20.—Jet diffuser.

---

* Clark, I. T., Air Conditioning the Usher Hall, Edinburgh. *J.I.H.V.E.* 1977. 45.125.

available or plain butterfly dampers in the ducts may be used. Any damper is a potential source of noise and its location and duty require careful consideration if such is to be avoided.

Discharge Velocities—The selection of the best number and size of inlet grilles, diffusers and the like, involves a choice to meet a number of variables at the same time:

(1) Air velocities produced at head or foot level must not exceed 0·15 to 0·20 m/s if draughts are to be avoided.

(2) If the air is entering cooler than the room, as in an air-conditioning installation, there will be a tendency to fall, best avoided by keeping a reasonable entering velocity with the object of inducing air entrainment from the room; the air mixture then being warmer will have less tendency to fall.

(3) The velocity of inlet must not be so high that the air will impinge on the wall opposite, thereby causing undue turbulence.

(4) The distance apart of inlets, particularly in the case of ceiling diffusers, must be such that the streams from two adjacent units do not collide with such force that a strong downward current results.

(5) The velocity selected for the grille or diffuser must be such that the sound level produced therefrom is below the design standard for the room.

(6) Appearance, layout and pattern are all architectural matters, which must also as a rule be taken into account and this sometimes creates a difficulty where the desired spacing or size is not acceptable on these grounds.

TABLE 22.1

APPROXIMATE SIZES OF AIR INLETS (REPRESENTATIVE ONLY)

Type	Air quantity litre/s				
	50	100	200	300	500
	dimensions in mm.				
Perforated ceiling panels (face size) - - - - -	300 × 150	300 × 300	500 × 500	600 × 600	1200 × 600
Line diffuser (duct size) - -	750 × 50	1000 × 90	2000 × 90	3000 × 90	2400 × 165
Ceiling diffuser					
(neck dia.) - - - -	100	150	200	250	380
(overall dia.) - - - -	330	330	450	600	860
Side Wall grille double deflection type (face size) - - -	200 × 150	300 × 200	450 × 250	450 × 400	660 × 450

Note: Above are all selected at approximately the same sound level of 40 dB.

Primary selection should be by use of the design guides published by B.S.R.I.A.* prior to reference to manufacturers' data. These latter are now freely available for all types of equipment and by a study of them, a variety of alternative solutions to a particular problem can be put down with the object of a final selection being made best suited to meet all the other conditions. The reader is referred to such data for precise information, but a few sample sizes are given in Table 22.1. It will be clear that with the large number of variables, air-distribution design is perhaps more an art than a science, but on it, to a large measure, as has previously been emphasised, depends the success or otherwise of any air handling installation, particularly in the case of air conditioning.

High-Velocity Supply Fittings—The use of high velocity air-distribution in ducts has been referred to in connection with the induction, dual-duct and variable air volume systems. High velocity air-distribution may also be the most practical and economical method both in cost and space for use with any otherwise normal ventilation or air-conditioning system where extensive ductwork is involved. This is particularly the case in multi-storey buildings where duct sizes are much reduced by use of higher velocities. Duct velocities up to 30 m/s may be used, although it is more usual to limit them to the range of 15 to 20 m/s.

At the terminal end, it is necessary to break down the high velocity to low velocity for introduction into the room, and a silencing or attenuating box in some form is required: this subject has been referred to previously and illustrated in Chapter 17.

Combined Lighting and Air Distribution—Mention has previously been made (p. 407) of the use of special fittings which enable some part of the heat generated by lighting to be dealt with at source before it enters the room. Early types of such fittings used 'boots' mounted to conventional lighting enclosures, as Fig. 22.21, but more sophisticated handling functions have been considered in an integrated design, as previously referred to, Fig. 22.17. The performance of the lighting apparatus can be improved by such arrangements due to temperature control of the tubes, and the heat entering the room may be reduced very considerably.

Extract or Return Air Grilles—The particular form of grille for extract is unimportant since air approach to a return fitting is no aid to distribution. It may, for instance, be of egg-crate plastic within an aluminium frame as Fig. 22.22, or louvred, or any design giving the required free area. Dust collects on extract gratings on the outside, and close mesh or closely placed slats are undesirable as they quickly block up and impede the air flow.

Where extraction takes place naturally into a corridor, as in an office building, it may be necessary to use a light-trap grille to avoid direct vision. Fig. 22.23 shows alternative fittings suitable for this purpose.

---

* Laboratory Reports 65, 71, 79, 81 and 83. The principal author is P. J. Jackman.

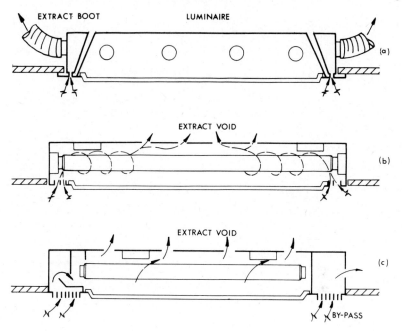

FIG. 22.21.—Air-handling luminaires.

A method of extracting room air which has been used with some success is illustrated in Fig. 22.24. It has been claimed that heat loss and gain to the occupied space is much reduced by passing room return air through the cavity of the double window. A glazing U-value equivalent to about 0·6 W/m²K is produced and, of course, hot and cold radiation, in summer and winter respectively, is reduced proportionately.

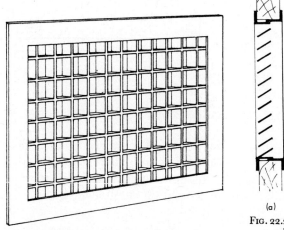

FIG. 22.22.—Standard egg-crate register.

FIG. 22.23.—Cross-section through transfer grilles (a) simple (b) vision-proof (c) light-proof.

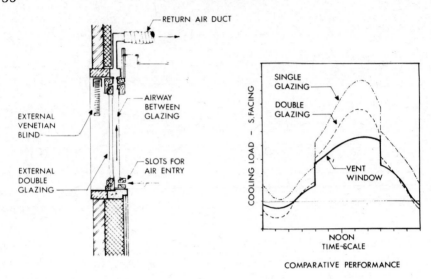

FIG. 22.24.—Ventilated double window.

Details of the mushroom ventilator and gallery-riser vent referred to earlier are shown in Figs. 22.25 and 22.26. These types are more commonly used as extracts, though they may be used as inlets at low velocities.

Toilet Extract—For this purpose, use may be made of the special wall-type mushroom ventilators shown in Fig. 22.27. These combine a facility for air volume regulation with a degree of acoustic attenuation.

UNORTHODOX ARRANGEMENTS

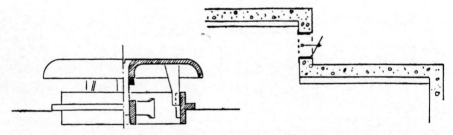

FIG. 22.25.—Ventilator in gallery stepping.   FIG. 22.26.—Mushroom floor ventilator.

Mention has previously been made (p. 517) of an experimental method of air distribution using a low level slot. In a quite different context, that of air movement and avoidance of temperature gradient in lofty buildings such as churches and high-bay factories, information has reached the authors of a technique used in the U.S.A.*

* Private communication. K. E. Robinson, late Chairman Ventilation Manual Committee, American Conference of Governmental Industrial Hygienists.

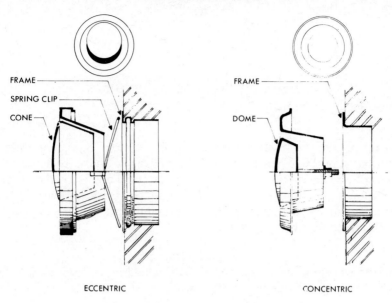

FIG. 22.27.—Mushroom wall ventilator.

Here, with conventional methods of space heating at low level, heating effect and air movement have been greatly improved by means of a fan circulation arrangement as shown in Fig. 22.28. By this means the warm air, convected to high level by natural currents, is displaced to the occupied zone: an air circulation equivalent to about 3 room volumes per hour has been found to be adequate to maintain the temperature gradient at about 2 K.

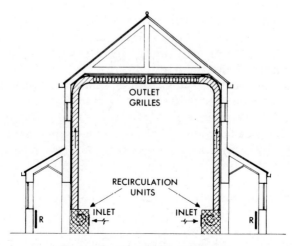

FIG. 22.28.—Air transfer arrangement.

CHAPTER 23

Tall and Deep Plan Buildings

UNTIL THIRTY YEARS AGO buildings in the United Kingdom were, by tradition, modest in scale and so designed as to make best use of the materials of construction as climate modulators. They were, furthermore, dimensioned generously so that they could be reserviced every quarter century to meet changing needs for accommodation and changing demands for standards of comfort.

TALL BUILDINGS

With a normal building height of 30 metres or so, some shelter from sun and wind can be expected from adjacent structures but this is not so in the case of an isolated tower block. A building 60 metres or more in height is exposed to sun and wind on various sides throughout the day through all seasons of the year and the effect of these climatic influences can cause marked fluctuations in the total demand for heating or cooling, shading and lighting throughout the day.

The force of wind at these higher levels is greater in strength than at ground level, rendering the opening of windows for ventilation well nigh impracticable under certain conditions. It has been found by experience that wind pressure on one side of a building and corresponding suction on the other is sufficient to cause undesirable pressure differences on doors to corridors, as well as discomfort in the rooms. It appears paradoxical that a tall building, which should have the freshest of ventilation from the upper atmosphere, cannot so avail itself as a general rule.

Solar Gains—Reference to the calculation of solar-heat gain was made in Chapter 18. Orientation is particularly important in tall buildings as will be evident from the following notes:

Fig. 18.2 is an azimuth and altitude chart for Latitude 50° N for various seasons of the year plotted by a computer. It is assumed that a horizontal surface at this latitude at sea level (or up to 300 metres above) in June will receive a maximum peak of solar radiation, under a clear sky, of 910 W/m². Intensities on surfaces other than the horizontal will be proportional to the cosine of the angle of incidence.

Fig. 23.1 shows the plot of solar incidence on the east, south and west faces throughout the day at the summer solstice. It will be noted that due to the high angle of the sun at midday (63°), the peak solar incidence is lower on the south face than the peaks on the east and west faces which occur at times when the angle is less. Fig. 23.2 is a similar plot for the

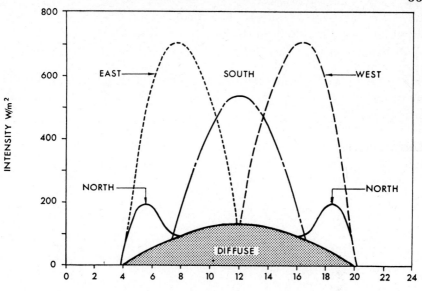

FIG. 23.1.—Solar heat on a vertical face (June 21).

S. and N. faces 24 hour mean =245 W/m².
E. and W. faces 24 hour mean =390 W/m².

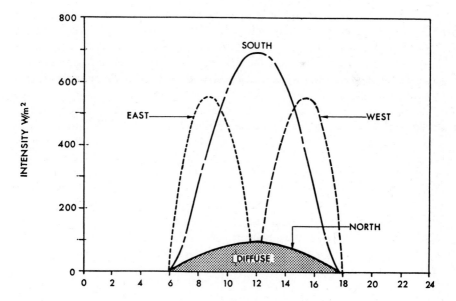

FIG. 23.2.—Solar heat on a vertical face (March 22 and Sept. 22).

S. and N. faces 24 hour mean =241 W/m².
E. and W. faces 24 hour mean =236 W/m².

spring and autumn equinoxes. The incidence on the south face at 40°
altitude now exceeds that on the east and west faces.

A measure of the relative incidence of solar heat gain is given by the
24 hour mean values, and these are stated on the figures.

Fig. 23.3 illustrates the plan of a rectangular slab block: Case 1 is with
the main axis east and west, Case 2 with the main axis north and south.
Assuming that length =3 × breadth:

Case 1, in June:

$$S \text{ and } N \text{ faces } 245 \times 3 = 735$$
$$E \text{ and } W \text{ faces } 390 \times 1 = 390$$
$$\overline{\quad 1125 \quad}$$

Case 2, in June:

$$S \text{ and } N \text{ faces } 245 \times 1 = 245$$
$$E \text{ and } W \text{ faces } 390 \times 3 = 1170$$
$$\overline{\quad 1415 \quad}$$

Case 1, in September:

$$S \text{ and } N \text{ faces } 241 \times 3 = 723$$
$$E \text{ and } W \text{ faces } 236 \times 1 = 236$$
$$\overline{\quad 959 \quad}$$

Case 2, in September:

$$S \text{ and } N \text{ faces } 241 \times 1 = 241$$
$$E \text{ and } W \text{ faces } 236 \times 3 = 708$$
$$\overline{\quad 949 \quad}$$

It is evident that at mid-summer the orientation of Case 1 produces
lower heat gain than Case 2, but in spring and autumn the difference
would be much less. In reality, in a slab block of this kind, the narrow
ends would be windowless which, when due allowance is made for gain
through glass, shows Case 1 to even greater advantage. From a planning
point of view, however, Case 1 may possibly be less attractive than Case 2,
and the consequences of the additional heat gain of the latter might have
to be accepted.

This example may serve to illustrate how other cases may be compared and Fig. 23.4 shows solutions for a number of different building

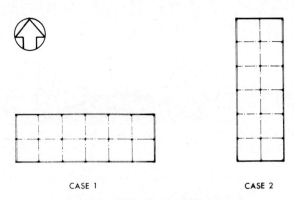

CASE 1   CASE 2

Fig. 23.3.—Plan of tall slab building.

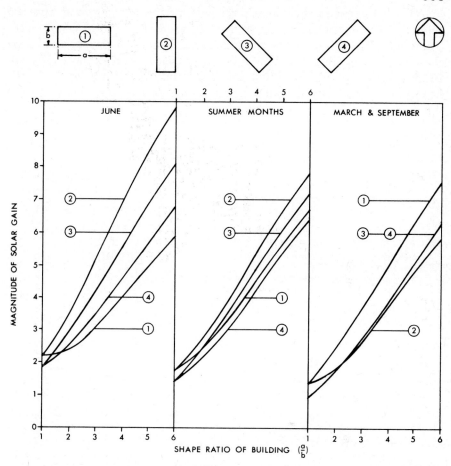

FIG. 23.4.—Solar incidence on buildings of different shapes and varying orientations. (All faces assumed to be glazed in equal proportion: floor areas assumed to be constant: hours considered are between 0900 and 1730 GMT.)

orientations in mid-summer, at the solstice and for the five summer months.
Glazing—U factors for single and double glazing are given in Chapter 2 but the values tabulated are for 'normal' exposure, i.e. for a wind speed at roof level of about 3 m/s, where the external surface resistance R_{S_2} is taken as 0·06 m²K/W. In the case of a tall building, with much higher wind speeds, this external surface resistance will be nil and the factors become:

$$\text{Single glazing in metal frames} \quad - \quad U = 8\cdot 1 \ \text{W/m}^2\text{K}$$
$$\text{Double} \quad ,, \quad ,, \quad ,, \quad ,, \quad - \quad U = 4\cdot 3 \ \text{W/m}^2\text{K}$$

These increased values concern us chiefly in calculating winter heat losses, as in summer winds may be assumed to be less and normal U values may be taken.

The reduction in solar-heat gain with double glazing, and also the reduction due to various forms of shading are referred to in Chapter 18. Table 23.1 lists a number of combinations of shading devices and ranks these in order of efficiency. It will be noted that it is much more effective to make use of blinds etc. fitted externally than to attempt to deal with heat gain which has penetrated glazing and entered the building.

TABLE 23.1

RELATIVE EFFICIENCY OF SHADING DEVICES IN REDUCING SOLAR HEAT GAIN
(PLAIN SINGLE GLAZING = 0)

Type of Solar Protection	Relative Exclusion Efficiency (%)	
	Single Glazing	Double Glazing
External dark green miniature louvred blind - -	83	87
External canvas roller blind - - - -	82	85
External white louvred sunbreaker, 45° blades - -	82	85
External dark green open weave plastic blind - -	71	78
Heat reflecting glass, gold - - - - -	66	67
Clear glass with solar control film, gold - - -	66	67
Densely heat absorbing glass - - - -	—	67
Mid-pane white venetian blind - - - -	—	63
Internal cream holland blind - - - -	57	61
Lightly heat absorbing glass - - - -	—	50
Densely heat absorbing glass - - - -	49	—
Internal white cotton curtain - - - -	46	47
Internal white venetian blind - - - -	39	39
Lightly heat absorbing glass - - - -	33	—
Internal dark green open weave plastic blind - -	18	26
Clear glass in double window - - - -	—	16

Curtain Walling—The infilling between continuous fenestration in tall buildings is frequently of some light-weight material, as in Fig. 23.5. The U factor for such a construction was considered in Chapter 2 (pages 19 and 21). If the external surface resistance is ignored due to exposure, as in the case of glazing, the values quoted in Table 2.1 will be increased by up to 20 per cent. Other constructions may be similarly evaluated, but the importance of bridging by concrete or metal ribs should be noted (as in the U values given in that Table). Such members are often in contact with both inside and outside air, without any allowance for discontinuity, and thus condensation can occur on the inner surfaces of the 'cold bridges' in winter. A further problem may arise due to interstitial condensation where well insulated infill panels have an impervious external finish and moisture from within the building is able to penetrate some part of the structural insulation material and start a condensation cycle leading to corrosion on the panel rear faces.

The effect of solar radiation on this form of construction is to raise the outer surface temperature rapidly. It can be shown that in this country the rise will be to about 50 K. The difference between this temperature and that to be maintained in the room in summer (say 21° C) multiplied by the U value gives the heat gain to be allowed per square metre in respect of this wall area.

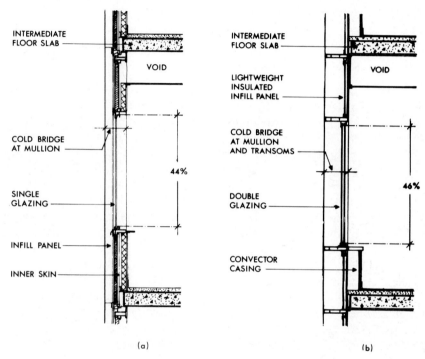

FIG. 23.5.—Examples of curtain walling.

Since the mass of some types of infill panel used in curtain wall construction is low, the insulation material being part of a pre-fabricated sandwich of some form, the time lag for heat transfer may be no more than an hour. Thus, if solar gain (say on a S.E. face) is at a peak at about 10.30 a.m., it may be assumed that the peak heat gain *via* such curtain walling will take effect internally at about 11.30 a.m.

Time Lag—It has been shown* that for British climatic conditions the effect of heat storage in a building is to reduce the effect of peak heat gains from all sources. The paper postulates the factors given in Table 23.2, below, as a reasonable basis for approximate design and states that they are 'based on 24 hour operation of air-conditioning plant during peak load

* Knight, J. C., and Knight, J. L., *The air conditioning of multi-room buildings*, J.I.H.V.E. 1962, 30.1.

TABLE 23.2

MULTIPLYING FACTORS FOR COOLING LOAD

(J. C. and J. L. Knight)

Instantaneous heat source	Multiplying factor for Cooling Load
(a) Peak solar radiation through bare glass	0·55
(b) Peak solar radiation but with venetian blinds on inside - - - -	0·75
(c) Peak solar radiation through double windows with venetian blinds in inter space - - - - - -	0·6
(d) Lighting and occupants - - -	0·6

conditions or pre-cooling. They should also be suitable for 12 hour operation providing an upward momentary swing in air temperature is allowed.'

Obviously, the application of empirical factors such as these is inadequate for detailed design. A rigorous analysis of the available data, using a computer based method of calculation, as described on page 401 and illustrated in Fig. 18.3, should be sought at this final stage.

It will be clear that the effect of solar gain through glazing, in so far as effect on temperature rise of the structure and contents is concerned, is less than the peak, but this and the effect of other heat gains will depend on many factors, such as the nature of the materials of the building and its contents. Another approach to this same problem of the actual effect of the time lag on the cooling load is that referred to on page 398, Tables 18.3 and 18.4, where the choice of two factors is given depending on the mass of the internal construction.

Diurnal Cycle—The diurnal cycle of the sun's motion subjects the various faces of a building to more or less solar incidence at the various hours of the day. When considering the effect on a tall building, it will be apparent that changes can be quite rapid. Thus, although for some hours in the morning an easterly face may receive solar radiation, by about 10.30 a.m. it will begin to lose the benefit and, by 11.00 a.m., will be almost in shadow.

If the building contains wings, such as an L, T or cruciform shape, the degree of solar gain is more complicated and more violent changes are possible, bearing in mind, also, angular incidence over roofs. A simple way to study these effects is to set up a small model of the building at the centre of a chart, such as Fig. 18.2. If external shading or sun-breaks in some form are being considered, a process such as this will be particularly necessary.

Wind Effects—The pattern of wind-flow over a tall building is of the form shown in Fig. 23.6. The positive pressure on the windward side can cause peculiar and undesirable effects within the building if leakage through windows occurs. Infiltration on the windward side may account for a high rate of air-change, rendering the maintenance of temperature difficult, whilst on the leeward side air movement tends to be outward, the only ventilation being by second-hand air from the corridor.

Wind speeds up to 25 km per hour account for up to 30 per cent of frequency of winds in the prevailing direction. Higher speeds are a less frequent phenomenon. Wind speed data as published by the Meteorological Office are at 10 m above ground in open country, but the greater

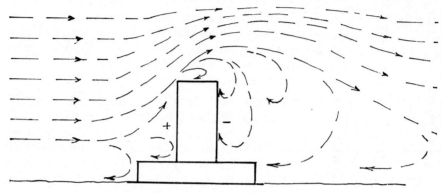

FIG. 23.6.—Wind currents about a tall building.

the height the greater the wind speed. The Meteorological Office formula for correction for height is:

$$V_h = V_{10} \left(\frac{h}{10} \right)^\alpha$$

where V_h = Velocity in m/s at a height of h metres.
V_{10} = Velocity in m/s at 10 metres.
α = a function depending on location, 0·14 in flat open country, 0·29 in suburban areas and 0·5 in a city centre.

If a wind speed of 5 m/s be taken at 10 metres, it will be found that the velocities at other heights above ground in a city centre would be:

at 30 m	8·7 m/s
at 60 m	12·3 m/s
at 90 m	15·0 m/s.

For a tall building, an upper limit may perhaps be taken at 20 m/s for winds of significant frequency, corresponding to a velocity pressure of about 0·24 kPa.

The leakage of air into and out of tall buildings through cracks around windows is dealt with in detail in the *Guide, Section A4*, the basis being work at H.V.R.A.* It is concluded that wind pressure is the principal

---

* Jackman, P. J., *A Study of the Natural Ventilation of Tall Office Buildings.* H.V.R.A. Lab. Report No. 53, 1969.

factor, the pressure differences due to stack effect being small in comparison except in certain unusual circumstances. Some mention of this subject has already been made in Chapter 2 (page 29), but the following few comments are appropriate in the present context.

Air flow through cracks may be calculated from the formula

$$Q = C(P_1 - P_2)^{0.63}$$

where Q =volume of air in litre/s per metre run of window crack.

C = a constant depending on the window type, 0·25 where there is no weather-stripping and 0·05 or 0·13 for weather-stripped pivoted and sliding windows respectively.

$P_1 - P_2$ = pressure difference across the window in Pa.

Solutions for this equation are, however, strictly applicable only when the building is in open plan form and the wind entering on one side has free access to a similar escape route on the other. Where the building has many internal partitions which impede the cross flow of air, then the total infiltration may be reduced to as much as 40 per cent of the calculated value. An average figure for normal cases might be 70 per cent.

For a twenty-storey building, 60 m high, the *average* infiltration rate (which is based upon the *total* height to allow for disturbed wind currents as shown in Fig. 23.6), with a prevailing local wind speed of 3 m/s and windows without weather-stripping, would therefore be:

$$\text{Speed at 60 m height} \qquad 3 \times \left(\frac{60}{10}\right)^{0.5}$$

$$= 7·35 \text{ m/s}$$

$$\text{Equivalent velocity pressure} = 0·6v^2$$

$$\text{(see also Fig. 21.1)} \qquad = 0·6 \times 7·35^2$$

$$= 32·5 \text{ Pa}$$

Thence,

$$Q = 0·25 \times 32·5^{0.63} = 2·24 \text{ litre/s per metre run of crack}$$

The *Guide* proposes that, for the topmost floor, the average result should be increased by 8 per cent and thus, for application to a situation where average partitioning exists, the infiltration to a private office 5 m × 5 m × 3 m with opening windows manufactured to a one metre module would be:

$$\text{Crack length} = 5 \times 4 \text{ m} = 20 \text{ m}$$

$$\text{Infiltration} = 2.24 \times 1.08 \times 0.7 \times 20$$

$$= 33.8 \text{ litre/s}$$

Since the volume of the office is 75 m³, this infiltration rate *with the windows closed* amounts to 1·6 air changes per hour, or a ventilation loss of 40 W/K.

Stack Effect—A tall building, warm inside and cold outside in winter, acts like a chimney. Cold air will enter through entrance doors and other apertures at ground level, rise through lift shafts, staircases and other vertical communications to the upper storeys whence it will tend to be exhausted through smoke vents, crackage and other openings at high level. During spells of warm weather, a reversal of the stack effect will lead to a downflow of infiltrated air.

A theoretical approach to this problem is valueless on account of the unknowns—e.g., what is the resistance to air-flow through a complex of shafts and corridors? Research work on this subject has been published* but the results have not yet been accepted into practice since they are inconclusive. In general terms:

(*a*) At ground floor, wind pressure is found to be insignificant and temperature forces dominate.

(*b*) For higher levels wind effect is more important than stack effect.

(*c*) It is not economically feasible to use excessive ventilation to pressurize a tall building in an endeavour to eliminate infiltration due to wind.

(*d*) The criteria of tightness of the envelope of a building (such as curtain walling) have not been established for quantitative approach.

(*e*) Swinging door entrances—for a particular set of conditions — infiltrate about 25 m³ per person entering or leaving for a single door; 15 m³ per person for a vestibule-type entrance with air lock.

(*f*) Manual revolving doors under the same conditions infiltrate about 2 m³ per person, and motor driven doors about 1 m³ per person.

The conclusions of this study are:

(i) that the greater part of the stack effect problem concerns the ground floor entrances and vestibules where adequate heating in winter is required either by warm air curtain just inside the door, or by forced convectors augmented by floor panel heating.

(ii) that in summer the reverse stack effect is not serious, but some form of air-lock at entrances is necessary if control is to be maintained.

Reference must be made to significant problems encountered in tall buildings clad in curtain walling. Consideration of Fig. 23.5 shows that if particular care has not been taken to seal the structural joint where the

* Min, T. C., *Engineering concept and design of controlling ventilation and traffic through entrances in tall commercial buildings*. I.H.V.E. Conference, 1961.

cladding passes the edge beam of the floor slab, leakage can occur and a stack effect cause an air movement upwards, floor to floor in cascade, thus disrupting the planned method of air movement handled by mechanical ventilation plant. A further example of unexpected air flow can arise where wind pressure applied to small openings in cladding, drillings in members provided to drain away condensation etc., leads to uncontrolled air flow through voids above false ceilings and considerable heat loss from room to void in consequence.

Control of Smoke—The movement of smoke in or towards escape routes is a hazard in all circumstances but one of particular importance in tall buildings. It is necessary to ensure that mechanical ventilation systems are so arranged or controlled that they do not act to reinforce the air distribution influences of wind and stack effect or those produced by thermal pressures originating at a fire source.

Systems may be controlled in the negative sense such that they are closed down in the event of fire or used positively by selective close down of some plants and boosting of others to prevent smoke entry to escape routes. There are differences of opinion between Fire Authorities on this subject but the positive approach is gaining favour.*

The Guide, Section B3, postulates that the maintenance of a positive pressure of about 50 Pa in staircases and lobbies etc., by means of injection of outside air, is adequate to counteract pressure conditions produced in the U.K. due to wind, stack and thermal effects, in most circumstances. This pressure may be produced by boosting staircase supply plant in the event of fire, retaining normal extract systems in use and isolating normal supply systems. Each case must, of course, be considered individually since particular circumstances will influence this general proposition.

To produce this pressure differential of 50 Pa, the outside air supply to the means of escape may be calculated from an assessment of the leakage areas to adjoining spaces and use of the expression:

$$Q = 6000 \, A$$

where   Q = volume of air supplied, litre/s
        A = total of leakage areas through closed doors,
             $0.013 \, m^2$ for single doors, $0.026 \, m^2$ for double
             swing doors and $0.06 \, m^2$ for lift doors.

For one building where these principles were applied,† the required air volume was equivalent to 16 air changes per hour, four times that supplied in normal circumstances.

Perimeter and Core Areas—Tall buildings are sometimes planned in relative depth but cannot, in the context of this Chapter, be considered as

---

* Butcher, E. G., Fardell, P. J., and Clarke, J., *Pressurisation as a means of controlling the movement of smoke and toxic gases on escape routes*. Fire Research Station. H.M.S.O. 1971.

† Hogg, A. E., *Pearl Assurance House, Cardiff*, J.I.H.V.E., 39, 1971, 39.201.

fully 'deep planned'. The perimeter areas will be subject to solar gains, and to exposure to wind. The inner and core areas will be subject to internal heat gains only and ventilation will be essential. In tall buildings, the intensity of daylight at upper levels is often such that shades are left down and artificial light is thus required at all times. The intensity of lighting for inner and core areas which receive little or no daylight must be such as to provide an acceptable visual balance between perimeter and core, and this has led to some very high loadings—as much as 50 W/m².

DEEP PLAN BUILDINGS

Deep plan arrangements, generally single storey, have been a commonplace approach to industrial building for many years, Figs. 10.10 (page 217) and 16.22 (page 372) show examples. In most cases room heights are generous, daylight is available from roof glazing and occupancy levels per unit of floor area are low. Simple heating and ventilating systems will thus, in such cases, provide adequate service. Exceptions to this general statement arise of course where the industrial process requires close control of temperature and humidity or where, in light industries such as assembly of micro components, the necessity for a particularly high standard of illumination leads to a requirement for cooling. Plate XLIII illustrates a ceiling arrangement* which integrates lighting with a radiant heating and cooling system.

In the context of this present Chapter however, we are concerned with the deep plan office building in contrast to the multi-storey slab block discussed in previous paragraphs. A not untypical example might be a four storey building having each floor planned as shown in Fig. 23.7. For equal floor area overall and assuming reasonable proportions, length to width, a tall building of twenty storeys would be required: the perimeter

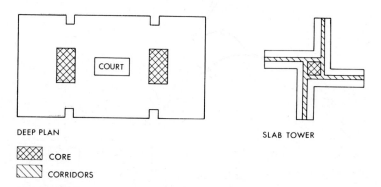

DEEP PLAN

SLAB TOWER

CORE

CORRIDORS

Fig. 23.7.—Arrangement of deep plan building (4 floors) and tall building of same area (20 floors).

* Jamieson, H. C., *A variety of Engineering Experiences*, J.I.H.V.E., 1959, 27.245.

surfaces exposed to wind, frost and sun would then have an area 50 per cent greater than for the lower building.

Solar Gains—Whilst of some significance, the effect of orientation is less marked in the case of deep plan buildings since shape ratios tend to be low. Fig. 23.4 shows that least solar gain will arise throughout the year if the longer faces of a rectangular building were arranged to face north-east and south-west.

Insulation to large areas of roof is of importance as is the use of a light coloured surface finish. Where practicable a light-weight construc-tion is marginally to be preferred. In many cases, accommodation for lift motors, water tanks, cooling towers and air handling plant can be arranged so as to provide some shading of the roof surface.

Glazing—In a deep plan building, windows are provided more to ensure that occupants retain some visual contact with the outside than as a means of illumination. The use of internal or external blinds or the pro-vision of tinted or reflective glass to reduce solar heat gain may not be acceptable. Protection must therefore rely upon suitable elevational treat-ment of the building in the form of vertical fins or horizontal projections.

Vertical fins, for the latitude of Great Britain, are useful only on east and west faces. The reduction in *peak* cooling load for such an application will not be large but their use may have significance with respect to the maximum co-incident load for the building as a whole. Horizontal pro-

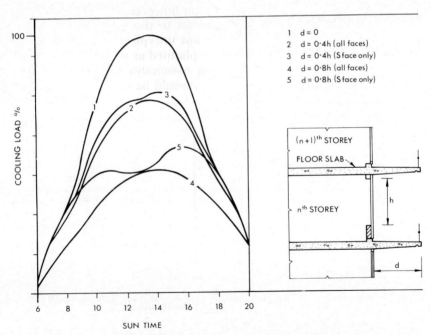

FIG. 23.8.—Effect of horizontal projections upon cooling load.

jections formed by extension of floor slabs above windows are more effective, particularly on a southern face, as shown in Fig. 23.8.

Illumination—Inherent to the concept of deep plan buildings is the provision of a high standard of artificial lighting and, prior to the energy crisis of 1974, the trend was towards a level overall of 1000 lux. This, dependent upon the planning grid and the consequent choice of luminaire, led to an electrical power loading of between 40 and 50 W/m². Present practice suggests that levels of between 500 and 750 lux are adequate, augmented in the former case by local task lighting as may be necessary, which equate to a power loading of between 25 and 30 W/m².

Landscaping—Whilst deep plan offices are not necessarily arranged to follow 'landscape' or even 'open plan' principles, such is often the case. With open plan arrangements, the entire floor area requires to be air conditioned there being no formalised arrangement of communication routes: space utilisation is thus greater and occupancy per m² gross is likely to be higher by about 15 per cent than in the case of a conventional spine corridor layout.

Some demand normally remains, however, for cellular office accommodation within the total floor area and the arrangements for air distribution must be sufficiently flexible to permit adjustments in this respect as requirements change. The choice of planning grid is particularly important in this respect since this will dictate the choice and siting of luminaires and thus affect not only the resultant power loading but also the disposition of air supply terminals. Fig. 23.9 illustrates this proposition

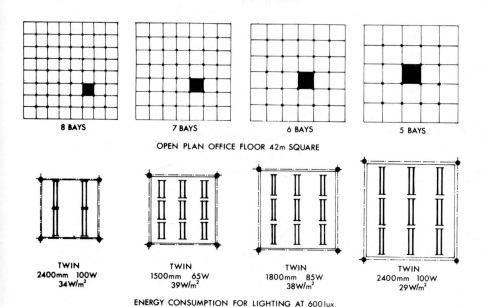

FIG. 23.9.—Alternative arrangements of luminaires: effect upon planning grid.

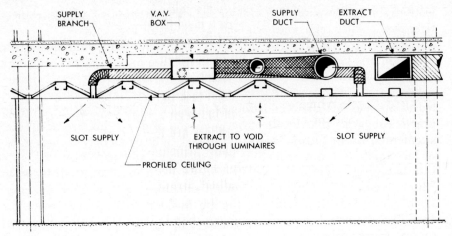

SUPPLY
BRANCH

V.A.V.
BOX

SUPPLY
DUCT

EXTRACT
DUCT

SLOT SUPPLY

EXTRACT TO VOID
THROUGH LUMINAIRES

SLOT SUPPLY

PROFILED CEILING

FIG. 23.10.—Co-ordination of luminaires with air terminals and suspended ceiling.

and Fig. 23.10 shows arrangements which integrate lighting and air distribution terminals with ceiling constructions.

The acoustical treatment of boundary surfaces for landscaped offices is particularly important since, relative to the floor area, heights from floor to ceiling are low. Floors are commonly carpeted and the ceiling is therefore the principal available surface which can be treated to control

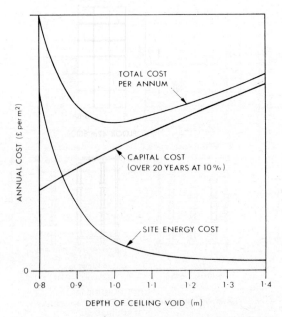

TOTAL COST
PER ANNUM

ANNUAL COST (£ per m²)

CAPITAL COST
(OVER 20 YEARS AT 10%)

SITE ENERGY COST

0

0·8   0·9   1·0   1·1   1·2   1·3   1·4

DEPTH OF CEILING VOID (m)

FIG. 23.11.—Economic depth of void over suspended ceiling.

the ambient sound level. Since the ceiling construction and support system are integrated to such an extent with the lighting, air distribution and acoustic design, it is a growing practice to arrange for the whole to be treated as an engineering exercise. Simulation on a water table of air flow patterns beneath profiled ceilings and full scale mock up facilities can be made available by B.S.R.I.A.

Determination of the void space necessary above the suspended ceiling structure requires careful analysis of capital and plant-running costs, a typical result being as plotted in Fig. 23.11.

CHOICE OF SYSTEMS

To summarise the alternative situations with respect to the two building forms:

The tall building is:

(a) subject to considerable solar gain varying in intensity throughout the day as to the various aspects of the building.
(b) subject to wind effects such that infiltration or exfiltration through openable lights are liable to cause excessive uncontrolled ventilation.

The deep plan building is:

(a) Subject to considerable solar gain on periphery and roof, varying in intensity throughout the day.
(b) Subject to high internal heat gains from artificial lighting, often of high intensity.

Heating and Ventilation (Tall buildings only)—Consider, first, heating for the perimeter. From the foregoing it will be clear that the system must be quickly responsive to change. Any system having a slow rate of response to control will be liable to cause overheating when solar gains occur. Embedded systems, such as floor heating or panel heating, are unsuitable for this reason. Convectors in any form, or heated metal acoustic ceilings having low thermal mass, are suitable provided control is adequately zoned or sectionalised room by room.

Reliance on opening of windows being inadvisable due to the increased wind effect, a heating system needs to have a ventilation system in combination with it. This will introduce fresh air warmed in winter to room temperature, or slightly below, and means for extraction would be necessary. This arrangement would suffice for that period of the year when outside temperatures are low enough to cause sizeable heat loss, say 7° C and below, since control of room temperature may then be achieved by adjustment of the amount of heat supplied by the heat emitters. At higher external temperature, solar gain plus internal heat gains may cause overheating, even with the heat supply shut off.

In the core areas not subject to variations in weather, heating is not required, but ventilation only. The ventilation air will be the vehicle for the removal of heat gains and, when air can be introduced from outside some 5 K lower than room temperature, results will be satisfactory. At higher external temperatures the same 5 K temperature rise will occur, so that when the outside temperature is 21° C and over temperatures of 27° C and over may be expected internally (see also page 401 with reference to this).

Air-Conditioning—For the tall building, therefore, 'straight' heating and ventilation cannot provide for a truly satisfactory service. A means for cooling becomes a necessity as is the case for all deep plan buildings.

Air-conditioning provides for such circumstances and includes also facilities for heating. Thus, one system replaces two and, in addition, supplies the means for dehumidification in summer. Windows may be sealed if cleaned externally, or be openable by key for cleaning from inside. Temperatures may be controlled individually floor by floor, aspect by aspect or room by room.

Air-conditioning is immediately responsive to control; air, being the medium, can be varied in temperature rapidly to deal with sudden changes of heat gains or heat losses. Core areas in tall buildings may be maintained at a constant condition regardless of the season of the year, as may the whole occupied space of deep plan buildings.

Systems of Air-Conditioning—In view of the differing needs, it is necessary to consider tall slab and low deep plan buildings separately. Systems in general were described in Chapter 17 and the following would be suitable applications:

Tall buildings.

Perimeter

Air-water	Induction, zoned to aspect.
	Fan coil with ducted outside air.
	Heat recovery with auxiliary outside air supply at ceiling or side-wall.
All-air	Dual duct, zoned to aspect (for special applications).

Core of limited area

Air-water	Fan coil with ducted outside air.
All-air	Dual duct.

Core of larger area

Air-water	Fan coil with ducted outside air and ducted discharge.
All-air	Dual duct (for special applications).
	Variable volume.

Deep Plan buildings.

Perimeter

Air-water	Induction, zoned to aspect.
All-air	Dual duct, zoned to aspect (for special applications).
	Single duct constant volume, zoned to aspect.
	Variable volume.

Core

All-air	Dual duct (for special applications).
	Variable volume.

A schedule such as that given above cannot, of course, be wholly comprehensive since particular circumstances may give rise to a demand for equipment application which is unusual. For instance, were a tall building to be a hospital, induction units would be unsuitable for service at the perimeter and either all-air single duct terminal re-heat or, more likely, all-air dual duct would be selected.

The system of ducting for the primary air and water circulation to the perimeter zones of a slab block would be determined by the plant location. Fig. 23.12(a) and (b) shows some alternatives which will be self-explanatory: in the case of a deep plan building, air handling plant would be arranged at either roof or basement level as most convenient.

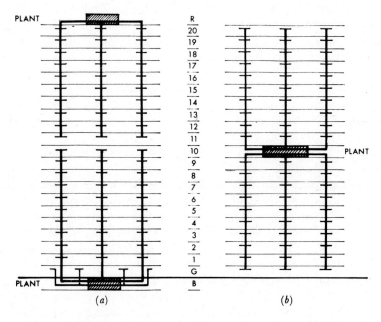

Fig. 23.12.—Disposition of plant rooms (a) basement and roof (b) intermediate.

Zoning of distribution systems, both air and water, is important as far as perimeter service is concerned. In a tall building, for instance, three or four compass aspects might require individual service in addition to the vertical separation illustrated in Fig. 23.12(a).

The arrangement of ducts to serve the core areas in each case would probably concentrate the main risers at one or two points in the core, and to run out floor by floor in a ring main fashion to serve terminals or attenuating boxes as appropriate.

The ring main system has the advantage that the air handling capacity of the ducts is reduced, due to the fact that maximum loads at each zone do not occur simultaneously. Hence the ring need only handle the average load, whereas individual ducts to each section must handle 100 per cent.

The *water-chilling plant* would be common to all systems and could be located in one or other of the plant spaces indicated, such as the basement, roof or intermediately, due regard being paid to reduction of noise and vibration.

The *air heater batteries* would be served from boilers either at the bottom of the building, or on the roof. The latter avoids a flue, which is an expensive item in a tall building as well as taking up valuable space.

INTERMEDIATE PLANT ROOMS

Apart from the areas served by air-conditioning, there will be staircases, entrance halls, lavatories and the like where direct heating in some form may be necessary. In order to avoid undue static pressures on apparatus such as radiators or convectors, valves and fittings, a tall building of over 50 metres height should preferably be divided into two or more vertical zones.

The same applies to hot and cold water services where a pressure of little over 30 metres is the desirable maximum on taps and other fittings.

An intermediate plant room half way up the building, as shown in Fig. 23.12(b) is thus an advantage. The lower part of the building is served from tankage at this level, with calorifiers in the basement. In the intermediate plant room are accommodated calorifiers serving the upper half of the building. The tankage for the upper half is of course on the roof, as shown in Fig. 23.13.

A building 100 metres high would require two such intermediate plant rooms, and so on and the same intermediate level(s) would also be convenient for air plants serving up and down, as previously discussed.

At least some sections of the chilled-water system must, for other than exceptional circumstances, be designed for the full head of the building, as any heat-exchange surface for breaking static head would introduce undesirable temperature drops probably necessitating the use of brine or introducing similar complications. The introduction of water chillers at an intermediate floor level, or on the roof, can be considered as a means of overcoming this problem as far as plant items are concerned. Plate

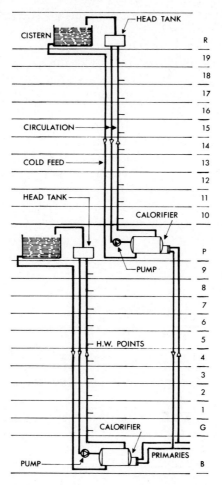

Fig. 23.13.—Diagram of hot water supply system for tall building.

XXXVI illustrates water chillers being installed on the roof of a tall block. **Turbine and Absorption Plant in Combination**—One interesting development of the last named method is to provide in the basement a centrifugal compressor plant driven by steam turbine, and to provide at the upper plant room an absorption cooling plant. The absorption system of refrigeration can make use of the exhaust steam from the turbine, so that a highly efficient thermal balance is obtained. To obtain the optimum efficiency, it is necessary for the two plants to be in a similar state of load, although in practice this may not always be possible. The steam is obtained from boiler plant serving heating, in the winter, as well as other requirements. It is probably true to say that this system has arisen chiefly where district steam is available, as in N. America.

AIR-CONDITIONING AND HEAT RECOVERY

Both tall and deep plan buildings present problems inherently favour-ing the use of some form of heat-recovery system (see p. 576). The inner and core areas of such buildings are heat producing on a massive scale due to high lighting and other electrical loads, whilst the perimeter areas are, in winter, heat losers. Thus, if the system of ventilation is such that the exhaust air from the centre can be used as the low grade heat supply to a heat-recovery system, a great deal of energy may be saved.

For air-conditioning in summer, cooling plant is required in any event, and thus it is obvious that the same plant can be put to good use, with the cycle reversed in the winter, without significant extra capital costs.

This principle is used in the specialised system referred to on page 394, but may equally be applied to conventional central plant, provided the additional complications are accepted. It is obvious that running costs for the year-round conditioning of the building as a whole should be con-

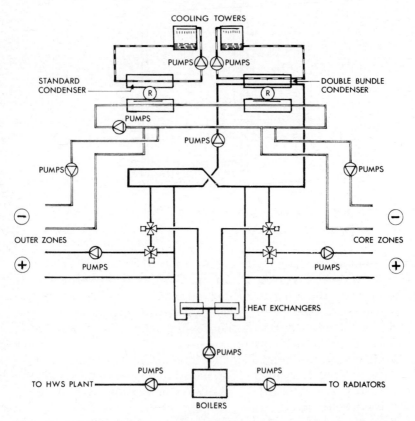

FIG. 23.14.—Large scale heat recovery system. Diagram of water pipework and plant connections.

siderably less with a conventional approach. Fig. 23.14 illustrates such an arrangement as applied to a large deep plan building.

Balance Point—This term has of recent years acquired a specific meaning following on the publicity given to the concept of Integrated Environmental Design (IED). It may be defined as being that outside dry bulb temperature at which, if all heat generated within the building (from lighting, occupancy and machinery etc.) is distributed usefully, no external heat source is necessary. A low balance point is not necessarily a criterion of excellence since it implies a requirement for cooling at any higher outside temperature.

CONCLUSION

Other aspects not relevant to the present book concern water supplies, sanitation, lifts, electrical distribution, and fire protection, all of which fall within the province of the building services engineer. These require the most careful co-ordination if the structure as a whole is to provide a tolerable environment for the occupants and a satisfactory investment in both capital and running costs to the building owner.

Tall and deep plan buildings, as will be seen, present a number of unusual problems. For the purpose of this discussion, it has been convenient to treat these in the context of a narrow slab tower block and a low rise open plan structure. This separation is, of course, artificial since there is no reason why a deep plan building should not be multi-storey also. The most economic building shape, in all respects, is the cube which brings us back very nearly to the traditional form mentioned in the preamble to this Chapter!

CHAPTER 24

Energy Conservation

IT HAS BEEN ESTIMATED that between 40 and 50 per cent of national annual consumption of primary energy in the United Kingdom is used in building services and that over half of this relates to the domestic sector.* By the introduction of economy measures, without real detriment to environmental standards, it should be possible to reduce this proportion by as much as 15 per cent of the gross, which roughly equals the current usage by the whole transport system, private and public.

Primary energy is that contained in fossil fuel, coal, oil and natural gas etc., but an overhead must be deducted from this to cover consumption in production and distribution in order to arrive at the net supply received by the actual consumer. Data extracted from the latest available edition of the H.M.S.O. publication, *United Kingdom Energy Statistics*, show that after allowance has been made for combustion inefficiency the proportion of primary energy actually remaining for use in building systems is approximately as follows:

Electricity	25%
Oil and Coal	58%
Natural Gas	62%

The order of these figures must be borne in mind when considering the merits of conservation measures and those of alternative sources of energy.
Alternative Energy Sources—A whole separate literature exists covering alternative sources of energy, wind and wave power, geothermal potential from steam or hot water as occurring naturally in Iceland and New Zealand or by bores in California, France, Italy and Japan. It is not within the scope of this Chapter to consider these nor to enter into the nuclear debate. It is rather to deal with the application of established conservation techniques to buildings and to the heating, hot water and air-conditioning systems associated with them: even these subjects can be covered only in *précis* form.

ENERGY USE IN BUILDINGS

The level and pattern of energy use in a building starts to be established when the architect accepts his client's brief and develops his initial sketch plans, with or without the advice of an engineer. They are subse-

---

* B.R.E. Working Party Report. *Energy conservation: a study of energy consumption in buildings and possible means of saving energy in housing.* CP56/75, 1975.

quently affected throughout the design and construction process and until the day of building demolition by a host of influences including:

(a) *Building exposure,* orientation, shape, modules, mass, thermal insulation, glazing, solar shading, plant room siting, space for service distribution and construction programme;

(b) *Plant and system design* to match the building characteristics and to meet the needs, known and unknown, of the ultimate occupants;

(c) *Commissioning and testing* of the completed plant and adjustment to ensure that it operates as designed in all respects;

(d) *Operational routines* as adopted to match the building and plant to the continuing or changing pattern of weather conditions, working hours and usage generally;

(e) *Level of maintenance* provided to both building and plant, preservation of adequate records, use of energy audit techniques and replacement or updating of worn out or obsolete components.

Some of the characteristics of building structures which lead towards energy conservation have been dealt with previously—see Chapters 2 and 23 in particular. Fig. 24.1 illustrates, in terms of primary energy, how building form may affect consumption. The three blocks in the histogram identified as A, B, and C are taken from published work* and relate to consumption per m² of floor area. If, however, consumption is related to

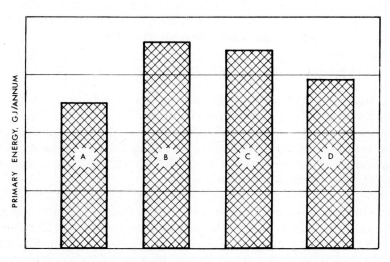

FIG. 24.1.—Primary energy consumption for three built-forms of tall offices. A =naturally ventilated, naturally lit, shallow plan: B =air conditioned, naturally lit, shallow plan: C and D = air conditioned, artificially lit at 750 lux, deep plan.

* Dick, J., *Ambient energy in the context of buildings.* RICS/CIBS/ISES/DOE Conference, 1977.

occupancy and the advantage of open planning thus taken into account, block D replaces C and the penalty over block A, for a greatly improved environment, is halved. This example has been introduced to illustrate the fact that energy use per unit area may not necessarily be the appropriate criterion in all cases.

The last three of the influences included in the list are only in part within the control of the designer. Commissioning and testing of plant is increasingly important as components become more sophisticated, but all too often the contract period planned to be available for these activities is curtailed due to over-run in construction time. Little recognition is given to the fact that although the systems in any building are made up from standard production line components, the combination is a prototype in each case. As to operational routines and maintenance, these are almost always given low priority by building owners but this neglect may result in part from the absence in many instances of an adequately written and illustrated 'driver's handbook'.

The remaining item, that of good practice for plant design, has been covered to a large extent in previous Chapters. There are, however, certain aspects which require re-emphasis and certain plant components or arrangements which are particularly suitable for inclusion in energy-economic designs.

HEATING AND HOT WATER SYSTEMS

As far as space heating plants are concerned, past bad practices have related principally to designs based upon continuous rather than inter-mittent operation. With the advent of more sophisticated control equip-ment, full advantage may be taken in commercial buildings of the economies which result from plant shut-down outside occupied hours. For the public sector,* areas of attack, and the potential savings, are listed in Table 24.1.

Optimum Start Control—This system, developed jointly by the Pro-perty Services Agency and control manufacturers, was mentioned in brief on page 202. The components include a conventional indoor/outdoor temperature controller, with three-way mixing valve etc., plus a time and temperature programmer associated with an outside thermostat and a room thermostat sited in a typical space.

The additional programme unit is commissioned to match the thermal inertia of each individual building and, taking this into account, computes plant start-up time to suit prevailing external conditions. The unit also provides for boost temperature control and frost protection. The opera-tional characteristics are shown in Fig. 24.2 and data available to date suggest that significant energy economies will result from its use.

---

* Johnston, H. P., and Gronhaug, A. C., *Energy economy in government buildings.* IHVE/IES/ InstF/DOE Conference, November 1975.

TABLE 24.1

ENERGY CONSERVATION MEASURES FOR EXISTING BUILDINGS

Type of conservation measure	Potential savings in % of sectorial consumption (years)	Costs/ benefits	Type of policy and action required
Running boiler plants at optimum efficiency	5% of fuel	Small cost compared to savings	Information, training, encouragement, monitoring performance, accountability
Less non-productive idling of boiler plant	5% of fuel	Small cost compared to savings	Information, training encouragement, monitoring performance, accountability
Introduction of optimum start control to Defence Estate	10% of fuel	Return on investment 1 to 2 years	Identifying suitable buildings for conversion contract, monitoring performance
Adjustment to existing heating controls— Defence and Civil Estate	10% of fuel	Small cost compared to savings	Monitoring performance to highlight large wastage. Corrections to controls. Training staff in control technology. Accountability
Installation of heating controls to inadequate system—Defence and Civil Estate	10% of fuel	Return on investment less than one year in most cases	Monitoring performance to highlight large wastage. Installations of controls. Training of staff in control technology. Accountability
Improve thermal insulation and temperature control	20 to 25% of sectorial consumption	Return on investment 5 to 10 years	Revised insulation and control standards. Implementation of schemes giving greatest cost benefits

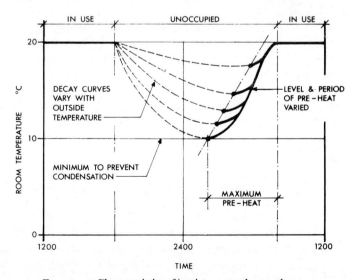

FIG. 24.2.—Characteristics of 'optimum start' control system.

Domestic Heating—Such evidence as is available suggests that the most effective method for control of domestic heating is the thermostatic radiator valve coupled, of course, with full use of a time programmer to provide for intermittent boiler shut-down at night or during other periods when the dwelling is unoccupied.

Proposals have been made for energy saving by reduction in the ventilation rate and, provided that these are not carried to the level which would lead to a recurrence of the condensation problems so prevalent in the late 1960s, economies must result. Double glazing of domestic windows is a measure which will undoubtedly save energy and increase comfort but, using conventional methods, cannot be considered economic: the well publicised concept of using cheap plastic wrapping film for this purpose will serve to produce an insulating air gap on a strictly temporary basis but fire risk and fume generation cannot be ignored.

Long term, the provision of a controlled supply ventilation system to dwellings, possibly taking advantage of solar heating effects upon roof areas or purpose provided surfaces, may be advantageous. Such arrangements are commonplace in Scandinavia in multiple housing.

Proposals have been made that dwellings be compartmented to permit temperature zoning, but this savours of the 'selective' warm air heating systems which had a brief phase of popularity some twenty years ago on grounds of capital economy. *Parker Morris* standards for home heating are mentioned on p. 601 and some hidden disadvantages of the zone concept are exposed in Fig. 25.8.

Hot Water Supply—The use of local as distinct from central systems is discussed in Chapter 13, as are the economies resulting from the use of instantaneous heaters, of all types, and spray taps in substitution for the more usual bib pattern.

In the interests of energy saving, more attention should be given to the provision of positive means for prevention of secondary hot water circulation overnight. Despite the fact that circulating pumps may be stopped by time controllers, some thermo-syphon water movement may continue during the shut-down period through a pipework system with the result that a storage vessel left full of hot water in the evening may be quite cold next morning, however well insulated. This comment applies equally to domestic and commercial premises.

AIR TO AIR HEAT EXCHANGERS

Any ventilation or air conditioning system which takes in outside air, heats and/or cools it and then discharges an equivalent or lesser quantity to waste, offers potential for energy saving. The simplest of plenum ventilation plants, arranged to recirculate 60 or 70 per cent of the air handled, will nevertheless require a heat source to raise the temperature of the remaining outside air supply to whatever level the application may demand. At the other extreme, a sophisticated air conditioning plant will consume energy in pre-heaters, re-heaters and zone-heaters plus that which will be required as a result of a humidification process. In either case, some proportion of the treated air delivered to the building will, by design, quite properly be discarded.

The opening paragraphs of Chapter 16 showed that the admission of certain minimum quantities of outside air were necessary for odour control and prevention of condensation and this subject was further discussed, with respect to design of air conditioning plant, on p. 404. For commercial and industrial premises where noxious fumes are generated, in however low a concentration, additional outside air quantities must be supplied beyond the minimum and an equivalent volume collected for discharge back to atmosphere.

To bring the quantities of heat so wasted into perspective, consider a small office block with a floor area of 1000 m² which has a mechanical ventilation plant to provide the minimum recommended outside air quantity of 1·3 litre/s per m², the plant running for 60 hours per week. Over an average winter season, a heat supply of about 140 GJ (equivalent to that provided by burning about 4 tonnes of oil) would be necessary to do no more than raise the ventilation air supply to a degree or so *less* than room temperature. The associated extract ventilation plant would then reject a similar air quantity, certainly at room temperature, back to outside.

In order to overcome the nuisance of heat gain to rooms from lighting, modern practice allows extracted air to pass over and through luminaires, see Fig. 22.21, taking with it up to 70 per cent or so of the electrical input thereto. In our hypothetical example therefore, the temperature of the air discharged might well be several degrees *above* that held in the office space proper—say 22 or 23° C.

For applications such as hospitals and similar buildings, which ventilate or air-condition without recirculation, in order to avoid contamination, and swimming pools which are similarly served in order to reduce risk of condensation, the air quantity rejected is far greater per m² of floor area. Further, in industrial premises where process heat gain may be high, air will be exhausted at temperatures much greater than those quoted above.

It will be obvious that great scope for energy conservation exists if the heat in exhaust air can be reclaimed and applied, in part at least, as a source of energy to raise the temperature of the outside air used in the parallel supply ventilation plants. These same comments apply of course to economies to be achieved in cooling capacity during the summer months since the temperature of air discharged from an air-conditioned building may then be *less* than that of the outside ambient: it may, furthermore, carry less moisture. This aspect assumes more importance in climates which produce extremes of summer temperature.

Available Equipment—A variety of types of equipment is available for air-to-heat exchange in ventilation plants. They fall under one of the following headings:

(a) Plate type heat exchangers
(b) Ljungstrom and Munter thermal wheels
(c) 'Heat pipe' heat exchangers
(d) Run around coils with water circulation
(e) Run around coils using refrigeration.

Of these five types, the first two effect heat exchange directly from air to air whereas the remainder employ an intermediate circulating medium. **Efficiency of Heat Reclaim**—Before describing the alternative types of equipment in more detail, the matter of their efficiency in operation requires definition. The data normally quoted derive from the expression:

$$\eta = \left(\frac{t_3 - t_1}{t_2 - t_1}\right) 100$$

where t_1 is the temperature of the outside air, t_2 is the temperature of the exhaust air from which heat is to be reclaimed, t_3 is the temperature of the supply air after it has been passed through the heat exchanger and η is the efficiency, per cent.

As will be noted, the expression makes use of dry bulb temperatures and thus, strictly speaking, relates only to sensible heat recovery. Some manufacturers emphasise this in their literature whilst others do not.

Consider the case of a heat exchanger unit having a quoted efficiency of 76 per cent which is applied to a winter reclaim situation, handling equal quantities of outside air at $-1°$ C saturated and exhaust air at $22°$ C DB, $15.5°$ C WB. Thus:

$$t_3 = 0.76(22 + 1) - 1 = 16.5° \text{ C}$$

For use in summer, when the stated efficiency might have fallen to say 66 per cent,* and taking the outside air supply to be at $28°$ C DB and $20°$ C WB

$$t_3 = 28 - 0.66(28 - 22) = 24° \text{ C}$$

If these data are plotted on a psychrometric chart as Fig. 24.3(a), the processes may be then represented in an idealised form as shown and the various significant quantities read off as summarised in Table 24.2 (p. 568). In terms of enthalpy, it may be seen that the efficiencies may be calculated as:

Winter

$$100\left(\frac{S_3 - O_1}{R - O_1}\right) = 100\left(\frac{25.3 - 7.7}{43.4 - 7.7}\right) = 49.2\%$$

Summer

$$100\left(\frac{O_2 - S_4}{O_2 - R}\right) = 100\left(\frac{57.8 - 48.4}{57.8 - 43.4}\right) = 29.2\%$$

---

* In winter, as may be seen from Fig. 24.3 (a), exhaust side of the heat exchange surface will be wetted for part of the process. This situation may not occur in summer if the heat exchanger type is such that the surface remains dry.

In instances where a heat exchanger has equal ability to recover both sensible and latent heat, the situation is somewhat better, as illustrated in Fig. 24.3(b). In this case the efficiencies in enthalpy terms will be as those quoted for temperature:

Winter

$$100\left(\frac{S_3-O_1}{R-O_1}\right)=100\left(\frac{34\cdot8-7\cdot7}{43\cdot4-7\cdot7}\right)=76\%$$

Summer

$$100\left(\frac{O_2-S_4}{O_2-R}\right)=100\left(\frac{57\cdot8-46\cdot9}{57\cdot8-43\cdot4}\right)=76\%$$

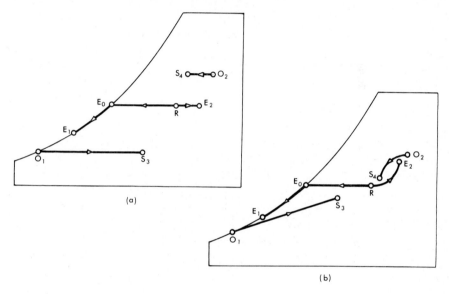

FIG. 24.3.—Air-to-air heat exchanger performance (a) sensible heat reclaim only (b) enthalpy reclaim.

TABLE 24.2

PHYSICAL QUANTITIES APPROPRIATE TO STATE POINTS INDICATED IN FIG. 24.3

Position from Fig. 24.3		DB Temp. °C	Per kg dry air	
Point	Statement		Enthalpy kJ/kg	Moisture kg/kg
O_1	Winter, outside	-1	7·7	0·0035
O_2	Summer, outside	28	57·8	0·0116
R	Room exhaust	22	43·4	0·0084
S_1	Unit output, winter	16·5	23·5	0·0035
S_2	Unit output, summer	24	53·6	0·0116
S_3	Unit output, winter	16·5	34·8	0·0072
S_4	Unit output, summer	23·4	46·9	0·0092

Some types of equipment have different characteristics with respect to sensible and latent heat exchange. Taking the previous example and allowing for latent transfer to be at 50 per cent efficiency, the winter and summer results in terms of enthalpy, would be 67 and 58 per cent respectively.

Where supply and exhaust air quantities are not equal, the notional 'efficiency' will change *pro rata* to the mass ratio. That is, if the supply air quantity were twice that of the exhaust air, the efficiency would be halved: in the converse case, the efficiency would increase by about 25 per cent. **Plate Type Heat Exchangers**—Having no moving parts, this is probably the most simple type of equipment. The two air streams are directed in cross- or counter-flow through a casing which is compartmented to form narrow passages carrying, alternately, exhaust and supply air. Energy is transferred by conduction through the separating plates and contamination of one air stream by the other thus avoided. The form of the casing is arranged to suit the configuration of the transfer surface and to provide for convenience of air duct connections: examples are shown in Fig. 24.4.

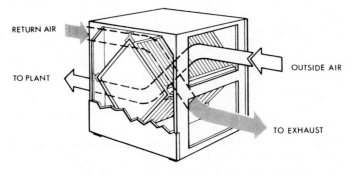

RETURN AIR

TO PLANT

OUTSIDE AIR

TO EXHAUST

Fig. 24.4.—Plate type heat exchanger: note air flow pattern.

Since the separating plates are normally of metal (aluminium or stainless steel) or of glass, moisture transfer is not possible and sensible heat only is exchanged. A departure from this principle occurs in the case of one type where treated Japanese paper is used in a honeycomb formation to separate the air paths and latent heat exchange through the material results in consequence.

Units are available in module form to handle air quantities in the range of 500 to 20 000 litre/s and may be built up to suit individual requirements. Due to the relatively low rate of heat transfer per unit area, the plate surface necessary is large but unit sizes are reasonably compact since air flow passages are kept to minimum width. Efficiencies in the range of 50 to 75 per cent are claimed, and resistance to air flow is from 150 to 300 Pa at a face velocity of 3 m/s.

Thermal Wheels—Constructed on the lines illustrated in Fig. 24.5, the thermal wheel or regenerative heat exchanger consists of a shallow drum containing appropriate packing which is arranged to rotate slowly between two axial air streams, picking up energy from exhaust air and releasing it to supply air. The wheel is mounted in a supporting structure and motor driven at approximately 10 rpm maximum, the speed sometimes being made to vary as a means of controlling output.

The media and the form of the heat transfer surface varies as between manufacturers. The original Ljungstrom type, first produced sixty years ago for industrial applications, has a random mass of stainless steel or aluminium wire, which may be knitted as for a pan scourer, as the packing. Other manufacturers use a matrix made up from alternate flat and corrugated sheets of aluminium foil, rolled concentrically to form a multitude of axial passages.

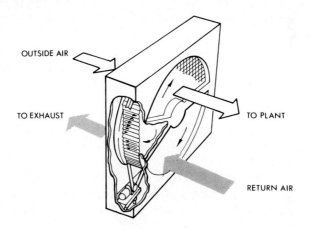

OUTSIDE AIR

TO EXHAUST

TO PLANT

RETURN AIR

Fig. 24.5.—Thermal wheel (Curwen and Newbery).

The later Munter type of wheel uses packings which have hygroscopic properties. These may be similar to the metal mesh described above but coated with a desiccant such as lithium chloride. Alternatively, an inorganic fibrous paper may be used, made up as for the aluminium foil arrangement mentioned previously but impregnated with desiccant.

In operation, the difference between the two types is that heat interchange by the former will be largely sensible whereas the latter, by successive attraction and rejection of water vapour at the dessicant, will recover latent heat also.

Cross contamination between the two air streams is minimised by so arranging the respective fan positions that the supply air pressure at the recuperator is greater than that of the exhaust stream. By using suitable labyrinth seals and incorporating a purge sector which allows for the matrix to be scavanged before supply air passes to the building, it is claimed that contamination is kept to less than 1 per cent. Lithium bromide as used for treatment of the hygroscopic type matrix is stated to be bacteriostatic, i.e. it inhibits the *propagation* of bacteria.

Wheels are available in sizes up to about 4 m diameter to handle air quantities in the range of 300 to 20 000 litre/s but multiple units in the middle of the range are often more physically convenient to handle for large air quantities. Efficiency may be as high as 80 per cent in sensible heat reclaim and about 65 per cent in enthalpy terms for the hygroscopic type. Resistance to air flow at a face velocity of 3 m/s will be between 200 and 300 Pa. The power required to rotate the wheel is quite small being between 400 and 800 W for the largest size handling 20 000 litre/s.

'Heat Pipe' Heat Exchanger—As in the case of the plate type, heat pipe units have no moving parts and are simple in concept. A working fluid is however employed to effect heat transfer. Construction consists of a box enclosure having a dividing partition to separate the supply and exhaust air streams, through which an array of finned heat pipes is assembled.

The 'heat pipe' itself is a by-product of nuclear research developed further in connection with the space programme: in essence it is no more than a super-conductor of sensible heat. Each individual conductor is a sealed tube, pressure and vacuum tight, provided with an internal wick of woven glass fibre normally as a concentric lining to the tube. During manufacture, a working fluid is introduced in sufficient quantity to saturate the wick. The actual fluid used is selected to suit the temperature range required and, for the application here considered, Freon (R12) is commonly chosen.

In operation, heat applied to one end of the pipe will cause the liquid to evaporate and the resultant vapour will travel to the 'cool' end where it will condense, surrendering energy, and the liquid will return through

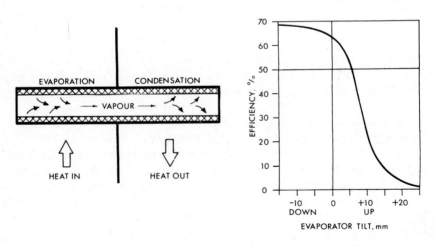

FIG. 24.6.—Heat pipe operation and performance (Q-Dot Corporation).

the wick by capillary action to the 'hot' end. Fig. 24.6 illustrates this process and shows also how the heat transfer capacity of a pipe may be adjusted by varying the angle to the horizontal at which it lies. This characteristic may be used to match capacity to a given application or, by automation, to provide a means of control. Where the angle of tilt is used in this way, a facility must be provided to reverse the action when the season changes, winter to summer.

The capacity of a built-up heat exchanger for given overall dimensions will vary according to the number of rows of heat pipes, the fin spacing and the air velocity. A six row unit by one manufacturer having 55 fins per 100 mm would, for equal supply and exhaust air quantities, have an efficiency for sensible heat exchange of 61 per cent at a face velocity of 3 m/s and with a resistance to air flow of 250 pa. Module sizes range from 0·2 to 4·5 m² which at that velocity would indicate capacities from about 500 to 13 500 litre/s.

Run-Around Coils (**water circulation**)—This approach to the problem has the merit of extreme flexibility and is, moreover, founded upon a well understood technology. As shown in Fig. 24.7, the basis of the system is a pair of conventional finned tube heating/cooling coils, one fitted in each air stream, connected by a pipework system for pumped circulation of the working fluid, often a 25 per cent solution of glycol antifreeze in water.

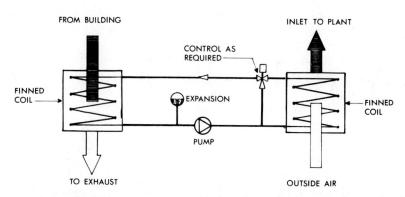

FIG. 24.7.—Run-around coils (water circulation).

The flexibility of the system derives from the obvious ease by which the coils may be connected together: there is no need to disturb the routes of what may be large air ducts to bring them, inlet and outlet for both supply and exhaust, to the heat exchanger. Furthermore, coils may be fitted to any number of exhaust ducts and the heat therefrom collected and distributed to any number of similar coils fitted to supply air ducts. Diversity of energy availability and energy demand between air handling plants may thus be used to best advantage.

There are, of course, compensating disadvantages, those of most consequence being the need for double heat exchange (exhaust air to fluid and fluid to supply air), the relatively small temperature differentials available for such energy transfer, the need for water pumping power and the matter of heat loss and gain from and to the pipework system.

Direct transfer of latent heat is not possible with this system, but in winter the coil in the exhaust air stream would run wet as would that in the supply air stream during some summer conditions: energy transfer would thus be assisted and efficiency improved as in the case of plate type heat exchangers. In the context of what has been said before, however, heat transfer would be sensible only.

Little more needs to be added with regard to this type other than to emphasise that the small temperature differentials between either air stream and the working fluid will result in deep coils and high resistance to air flow with consequent penalties in fan power requirement. The over-all efficiency, ignoring fan and pump power, some of which will be re-covered in winter but will be a penalty in summer, is not likely to be more than 40 to 60 per cent at best.

Run-Around Coils (using refrigeration)—If one considers an exhaust and a supply air duct, separate but not too distant, it is obvious that the evaporation and condensing elements of a refrigeration plant could be fitted within the respective air streams, Fig. 24.8. By such means, one of the disadvantages of a water circulating system, i.e. small temperature differentials, could be overcome: in fact, using this 'heat pump' principle (see p. 569), the supply air temperature may be raised above that of the exhaust air. The energy required to drive the compressor imposes a penalty but much of this would be recovered as heat to the supply air stream.

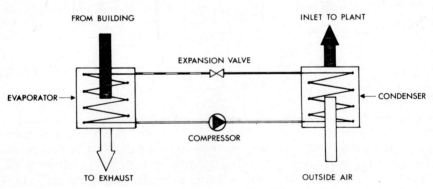

FIG. 24.8.—Run-around coils (using refrigeration).

Many types of packaged air conditioning plant designed for roof mounting incorporate not only supply and exhaust fans but also air-cooled refrigeration equipment for summer use. In some cases facilities are provided whereby air paths may be re-directed during the winter and some part of the refrigeration capacity used as a heat pump drawing energy from exhaust air.

Systems compared—The principal disadvantage of any air-to-air heat recovery system which does not make use of an intermediate fluid for heat transfer is, as previously explained, that four air ducts which may be quite large must be brought together at the heat exchanger. Other problems occur also in analysis but these are amenable to technical rather than spatial solution:

(1) Any heat exchange element which may at some periods of the year operate with wetted surfaces must be provided with a condensation collection tray and drain facilities. It is important that the configuration of the exchange surfaces permits flow and collection of moisture.

(2) In extreme weather conditions, a small amount of pre-heat may be required to prevent freezing of condensed moisture since this could lead to damage to the equipment and an unacceptable resistance to air flow.

(3) Heat exchangers which have small passages presented to air flow will soon be clogged with dirt unless pre-filters are provided: wash down may be required at intervals. Some manufacturers of thermal wheels claim that the reversal of air flow which occurs in normal operation acts to maintain cleanliness.

In considering the relative economics of alternative methods, account must be taken of annual mean rather than peak load efficiencies and of available means for control of heat exchange. As has been mentioned, the performance of a thermal wheel may be varied by speed change and that of a heat pipe unit by automation of the angle of tilt. The water circuit of run-around-coils may be fitted with motorised valves as required, but for plate heat exchangers an arrangement of face and by-pass dampers will be necessary.

Resistance to air flow and consequent increases in fan power must be considered as must the energy consumption by auxiliaries, drive motors, pumps, etc. These, of course, are likely to remain running even when the enthalpy of the supply and exhaust air streams is so close as to lead to minimal interchange. Table 24.3 presents a summary comparing the performance of the various types of equipment based upon the design conditions listed in the first three lines of Table 24.2, with supply and exhaust air quantities equal at a rate of 5000 litre/s.

HEAT PUMPS

The operating principles of refrigeration equipment are described in Chapter 20 in their application to the proper performance of an air-

T   K.H.A.C.

conditioning plant. In that context, the capability of the evaporator is utilised as a source of cooling whilst heat produced at the condenser is rejected. When similar equipment is utilised in the reverse sense, drawing energy from a low temperature source (a heat 'sink') at the evaporator and making use of the higher grade output at the condenser, the apparatus is described as a 'heat pump', see Fig. 24.9. Brief mention of simple small scale units is made in Chapter 12, in their role as unusually efficient means of converting primary energy to space heating use *via* an electrical supply.

As may be seen from Fig. 24.9, mechanical equipment is used to compress a gas and in doing so raises its pressure and temperature; that is, energy is added to the gas. The compressed hot gas is passed to the condenser where heat is removed for use and the gas condensed into a hot

TABLE 24.3

COMPARATIVE PERFORMANCE OF VARIOUS TYPES OF AIR-TO-AIR
HEAT EXCHANGER*

Type of equipment	Make	Energy reclaim efficiency (%)			
		Sensible heat		Enthalpy	
		Winter	Summer	Winter	Summer
Parallel Plate					
Metal	A	70·9	66·7	45·9	28·0
Glass†	B	70·0	66·0	46·2	29·5
Paper†	C	73·0	73·0	68·0	64·0
Thermal Wheel					
Non-hygroscopic	D	77·8	77·8	49·9	34·3
Hygroscopic	D	77·8	77·8	77·8	77·8
Hygroscopic	E	74·0	74·0	74·0	74·0
Heat Pipe					
6 row, 1·8 mm fin pitch†	E	62·1	62·1	41·4	27·9
Run-around Coil					
Water	D	60·4	55·0	38·9	26·5
Water/glycol	F	45·7	41·7	29·4	17·5

* For the winter and summer design conditions listed in Table 24.2. Supply and exhaust air quantity = 5000 litre/s.
† Calculated from catalogue data. Base values for the remainder kindly provided by various manufacturers.

liquid, surrendering latent heat. The refrigerant then expands through an expansion valve where the pressure drops and the liquid enters the evaporator at a reduced temperature. In this state the refrigerant can absorb heat at a relatively low temperature from the heat sink and in so doing boils, or vaporizes, the resultant vapour returning to the suction of the compressor and the cycle then being repeated.

It will be appreciated that it is practicable for the same group of

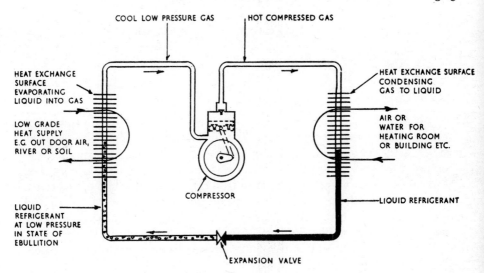

COOL LOW PRESSURE GAS   HOT COMPRESSED GAS

HEAT EXCHANGE
SURFACE
EVAPORATING
LIQUID INTO GAS

LOW GRADE
HEAT SUPPLY
E.G. OUT DOOR AIR,
RIVER OR SOIL

HEAT EXCHANGE SURFACE
CONDENSING
GAS TO LIQUID

AIR OR
WATER FOR
HEATING ROOM
OR BUILDING ETC.

COMPRESSOR

LIQUID
REFRIGERANT
AT LOW PRESSURE
IN STATE OF
EBULLITION

LIQUID REFRIGERANT

EXPANSION VALVE

Fig. 24.9.—Diagram of heat-pump.

equipment to have a dual function, providing cooling output in summer and heating output in winter. A unit such as that shown in Fig. 12.11 may very well be designed to operate in this way and such a change-over facility is fundamental to the operation of the heat recovery terminal units described on p. 394 and shown in Fig. 17.22. The sole technical reservation is that the operating temperatures and pressures of the working fluid (refrigerant) chosen for optimum operation in one cycle may not be those best suited to the other, the condensing and evaporating conditions changing with the seasons.

Coefficient of Performance (C.O.P.)—This is the term used to describe the 'advantage' offered by operation of a heat pump, use of the parallel term 'efficiency' not being truly appropriate when the ideal of 100 per cent is exceeded, such a condition being impossible of achievement. In strict theoretical terms, the coefficient of performance is defined as:

$$\text{C.O.P.} = \frac{T_1}{T_1 - T_2}$$

where T_1 is the condensing, and T_2 the evaporating temperature of the thermodynamic cycle, in degrees Kelvin. It is more usual however, in practical terms, to express this coefficient as the simple ratio of energy output to energy input but it is necessary in this respect to be aware whether the consumption of all appropriate auxiliaries is included in the calculation.

It must be remembered of course that, in comparison with a cooling application, the energy in driving the compressor is an asset as far as a heat pump is concerned; the equation being transposed algebraically:

Cooling

$$\frac{\text{Evaporator}}{\text{Capacity}} = \frac{\text{Condenser}}{\text{Capacity}} - \frac{\text{Compressor}}{\text{Power}}$$

Heating

$$\frac{\text{Condenser}}{\text{Capacity}} = \frac{\text{Evaporator}}{\text{Capacity}} + \frac{\text{Compressor}}{\text{Power}}$$

As may be seen from Fig. 24.10 which represents the output of a reciprocating semi-hermetic machine using refrigerant R22, performance varies with evaporating and condensing temperature, the smaller the temperature difference, the better the performance characteristic. The properties of various refrigerants are listed in standard works of reference and pressure/enthalpy charts appear in the *Guide, Section B14.* In the application of these to practical problems, it must be remembered that the theoretical performance shown in Fig. 20.2 is distorted in practice (Fig. 24.11) by superheat and sub-cooling effects: these affect predicted per-

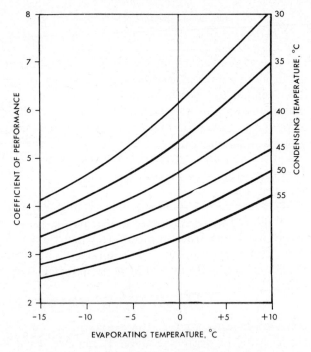

Fig. 24.10.—Heat pump performance using R22.

formance which may well be only two-thirds of the theoretical.

In consequence of the requirement that the temperature difference between the condensing and evaporating temperatures (and hence between heat utilization medium and heat sink) be minimised, applications to space heating are restricted. In winter, where heat output is required

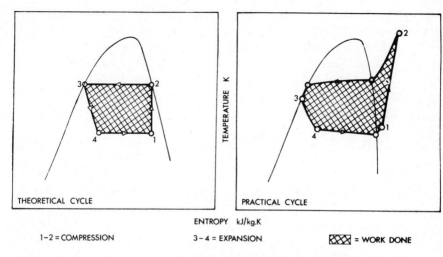

THEORETICAL CYCLE   PRACTICAL CYCLE

ENTROPY   kJ/kg.K

1–2 = COMPRESSION   3 – 4 = EXPANSION   ⬚⬚⬚ = WORK DONE

FIG. 24.11.—Practical Rankine-cycle heat pump.

to be at maximum, heat sinks such as outside air, river water or the earth's crust are likely to be at their annual minimum temperature. Industrial effluents and power station waste are, however, relatively constant as to temperature throughout the year and it is to these that attention should be directed as well as to any advantage that may arise from collection of solar energy.

Compressor Drive—Whilst it is common practice to consider the heat pump as being electrically driven, this is not necessarily the case except where small production units are concerned. If a broad approximation of 25 per cent be considered as representative of the conversion of primary energy to electrical power, it is obvious that a heat pump so driven, having a practical coefficient of performance of say 2·5, will equate as follows to, say, a boiler fired by natural gas:

Electrical Heat Pump
C.O.P. (primary energy) = 0·25 × 2·5 = 0·625
Gas Fired Boiler
C.O.P. approx. (primary energy) = 0·62

In consequence, it is proper to consider the case of a heat pump driven by a prime mover such as a gas or diesel engine in circumstances where the waste heat from the power source may be recovered. Commonly quoted data for prime movers are, at full load:

Shaft power   33%
Waste heat to oil
and jacket coolers   30%
Waste heat to exhaust 32%

If a recovery potential of 50 per cent is applied to waste heat and an average 5 per cent energy overhead considered for oil or gas supply, then the following equation results:

Motive Power 0.33×2.5 $= 0.825$

Waste Heat $0.62 \times 0.5 \times 0.95$ $= 0.295$

Combined C.O.P. $= 1.12$

In consequence, a clear case exists for further examination of the use of heat pumps driven by other than electricity, the ratio of advantage being of the order of 1.8 to 1 in favour of this approach.

Packaged Heat Pumps—Domestic scale installation of heat pumps has grown considerably in volume over recent years. Various types of packaged units of this sizé are now available in the U.K. and a number of enthusiasts has applied a variety of experimental concepts.* Most available equipment operates on an air to air cycle and, in consequence of winter freezing problems at the evaporator, manufacturers often recommend that the heat pump be put out of use when the outside air temperature falls below 5° C. Hence, a full supplementary heat source will be required by way of boiler power or electric resistance heaters and the annual economics of operation are thus eroded.

Investigations are in hand at the Building Research Establishment into combination heat pump/solar collector projects which have been well reported in the technical press.† A problem which has arisen relates to the fact that most commercially available equipment, even of this size, requires a three-phase electrical supply which local Supply Boards appear reluctant to provide except at an exhorbitant capital charge.

Purpose Designed Heat Pumps—As far as can be established, only three installations in the U.K., of significant size, have been reported upon in detail. Each was a prototype and for explicit information the reader is referred to detailed published accounts. In brief however:

> **Norwich.** This unit owed its existence to the enthusiasm of the designer‡ since it was constructed during the period 1940–45, largely from second-hand or salvaged materials. The source of heat was the river Wensum in Norwich and the building served was the local electricity department offices and workshops. Sulphur dioxide was chosen as the refrigerant, R12 not being available at the time, and the second-hand reciprocating compressors served a demand of about 235 kW *via* shell and tube heat exchangers.
>
> The average 'seasonal reciprocal efficiency' during the winter of 1945–6 was reported to be 3.45 and the comparative unit cost of heat supplied was 94 per cent of that provided by a parallel

---

* Keable, J., *The application and economics of heat pumps*. CIBS/ISES/DOE Conference, April 1977.

† Freund, P., Leach, S. J., Seymour-Walker, K., *Heat pumps for use in buildings*. BRE Report CP19/76, 1976.

‡ Sumner, J. A., *The Norwich heat pump*. Proc.I.Mech.E 1947, 156.22.

boiler plant (burning coal at £3.25 per ton!) including capital charges. Unhappily, the plant was demolished on the retirement of the designer.

Festival Hall—Designed in 1949* to serve the new building complex on the south bank of the river Thames, this plant included high speed (over 2000 rpm) centrifugal compressors driven by Merlin aircraft engines converted to run using Towns gas. The output of the plant was approximately 2·64 MW to serve, in parallel with boiler plant, a connected load estimated at 7·33 MW: this disparity was a deliberate design feature selected to ensure that a 'base load' running condition for the heat pump would exist. The heat sink was the river Thames, using 136 litre/s through the evaporator at a design temperature drop of about 1 K. The working fluid was R12.

In the event, the maximum load available to the heat pump did not exceed about 2 MW and operation was possible only for pre-heat periods of short duration. On test the heat pump achieved a C.O.P. of 2·8, the ratio of heat supplied to gas consumed being 1·21. The plant has since been dismantled.

Nuffield College—This installation was designed by the authors to serve the new college in Oxford of this name.† It was built at the time of the construction of the college and the opportunity was thus available to match the characteristics of the heat pump to the heating systems and *vice versa*. The heat source was the main Oxford sewer which records showed to run at not less than about 16° C even with snow water entering. The two twin cylinder reciprocating compressors were works packaged for drive from a 5·1 litre diesel engine which also powered three circulating pumps by belt drive from the tail shaft of the engine.

The capacity overall was 127 kW, 24 per cent of which derived from engine waste heat. The heating systems within the college proper, having a design output of approximately 400 kW, incorporated three separate types of emission surface, low temperature embedded panels (50° C flow), medium temperature radiators (60° C flow) and normal temperature radiators (70° C flow). The automatic (weather sensitive) control systems to the three water circuits concerned are so sequenced as to provide the best possible annual load factor. The boiler plant was arranged to provide the necessary 'top up' capacity and to serve the domestic hot water system.

Test results in 1962, as recorded in Fig. 24.12, showed a heat pump C.O.P. of 3·98 and an overall heat supply to energy con-

---

* Montagnon, P. E., Ruckley, A. L., *The Festival Hall heat pump.* J.Inst.F., 1954, 27.170.

† Kell, J. R., Martin, P. L., *The Nuffield College heat pump.* J.I.H.V.E., 1962, 30.353.

sumed ratio of 1·54. The principal operational problems have related to the diesel engine which is currently in need of replacement. In retrospect, a diesel alternator providing electric drive to compressors, with waste heat recovery, would probably have been a more elegant design solution, bearing maintenance facilities in mind. (See Plate XLIV, facing p. 577.)

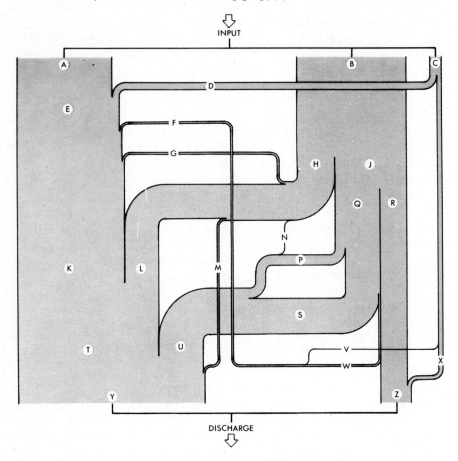

KEY		kW	KEY		kW
A	SEWAGE	65·5	N	WASTE HEAT PUMP	0·1
B	FUEL OIL	75·3	P	WASTE HEAT BOILER	8·0
C	ELECTRICITY	7·4	Q	ENGINE HEAT RECOVERED	30·8
D	SEWAGE PUMP	5·1	R	ENGINE HEAT LOST	18·6
E	ANNULAR HEAT EXCHANGER	70·6	S	ENGINE JACKET	22·1
F	AMBIENT AIR	0·8	T	CONDENSER	96·7
G	EVAPORATOR PUMP	1·0	U	WASTE HEAT EXCHANGER	30·2
H	POWER	25·9	V	FAN	0·1
J	ENGINE WASTE HEAT	49·4	W	RADIATION	0·7
K	EVAPORATOR	72·4	X	LOSSES	2·2
L	COMPRESSORS	24·3	Y	OUTPUT TO COLLEGE	127·4
M	MAIN CIRCUIT PUMP	0·5	Z	TOTAL WASTE	20·8

FIG. 24.12.—Sankey diagram of heat pump performance, Nuffield College.

Templifier Heat Pumps—An interesting new development is the Templifier* machine which uses single or two-stage centrifugal compressors having hermetic motor drive. These units are factory packaged with the associated shell and tube evaporator and condenser in a range of four units rated from 250 kW to 1·36 MW. The performance, as shown in Fig. 24.13, covers a wide range of temperatures for source and output. The ability of this machine to provide output at temperatures suited to conventional space heating systems should be noted.

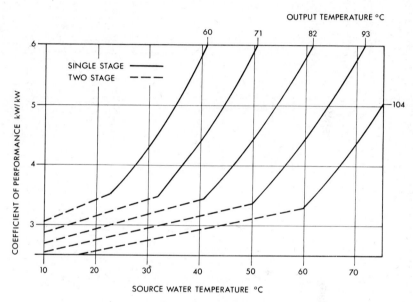

FIG. 24.13.—Performance of Templifier heat pump (Westinghouse).

CENTRALISED HEAT RECOVERY

The use of a run-around coil system with refrigerant as the working fluid has already been mentioned in the limited context of simple heat transfer between adjacent exhaust and supply air ducts. In the case of a fully air-conditioned building and in particular one where heat gains due to process or lighting are high, the refrigeration plant provided to supply chilled water for cooling purposes may be used as a heat source.

The practicalities of such a design develop through the stages shown in Fig. 24.14. Part (*a*) represents the conventional separatist approach where surplus heat extracted at the condenser of a refrigeration plant is rejected by a cooling tower, a simultaneous demand for heating being provided by a boiler plant. For circumstances where heat rejection plus compressor power matched heat demand, the cooling tower and boiler

* Derived from *temperature-amplifier*.

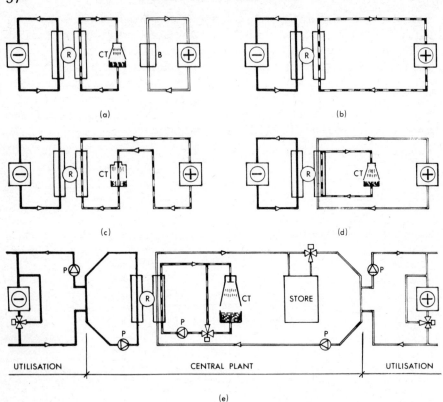

(a)   (b)

(c)   (d)

(e)

FIG. 24.14.—Centralised heat recovery systems, development of the concept.

plant could be done away with and the circuit of part (*b*) substituted: except for the addition of water circuits this arrangement is similar in function to that shown in Fig. 24.8.

In practice of course, heat rejection would not match heat demand and, in consequence, the components deleted would have to have alternatives substituted for them to cater for out-of-balance conditions, as shown in part (*c*) of Fig. 24.14. Since water from an open cooling tower carries atmospheric pollutants, it could not be circulated at large throughout the building and thus a closed circuit evaporative cooler is illustrated. The disadvantages of this type of equipment, in terms of efficiency and physical size, are such, however, that what is called a 'double bundle' condenser is introduced as illustrated in part (*d*). This consists of an over-size shell which contains two quite separate sets of tubes. Water to an open cooling tower circuit passes through one set in the conventional manner and to the heat recovery circuit through the other.

For the type of commercial building to which central heat recovery

Plate XLII. Downward air-distribution to concert hall (see p. 519)

Plate XLIII. Integrated ceiling in precision workshop (lighting/heating/cooling)
(see p. 543)

Plate XLIV. Diesel engine driven heat pump (above) and (left) sewage evaporator/heat exchanger (see p. 574)

Plate XLV. Array of solar panels for swimming pool water heating (see p. 581)

systems are applied, there is commonly a heat surplus over the 24 hours of operation and some means of storing the excess when available for those hours when a deficit occurs is thus needed. Heavily insulated water vessels may meet this need and could, in turn, be 'topped up' by immersion heaters or pipe coils from a boiler plant if necessary. Fig. 24.14(e) shows one form of the finally developed system, including automatic control valves, circulating pumps etc.

Integrated Environmental Design (I.E.D.)—The principles outlined above are those underlying the concept which has been widely publicised by the Electricity Council. Those areas of the building which require heat are serviced by those other areas which require cooling and the input of primary energy is thus, in an ideal case, reduced to a minimum. In terms of annual energy consumption however it is necessary that the summer condition be considered also when requirements for heat supply are negligible. A building environment which is so arranged that heat recovery will meet all winter needs may present demands for summer cooling which, in energy terms, erode any advantage to an unacceptable extent. Each individual case must be examined to ensure that the application of central heat recovery principles is in all respects the most energy-economic solution.

Use of Multiple Machines—In the case of larger scale projects, not all of the refrigeration plant provided for water chilling purposes would necessarily be equipped with double bundle condensers and used for energy recovery purposes; selection would depend upon circumstances. Taking for example a building with which the authors were concerned, the total capacity of the cooling plant was $11\cdot4$ MW but only one of the three centrifugal machines was used for heat recovery purposes. The equipment chosen produced chilled water at $5\cdot6°$ C from return at $12\cdot8°$ C and condenser water was made available at $41\cdot5°$ C from return at $32\cdot2°$ C. The power input was approximately 850 kW.

Water Cooling to Luminaires—A considerable proportion of the surplus heat available for recovery originates, in commercial as distinct from industrial buildings, as a result of electrical power consumption in lighting. Air handling luminaires are dealt with on p. 407 and illustrated in Fig. 22.21, but the use of such equipment means that return air is the medium used for movement of heat to the point of recovery.

An alternative method exists involving the use of water cooled luminaires. The units concerned, as illustrated in Fig. 24.15, incorporate a sinuous water-flow coil integral with the luminaire housing which the manufacturers claim will absorb approximately 70 per cent of the available heat. It is normal to arrange for series water circulation to groups of about five luminaires, flow to each such group being about $0\cdot08$ litre/s and output at about $27°$ C. The total collected flow is then available for use at a heat exchanger for reclaim as most appropriate.

Advantages claimed relate to increases in lamp efficiency; reductions in air circulation rate, with consequent savings in ceiling void space, and ease of heat recovery. It would appear however that the need for a water circulation system through areas which could otherwise be kept 'dry' and the particular care necessary with water treatment are not outweighed.

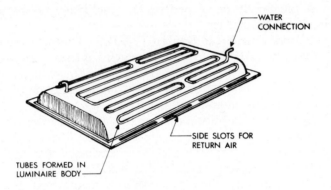

FIG. 24.15.—Water-cooled luminaires (Westinghouse).

SOLAR ENERGY

This Chapter would be incomplete if no reference were to be made to the use of solar energy in servicing buildings, but space will not permit that the subject be dealt with in depth. The reader must therefore refer to the many detailed works devoted to this subject alone and in particular to those publications originating from the U.K. Section of the International Solar Energy Society.*

Reference is commonly made in subject works to *passive* and *active* use of solar heat, the former term relating to building applications where the input is stored in some way in the structure and used internally without the aid of any associated mechanical equipment. A notable early example was that of St George's School Annexe in Wallasey which, despite much uninformed criticism, pioneered efforts in this field long before the present energy crisis, since it was completed and occupied as long ago as 1961. The Trombé solar wall system, illustrated in Fig. 24.16, has been developed in France and a number of adventurous experimental dwellings have been built in the U.K. It is perhaps pertinent to note that, in the majority of cases, such developments have been initiated by the prospective occupiers of the buildings and that they are continuing to live in and record the results of their initiative. The economics of the measures adopted in such cases are often confused by the fact that the designs were brought to fruition on a 'do-it-yourself' basis.

'Active' Systems—Large scale experimental work is being undertaken in various parts of the world and dramatic installations, using solar heat

* This Section is based at The Royal Institution, 21 Albemarle Street, London W1X 4BS.

to power heating and air conditioning plants, are reported in the technical press. It must be remembered that since the British Isles enjoy a temperate maritime climate, comparable opportunities are not available. Active systems using solar collectors of one form or another, often associated with heat pump equipment, are developing comparatively slowly and usually on a domestic scale. This is in no sense to decry such initiative but rather to suggest that the 'inevitability of gradualness' may be to our national advantage as the mistakes made will be confined to a finite scale. Table 24.4* summarizes the characteristics of some possible systems.

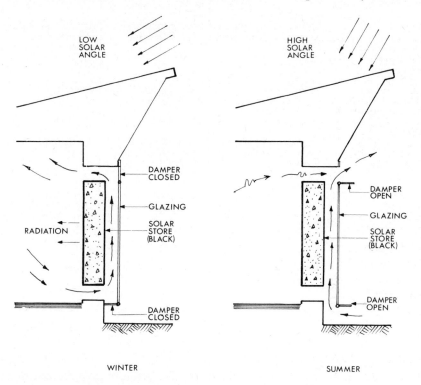

WINTER   SUMMER

FIG. 24.16.—The Trombe Solar wall. In hot weather a simple flap damper allows the air-ways to draw outside air into the house.

Solar Collectors—Active systems rely, in general, upon the use of a solar collector of some form. These consist of an extended surface bonded in any one of a variety of ways to a sinuous or parallel pipe coil which takes in solar energy for conveyance by a water circulation to the point of use. The surface and the pipe coil are contained within a casing, glazed at the exposed face and heavily insulated at the rear, advantage being taken of the diathermal property of glass (referred to on p. 5) to admit

* Energy Paper No. 16. *Solar energy: its potential contribution within the United Kingdom*. HMSO. 1976.

TABLE 24.4

SUMMARY OF THE CHARACTERISTICS OF SOME POSSIBLE DOMESTIC HEATING SYSTEMS
USING SOLAR ENERGY

Collection system	Storage	Energy collected	Additional energy input required to meet total load*
Air systems Air at $\geqslant 25°$ C. Partial heating	Short-term, within building structure	25% of heat load or 10 GJ/annum plus hot water in summer at 8 GJ/annum	30 GJ/annum for additional heating, 8 GJ/annum hot water
Air at $\geqslant 5°$ C. Air-to-water heat-pump transfers heat to store. Full heating	Short-term (days) e.g. 10 m³ of water at up to 70° C	70% of total load or 39 GJ/annum	30% of total load (electric drive for heat-pump) or 17 GJ/annum
Ditto, with solar powered heat-pump Full heating	Ditto	100% of total load or 56 GJ/annum	None
Water systems Water at up to 60° C. Full heating	Long-term (months) e.g. 200 m³ of water at 60° C max, 30° C min.	100% of heat load or 40 GJ/annum plus 8 GJ/annum hot water	8 GJ/annum for hot water
Water at up to 60° C. Water-to-water heat-pump transfers heat to and from store. Full heating	Long-term (months) e.g. 100 m³ of water at 70° C max, 10° C min.	70% of total load or 39 GJ/annum	30% of total load (electric drive for heat-pump) or 17 GJ/annum
Ditto With solar powered heat-pump. Full heating	Ditto	100% of total load or 56 GJ/annum	None

* Assumed annual heat loads = 40 GJ for heating; 16 GJ for hot water; 56 GJ total.

solar heat and retain it. The heat collecting surface is painted or coated dull black in order that it may absorb the maximum possible amount of available radiation. Arguments are advanced for the use of double glazing to collectors but, for the range of external ambient temperatures in the U.K., it would seem that a single sheet is enough.

The orientation and inclination of a solar collector has been much debated. Obviously, in an ideal situation, the collector would 'track' the movement of the sun such that it presented a face always normal to altitude and azimuth but in practice the mechanisms involved would be prohibitive in cost. Fortunately, for practical applications, adequate results may be obtained in the latitude of 51° N if the collector is arranged at an angle of between 30° and 60° to the horizontal, facing southwards. As may be seen from Table 24.5, the variation in annual incidence of solar heat received alters little between inclinations of 30° and 60° and in the same order of difference between orientations in the SE to SW quadrant. If limited use in terms of specific months of the year can be defined then, obviously, an optimum inclination and azimuth may be determined. Both are often a function of site arrangement and co-ordination with the pitch of the roof (Plate XLV, facing p. 577).

Various forms of collector are available, as shown in Fig. 24.17 which is self-explanatory: there are of course numerous other types but all are derivatives from those illustrated. The materials from which collectors are constructed are diverse; all copper; all aluminium; all plastic and any number of mixtures. Since, to avoid damage by frost, the working fluid circulated may contain a corrosive additive, metal mixtures are to be avoided. Copper piping in the system proper and an all-aluminium collector is a specification for disaster unless special steps are taken to ensure that the working fluid is treated appropriately.

In general, the efficiency of solar collectors is low with respect to the

TABLE 24.5

MONTHLY TOTALS OF INCIDENCE OF SOLAR RADIATION ON A FLAT PLATE COLLECTOR
FACING SOUTH*

Month	Monthly totals of solar incidence. MJ/m^2							
	Incidence on a horizontal collector		Effect of collector inclination from horizontal				Effect of change in orientation on collector at 45° inclination	
	Theory	Actual	30°	45°	60°	90°	SW	SE
January	120	45	90	105	105	100	105	95
February	250	75	135	145	140	125	145	130
March	450	215	295	310	295	240	310	280
April	620	310	360	345	325	235	350	310
May	785	495	480	455	415	285	460	415
June	810	535	515	475	425	280	480	430
July	785	465	475	450	400	265	455	410
August	640	355	435	420	380	275	425	380
September	435	235	350	350	335	265	355	320
October	275	125	235	250	245	210	250	230
November	155	55	120	130	135	125	130	120
December	120	35	85	95	100	95	95	90
Averages								
Winter	161	53	108	119	120	111	119	109
Mid-season	445	221	310	314	300	238	316	285
Summer	755	463	476	450	405	276	455	409
Annual	454	246	298	294	275	208	297	268

* Much of the data here listed are extracted or derived from papers by R. G. Courtney of BRE. See, in particular, CP 7/76 and CP 30/77.

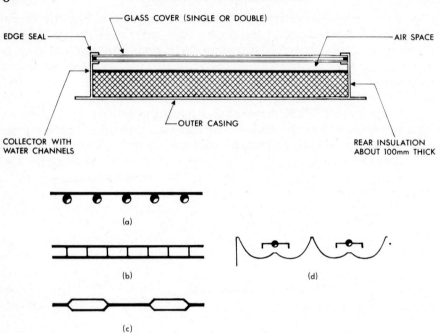

Fig. 24.17.—Various forms of solar collector (a) pipe and fin (b) water sandwich (c) semi-water sandwich (d) parabolic focusing channel.

energy available, the common level being between 20 and 40 per cent. Thus, although a collector may *intercept* up to about 3·5 GJ/m² of energy per annum, the most that can be expected to be received as useful heat is 1·4 GJ/m² which with an 'average' house having a collector surface of 5 m² will only provide about half of the energy used in producing domestic hot tap water.

As to the size of collector required for a given demand, the criteria are obscure. For the provision of solar heat to a domestic hot water system, one of the pioneers* set a rule of thumb, which seems still to be valid, is 1 m² of collector per 50 litre storage capacity (1 ft²/gallon). In domestic terms, this may well be transposed to mean 1 m² of surface per occupant —probably 4 to 5 m² for a normal family house.

Storage—For domestic hot tap water use, the most common application of solar heat is *via* pre-heating of the make up supply. This may be simple or complex, as shown in Fig. 24.18: simplicity, where circumstances permit, is obviously to be preferred for installations in private houses.

Since the incidence of solar heat is quite outside human control, steps must be taken to balance supply and demand. No guidance can be given in this respect since practice is not yet established. For space heating, as

* Heywood, H., *Operating experiences with solar water heating*. J.I.H.V.E. 197.39.63.

distinct from domestic water use, storage must be considerable if the maximum effect of climatic variables is to be taken to advantage. Water, in particular when stored in open tanks at temperatures of about 25 to 30° C, may prove to be a source for bacteriological growth. At lower temperatures, algae may form and, if the store be sealed, then an unsterilized nucleus of anaerobic bacteria may result.

Much useful experimental work is afoot in this area, with particular reference to phase change.* It is worth recapitulation here of the facts set out in Chapter 1 to emphasise that, approximately:

1 kg of water heated or cooled through 1 K	=	4 kJ
1 kg of water changed to 1 kg of ice	=	330 kJ
1 kg of water changed to 1 kg of steam	=	2300 kJ

If, in conjunction with solar heat supply and some form of heat pump operation, this ratio of energy storage or release can be harnessed then the

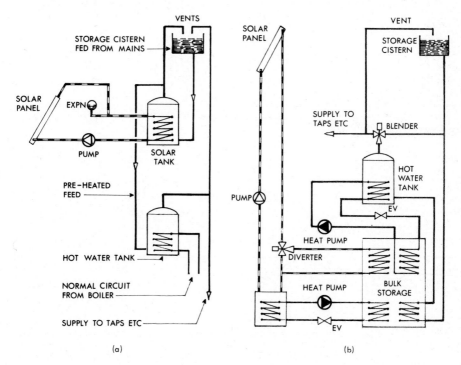

FIG. 24.18.—Solar hot water systems, simple and complex.

* Randell, J. E., *Long term heat storage*. CIBS/ISES/DOE Conference. April 1977.

present requirement for vast volumes and unacceptable insulation requirements may be overcome.

COMPUTER ANALYSIS

For large scale projects, facilities are now available for the detailed analysis of energy demands with respect to both building construction and the associated services. The first generation of such programs derived from trans-Atlantic sources and these were not adequate in scope or practice to deal with U.K. conditions and weather variables. Home produced successors are now available which will not only take account of British weather conditions, by reference to either a 'standard' or any chosen year, but will also analyse the relative annual energy consumption of a wide spectrum of systems as applied to a given project. Fig. 24.19 shows a plotted output from such a program* for two alternative system approaches. The printed output (not shown) divides the annual consumption between electrical, gas, oil or other energy input.

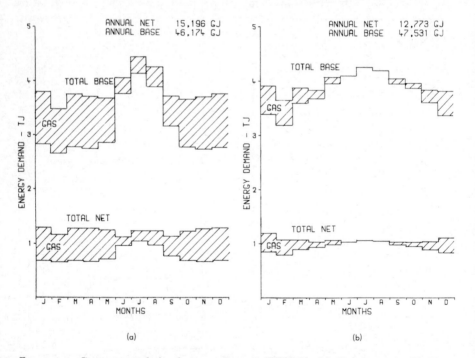

Fig. 24.19.—Computer analysis of energy demand (ENPRO). (a) Heat recovery (b) 'Free cooling'.

* The authors are indebted to Oscar Faber & Partners for this reproduction of output from an ENPRO program.

OTHER SUBJECTS

It has not been practicable to cover, in one short Chapter, many other aspects of Energy Conservation. Among these are the overall technicalities of storage systems insofar as these may affect the use of primary energy in the broadest sense; the use of electrical power for machinery drive at best efficiency; the question of Energy Targets for buildings and the various engineering systems within them and, last of all but by no means the least in importance, the desirability that all building owners, tenants and occupiers allocate time and staff resources to the completion of Energy Audits.*

ECONOMICS

Of all the equations which relate capital expenditure to revenue return, those appropriate to energy conservation measures are among the least amenable to solution. On the one hand the capital costs may relate to all scales and conditions of building, from the dwelling to the largest industrial complex and from present stock, having significant useful life remaining, to concepts as yet undeveloped from a basic client brief.

Similarly, revenue return may arise from *static* measures such as orientation, building form and materials of construction or from the *dynamic* characteristics, life cycle and maintenance needs of mechanical equipment. Obviously, the effect of corporation and other taxation cannot be ignored. Nor indeed can the impact of investment grants, capital allowances, accelerated depreciation allowances and so on. Each of these facets of the problem acts and interacts with the remainder.

Life Cycles of Conservation Measures—The financial return from 'static' energy conservation inclusions, orientation, building form and shape, solar exclusion, thermal insulation and the like may be assumed to relate directly to the remaining whole-life of the building. In the case of 'dynamic' action in the interests of energy conservation however, different criteria obtain. There are, of course, wide variations between varieties of plant ranging from life expectancies of say five to ten years for heat pumps, fifteen to twenty years for thermal wheels and twenty to thirty years for runaround and plate type heat exchangers.

Payback Period—Probably the most common term used when assessing the viability of energy saving methods is 'Payback Period', i.e. in the simplest form, the number of years required for a capital expenditure to be recovered through annual income or, in the present context, annual savings. Nevertheless, using raw cost data, payback is a rather crude concept for use in decision making since it takes no account of the fact that capital, if invested elsewhere, would earn interest.

The concept of *Present Value* has much to commend it for the type of exercise we are concerned with here, since it is easy to understand and can handle staged expenditure or known changes in the pattern of annual

* Fuel Efficiency Booklets I and II. *Energy Audits*. Department of Energy. HMSO. 1977.

saving. In brief, Present Value analysis converts all outgoings—capital expenditure in this case—and all income—annual energy savings in this case—to their equivalent values as measured at a single point in time, usually the present. The analysis relies upon the fact that a pound today is worth more than a pound tomorrow—inflation aside—since if today's pound were invested it would have earned interest by tomorrow. Thus, stating the converse, a pound at some future date is worth less than a pound today: the value will have been reduced—or discounted—in proportion to the rate of interest assumed.

Most relevant works of reference contain tables listing Present Values over a wide range of time periods for a selection of interest- or discount-rates. The public sector, for instance, uses a discount rate of 10 per cent but this may not be appropriate to private industry and, nearer to our present subject, the Department of Energy Loan Scheme for Industry currently uses a rate of $11\frac{1}{2}$ per cent. There are two important aspects which require emphasis, the first being that the method assumes constant money value in real terms over the period considered, i.e. that there is either zero inflation or that inflation is at a common level across the board. The second is that the analysis is sensitive to both the discount rate and the life cycle of the energy saving purchase.

Viability Charts—It is possible to use the tabulated figures to produce what, for want of a better description, we will call Viability Charts, plotting the ratio between capital cost and annual savings due to conservation measures against a time base. Fig. 24.20(a) shows how such a chart would appear for a number of different discount rates from zero to 20 per cent. Fig. 24.20(b) takes a discount rate of 10 per cent as a base and shows how the relationship will vary if annual costs (i.e. energy) inflated dispropor-

TABLE 24.6

PRESENT VALUES OF A SINGLE SUM, DISCOUNTED AT 10 PER CENT, AND LIFE CYCLE MULTIPLIERS FOR CAPITAL SUMS.

n (years)	Present Value of a Single Sum	Life Cycle (years)	Multiplier for Capital
Present	1·0	5	2·629
5	0·621	10	1·622
10	0·386	15	1·310
15	0·239	20	1·171
		25	1·101
20	0·149	30	1·057
25	0·092		
30	0·057		
35	0·036		
40	0·022		
45	0·014		
50	0·009		
55	0·005		

tionately to the general pattern. Three curves are shown, representing zero, plus 3·5 and plus 7 per cent per annum. An excess rise of 3·5 per cent per annum represents a doubling in 20 years and an excess rise of 7 per cent per annum represents a doubling in 10 years.

To cater for the situation where the energy saving material or equipment has a life shorter than that of the building to which it is provided, the capital outlay figure may be adjusted using data extracted from a further set of calculated factors as in Table 24.6. The first and second columns here show how the Present Value of a single sum varies with the passage of time: the fourth column lists multipliers for capital sums according to the life cycle of the material or equipment. These multipliers are no more than successive additions from the second column, e.g. the 20 year cycle factor is the sum of the initial outlay, unity, plus the 20 and 40 year values: 1·0 +0·149 +0·022 =1·171.

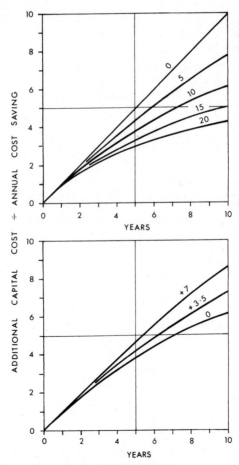

Fig. 24.20.—Viability charts showing effect of changes in (a) discount rate and (b) energy cost premium.

A developed chart for a discount rate of 10 per cent, plotted to a logarithmic base scale, is shown in Fig. 24.21 with added curves for excess premium in annual energy costs. The scale on the right hand side is provided as an aid to interpolation between the curves. The broken lines illustrate the following applications to practical problems:

Example 1

Capital cost of energy saving measure
having a 60 year life as for the building   £8,000
Estimated annual saving at present day prices   £1,540

$$\text{Ratio} = \frac{8,000}{1,540} = 5 \cdot 2$$

From the chart, even if energy costs do not rise disproportionately, the measure will be viable at 7·6 years and profitable thereafter.

Example 2

Capital cost of energy saving equipment
requiring renewal every fifteen years   £2,000
Multiplier from Table 24.6   1·31
Estimated annual saving at present
day prices   £ 236

$$\text{Ratio} = \frac{2,000 \times 1 \cdot 31}{2 \cdot 36} = 11 \cdot 1$$

From the chart it may be seen that:

(a) The equipment will never be viable unless energy costs rise disproportionately to the general level of inflation.

(b) If energy costs rise disproportionately at 7 per cent *per annum*, the equipment will be viable at 13·5 years.

(c) The equipment will be viable at the end of the building life (60 years) if energy costs have risen disproportionately by about one per cent *per annum* over the whole period.

To conclude this brief attempt to explain the impact of economic factors upon energy conservation measures, the authors feel bound to add that they cannot accept that cost equations are the end of the story. These take no account of amenity values, thermal and visual comfort, aesthetics, contentment, productivity and quality of life. Some at least of these factors should be quantified and introduced as weighting to the results produced by soul-less mathematics. There remains, furthermore, the fundamental question of the national and international importance that remaining reserves of energy be husbanded. Economic considerations should perhaps be regarded as of secondary importance.

Conclusion—In concluding this Chapter which has inevitably been somewhat discursive, the authors wish to quote from an article which appeared in an American journal.*

> 'It must be realised that the energy problem is a *people* problem. It is *people* who waste energy: it is *people* who design energy wasting systems: it is *people* who operate systems in a wasteful manner and it is *people* who fail to do the things they know how to do and use the tools that are available to save energy.'

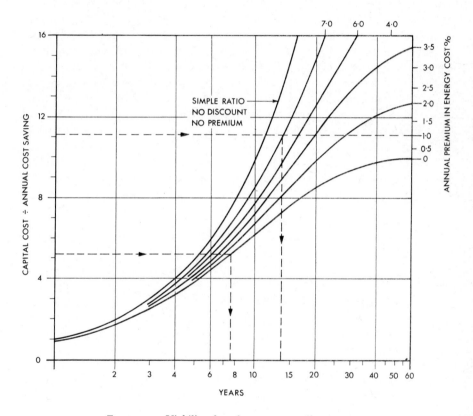

FIG. 24.21.—Viability chart for 10 per cent discount rate.

* Bridgers, F. H., *How new technology may save energy in existing buildings.* H.P.A.C. Aug. 1975.

CHAPTER 25

District and Group Heating

DISTRICT HEATING, PIONEERED IN THE USA and on the Continent on a small scale late in the last century, was later developed to such an extent that over 400 schemes now exist in Denmark alone. In Great Britain, however, probably due to the more equable climate and the traditional plentiful supplies of cheap fuel, little interest was shown in such schemes until the period immediately following the Second World War, although quite large institutional and factory sites were served from a single central boilerhouse. Fig. 25.1 shows the layout of such a heat distribution scheme for a university, having a site area of just over 1·1 km², the boiler capacity being 40 MW.

With the development of the New Towns and the intensive building programme after 1945, there was a growing awareness in this country of

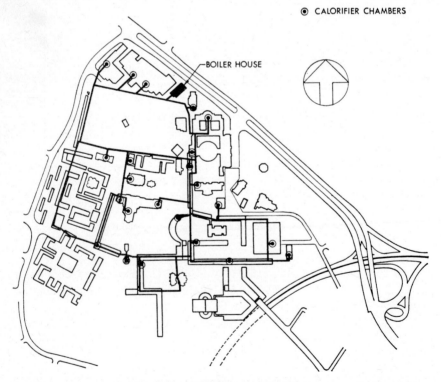

FIG. 25.1.—Site plan of a university.

DISTRICT AND GROUP HEATING

the advantages of district heating and many design projects were initiated, although regrettably few came to fruition. The publication in 1953 of *Post War Building Studies, Nos. 31 and 32*, provided a mass of detailed information but, by this date and in the prevailing economic climate, interest was receding.

Over the last fifteen years, with the tendency to plan for compact multi-storey building complexes, many heating schemes serving building groups —some of considerable size—have been constructed. But these do not fall strictly within the definition of 'district heating', since the groups are usually in common ownership and are often so arranged that the pipe-distribution system runs totally (and economically) in basements, car parks, interlinking corridors or other structural elements, thus avoiding such problems as underground networks and road crossings. Such plants are more properly classed as 'group heating'.

The advantages arising from district-heating systems lie in the avoidance of atmospheric pollution with the accompanying hazards to health, a reduction in the road traffic inherent to retail fuel delivery and the provision to building owners and occupiers of a convenient and trouble-free utility service. Where the heat source is an electrical generating station, economic use of the world's energy resources may be added to this list.

The principal disadvantages relate to matters of economics, the capital outlay necessary to initiate systems being large. The cost of central plant and distribution mains must be met before any income is produced and considerable disruption to work progress under public roadways is inevitable. Potential consumers in a commercial/industrial area of an existing town may have near-new boiler plant and thus be reluctant to use the new service for marginal cost benefit.

FIRST CONSIDERATIONS

The primary elements of a district-heating system are a source of heat supply, a system of distribution and a means of utilisation.

Heat Supply—The first and most obvious source of heat supply is the electrical generating station which, at best, operates at only about 35 per cent efficiency. The second is the municipal refuse incineration plant and the third is the conventional boiler plant burning one fuel or another.

Difficulties of load balancing (heat supply to electrical supply) and some technical complications have tended in the past to militate against the use of thermal electric-generating plant; in fact, less than two per cent of the European systems are so arranged. One notable exception exists, however, in the centre of London where part of Pimlico is heated by waste energy from Battersea Power Station. Also, Aldershot has a district-heating plant using waste heat from a diesel generator station.

Much has been written in recent months on this subject and the reader is referred to a discussion document published in February 1977 by the

Department of Energy.* Much of the content is superficial and not to the standard of the *Post War Building Studies* previously mentioned, but the summary poses a number of important questions for informed debate.

Refuse incineration as a source of heat is widely used on the Continent; and new plants operated by the GLC and by Nottingham City have recently come into service, so that British experience should soon be available. Domestic refuse has a calorific value of about 12 MJ/kg, about half of which can be made available by combustion. Since refuse output is rising to about 300 kg per person per annum it would appear that dwellings could be provided with about ten per cent or so of their annual heat requirements by 'fuel' otherwise wasted.

In context with the present time, however, it would seem that the conventional boiler plant (coal- or heavy oil-fired) must be considered as the most usual heat source for district-heating plants. In the opinion of the reviewers, the use of natural gas for this purpose should be banned completely and licence to use light grade fuel oil granted in only the most special circumstances. These refined fuels can be burnt in local domestic and other boilers at efficiencies comparable with those available at centralised plant: the effect of heat losses from distribution mains is thus an unnecessary penalty and is counter productive in terms of energy conservation.

Boiler plants of this type have been discussed in some detail in earlier chapters and it is necessary only to emphasise that group- or district-heating systems, being large by their very nature, can be equipped to burn the lower and cheaper grades of coal and oil with least disadvantage to the environment. Plate XLVIII (facing p. 609) shows an automatic coal-fired boiler plant serving a small group-heating system.

Heat Distribution—Generation of heat in a supply station is, in many ways, the least part of the design problem associated with the inception of a district-heating scheme.

The heat there produced must be distributed to the consumers in the most economical manner possible, from the viewpoint of both capital and recurrent costs. As is often the case in engineering economics, these two aspects are fundamentally opposed—the requirement for minimum capital costs militating against the need for excellence in choice of materials for pipework, techniques for heat insulation and structures for underground containment of both.

Traditional British methods of pipe insulation and enclosure have of recent years come under scrutiny, particularly since practice in the USA and on the Continent is known to be at variance with them. Following an expected pattern, the North American developments have been towards the production of relatively complex factory-built units in order to reduce site works to a minimum; conversely, European tendencies have been

* Energy Paper No. 20. *District heating combined with electricity generation in the United Kingdom.* H.M.S.O.

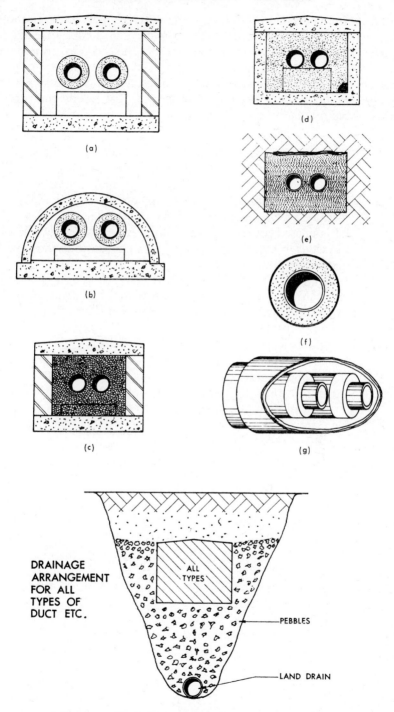

FIG. 25.2.—Typical constructions for underground mains.

towards simplification in the choice of materials and their containing structures whilst retaining meticulous attention to detail. In each case, however, the result of these advancing practices has been to reduce capital costs.

So much information has recently been published on this subject that it would be out of place here to do other than list briefly the principal methods used, all as shown in Fig. 25.2:

(a) The 'traditional', where excavation is made and an in-situ brick or concrete passage formed, in which pipes are installed and subsequently insulated with glass fibre or like material. A concrete cover is then fixed and the ground surface put back.

(b) The 'semi-traditional', where the in-situ passage is replaced by preformed channel, circular or semi-circular concrete sections.

(c) The 'enclosed loose fill', where the segmental insulation of the previous methods is replaced by a loose granular material packed around the pipes.*

(d) The 'solid fill', where a preformed channel section is laid in an excavation and the pipes are insulated by means of a light-weight aerated concrete material poured round them to fill the channel completely.*

(e) The 'open loose fill', where bare pipe lengths are laid in an excavation prior to insulation by means of a powdered mass of hydrophobic material, which cures on the application of heat to form a protective skin on the outer surface of the pipe, no structural enclosure of any type being provided.*

(f) The 'simple preformed', where single steel pipe lengths are factory-set concentric within a larger plastic tube and the annulus filled with a hard-setting plastic foam, prior to burying directly in an excavation without further protection or structural surround.

(g) The 'complex preformed', where single or multiple steel pipe lengths are factory-insulated and enclosed within an outer steel tube (protected by up to ten laminates of various materials) or an asbestos cement tube, prior to burying directly in an excavation without any structural surround.

Long term experience has shown that in this country, despite much ingenuity and the most careful design, small size in-situ or preformed structural ducts, as described in (a) and (b) above, are sooner or later subject to leakage such that ground water enters. Provision for drainage can in many instances prove disastrous, since at times of heavy rainfall surface water will actually enter the ducts *via* the drains, which are not normally allowed to be connected to a deep outfall system but only to

---

* This method and also those under (d) and (e) are reported here to complete the list of past practices, but they are not recommended (see Plate XLVIII, facing p. 609).

soakaways. Water entry having been accepted as an unfortunate fact of life has resulted in attempts being made to keep the insulation dry by wrapping. It has been found, however, that tiny failures occur in due time and that the insulation in the wrapping becomes poulticed on to the pipe by moisture unable to escape and quickly causes corrosion. The attempted cures have thus often proved more fatal than the diseases they were meant to remedy.

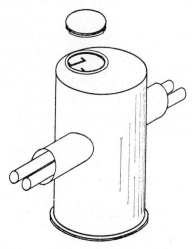

Therefore, unless generously dimensioned walk-in tunnels and subways can be formed (an unusual circumstance in terms of modern building economics), one of the preformed systems is to be preferred, and these may be associated with preformed access chambers for inspecting valves etc., as shown in Fig. 25.3. For these, many novel methods have been proposed to monitor possible failures: an air pressure may be maintained within the protecting enclosure tube, or tracer cables laid so that moisture

FIG. 25.3.—Factory-built access chamber.

causes current flow to provide both indication and fault location. Mr A. E. Haseler, first chairman of the District Heating Association, has written

TABLE 25.1

HEAT LOSS FROM PREFORMED 'PERMAPIPE'
BURIED 600 mm BELOW GROUND LEVEL

Size of service pipes (mm)	Insulation thickness on pipes (mm)	Diameter of enclosing conduit (mm)	Heat loss, W/m run of enclosing conduit for following mean water temperatures		
			75 °C	100 °C	125 °C
Two × 20	25	250	39	53	63
Two × 25	25	250	42	58	73
Two × 32	25	300	50	65	83
Two × 40	25	300	54	71	89
Two × 50	25	300	58	79	99
Two × 65	25	350	69	92	117
Two × 80	37	350	62	84	108
Two × 100	37	400	73	98	123
Two × 125	37	450	83	110	140
Two × 150	37	600	96	130	165

These data have been obtained by converting manufacturers published figures into SI units. The published values were calculated according to the methods of BS 4508, Part 1.

extensively on this subject,* and the Building Services Research and Information Association is actively initiating developments.

Table 25.1 lists data regarding dimensions and heat losses from one proprietary type of preformed system ('pipe-in-pipe'—to use a common description). Reference should be made to the *Guide, Section C3,* for further information on heat loss from buried pipes.

In recent City Centre and housing developments, there has been a tendency towards separation of pedestrian and vehicular traffic by elevation of the former to bridges, overhead walkways or upper levels of a podium. These proposals offer further solutions to the problem of heat distribution since it is often possible in such instances to suspend the necessary pipework below the pedestrian ways and thus avoid the problems inherent in any form of excavation and burying; protection against

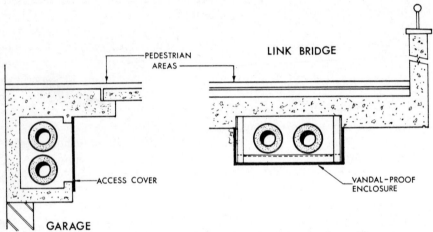

FIG. 25.4.—Heating mains suspended below elevated pedestrian walkways.

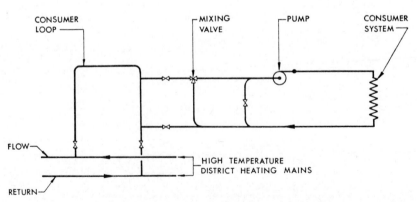

FIG. 25.5.—High-temperature/low-temperature mixing arrangement.

* *New Heat Mains Techniques for Telethermics. J.I.H.V.E.* 1970, 38.194.

vandalism is all that is necessary. Fig. 25.4 shows two instances of application in this way. (See Plate XLIX, facing p. 609.)

Heat Utilisation—Heat energy utilisation, at the consumer end, can follow any of the methods described in previous chapters. Initially, of course, much will depend upon the medium used for distribution since if steam or high-temperature water be employed, these may require to be fed into a 'transformer' station to reduce temperature potential to that which may be employed safely by the consumer. Such a transformer station would incorporate heat exchangers—storage or non-storage according to the service required—or, in the particular case of high-temperature water, some form of mixing arrangement, as illustrated in Fig. 25.5.

HEATING MEDIA

Basic design decisions for a district- or group-heating system are determined by the nature of the load to be served. Domestic and commercial premises rarely require other than water at low to medium temperatures, whereas steam or some other high-temperature energy source may be necessary for industrial processes. Each case must be considered on its merits, some of the alternatives being as follows:

Mixed Areas

(1) Primary distribution at high temperature (water at say 180°C or steam at an absolute pressure of 1 to 1·2 MPa): Industrial buildings would be connected directly to the primary mains. Domestic and commercial building systems would be connected either individually or in groups to the primary mains but served thence, via transformer stations, at low temperature.

(2) Dual primary distribution, at high temperature, as above, and with low temperature (water at say 90° to 100°C) in parallel: Industrial buildings would be connected to the high temperature mains. Domestic and commercial building systems would be connected to the low temperature mains.

(3) Primary distribution at low temperature (90° to 100°C as above): Industrial processes would be served by some local heat generating plant. Industrial, domestic and commercial building systems would be connected to the primary mains.

Domestic and Commercial Areas

(1) Primary distribution at medium temperature (water at say 120 to 150°C): Building systems would be connected in groups to the primary mains but served thence, via transformer stations, at low temperature. Building systems, if of warm air type, would be connected directly to the primary mains.

(2) Primary distribution at low temperature as above: Building systems of any type would be connected directly to the primary mains.

Obviously, there are any number of combinations of these principles which could be adopted and space will not permit further discussion. It must suffice to say that the essence of the matter lies in the selection of the most economic arrangement in both capital and operational cost, since these are the penalties which have to be set against the advantages which may be gained from centralised heat generation.

PIPING DISTRIBUTION SYSTEMS

Where district- or group-heating is to serve domestic and commercial premises only, there are four principal methods which may be used in arranging the distribution system, as shown in Fig. 25.6. These are:

The 4-Pipe Primary/Secondary System (a)
>In this case, similar in many ways to that which might occur in a single large building, one pair of distribution mains serves the various heating appliances in the buildings connected, and a parallel pair of distribution mains provides domestic hot water to draw-off fittings from a central storage vessel or vessels. There are obvious cost disadvantages in that the domestic water mains may well have to be in copper and, in any event, will be subject to scaling and the other varieties of attack inherent to any pipe distribution network conveying a changing water content. Further, and perhaps more important, over-use or wastage in any building connected will limit the supply of hot water available to other buildings.

The 4-Pipe Primary System (b)
>Here, one pair of distribution mains serves the various heating systems in the buildings connected and the parallel pair serves the

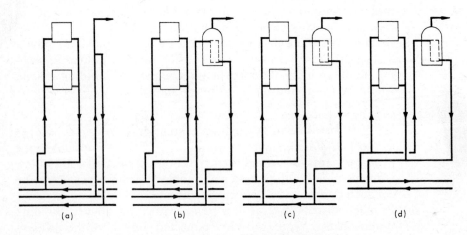

FIG. 25.6.—Alternative methods for distribution.

primary coils of hot-water storage vessels, also, in the various buildings. The latter mains may now be in steel and are not subject to scaling or internal attack. Any building requiring year-round service could demand *heating* connection to the *hot-water* primary mains.

The 3-Pipe Primary System (c)

This is a refinement of the last system whereby two flow mains, one for heating and one for hot water primaries, have a common return. The advantages of the 4-pipe primary system are retained and there is some saving in capital cost. This system is not recommended, however, due to difficulties which arise in isolation of the heating circuit during summer.

The 2-Pipe System (d)

This arrangement, which for all practical purposes has superseded the previous three, has one pair of distribution mains serving both heating and domestic hot-water primary coils in the buildings connected. Heat losses and pumping power are saved during the summer months either by use of alternative smaller-duty pumps or by arranging for the main pumps to have variable speed. The latter alternative has many advantages since, with reasonable operating skill, the pumped output may be governed to meet the actual load at any time of the year.

The four methods mentioned are those which have been, or are, appropriate to this country. In the USSR, where distances are immense, and in Iceland, where the heat source is geothermal springs, single pipe 'open' systems have been used.* In such cases the single flow pipe conveys heated water to the buildings connected, the maximum usage is made of the energy potential by balancing heating and domestic hot-water demands where possible, and the residual is sent to the cold-water mains, to swimming pools or to waste.

As emphasised previously, there are many possible combinations of piping systems. It may be desirable, for instance, to use a 2-pipe distribution for sixty per cent of the connected load and a 4-pipe primary/secondary system for groups of buildings making up the remaining forty per cent. Once again, it must be stressed that the most economic solution for a given set of circumstances will determine the 'right' choice of system.

SERVICE TO DWELLINGS

A number of ways exist in which service to individual dwellings may be arranged and these are shown in Fig. 25.7. They relate to the manner

---

* Kristinsson, G. H. and Jonsson, K. O., *Geothermal energy and its use for District Heating in Iceland. J.I.H.V.E.* 1971, 39.103.

in which service is provided and fall into two groups for 'unlimited' and 'limited' supply:

Unlimited—Here, as illustrated in parts (*a*) to (*d*) of the figure, the space heating and the hot water supply arrangements are to a large degree independent of one another in each dwelling.

(*a*) In this case two totally separated circuits supply the two services from a central point per block.

(*b*) As a marginal simplification, a common flow pipe is used, but separate returns. This arrangement gives rise to difficulty in providing heating control.

(*c*) The converse of (*b*), a better arrangement with separate flows and a common return but still presenting problems in hydraulic balance.

(*d*) Similar to (*c*) but incorporating an arrangement whereby a simple reversal of circulation through the hot water circuit would isolate the heating surface during the summer (black arrows in winter, open arrows in summer).

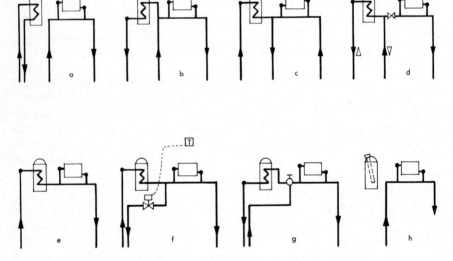

Fig. 25.7.—Alternative methods of service connection for dwellings.

Limited—Here, as illustrated in parts (*e*) to (*h*) of the figure, the two circuits are to a greater or lesser extent dependent upon one another. This would mean that a specific allocation of heat energy would be made by design to each dwelling and advantage taken of the thermal capacity of the building structure to even out variations in room temperature at times when exceptional demand for hot water gave rise to lower radiator temperatures.

(e) This is the crudest form with the hot water storage cylinder in simple series with the space heating. At times of heavy hot water demand the radiators would be cool but, when no hot water had been used, the radiators would overheat.

(f) A minimal refinement, a by-pass to the radiator circuit is opened automatically by a thermostat in the living room, thus in part overcoming the disadvantage of method (e).

(g) An alternative refinement whereby a manually operated diversion valve allows the tenant choice of usage of his heat allocation to radiators, to hot water or in some proportion between them.

(h) Here the hot water system is entirely separated from the central plant and provided with an electric immersion heater.

Of these alternatives, the choice would, in engineering terms, best lie between the simplest form of the unlimited and one or other of the refined limited systems (either parts (a) or (f) or (g) of the figure).

DOMESTIC BUILDINGS

District heating, or often group heating as applied to large housing developments, is currently in the public view. This presents certain very special problems insofar as heat requirements and heat consumption become a matter for the householder's pocket. Many such schemes are initiated by local authorities who obtain guidance from the *Housing Subsidies Manual* (Appendix XIII: *District Heating*) published by HMSO for the Department of the Environment.

Standards for Internal Temperatures—The Parker Morris Report,[†] in advance of contemporary thought but now sadly outdated, includes the following recommendations with regard to standards:

'*Minimums*'
'The heating installation should be one capable of heating the kitchen and the areas used for circulation to 13° C (55° F) and the living and dining areas to 18° C (65° F), both temperatures attainable when the outside temperature is −1° C (30° F).

'*Recommended additions*'
'Where daytime and evening use of bedrooms is likely to be considerable, it will be worthwhile to adopt a higher standard and provide an installation capable of heating the bedrooms to 18° C (65° F) as well.'

These recommendations have been variously interpreted by local and other authorities, as shown in Table 25.2. Dwellings for old persons are required to be heated to 21 °C throughout.

Undesirable Standards—It is in context to make particular mention here of the fallacy which exists in attempting to warm only the living room

† *Housing for Today and Tomorrow.* HMSO: 1961.

TABLE 25.2

STANDARDS FOR INTERNAL TEMPERATURE

Authority	'Parker Morris standard'	'Full Parker Morris standard'
A	18 °C in all living areas (not bedrooms!)	—
B	18 °C in living room 13 °C in hall. No heat in bedrooms	—
C	18 °C in living room, 13 °C in hall, kitchen and one bedroom	—
D	—	21 °C in living room, 16 °C in all other rooms
E	—	21 °C in living room, 18 °C in all other rooms

and hall to temperatures of 18°C and 13°C respectively, leaving the bedrooms unheated. Fig. 25.8 shows a typical flat, the shaded areas being those directly heated. The sinuous arrows show the route of the heat loss from the flat to the outside air, whereas the straight arrows show the heat transfer between the heated and unheated rooms of the flat.

The figures circled in each room show the temperature levels which would be achieved in steady state conditions, with the outside temperature

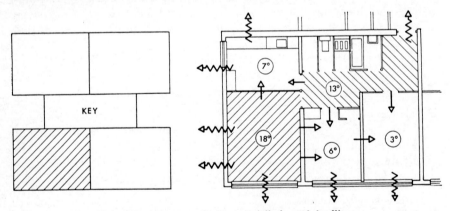

FIG. 25.8.—Heat transfer in a partially heated dwelling.

at −1°C; and it will be noted that although kitchen and bedrooms are excluded, these are in fact heated to some extent, second-hand. The point that requires emphasis here, of course, is that this condition is only achieved by putting excess heat into the living room and hall with the inevitable result that, if kitchen and bedrooms are provided with local heaters, such as electric fires, the living room and hall will be *overheated* by the fixed plant.

Quantitatively, for the case considered, energy required to heat the living room and hall only is 2·4 kW. To 'overheat' them as necessary 3·7 kW is required whereas only 5·4 kW would be needed to heat the whole dwelling properly. This goes towards proving that it is more reasonable to design for whole-dwelling heating than to use any half-baked, half-way measure.

Standards for Consumption of Domestic Hot Water—The more important of the standards laid down with regard to provision for domestic hot-water consumption are:

> *The Parker Morris Report* (*5-person house*)
> 'All systems should allow for a supply of 1150 litres (250 gallons) of hot water per week at 60° C (140° F).'
>
> *The I.H.V.E. Guide* (1970)
> For low rental flats, consumption is quoted as 70 litres per person on the day of heaviest demand. This is usually interpreted to mean 280 litres per person per week.

Records exist which indicate that the recommendations set out above are too high if the system incorporates some in-built means for the prevention of water wastage. With district heating, the provision of a local storage vessel in each dwelling, with a limited facility for re-heating, has been found to produce such economies. From data available from several sources, it is considered that, if separate facilities are provided for bulk laundry use, hot-water consumption may be limited to approximately 180 to 230 litres per person per week, on the basis of nominal design occupancy.

Typical Consumptions of Heat Energy—For domestic premises, the peak requirement for space heating will vary considerably according to the construction, size and isolation, or otherwise, of the dwelling. A detached four-bedroom house might require about 18 kW, whereas similar accommodation within a multi-storey block would require only about 10 kW. Most available statistics relate to local authority housing and suggest a range between 5 and 8 kW, with an average of about 6·5 kW for a two- to three-bedroom (four-person) dwelling in a low rise block, fully heated.

Annual consumptions of heat energy vary considerably according to the method of heating and the fuel used. Electricity Boards are able to produce evidence that annual usage for electric warm-air heating in a dwelling is much less than that which would normally be calculated for the connected load. The aspects of ease of metering and the relatively high fuel cost are important factors here.

Again, for local authority dwellings, statistics show that, where district heating is available, a total annual consumption of between 53 GJ and 80 GJ may be anticipated, with an average of about 65 GJ. Of this total, some 13 GJ per annum may be considered as usage for domestic hot water, leaving about 52 GJ per annum for space heating. Current rising standards in demand for comfort in the home suggest that future total requirements may well be of the order of 65 GJ per annum for space heating plus 15 GJ for domestic hot water, making a total of 80 GJ per annum for the 'average' dwelling.

CHARGES FOR HEAT SUPPLY

Individual occupiers may be charged for the supply of group-heating to their premises in four principal ways:

By a flat-rate charge included in the total of the rental where appropriate.
By a separate flat-rate charge for heating service.
By a unit charge derived from metering.
By a fixed charge plus a unit charge derived from metering.

It is usually thought to be important that an impression should not be created that the individual can wholly 'contract out' of receiving service, since the economics of the case overall are at a maximum with universal supply.

The *Parker Morris Report* mentioned previously states that, for dwellings, some form of heat metering is desirable. This conclusion is no longer accepted, since metering inevitably introduces the necessity for meter reading, preparation of accounts and collection of debts.

Metering—In the early days of district heating in the USA, heat metering was comparatively easily achieved. Energy was distributed by steam mains and a water meter on the condense line provided a measure of the heat consumed.

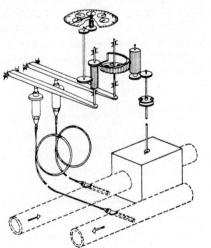

(a)

Fig. 25.9 (a).—Mechanical heat meter.

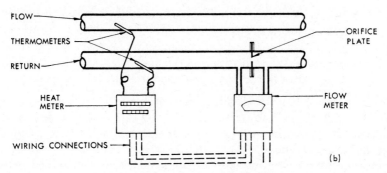

FIG. 25.9 (*b*).—Electrical heat meter.

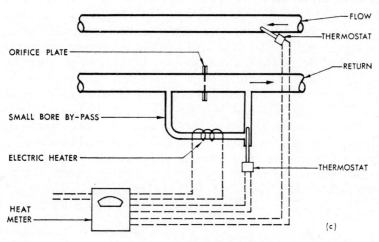

FIG. 25.9 (*c*).—Shunt heat meter.

With the advent of water distribution, matters became more complex since it was then necessary to meter not only the water quantity flowing to each consumer but to integrate this with the temperature difference between the supply and return mains.

Heat Meters—Complex Btu (now presumably GJ) meters have been produced and proved to be successful. They depend for their operation upon some means which will measure water flow and integrate this with temperature difference. Three principal types have been evolved:

(*a*) *Mechanical.* In this case (the principle of which is shown in Fig. 25.9 (*a*)), the heat-energy reading is produced by biasing the output drive from a water meter mechanically, by means of temperature sensitive elements in the supply and return mains.

(*b*) *Electrical.* Here, as shown in Fig. 25.9 (*b*), both flow rate and temperature difference are measured electrically and are integrated by an electronic device.

(c) *Shunt.* In this case (Fig. 25.9(c)), a known small proportion of return water is electrically heated back to the supply temperature, as sensed by a thermostat. The heating current is recorded by an ordinary kWh meter to provide an inferential measure of the energy abstracted from the circuit proper.

Such meters have the disadvantage of being quite large, relatively expensive and subject to error if not regularly maintained and re-calibrated. Recourse has thus been made to simpler devices, but these meter by apportionment rather than directly: i.e. they provide data whereby the total operating costs of the system may be divided between the users without actually measuring the energy consumed.

Inferential Meters—For radiator systems, the pattern shown in Fig. 25.10 has been evolved. This consists of a housing bearing a scale which is clamped to each radiator and which conceals a removable open phial containing a volatile liquid. The amount of liquid evaporated by radiator heat is a broadly approximate measure of usage, and this may be read against the scale each season or other period. The scale readings are then totalled and the cost per radiator apportioned. Obvious disadvantages exist here in that 'informal adjustment' of meters is easily achieved. Further, access to individual radiators by meter readers is necessary, and this is not always easy to achieve in domestic premises. The manufacturers, however, offer an all-in meter reading and accounting service.

An alternative method of metering for radiator systems, which could be described as 'part-inferential', relies upon the use of a type of circuit control known as a 'temperature limiting' valve. This, fitted to the return

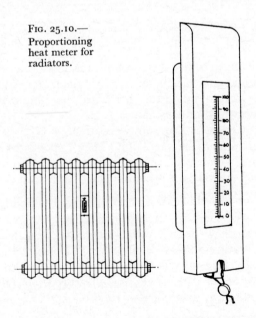

FIG. 25.10.—
Proportioning
heat meter for
radiators.

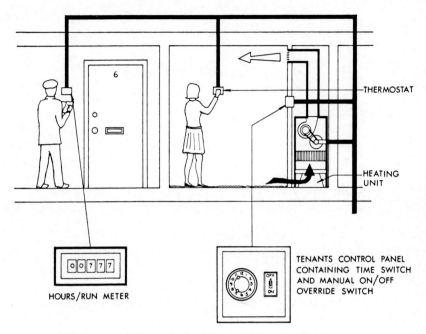

THERMOSTAT

HEATING
UNIT

HOURS/RUN METER

TENANTS CONTROL PANEL
CONTAINING TIME SWITCH
AND MANUAL ON/OFF
OVERRIDE SWITCH

FIG. 25.11.—A proportioning heat meter for warm-air systems.

pipe leaving a dwelling, acts to maintain a constant *outlet* temperature and thus restricts water flow relative to heat demand. A simple water meter fitted in series with such a valve will record an approximation of heat use.

For warm-air systems, inferential readings are easily achieved. A cyclometer type 'hours run' meter may be wired in parallel with the fan control, as Fig. 25.11, and read from outside the premises. The products of hours run and unit size may be totalled and then used to apportion the cost to users.

Flat-rate Charges—Some system of producing flat-rate charges—per m² of dwelling area, for example—is an obvious alternative to metering, and this, for domestic premises at any rate, seems to be the preferred method on the Continent. For local authority developments, this approach has advantages in that the tenant who is unfortunate enough to live on the twenty-fifth floor of a tower block, facing north east, is not penalised *vis-à-vis* his near neighbour living in an identical flat on the ground floor, facing south. The disadvantage, of course, is that no incentive exists to be economical in heat usage and that the frugal tenant is penalised at the expense of his extravagant equivalent, who opens the windows rather than turning the radiators down and leaves all equipment running at full blast whilst he is away on holiday.

Flat-rate plus Unit Charges, etc.—A system of flat-rate plus unit charges would seem to have advantages over both the alternatives: the economics of the system would be assured and equity would be preserved. It would seem reasonable that charges should fall into three categories:

(*a*) Connection.
(*b*) Floor area, cube or loading.
(*c*) Usage.

The first of these would, not unreasonably, cover the actual cost of service provision, and the second would be related to the physical aspects of each building or dwelling connected. The last component, heat metering aside, could be related quite simply to the product of the connected load and the hours of operation: a sealed time-switch could provide a simple means of measuring the latter, or even a 'white-meter' assessment related to the operation of local circulating pumps or thermostatic-control devices.

SOCIOLOGICAL ASPECTS

Much has been written upon the incidence of ill health and even death, in the case of elderly persons, due to hypothermia, a matter directly associated with the availability of low-cost district heating. One Lloyds insurance firm has, in fact, offered a ten per cent reduction in premiums to persons living in centrally-heated houses. The social costs of the 'four damp cold walls which are some peoples environment' is not inconsiderable in terms of loss of production and drain on the Health Service.

With particular respect to the viability of district- and group-heating schemes, however, and specifically to those which provide a metered supply, it is necessary to consider the occupation incidence of homes in Britain

TABLE 25.3
TYPICAL POPULATION DISTRIBUTION IN A
HOUSING ESTATE

Dwelling		Occupancy	
Type	No.	Adults	Children
Bed-Sitting Rooms	6	6	—
One-Bedroom Flats	112	200	12
Two-Bedroom Flats	84	169	76
Three-Bedroom Houses	27	77	51

TABLE 25.4
OCCUPANCY DURING WORKING WEEK

Dwelling	Assumed condition 9.00 to 1700 hrs (%)		
	Empty	Occupied	
		In part	In full
Bed-Sitting Rooms	100	—	—
One-Bedroom Flats	75	10	15
Two-Bedroom Flats	55	20	25
Three-Bedroom Houses	25	40	35

Plate XLVI. Boiler plant with automatic coal firing to group heating system (see p. 592)

Plate XLVII (above left). Pre-formed factory-insulated 'pipe-in-pipe' mains connecting to covered crawlway (above right)

Plate XLVIII (above left). Excavation necessary to locate a pipe leakage and (above right) corroded pipework when uncovered

Plate XLIX. Elevated pedestrian walkway providing protected route for distribution pipework *(see p. 596)*

today. Statistics collected and published in relation to one housing estate show that the occupancy set out in Table 25.3 may be anticipated for a local authority housing development. This leads, *via* certain empirical assumptions, to the conclusions listed in Table 25.4.

Such data are very pertinent when it becomes a matter of considering the annual heat consumption which may be expected from a district- or group-heating scheme (particularly when the heat supply may be metered).

TYPICAL HOUSING DEVELOPMENT

In order to bring the subject into perspective, a housing development with ancillary amenity buildings (as shown in Fig. 25.12) may be considered. There are 1200 dwellings with a notional occupancy of 4500 persons.

Assuming the average peak heat requirement per dwelling to be 6·0 kW, the annual requirements per average dwelling might be:

Space heating	56 GJ
Hot-water supply	14 GJ
Total	= 70 GJ

It would be necessary to consider the comparative capital costs of equipment and plant within and outside the dwellings, as set out in the first two columns of Table 25.5, for group and local heating. These costs

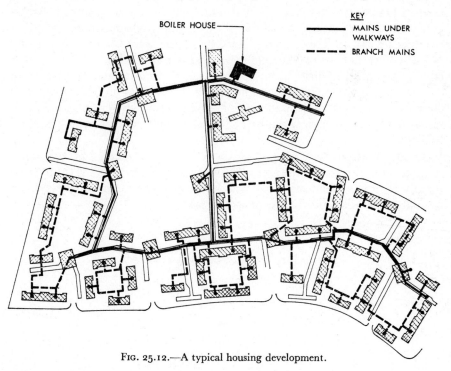

Fig. 25.12.—A typical housing development.

TABLE 25.5

CAPITAL AND ANNUAL COSTS OF EQUIPMENT FOR ALTERNATIVE
HEATING METHODS

Item	Capital Cost £ per dwelling*		Annual Cost† £ per dwelling*	
	Group Heating	Individual Unit	Group Heating	Individual Unit
Equipment				
Boilers, Pumps, etc.	180	—	15·3	—
External Mains	250	—	23·6	—
Internal Mains	200	—	18·9	—
In dwellings	580	870	49·4	74·1
Builders Work				
Boilerhouse and Flue	60	—	6·0	—
External Mains	20	—	1·9	—
Internal Mains	10	—	1·0	—
In Dwellings	20	50	1·9	4·7
Total	1320	920	118·0	78·8

* Based on a 4 person 2 bedroom flat.
† Interest rate taken at 10% over terms of 10, 20, 30 and 60 years as
appropriate.

would then be amortised over varying periods, using current rates of
interest, to produce the annual owning costs listed in the last two columns
of the same Table.

Group Heating—Assuming that the fuel used is either coal or a heavy
grade oil, the operating cost of the group-heating system would be calculated as set out in Table 25.6, if raw fuel costs were about 10p per 100
MJ. It will be noted that the ancillary buildings are included here as being
equivalent to an additional 100 dwellings. The individual costs may be
viewed in terms of percentage of the total as shown. It will be seen that the
charges for interest and repayment of capital make up very nearly half of
the annual cost—which situation reflects the high interest rates currently
in force.

Individual Heating Units—For the individual local units, assuming
usage of fuel purchased at a 'retail' price of say 28p per 100 MJ (approximately typical of domestic-tariff gas burnt at 80 per cent efficiency), an
annual cost-in-use could then be as follows:

	£
Owning cost	79
Standing charge (say)	20
Maintenance	15
70 GJ × 2·8	196
	£310

or £5·96 per week.

However, if it were assumed—and this is the situation reflected by statistics quoted by the supply authorities—that the occupier economises in energy usage at the inevitable expense of comfort conditions in the dwelling, then the consumption might only be 40 GJ per annum. The resultant cost-in-use would then be £226 per annum and the weekly charge would fall to £4·35 approximately.

TABLE 25.6

COST-IN-USE FOR GROUP HEATING

Item	£	£	%
Fuel			
Heat used/annum			
$\quad$ = 1300 × 70 = 91 000 GJ			
Heat losses, fuel pre-			
heating (where appro-			
priate) etc. $\quad$ = 19 000 GJ			
Heat generated $\quad$ = 110 000 GJ			
Fuel used at 80% eff. = 137 000 GJ			
Cost = 137 000 × £1.00	137,000		
Electricity to Plant			
Approximately 10% of fuel	13,700	150,000	45·7
Salaries of Operators, Spares etc.			
Operators	14,000		
Insurance and holidays (25%)	3,500		
Management	3,500		
Spares, 0·5% of equipment	5,200	26,200	7·9
Insurance and Rates			
Insurance, 0·1% of equipment cost	1,450		
Rates, 0·25% of equipment cost	3,630	5,080	1·5
Interest and Repayment of Capital			
Boiler Plant	18,390		
External Mains	28,280		
Internal Mains	22,620		
Dwelling Equipment	59,260		
Builders Work	13,860	142,410	43·2
Sundry Items			
Say		5,610	1·7
Total cost-in-use/annum		£330,000	100%
Cost-in-use/dwelling per annum		£275	
Cost-in-use/dwelling per week		£ 5·29	

Comparison—It will thus be seen that for comparable service, group heating is 67p per week cheaper than individual units. Parity of cost will be achieved only if service from the individual units is reduced to 82 per cent of that provided by the group system.

Other Factors—It will be appreciated by the reader that this fiscal examination is necessarily superficial and refers to a hypothetical development with a boiler plant fired by coal or heavy oil. Fuel costs are constantly varying, as are interest rates, so that any precise comparisons would quickly be out-dated. The picture presented is, however, fairly typical of the results of a number of surveys completed during the period 1968 to 1971 in relation to real situations, up-dated to present day prices. Were firing to be by light oil or gas however, the comparison would be far less favourable to the district heating case—and unacceptable in the sense of conservation of dwindling resources.

As examples of circumstances which may confuse matters, the following are typical:

Some electrical supply authorities claim, not unreasonably, that in order to provide a supply to a new housing development it is necessary for a considerable capital investment to be made by way of transformer stations, distribution networks and so on. If group heating is the capital cost. Such a sum, if capitalised over fifty years, could be considered as *adding* to the cost of group heating per week for each dwelling.

Gas-supply authorities have been known to suggest that, if a group heating boilerhouse uses that fuel, then a reduction in the domestic cooking tariff will be made available to the individual dwellings. This situation would produce a saving per week for each dwelling, which could be considered as *reducing* the cost of district heating.

HEAT SERVICE OPERATORS

A development in this country which has grown in popularity during the past decade has been the formation of heat-service organisations which will undertake the operation of district- and group-heating plants and will offer expertise imported from the Continent, where such arrangements have been in being for some decades.

Such arrangements relieve the owning authority of operational responsibilities for heat generation and include the cost of fuel, insurance, maintenance and spares, etc., in the boilerhouse. In return, the authority will be required to enter into a long term contract with the service organisation. Terms of payment vary: that most common in the immediate past being a standing charge plus a unit charge for heat as metered at the 'boilerhouse wall', the owning authority retaining all responsibility for external and internal mains, dwelling equipment, etc.

There are current indications that operators are now prepared to accept responsibility for the operation and maintenance in good order of external mains up to the entrance of each building, thus placing themselves on all fours with other public utilities.

CONCLUSION

This brief survey of district and group heating has necessarily omitted mention of many technical aspects implicit in the design of installations, pressure, and temperature-limiting controls, volume regulators and the like. These are beyond the scope of this volume. But it is worthy of note that among the most obvious features of successful plants visited on the Continent is the severe simplicity of the arrangements overall: maintenance skills for complex equipment are at a premium in present-day circumstances, and this would seem to dictate the avoidance of unnecessary 'gadgets' all too prone to failure.

CHAPTER 26

Total Energy

THE FIRST, SECOND AND THIRD EDITIONS of this book, published in 1936, 1943 and 1957 respectively, included a chapter entitled *Combined Electrical Generating Stations*. The subject matter of these earlier inclusions reappeared in the Fifth Edition, suitably modernised, under a new title: *Total Energy*.*

In 1967, the *Journal of the Institute of Fuel* proposed a definition for the total energy concept as being 'a scheme for the energy supply to a site in which the whole of the electricity and heat requirements are derived from only one form of energy'. This definition has been accepted in principle for the present chapter, with the rider that the equipment concerned should be an assembly of heat-producing, heat-accepting and heat-rejecting units arranged in a manner which derive the maximum conversion of energy in a fuel to useful energy forms over extended periods. In short, with a minimum of waste.

BASIC CONSIDERATIONS

Heat generated in a conventional boiler plant is normally obtained at high thermal efficiency, whilst the generation of electrical power from heat is seldom obtained at overall thermal efficiencies exceeding 35 per cent. The intentional wastage of heat through cooling towers is a feature of generating stations: some small attempts, such as the Pimlico scheme referred to in Chapter 25, have been made to use the cooling water for heating, but nothing in this country compares with the co-ordinated use of energy resources by other countries. It is sufficient to say that the least thermally efficient of the existing British power stations could be made the most efficient if use were made of the waste heat. Further, it is obvious that, when an appreciable proportion of the nation's electricity supply has to be generated in stations which have efficiencies as low as 20 per cent, the cost to the consumer will inevitably be high. Fig. 26.1 illustrates the comparative efficiency of various forms of electrical, thermal and thermal-electric energy production.

A total-energy installation will necessarily comprise a fuel supply, a prime mover, an electrical power generator, heat-recovery equipment and

---

* Much of the content of this chapter has been taken from a paper read by one of the Authors, in conjunction with Mr D. G. Axford, to the 1970 Conference of the Combustion Engineering Association.

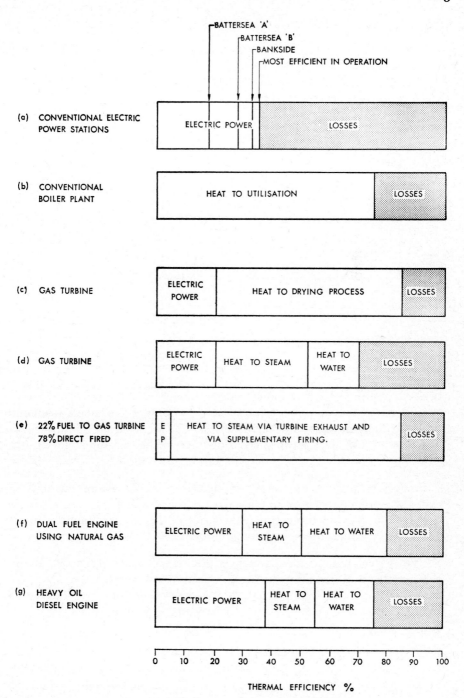

FIG. 26.1.—Comparative efficiencies for various methods of energy production.

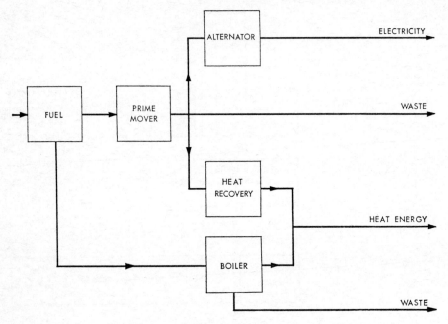

FIG. 26.2.—Components of a total-energy system.

perhaps a supplementary boiler—all as shown in diagram form in Fig. 26.2. It is true that components in similar groupings have been known and built for many decades, and the idea of independence from public electricity supplies has always appealed to planners. When local generation has been employed, it has usually been for one of two reasons:

(*a*) To ensure a continuity of electrical supply where essential to some process or functional use of the building.

(*b*) To provide the complete security of an electrical supply in all circumstances, including national disaster or insurrection.

Two developments have advanced the use of total-energy schemes in this country and the USA during the past twenty years, these being the discovery and exploitation of natural gas reserves and the development of the industrial gas turbine, for which natural gas is an ideal fuel.

EQUIPMENT CHARACTERISTICS

At this stage it would be as well to examine the form and availability of the heat rejected from some of the normally encountered prime movers, since it is an essential to the design of total-energy systems that the marked differences in the characteristics of prime movers are not only appreciated but are deliberately put to use. The more important of these characteristics are:

(a) the heat-to-power ratio at both full and partial load;
(b) the level or levels of heat rejected;
(c) the heat input rate throughout the operating range.

The first of these has the greatest influence on equipment selection, since the exercise will be centred upon the matching of the heat-to-power ratio of the total-energy installation to the building energy-demand profiles. However, it is not sufficient to ensure a match at full or at any one loading, but rather to maintain the balance as far as possible throughout the year. The object, ultimately, is not to create a match of maximum energy rates but to transfer the greatest possible quantity of fuel energy into useful energy forms.

PRIME MOVERS

Prime movers fall into three general classifications:

1. Steam turbines.
2. Gas turbines.
3. Reciprocating engines.

Of these, the steam turbine has the highest heat-to-power ratio at some ten or more to one. It follows that installations using such equipment are most suited to applications which have large thermal demands in relation to electric-power requirements: these will normally lie in the industrial field where a high process heat load may be accompanied by an associated electric-power demand in a ratio particularly suited to these machines. Gas turbines have heat-to-power ratios of the order of three to one, but the manner of application can materially increase this ratio by adding supplementary direct firing to the waste-heat boiler. The third group includes gas ignition engines, diesel engines and dual-fuel compression engines having heat-to-power ratios of approximately one to one.

On occasions, of course, the thermal and electrical demands of the building complex can be matched with the heat-to-power characteristics of a total-energy system by employing a combination of prime movers.

Steam Turbine Generator Sets—High efficiency electricity generating stations on a national scale employ high-pressure boilers and sophisticated techniques to obtain the maximum electric power from the fuel input. However, thermal-electric stations of comparable generating capacity, designed to produce useful heat as well as electric power, would be arranged for quite different pressure and operating conditions and would, in general, tend to employ less complex and cheaper steam-raising plant. In such cases, back-pressure turbines are most likely to be used or, where some proportion of higher pressure steam is required for process use, this may be obtained either directly from the boiler or by pass-out from the

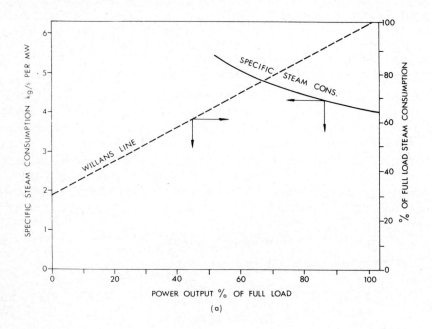

(a)

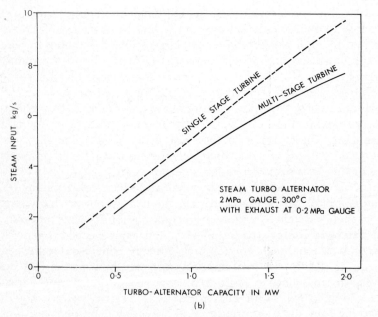

(b)

FIG. 26.3.—Typical energy consumptions of steam turbines.

turbine at an intermediate pressure. Where the generator load is less than say 700 kW, consideration would be given to reciprocating engines which in this range may be more economical of steam.

Turbine steam consumption per kilowatt produced varies widely according to the load and capacity of the machines. Fig. 26.3 (a) shows the variation in steam consumption for a 1 MW set under a varying load. Typically, 4 kg/s of steam per megawatt would be available at a gauge pressure of say 200 kPa. This represents a heat-to-power ratio of ten to one. Fig. 26.3 (b) gives full-load data for other sizes of machines and for single and multi-stage turbines.

Gas-Turbine Generator Sets—The full-load thermal efficiencies of typical open-cycle gas-turbine sets are illustrated in Figs. 26.1 (c) and (d) This shows that the generation of electricity without subsequent heat recovery produces an efficiency of only 19 per cent. In fact, industrial gas-turbine sets have full-load heat ratios varying from about 5 to 7, as shown in Fig. 26.4. These represent an efficiency range of between 14 per cent and 20 per cent and hence, as straight generators of electricity, such machines are not very efficient units.

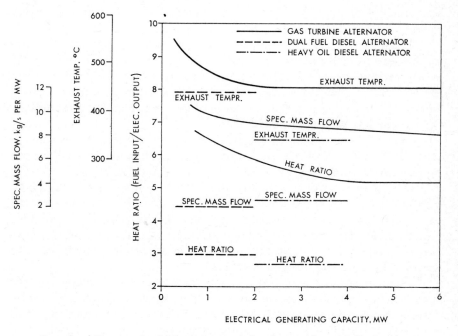

Fig. 26.4.—Comparative full-load characteristics of gas turbines and diesel engines.

However, the exhaust gas temperature at full load will normally be between 450° C and 550° C and will thus permit relatively simple heat exchange from a single unit; further, the thermal efficiency can be

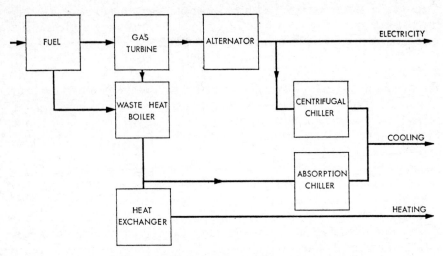

FIG. 26.5.—Line diagram of a gas turbine total-energy system.

increased by arranging two or more stages of heat transfer. Fig. 26.5 represents a flow diagram of a typical arrangement.

Occasionally, very high efficiency can be achieved by allowing the exhaust gases to fall to quite low temperatures: such an application might involve a direct drying process as shown in Fig. 26.1 (c). In practice, there are remarkably few problems associated with low final gas temperatures when the fuel used is good quality natural gas and the inlet air supply to the turbine has been adequately filtered.

Another characteristic of gas turbines has an important bearing on overall efficiency, this being the requirement that large volumes of air are handled by the turbine in order that the temperature of the turbine blades is limited. The limitation is usually about 850° C for industrial machines. This results in substantial quantities of excess air which pass through the turbine and into the heat exchanger, which large volumes of preheated air clearly invite supplementary firing. Fig. 26.1 (e) shows the increase in overall efficiency of gas-turbine-boiler combinations which can be achieved by adopting this arrangement.

When considering gas turbines, it is important to take account of the condition under which the machine is to produce maximum electrical output. Machines are normally rated at N.T.P. and deviations from this with respect to inlet-air temperature affect the rating, as illustrated in Fig. 26.6. Similarly, exhaust temperature, inlet pressure loss and exhaust-system pressure must be considered when assessing the actual output of a gas-turbine set under specific conditions. Of particular importance is an appreciation of how the heat input rate varies with load. Below 50 per cent of full load this rate increases sharply, as shown also in Fig. 26.6.

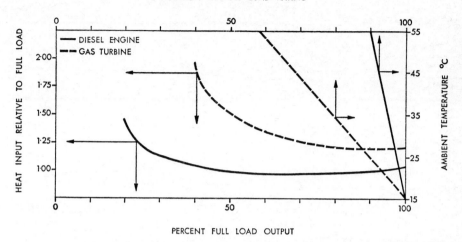

FIG. 26.6.—Variation in gas turbine and diesel engine performance at part-load and with ambient temperature.

Reciprocating Engines—Whilst the open cycle gas turbine has the great merit of rejecting waste heat in a single fluid stream and at a relatively high temperature, it has a high proportion of fuel energy rejected as heat, and the shaft power for generation of electricity is comparatively low.

The various forms of reciprocating engine, on the other hand, can convert an equivalent amount of fuel into approximately double the shaft power obtained with gas turbines. Among the reciprocating machines generally considered, and available in Britain for total-energy systems, are:

(a) Diesel engines using light diesel oil. These may be either naturally aspirated or turbo-pressure charged.

(b) Diesel engines using heavy fuel oil, up to 70 centistokes. These engines are normally turbo-pressure charged.

(c) Dual fuel engines capable of using either diesel fuel oil or natural gas. These may be naturally aspirated or turbo-pressure charged.

(d) Spark ignition gas engines.

Heat recovery from reciprocating machines may be better appreciated by reference to Fig. 26.7, which is a typical flow diagram for an installation capable of utilising heat at the various thermal levels at which it is available. Low grade heat may be obtained from the engine jacket, lubricating oil cooler and, where fitted, the inter-cooler. In addition, a higher grade heat is available from the exhaust gas *via* a waste-heat boiler. British practice is to take the waste heat from the engine jacket in the form of low-temperature hot water, whilst the exhaust gas is used to generate steam or

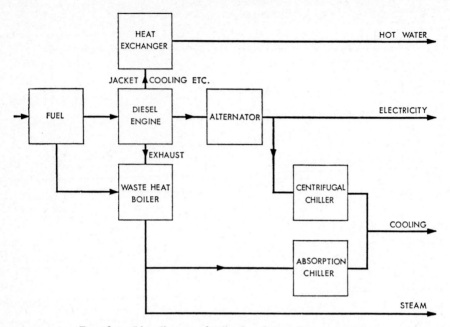

Fig. 26.7.—Line diagram of a diesel engine total-energy system.

to heat water to higher temperatures. In the United States, however, it is more usual to employ ebullient cooling which virtually generates steam within the cylinder jacket. This steam is not allowed to collect in the water-ways but is removed by a steam separator above the engine. It may be that, in due course, this practice will extend to this country.

The various waste-heat sources not only have different heat potential but also have differing limiting factors. For example, the inter-cooler water-outlet temperature is usually limited to 30° C, and hence it is often connected in series flow with the oil cooler which can allow the oil tempera-ture to approach 90° C and so raise water to a temperature of say 60° C.

Full-load thermal efficiencies of diesel and dual-fuel engines are shown diagrammatically in Fig. 26.1 (*f*) and (*g*). The heavy-oil diesel is shown to convert some 38 per cent of the heat input into shaft power for producing electricity and, provided that the grades of heat transferred to water and low-pressure steam are acceptable, then overall efficiencies of nearly 80 per cent are obtainable. The dual-fuel engine heat recovery is similar to that of the heavy-oil diesel and, although heat input con-version to shaft power is a little lower, higher overall efficiencies can be obtained. Notwithstanding the lower potential of the heat available, as compared with the gas turbine, satisfactory and economically sound installations are quite possible by using separately fired boilers to provide higher pressure steam or heat demands in excess of the heat recoverable from the engine.

Further, the diesel or dual-fuel engine has the merit of a substantially constant heat ratio over a wide range of operating loads, as shown in Fig. 26.6, and suffers much smaller changes in output in relation to intake-air temperature as compared with the gas turbine. Fig. 26.6 also illustrates the variation in output for diesel engines in relation to air-intake temperature. Changes in heat-input rate with varying load conditions are also shown in this Figure.

BUILDING LOAD CHARACTERISTICS

To ensure maximum utilisation of the potential energy in the fuel, the heat and power demands of the building load must be matched by the output of the prime movers quantitatively and in an acceptable ratio of heat to power. This match must be considered not only at full load but in accordance with load variations.

Electrical Supply—Total-energy systems set out to provide all the on-site energy requirements, including electrical power: i.e. including an electrical supply which would otherwise be obtained from the national network or public utility. It follows, therefore, that the on-site power supply should provide no less security than that obtained from public supplies. In some circumstances, of course, a guaranteed minimum supply may be the major reason for considering on-site generation. This security factor will affect the number, sizes and, on occasion, the type of prime mover and generator combination.

All machinery requires periodic maintenance, and it may not always be convenient to arrange for a machine to receive the service during light-load periods of the year. Therefore, where the load is substantially constant all the year round, provision should be made for at least one set in excess of maximum requirements. Such an arrangement, however, would do no more than allow a spare machine to be itemised on the maintenance schedule and, whilst this would reduce the possibility of breakdown, it could not be depended upon to obviate it. It is quite possible, then, that security of supply at full load may demand two machines in excess of maximum requirements.

It is on this aspect that the economic viability can turn so easily. The choice will lie between a small number of large machines with consequent large and costly excess capacity, or a larger number of small machines with lower capital cost for provision of standby but with higher cost per installed kilowatt of capacity. Frequently, too, the smaller machines would provide flexibility in load matching which would allow longer periods of running at part total-load at relatively low heat-rate and without by-pass and associated heat wastage. Reference to Fig. 26.6 will show that the heat rate increases as the power output is reduced, and this characteristic is particularly pronounced with gas turbines.

Electrical Load—The electrical load will be made up from components of power and lighting and will vary considerably between buildings put to different usage. In commercial buildings, offices and the like, the power element (apart from a small amount of power to socket outlets) is likely to be mainly for motors driving mechanical building-systems plant, such as pumps, fans, compressors and lifts. An industrial complex, on the other hand, may have a similar type of load confined to the office section of the site combined with a predominant process-power load elsewhere.

The lighting, too, can vary considerably, even when comparing buildings of similar usage. Deep planning will tend to show that the lighting load is substantially constant the year round, although the peak demand may be close to that of a shallow plan building of equal area and lit to comparable standards. Lighting may also be deliberately employed as a heating element during the pre-warm periods of an intermittently heated building.

Space and Process-Heat Load—Space- and process-heating loads for any one complex having such a dual demand may not only vary considerably quantitatively but are frequently of different character. For example, heated water at the relatively low temperature of 95° C will ordinarily satisfy space heating and domestic hot-water requirements. A process, on the other hand may well require higher temperature water or steam at a level such as 200° C for moulding presses. The type of demand will therefore influence the choice of prime movers and supplementary plant, but process applications are not considered in this Chapter.

Air-Conditioning Load—A desirable feature of any installed plant is that it shall have high utilization and this aspect will considerably influence feasibility in a total-energy study. It follows that substantially even demands, sustained over long periods, for both heat and electrical power have the highest acceptability. By its very nature, space heating will be seasonal, so that in Britain, between May until late September, the requirement is likely to reduce progressively to zero and then re-build similarly. However, an air-conditioned building will have a cooling demand which can be expected to increase inversely with movements in outside temperature. Cooling will not of course be entirely related to outside temperature, but this variable will nevertheless influence changes in demand considerably.

But a cooling requirement can become a heat demand if considered in conjunction with absorption refrigeration. This type of refrigeration plant not only accepts heat directly but at a relatively low level. Steam at a gauge pressure of 100 kPa is more usual, but hot water at the equivalent temperature of about 110° to 120° C is also used. This temperature requirement is significant since there is no restriction in choice of type of prime mover, all of which can easily be arranged to transfer rejected heat at this level.

Total Load—Fig. 26.8 illustrates how load factors, expressed there in heat-to-power ratios, may vary for two typical applications. The matching of these with the characteristics of the prime movers is essential to the

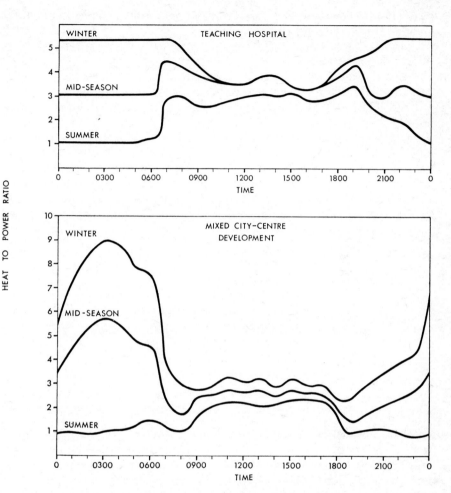

FIG. 26.8.—Heat-to-power demands for typical applications.

success of any total-energy plant. The process of carrying out a detailed examination of the electrical and thermal loads, hour-by-hour and season-by-season, is tedious, but the economic viability and practicality of the plant depends upon such an examination in great detail. A superficial approach and a cursory treatment of load profiles can lead to highly misleading conclusions.

ECONOMIC FEASIBILITY

Where a total-energy installation is to be considered as an alternative to conventional methods of obtaining power and heat, the *feasibility* must be founded on economic viability and not on the practicability of installing the equipment and meeting the energy demands. Thus a comparison must be made between the cost of installing and operating a conventional plant and the costs of the total-energy system.

The cost analysis will devolve upon two principal evaluations:

(a) The additional capital expenditure to install the total-energy installation compared with a conventional system.

(b) The annual savings effected by operating the total-energy system.

The building owner (or operator) will not only need to raise the additional capital but will not unreasonably expect returns at least as great as the marginal cost of raising that finance. To offset this added finance there will, of course, be an annual operating saving. Both of these factors must be taken into account in the preparation of cash flow analysis. Computations, similar to those described in Chapters 15 and 25, will involve the following:

Fuel costs.
Electricity charges.
Cost of maintenance.
Cost of operating staff.
Capital cost of equipment.
Capital cost of space.
Insurance charges.
Cost of finance.

Energy Demands—A first step in the computation of annual operating costs is clearly the assessment of annual energy demands. Throughout the cost analysis, the conventional scheme must be compared with the total-energy installation. Differences in the form of refrigeration equipment will affect the energy demands of the two installations. For a conventional scheme it may be assumed that electrically driven compressors would be employed: the total-energy installation on the other hand may use absorption type water-chilling apparatus.

The conventional plant would incur energy costs in two parts. Firstly, the provision of heat by gas-fired or oil-fired boilers and, secondly, the cost of bought-in electricity to supply power and lighting together with the power to drive the refrigeration plant. The total-energy system, conversely, would buy-in fuel to enable the prime movers to meet the electrical requirements and, in addition, fuel would be required to fire any supplementary boiler plant.

Fuel—Since the case for total energy may largely turn upon the cost of fuel it is imperative that the annual energy demands are accurately assessed; it is important, too, that the timing of these demands is carefully related to fuel costs where more than one fuel may be involved. For example, if advantage is to be taken of an interruptible supply tariff for natural gas, it is not sufficient to adjust running costs by the number of hours of gas cut-off. Full consideration must be given to the loading during those hours.

Bought-in Electricity—These outgoings will only apply to the conventional plant, although, on occasion, a case can be made for *partial energy* schemes. Negotiations with the Area Supply Authority for the bought-in portion of electric power must, however, be firmly concluded before carrying out the feasibility analysis. The supply of power may sometimes involve a contribution towards the cost of switchgear and transformers or the laying of cables, and conventional system costs should not fail to include such charges where these occur.

Capital—Since the total-energy system sets out to generate electric power on site rather than to buy-in this commodity, it follows that space requirements will be greater than those for a conventional plant. Space inevitably costs money and in most cases there will be the added cost of enclosing that space, and comparative costings must take full account of this. Equipment costs, too, must be carefully examined so that every item which is not common to both types of installation is properly listed in the analysis.

Maintenance and Staff—Whilst maintenance will generally not be a major component of total cost, there will be marked differences between the different types of prime mover. In general, reciprocating engines with their relatively large number of moving parts will require more frequent maintenance routines than turbines—which latter will normally run for 8000 to 20 000 hours between overhauls on good quality fuel. The maintenance costs will not only include the wages of operating staff and overheads but also the replacement of components, small expendable tools, lubricating oil, consumable stores and administration costs.

INVESTMENT APPRAISAL

There are several methods of appraising the economic soundness of any project, and difficulties in comparing the different methods have been shown to lead to wrong decisions. It is for this reason that the *discounted cash flow* method is recommended in examining the case for total energy. This process takes account of variation in earnings over the expected life of a project and allows investment incentives and taxation to be taken fully into account.

The various means of setting out the economic case must in one way or another recognise that:

(a) equipment, and to a lesser extent buildings, will deteriorate and depreciate in value;

(b) any additional capital expenditure involved in installing a total-energy system must not only be met but must show realistic returns.

Notwithstanding a high yield when correctly assessed, commercial circles may be unwilling to countenance pay-out periods in excess of five years. Very often in practice, therefore, the paramount issue becomes return on capital expenditure within periods which tend to swamp the effect of depreciation of equipment. Space for housing the installation must also be financed, but the returns for this are usually long term. It is not unusual, for example, to amortise the cost of additional space over periods of forty to sixty years.

CONCLUSIONS

Feasibility studies carried out before the current fuel crisis (the results of three of which are summarised in Table 26.1), have shown some quite remarkable returns on capital. Re-examination, in energy terms, suggests that conservation ratios for the cases examined are of the same order as those quoted for cost. Other studies, whilst less exciting, have demonstrated that total energy, if given the right conditions, can not only be shown to be economically viable for the managing authority using thoroughly realistic commercial rates of return, but can be considered as a step towards reducing the enormous waste of energy resources throughout the World and, in consequence, reducing the pollution of the atmosphere by waste products from unnecessary fuel consumption.

TABLE 26.1

COMPARATIVE TOTAL RECURRENT COSTS FOR
CONVENTIONAL AND TOTAL ENERGY SYSTEMS

Central Plant	Cost Ratios for following cases		
	Case A	Case B	Case C
Conventional System Boiler plant plus bought-in electricity for all power and lighting etc.	100	100	100
Total Energy Steam Turbines	94	—	—
Gas Turbines	76	66	81
Diesel Engines	73	—	—
Dual Fuel Engines	62	—	72

Note
Cases A and B include major air-conditioning components.

APPENDIX I

AN INTRODUCTION TO SI UNITS

The Old System

IN THE PAST, ENGINEERS have been accustomed to using a variety of disconnected and unrelated units. Thus we had temperatures measured in degrees Fahrenheit, which started from the freezing point of water at 32 degrees and divided the space between this and boiling point into 180 degrees. Heat quantities were based on the properties of water—the *British Thermal Unit* in the British and American system, or the *Calorie* in the Continental system. But the properties of water vary according to temperature, so this basis had to be defined in relation to temperature.

Energy was measured in horsepower using an odd figure of 33 000 foot pounds a minute to one h.p. Its equivalent in terms of heat, originating in the experiments of Joule (the son of a brewer) was the even odder figure of 2544 Btu/hour. For pressures there were various measurements to choose from—pounds per square inch or square foot, inches of water gauge, inches of mercury (or their metric equivalents: millimetres of mercury).

This whole edifice of engineering data was built up on the foundations of the inch, foot, yard and rod, pole or perch for linear dimensions; the square foot, square yard, pole, rood and three varieties of acre for areas; and the cubic inch, cubic foot, gallon and so on for volumes. For weight (avoirdupois) we had the pound, hundredweight, and ton, and for small weights the grain, of which 7000 went to the pound. A gallon of water weighed 10 lb which was convenient enough, but confusion with the American gallon at $8\frac{1}{3}$ lb had to be avoided, as likewise the American short ton of 2000 lb compared with the British 2240 lb.

This multifarious array of archaic tools—for means of measurement and calculation are no more than tools—has been likened to the Chinese alphabet. Those who mastered it over the years were proud of their great feat of memory, but it could not be regarded as suited to the present age of rapid scientific and technical development. Indeed, the scientists had long abandoned it, so that a student changing from the laboratory to applied engineering was faced with a completely new set of unfamiliar symbols and values to be learnt.

The New System

In the United Kingdom it has been decided to adopt the *Système International*, or SI. This is based on the metric system commonly in use on the Continent in so far as it is decimal, but it differs in that it is coherent throughout, which the old metric is not. For instance, in the metric system the unit for measurement of heat is, as explained, calorimetric—relying on the properties of water as in the British Thermal Unit. In the SI system the basis is energy.

SI units have already been accepted by a large number of other countries throughout the world and are now well on the way to becoming completely international.

It is not proposed here to enter into a complete dissertation on SI, which can be found elsewhere, but to summarize its principal features as they affect the subject of this book.

There are, in fact, only six basic units:

Length	the metre	(m)
Mass	the kilogram	(kg)
Time	the second	(s)
Electric current	the ampere	(A)
Absolute temperature	the kelvin	(K)
Luminous intensity	the candela	(cd)

All other units are derived from these: such as the units of force, energy, pressure, volume, area, motion, viscosity, etc.

Force

The unit of force is the *newton* (N), which is the force necessary to accelerate a mass of one kilogram to a speed of one metre per second in one second,

$$\text{or } 1 \text{ N} = 1 \text{ kg m/s}^2.$$

When a mass of one kilogram is subjected to acceleration due to gravity (which is $9 \cdot 81$ m/s^2), the 'force' or 'weight' exerted is $9 \cdot 81$ N. Weight is not otherwise referred to in SI terms.

Energy: Heat

Energy may take a number of forms which are mutually convertible; for instance, chemical energy in a battery may produce electrical energy which in turn, in passing through a resistance, will produce heat energy.

Heat energy in fuel, in a boiler producing steam, may drive an engine producing mechanical energy and so on.

The unit of energy, or quantity of heat, is the *joule* (J) which is equal to a force of one newton acting through one metre, or

$$1 \text{ J} = 1 \frac{\text{kg m}^2}{\text{s}^2}.$$

Heat-Flow Rate

The unit for rate of heat flow or power is the *watt* (W). One watt is equal to one joule produced or expended in one second:

$$1 \text{ W} = 1\frac{\text{J}}{\text{s}} = 1\frac{\text{Nm}}{\text{s}}.$$

The watt is already familiar in its narrower electrical context:

$$1 \text{ W} = 1 \text{ ampère} \times 1 \text{ volt}.$$

Thus the volt can be derived from the newton.

The amount of energy flowing through unit area in unit time is to be known as *the density of heat-flow rate*—W/m².

Temperature

The unit of thermodynamic temperature is the kelvin (K). It is defined as $1/273 \cdot 15$ of the thermodynamic temperature of the triple point of water. Use is also made of Celsius (°C) temperature defined as:

$$t = T - T_0$$
$$t = \text{Celsius temperature}$$
where $\quad T = \text{Thermodynamic temperature}$
$$T_0 = 273 \cdot 15 \text{ K}$$

Celsius was a Swedish professor and, as the inventor of the scale, his name is used in preference to 'centigrade' to avoid confusion with a French term concerning angular measurement.

A difference of $\quad 1°$ Celsius $= 1\cdot8°$ Fahrenheit $= 1$ kelvin
A temperature of $\quad 0° \quad ,, \quad = 32° \quad ,,$
$,, \quad ,, \quad ,, 100° \quad ,, \quad = 212° \quad ,,$

Volume

The cubic metre (m³) is the preferred unit, as is the cubic millimetre (mm³). In view of the disparity in magnitude between these units, however, the cubic decimetre (dm³) and cubic centimetre (cc or cm³) are acceptable. The litre (originally $1\cdot000\ 028$ dm³) has been redefined such that

$$1 \text{ dm}^3 = 1 \text{ litre} = 0\cdot001 \text{ m}^3.$$

A difficulty arises as to the symbol for *litre* (l), which when typed is indistinguishable from the figure *one*. Hence '1 l' is easily mistaken for eleven. This matter is not finally resolved. 'L' has been suggested, but could be confused with the sign for the Italian Lire. In this book, *litre* in full has been used.

At standard atmospheric pressure and 4° C:

> 1 litre of water has a mass of 1 kilogram
> 1 cc   ,,   ,,   ,, ,, ,,   ,, 1 gram
> 1 m³   ,,   ,,   ,, ,, ,,   ,, 1 tonne (1000 kg)

Time

The *second* is the only acceptable unit in SI. This should avoid error sometimes liable to creep in when using a formula in which some quantities are in one unit and others are in a different unit. The *hour, day, week* and *year* remain, but it will be important in calculations to convert them to seconds:

$$1 \text{ hour} = 3600 \text{ seconds}$$

The *cycle per second* is replaced by the term *hertz* (Hz), named after Heinrich Rudolf Hertz, the pioneer of the theory of radiation.

Pressure

The standard is the *newton per square metre* (N/m^2), also known as the *pascal* (Pa). This is for some purposes an inconveniently small unit, and hence the *bar* is permitted, which is approximately one standard atmosphere.

> 1 bar   = 100 000 Pa
> 1 pound/sq in =   6895 Pa
> 1 inch wg   =   249 Pa

In meteorological usage the *millibar* occurs,

> 1 mb   =   100 Pa.

Multiples and Sub-multiples

The preferred multiples are in threes:

> kilo (k), a thousand, 10^3
> mega (M), a million, 10^6
> giga (G), a thousand million, 10^9
> tera (T), a billion, 10^{12}.

Sub-multiples are:

> milli (one thousandth), 10^{-3}
> micro (one millionth), 10^{-6}.

Other smaller sub-multiples need not concern us. Reference has, however, been made to exceptions where use has had to be made of:

> deci   (one tenth), 10^{-1}
> centi (one hundredth), 10^{-2}.

The purpose of multiples and sub-multiples is to enable the values of units in common use to be whole numbers or single decimals, the preferred range to be 0·1 to 1000. For instance, as pointed out, the newton and the pascal are too small for most purposes, hence the *kilonewton* (kN) and the *kilopascal* (kPa) may be used. The same applies to the watt where, for most purposes, the *kilowatt* (kW), *megawatt* (MW), and *gigawatt* (GW) are used.*

The Decimal Point

In the British version of SI, the decimal marker remains as the familiar point above the line for writing and printing, and on the line for typescript —until special typewriters have become common.

The use of the comma to mark thousands *is not acceptable* owing to its liability to confusion with its use in other countries as the decimal marker. Instead, a gap is left to indicate thousands, but this does not apply where only four digits occur. Thus, for example,

six thousand will be 6000
sixty thousand will be 60 000.

Fractions are always preceded by a nought, and thus one-tenth is written as '0·1' (*not* '·1').

Coherence and Consistency

It will be clear that SI achieves the greatest possible measure of coherence and interchangeability as between one set of units and another. It is unfamiliar ground to regard heat as measurable in terms of force, distance and time, but this is what constitutes work; and work and heat have long had equivalents.

The units seemed strange for a time, but in use this disadvantage is beginning to be overcome as familiarity breeds content. What can never be overcome in the decimal system is, of course, its indivisibility by three. No longer can the yard be divided into feet or the gill into sixths. On the other hand, who could divide 2240 by three?

---

* See the list of SI symbols, Appendix II, for various multiples and sub-multiples, also, for conversion factors, Imperial to SI units.

APPENDIX II
Conversion Factors: Imperial Units to SI

Unit	Imperial	SI Exact	SI Approximate
Length	1 inch	25·4 mm	25 mm
	1 foot	0·3048 m	0·3 m
	3·28 feet	1 m	
	1 yard	0·9144 m	0·9 m
	1 mile	1·609 km	1·6 km
Area	1 sq. in	645·2 mm²	
	1 sq. ft	0·092 m²	
	10·77 sq. ft	1 m²	
	1 sq. yard	0·836 m²	
	1 acre	4046·9 m²	
Volume	1 cu. in	16·39 mm³	16 mm³
	1 cu. ft	28·32 litre	28 litre
	35·32 cu. ft	1 m³	
	1 pint	0·568 litre	0·6 litre
	1 gallon	4·546 litre	4·5 litre
Mass	1 pound	0·4536 kg	0·5 kg
	2·205 pounds	1 kg	
	1 ton	1·016 tonne	1 tonne
Density	1 lb/cu. ft	16·02 kg/m³	
Volume flow rate	1 gall/minute (g.p.m.)	0·07577 litre/s	0·075 litre/s
	1 cu. ft/minute (c.f.m.)	0·000472 m³/s	0·0005 m³/s
Velocity	1 foot/minute	0·0051 m/s	
	197 ft/minute	1·0 m/s	
	1 mile/hour	0·447 m/s	0·5 m/s
Temperature	1 Degree Fahrenheit	0·5556° C	
	t = 32° F	t = 0° C	
Heat	1 British Thermal Unit (Btu)	1·055 kJ	1 kJ
	1 'Old' Therm (100 000 Btu)	105·5 MJ	100 MJ
	1 'New' Therm* (947,867 Btu)	100 MJ	
	1 Unit of Electricity (kWh)	3600 kJ	
Heat flow rate	1 Btu/hour	0·2931 W	0·3 W
	1 horsepower	745·7 W	750 W
	1 ton refrigeration (12 000 Btu/hour)	3·516 kW	3·5 kW
Density of heat flow rate	1 Btu/sq. ft hour	3·155 W/m²	3 W/m²

*At the time of going to press, this is not an accepted unit.

APPENDIX II *(continued)*

Unit	Imperial	SI	
		Exact	Approximate
Transmittance (U value)	$\dfrac{\text{1 Btu}}{\text{sq. ft hour } ^\circ F}$	5·678 W/m² K	6 W/m² K
Conductivity (k value)	$\dfrac{\text{1 Btu inch}}{\text{sq. ft hour } ^\circ F}$	0·1442 W/m K	
Resistivity (1/k)	$\dfrac{\text{1 sq. ft hour } ^\circ F}{\text{Btu inch}}$	6·934 m K/W	
Calorific value	1 Btu/lb 1 Btu/cu. ft	2·326 kJ/kg 37·26 J/litre or 37·26 kJ/m³	2·5 kJ/kg
Pressure	1 pound force per sq. in (lb f/sq. in)	6895 Pa or 68·95 m bar	7000 Pa 70 m bar
	1 inch w.g. (at 4° C)	249·1 Pa or 2·491 m bar	250 Pa 2·5 m bar
	1 inch mercury (at 0° C)	33·86 m bar	34 m bar
	1 mm mercury	1·333 m bar	
	1 Atmosphere (standard)	101 325 Pa	1 bar
Pressure drop	1 inch w.g./100 ft	8·176 Pa/m	
Latent heat of steam (atmospheric pressure)	970 Btu/lb	2258 kJ/kg	2300 kJ/kg
Latent heat of fusion of ice	144 Btu/lb	330 kJ/kg	
Steam flow rate	1 lb/hr 8 lb/hr	0·126 g/s —	1 g/s
Heat content	1 Btu/lb	2·326 kJ/kg	
	1 Btu/gall	0·2326 kJ/litre	
	1 Btu/cu. ft	0·0372 kJ/litre	
Thermal diffusivity	1 ft²/hr	$2\cdot581 \times 10^{-5}$ m²/s	
Moisture content	1 lb/lb 100 grains/lb	1 kg/kg 0·014 kg/kg	

APPENDIX II (*continued*)

TEMPERATURE LEVEL
DEGREES CELSIUS TO DEGREES FAHRENHEIT

C	0°	1°	2°	3°	4°	5°	6°	7°	8°	9°
	°F	°F	°F	°F	°F	°F	°F	°F	°F	°F
0°	32·0	33·8	35·6	37·4	39·2	41·0	42·8	44·6	46·4	48·2
10°	50·0	51·8	53·6	55·4	57·2	59·0	60·8	62·6	64·4	66·2
20°	68·0	69·8	71·6	73·4	75·2	77·0	78·8	80·6	82·4	84·2
30°	86·0	87·8	89·6	91·4	93·2	95·0	96·8	98·6	100·4	102·2
40°	104·0	105·8	107·6	109·4	111·2	113·0	114·8	116·6	118·4	120·2
50°	122·0	123·8	125·6	127·4	129·2	131·0	132·8	134·6	136·4	138·2
60°	140·0	141·8	143·6	145·4	147·2	149·0	150·8	152·6	154·4	156·2
70°	158·0	159·8	161·6	163·4	165·2	167·0	168·8	170·6	172·4	174·2
80°	176·0	177·8	179·6	181·4	183·2	185·0	186·8	188·6	190·4	192·2
90°	194·0	195·8	197·6	199·4	201·2	203·0	204·8	206·6	208·4	210·2
100°	212·0	213·8	215·6	217·4	219·2	221·0	222·8	224·6	226·4	228·2
110°	230·0	231·8	233·6	235·4	237·2	239·0	240·8	242·6	244·4	246·2
120°	248·0	249·8	251·6	253·4	255·2	257·0	258·8	260·6	262·4	264·2
130°	266·0	267·8	269·6	271·4	273·2	275·0	276·8	278·6	280·4	282·2
140°	284·0	285·8	287·6	289·4	291·2	293·0	294·8	296·6	298·4	300·2
150°	302·0	303·8	305·6	307·4	309·2	311·0	312·8	314·6	316·4	318·2
160°	320·0	321·8	323·6	325·4	327·2	329·0	330·8	332·6	334·4	336·2
170°	338·0	339·8	341·6	343·4	345·2	347·0	348·8	350·6	352·4	354·2
180°	356·0	357·8	359·6	361·4	363·2	365·0	366·8	368·6	370·4	372·2
190°	374·0	375·8	377·6	379·4	381·2	383·0	384·8	386·6	388·4	390·2
200°	392·0	393·8	395·6	397·4	399·2	401·0	402·8	404·6	406·4	408·2
210°	410·0	411·8	413·6	415·4	417·2	419·0	420·8	422·6	424·4	426·2
220°	428·0	429·8	431·6	433·4	435·2	437·0	438·8	440·6	442·4	444·2
230°	446·0	447·8	449·6	451·4	453·2	455·0	456·8	458·6	460·4	462·2
240°	464·0	465·8	467·6	469·4	471·2	473·0	474·8	476·6	478·4	480·2
250°	482·0	483·8	485·6	487·4	489·2	491·0	492·8	494·6	496·4	498·2
260°	500·0	501·8	503·6	505·4	507·2	509·0	510·8	512·6	514·4	516·2
270°	518·0	519·8	521·6	523·4	525·2	527·0	528·8	530·6	532·4	534·2
280°	536·0	537·8	539·6	541·4	543·2	545·0	546·8	548·6	550·4	552·2
290°	554·0	555·8	557·6	559·4	561·2	563·0	564·8	566·6	568·4	570·2
300°	572·0	573·8	575·6	577·4	579·2	581·0	582·8	584·6	586·4	588·2

APPENDIX III
SI Unit Symbols

Quantity	Unit	Symbol	Common multiples or sub-multiples	Multiplier	Symbol
Length	metre	m	kilometre	$m \times 10^3$	km
			millimetre	$m \times 10^{-3}$	mm
Area	square metre	m^2	hectare	$m^2 \times 10^5$	ha
			sq. millimetre	$m^2 \times 10^{-9}$	mm^2
Volume	cubic metre	m^3	litre	$m^3 \times 10^{-3}$	(litre) *
			cu. millimetre	$m^3 \times 10^{-9}$	mm^3
Time	second	s	hour	$s \times 3600$	h
Velocity	metre per second	m/s			
Acceleration	metre/sec^2	m/s^2			
Frequency	hertz (cycle per sec)	Hz			
Rotational frequency	revolutions per sec	s^{-1}			
Mass	kilogram	kg	tonne	$kg \times 10^3$	t
			gram	$kg \times 10^{-3}$	g
			milligram	$kg \times 10^{-6}$	mg
Density		kg/m^3			
Specific volume		m^3/kg			
Mass flow rate		kg/s	litre per second	$(m^3/s) \times 10^{-3}$	litre/s
Volume flow rate		m^3/s			
Momentum		kg m/s			
Force	newton	N	meganewton	$N \times 10^6$	MN
			kilonewton	$N \times 10^3$	kN
Torque		Nm			
Pressure (and stress)	pascal	Pa	megapascal	$Pa \times 10^6$	MPa
			kilopascal	$Pa \times 10^3$	kPa
			bar	$Pa \times 10^5$	b
			millibar	$Pa \times 10^2$	mb
Viscosity					
Dynamic	pascal second	Pa s	centipoise	$Pa\,s \times 10^{-3}$	cP
Kinematic	centimetre^2/sec	cm^2/s	centistoke	$(cm^2/s) \times 10^{-2}$	cSt
Temperature	degree Kelvin	K			
	degree Celsius	°C			
Heat					
Energy	joule	J	gigajoule	$J \times 10^9$	GJ
Work			megajoule	$J \times 10^6$	MJ
Quantity of heat			kilojoule	$J \times 10^3$	kJ
Heat flow rate (**power**)	watt	W	gigawatt	$W \times 10^9$	GW
			megawatt	$W \times 10^6$	MW
			kilowatt	$W \times 10^3$	kW
Thermal conductivity		W/mK			
Thermal resistivity		mK/W			
Specific heat capacity		kg/kgK			
Latent heat		kJ/kg			

NOTE. For conversion factors, Imperial units to SI units, see Appendix II.

* This book does not accept *l* as a symbol (see Appendix I).

INDEX